Order and Chaos in Nonlinear Physical Systems

PHYSICS OF SOLIDS AND LIQUIDS

AMORPHOUS SOLIDS AND THE LIQUID STATE
Edited by Norman H. March, Robert A. Street, and Mario P. Tosi

CHEMICAL BONDS OUTSIDE METAL SURFACES
Norman H. March

CRYSTALLINE SEMICONDUCTING MATERIALS AND DEVICES
Edited by Paul N. Butcher, Norman H. March, and Mario P. Tosi

ELECTRON SPECTROSCOPY OF CRYSTALS
V. V. Nemoshkalenko and V. G. Aleshin

FRACTALS
Jens Feder

HIGHLY CONDUCTING ONE-DIMENSIONAL SOLIDS
Edited by Jozef T. Devreese, Roger P. Evrard, and Victor E. van Doren

MANY-PARTICLE PHYSICS
Gerald D. Mahan

ORDER AND CHAOS IN NONLINEAR PHYSICAL SYSTEMS
Edited by Stig Lundqvist, Norman H. March, and Mario P. Tosi

THE PHYSICS OF ACTINIDE COMPOUNDS
Paul Erdös and John M. Robinson

POLYMERS, LIQUID CRYSTALS, AND LOW-DIMENSIONAL SOLIDS
Edited by Norman H. March and Mario P. Tosi

SUPERIONIC CONDUCTORS
Edited by Gerald D. Mahan and Walter L. Roth

THEORY OF THE INHOMOGENEOUS ELECTRON GAS
Edited by Stig Lundqvist and Norman H. March

A Continuation Order Plan is available for this series. A continuation order will bring delivery of each new volume immediately upon publication. Volumes are billed only upon actual shipment. For further information please contact the publisher.

Order and Chaos in Nonlinear Physical Systems

Edited by

Stig Lundqvist
Chalmers University of Technology
Göteborg, Sweden

Norman H. March
University of Oxford
Oxford, England

and

Mario P. Tosi
International Center for Theoretical Physics
Trieste, Italy

Springer Science+Business Media, LLC

Library of Congress Cataloging in Publication Data

Order and chaos in nonlinear physical systems.
(Physics of solids and liquids)
Bibliography: p.
Includes index.
1. Order-disorder models. 2. Chaotic behavior in systems. 3. Nonlinear theories. I. Lundqvist, Stig, 1925- . II. March, Norman H. (Norman Henry), 1927- . III. Tosi, M. P. IV. Series.
QC173.4.073073 1988 003 88-15113

DOI 10.1007/978-1-4899-2058-4

Originally Published by Plenum Press, New York in 1988
MyCopy version of the original edition 1988

Contributors

F. T. Arecchi, Istituto Nazionale di Ottica and Department of Physics, University of Florence, Florence, Italy

M. V. Berry, H. H. Wills Physics Laboratory, Bristol BS8 1TL, England

C. E. Bottani, Istituto di Ingegneria Nucleare, CESNEF, Polytecnico di Milano, 20133 Milan, Italy; Gruppo Nazionale di Struttura della Materia del CNR, Unita di Ricerca 7, Milan, Italy; Centro Interuniversitario Struttura della Materia (CISM) del Ministero Pubblica Istruzione, Milan, Italy

P. N. Butcher, Department of Physics, University of Warwick, Coventry CV4 7AL, England

G. Caglioti, Istituto di Ingegneria Nucleare, CESNEF, Polytecnico di Milano, 20133 Milan, Italy; Gruppo Nazionale di Struttura della Materia del CNR, Unita di Ricerca 7, Milan, Italy; Centro Interuniversitario Struttura della Materia (CISM) del Ministero Pubblica Istruzione, Milan, Italy

F. Calogero Dipartimento di Fisica, Università degli Studi di Roma, "La Sapienza," 00185 Rome, Italy

P. Cvitanović, Institute of Theoretical Physics, Chalmers University of Technology, S-412 96 Göteborg, Sweden

A. Degasperis, Dipartimento di Fisica, Università degli Studi di Roma, "La Sapienza," 00185 Rome, Italy

P. De Kepper, Centre de Recherche Paul Pascal/CRNS, Domaine Universitaire, 33405 Talence Cedex, France

H. Haken, Institut für Theoretische Physik und Synergetik, Universität Stuttgart, 7000 Stuttgart 80, Federal Republic of Germany

Hao Bai-Lin, Institute of Theoretical Physics, Academia Sinica, Beijing, China

J. M. Hutchinson, Artificial Intelligence Laboratory, Massachusetts Institute of Technology, Cambridge, Massachusetts 02139, USA

C. Koch, Center for Biological Information Processing, Massachusetts Institute of Technology, Cambridge, Massachusetts 02139, USA. *Present address:* Division of Biology 216-76, California Institute of Technology, Pasadena, California 91125, USA

R. Loudon, Physics Department, Essex University, Colchester CO4 3SQ, England

S. Lundqvist, Institute of Theoretical Physics, Chalmers University of Technology, S-41296 Göteborg, Sweden

I. C. Percival, School of Mathematical Sciences, Queen Mary College, University of London, London E1 4NS, England

L. Pietronero, Solid State Physics Laboratory, University of Gröningen, 9718 EP Gröningen, The Netherlands

G. Rowlands, Department of Physics, University of Warwick, Coventry CV4 7AL, England

R. B. Stinchcombe, Department of Theoretical Physics, University of Oxford, Oxford OX1 3NP, England

A. Wunderlin, Institut für Theoretische Physik und Synergetik, Universität Stuttgart, 7000 Stuttgart 80, Federal Republic of Germany

Preface

This volume is concerned with the theoretical description of patterns and instabilities and their relevance to physics, chemistry, and biology. More specifically, the theme of the work is the theory of nonlinear physical systems with emphasis on the mechanisms leading to the appearance of regular patterns of ordered behavior and chaotic patterns of stochastic behavior. The aim is to present basic concepts and current problems from a variety of points of view. In spite of the emphasis on concepts, some effort has been made to bring together experimental observations and theoretical mechanisms to provide a basic understanding of the aspects of the behavior of nonlinear systems which have a measure of generality. Chaos theory has become a real challenge to physicists with very different interests and also in many other disciplines, of which astronomy, chemistry, medicine, meteorology, economics, and social theory are already embraced at the time of writing.

The study of chaos-related phenomena has a truly interdisciplinary character and makes use of important concepts and methods from other disciplines. As one important example, for the description of chaotic structures the branch of mathematics called fractal geometry (associated particularly with the name of Mandelbrot) has proved invaluable. For the discussion of the richness of ordered structures which appear, one relies on the theory of pattern recognition. It is relevant to mention that, to date, computer studies have greatly aided the analysis of theoretical models describing chaos. Indeed, important aspects of computer science are related to the theory of order and chaos.

The volume should, therefore, prove useful to a wide spectrum of readers: experimental and theoretical physicists, applied mathematicians, and chemists and biologists who are seeking deeper understanding of the appearance of patterns which may be either ordered or chaotic.

Finally, we note that the book had its origins in a College organized at the International Centre for Theoretical Physics, Trieste. We are, of course, grateful to all of the contributors. All but one chapter appears in its present form for the first time. Chapter 3 has, however, been reprinted, with permission, from an earlier book and we are grateful to the publisher.

Stig Lundqvist
Norman H. March
Mario P. Tosi

Contents

1

Chaos, Order, Patterns, Fractals—An Overview

S. Lundqvist

1.1. Introduction

This chapter is an introduction to order and chaos, as well as an overview of the field and little prior knowledge of the subject will be assumed. It will deal with the many phenomena that are caused by the inherent *nonlinear nature* that under some conditions leads to strong irregular behavior, *chaos*, and, at other times to characteristic *ordering phenomena*. Chaotic behavior occurs in a variety of systems: in classical situations such as mechanical systems, astronomy, hydrodynamics, dropping faucets, meteorology, plasma physics, electronic oscillators, and solid-state systems, and in physiology, an example being fibrillations of the heart. It also applies, for instance, to the growth of biological populations, economic theory, social theory, and to predictions of revolutions and war. The concept of chaos in quantum physics is a novel field of very special interest, but this chapter will focus primarily on the classical aspects.

Similarly, *ordering phenomena* also cover a wide range, e.g., the regular formation of cloud patterns, a variety of patterns in hydrodynamic flow, oscillatory patterns in chemical reactions, in the behavior of lasers, and pulse propagation in the Gunn diode. These patterns may be spatial, temporal, or spatio-temporal in nature and computing structures are of very special interest.

One purpose here is to introduce, in a broad elementary way, a number of important concepts, (limit cycles, attractors, strange attractors, Liapunov exponents, information entropy, self-organization), which will appear over

---

S. Lundqvist • Institute of Theoretical Physics, Chalmers University of Technology, S-41296 Göteborg, Sweden.

and over again throughout this volume. Many of these phenomena are described with the aid of a modern branch of mathematics called fractal geometry.

1.2. Classical Dynamical Systems

We start by recalling a few basic concepts from analytical dynamics and consider a system with N degrees of freedom, described by a Hamiltonian

$$H = H(\mathbf{q}, \mathbf{p}) \tag{1.2.1}$$

The equations of motion are

$$\frac{dq_k}{dt} = \frac{\partial H(\mathbf{q}, \mathbf{p})}{\partial p_k} \quad \text{and} \quad \frac{dp_k}{dt} = -\frac{\partial H(\mathbf{q}, \mathbf{p})}{\partial q_k} \tag{1.2.2}$$

These are $2N$ coupled nonlinear equations of first order in time with $k = 1, \ldots, N$. One can transform the equations to new variables $\mathbf{q}'$, $\mathbf{p}'$ such that the transformed equations of motions have the same form via a so-called *canonical transformation.* A particularly interesting case is when we can find a transformation to *action* and *angle variables,* $\mathbf{I}$ and $\mathbf{Q}$, so that the Hamiltonian can be expressed in the form $H = H(\mathbf{I})$. This means that the Hamiltonian only depends on half the variables. We can then trivially integrate the equations of motion to obtain

$$\mathbf{I}(t) = \mathbf{I}(0) \quad \text{and} \quad \Theta_\kappa(t) = \omega_\kappa t + \Theta(0) \tag{1.2.3}$$

and if we can find such a transformation, we have completely solved the problem and obtained $2N$ constants of motion $[\mathbf{I}, \theta(0)]$. The very simple time dependence of the solution is to be noted—simply a multiple periodic system. Thus, the problem of solving the original equations of motion has been shifted to finding the transformation to action and angle variables. If this problem can be solved, then it is said that the Hamiltonian system is *integrable,* and the method given in the textbooks to obtain the formal solution is usually referred to as the *canonical perturbation theory.*

The point of view traditionally presented in textbooks asserts that Hamiltonians for physical systems are generally integrable in the sense that the canonical perturbation theory converges and leads to well-defined results. However, as early as Poincaré, we were warned that most Hamiltonians may not have any well-behaved constants of motion except the energy itself, and that thus most dynamical systems are in fact *nonintegrable.* There are indeed very few examples of integrable systems but among such systems we can list:

1. All problems with only one degree of freedom.
2. All systems with linear equations of motions. In this case one can always find a normal-mode transformation such that the problem reduces to a set of independent one-dimensional problems. A well-known example is the theory of small vibrations around a stable equilibrium, such as the harmonic vibrations of a crystal lattice.
3. All nonlinear systems that can be separated into uncoupled one-dimensional systems. A timely and important example is given by the *soliton* waves in both continuous and discrete systems.

It is also worth noting that the theory of action and angle variables was used extensively in the quantization of physical systems even in the old quantum theory. In an integrable system the separation leads to N one-dimensional problems and thus one obtains N separate good quantum numbers, one for each action integral. By using the quantization rules of Bohr and Sommerfeld, one would simply put each action integral equal to $(n_k + \gamma)h/2p$. There is a very interesting paper by Einstein from 1917,[1] in which he presents a critical discussion of this procedure in view of the fact that many Hamiltonian systems are in fact nonintegrable. The theory of quantization of nonintegrable systems will be discussed later in this book.

In the study of Hamiltonian systems a standard method is to consider the *orbit* of the system in the $2N$-dimensional phase space of all the generalized coordinates q_k and their corresponding canonical momenta p_k. This is a very convenient way to hide all the real difficulties of the dynamics of real systems by formally reducing the problem to the motion of a point in $2N$-dimensional space—the *phase space of the system.* An important property of all systems in which energy is conserved is that the *volume* in phase space is also conserved. This means that if the orbit is defined within a volume element in phase space at a certain time, then this volume is unchanged during the subsequent motion. This is the famous Liouville theorem. One observes that only the volume will remain constant, while its shape may undergo dramatic changes.

As regards Hamiltonian systems, most of them are nonintegrable. The study of such systems has been intensified in the last three decades and has developed into a new multidisciplinary science, often called *nonlinear dynamics,* which forms a central part of the present volume. It has led to a wealth of new insights and explains a number of puzzling features of even very simple nonlinear systems. Systems of another class, even more important in condensed matter, are the *dissipative systems.* Owing to the dissipation, the volume in phase space is no longer conserved, and we have instead a reduction of the volume in phase space that often leads to a dramatic reduction in the dimensionality of the whole problem.

In the next section some very simple systems that can exhibit chaotic behavior will be mentioned briefly.

1.3. Some Simple Examples

1.3.1. The Simple Pendulum

We now introduce a few very simple models that exhibit nonlinear behavior and indeed will become chaotic if the nonlinear behavior is sufficiently strong. The well-known mathematical pendulum described by the equation

$$\frac{d^2\theta}{dt^2} + \omega_0^2 \sin\theta = 0 \tag{1.3.1}$$

$$\omega_0 = (g/l)^{1/2}$$

is considered first. For very small amplitudes, where we can set $\sin\theta \approx \theta$, the elementary solution is

$$\theta = \theta_0 \cos(\theta_0 t + \phi) \tag{1.3.2}$$

However, for larger amplitudes no such explicit solution exists. To obtain an understanding of the motion in the general case, it is convenient to study the position of the pendulum as a function of time, i.e., $\theta = \theta(t)$ together with the angular velocity $d\theta/\mathrm{d}t = \dot{\theta}$, namely the *phase portrait* of the system. In cases where the equations of motions cannot be integrated in closed form, a great deal of insight into the qualitative nature is gained by plotting the velocity as a function of the position, in this case the angle.

This example serves to introduce the important concept of *phase-space analysis.* In the case of the mathematical pendulum one just considers the trajectory of the pendulum in the plane (q, q). Since the mathematical pendulum without friction is a conserved system, i.e., the energy is a constant of motion, one obtains for this model the equation for the total energy:

$$E(\theta, \dot{\theta}) = \tfrac{1}{2}\dot{\theta}^2 + g/l(1 - \cos\theta) \tag{1.3.3}$$

The method is to exhibit the orbits in the phase plane for different values of the total energy, which is a constant of motion for conserved systems. The set of such curves gives a qualitative overview of the nature of the solutions of the dynamical model over a range of energies.

It should be noted that the motion is periodic and therefore only the interval $-\pi < \theta < +\pi$ in the phase plane need be considered. For small enough amplitudes (energies) one has a periodic motion, but a change occurs for the orbits connecting the points $(-\pi, 0)$ and $(+\pi, 0)$, and for higher energies an infinitely extended motion is obtained: in the upper region a clockwise motion, i.e. the angle increases indefinitely with time, while in the lower region the angle decreases with time.

For very small amplitudes the frequency of the pendulum is determined just by $(l/g)^{1/2}$. For higher amplitudes, i.e., stronger nonlinearities, the period changes and will depend on the energy of the pendulum.

1.3.2. The Damped Oscillator

The model in the previous section is too oversimplified to describe physically interesting situations for there is always friction in a mechanical system, which leads to a damping of the motion. The simplest assumption is that the damping is proportional to the velocity, and in the case of the pendulum, one obtains, for small amplitude oscillations, the equation in the linear approximation:

$$\frac{d^2\theta}{dt^2} + \gamma\frac{d\theta}{dt} + \omega_0^2\theta = 0 \tag{1.3.4}$$

Formally, the nature of the solutions depends on the sign of γ. For $\gamma > 0$ one has a damping of the motion, while in the case $\gamma < 0$ an unstable situation arises with an unlimited increase of the amplitude. The physically more common case is, of course, that of *dissipative systems* with positive damping, but there are also interesting cases where, e.g., nonlinear interactions lead to a motion with increasing amplitudes.

As for the simple pendulum, it is instructive to exhibit the motion in phase space. For positive damping, $\gamma > 0$, the motion is an inward spiral, converging toward the origin, but for negative damping the motion is an outward spiral. In the limiting case of zero damping, the motion is circular. In the case of small damping, i.e., $\gamma/\omega \ll 1$, one can solve the equation explicitly to obtain

$$\begin{aligned} \theta &\propto \exp(-\gamma\tau/2)\cos(\omega\tau + \phi) \\ \dot{\theta} &\propto \exp(-\gamma\tau/2)\sin(\omega\tau + \phi) \end{aligned} \tag{1.3.5}$$

The damped oscillator is an example of a general property of dissipative systems: *the contraction of phase space.* We consider a small surface element around the orbit. In the subsequent motion the surface element will decrease in size and tends toward zero when we approach the origin. In the general case there is not only a change in volume, but it is often associated with a drastic change in shape of the volume element so that the dimensionality is effectively reduced.

1.3.3. Oscillator with Nonlinear Damping

The assumption of a linear damping proportional to the velocity is often valid only for weak coupling and small amplitudes. In more general cases the interaction between the oscillator and its environment is more complex—the damping will depend on the amplitude so that the damping term will be nonlinear. The most studied case is the *Van der Pol* equation, replacing the damping constant γ by the function

$$\gamma(\theta) = -\gamma_0(1 - \theta^2/\theta_0^2) \tag{1.3.6}$$

For small amplitudes one gets amplification, i.e., negative damping, and the strength of the negative damping decreases with increasing amplitude, becoming zero at $\theta = \theta_0$. For $\theta > \theta_0$ the system becomes dissipative and the damping increases with the amplitude. Insertion of this formula into the equation for the damped oscillator yields the Van der Pol equation. It is convenient to change the units of amplitude and time and write the equation in the dimensionless form

$$\frac{d^2\theta}{d^2t} - (\varepsilon - \theta^2)\frac{d\theta}{dt} + \theta = 0 \tag{1.3.7}$$

By this change of units one can now study the properties of the solutions as functions of the dimensionless parameter $\varepsilon = \gamma_0/\omega$. There is a vast literature on the Van der Pol equation so we will only note a few points concerning the nature of the solution. It is recalled that for the damped oscillator with a constant damping constant one had a spiral inward with positive damping, $\gamma > 0$, and a spiral outward if $\gamma < 0$. In the limiting case $\gamma = 0$ one has a stable solution in the form of periodic circular motion in the phase plane. These properties make it easy to understand the solutions of the Van der Pol equation. If one starts rather close to the origin in the region of negative damping, the oscillator gains energy and spirals outward. However, as the negative damping gets weaker with increasing amplitude, the orbit converges toward a closed curve, a *limit cycle*, and the motion becomes periodic. If instead one starts at high amplitudes (energies), there is strong damping and the oscillator spirals inward, losing energy. As it loses energy, the damping decreases and the orbit converges toward one that is periodic, which is again the same limit cycle. The detailed nature of the orbits and the limit cycles depends on the choice of the parameter $\varepsilon = \gamma_0/\omega$. One can schematically distinguish three different cases:

1. The case of very small ε (damping). The motion here is dominated by the force and the limit cycle is very close to a circle. The motion in time at the limiting cycle is almost sinusoidal.
2. Rather strong damping, $\gamma_0 \cong \omega$. The limit cycle is now far from circular. The motion in time is of course still periodic but the time dependence is now represented by a Fourier series of the form $\sum X_n \sin(n\omega t + f_n)$.

3. Very large ε, $\gamma_0 \gg \omega$. The motion is now dominated by the nonlinear damping. As a result, the limit cycle now looks rather strange—in fact, not far from a rectangle. This also leads to a structure in the time-dependence, with essentially two time scales in the time-dependence of the amplitude.

1.3.4. The Parametrically Driven Oscillator

In the preceding section we discussed the effect of nonlinear effects in the damping term in the pendulum equation but even more interesting is the case where there is an external modification of the external driving force. Here one speaks of a parametrically driven oscillator, and we consider the example where the effective gravitational acceleration is changed. If one neglects the damping, this leads to an equation of the form

$$\frac{d^2\theta}{dt^2} + [g(t)/l] \sin \Theta = 0 \qquad (1.3.8)$$

1.4. Simple Systems with Two Degrees of Freedom

Section 1.3 dealt with a few simple examples of systems having only one degree of freedom, but still capable of showing strongly nonlinear behavior. Now some simple systems with two degrees of freedom will be introduced. First a system which is possibly the simplest nontrivial example of a Hamiltonian system, namely, two harmonic oscillators with a nonlinear coupling, will be considered. Then dissipative systems—in the simplest case a damped nonlinear pendulum driven by an external periodic field will be treated. In the latter case there is a competition between the two frequencies—that of the pendulum and that of the external field.

1.4.1. Hamiltonian System with Two Frequencies

Consider first an integrable system for which the Hamiltonian $H_0(\mathbf{q} \cdot \mathbf{p})$ possesses two degrees of freedom, i.e., in a four-dimensional phase space. The energy is conserved so that if no other constants of motion exist, the motion will be confined to a three-dimensional surface in the four-dimensional phase space. For an integrable system one has two constants of motion, J_1 and J_2, with frequencies ω_1 and ω_2. This means that the motion takes place on a two-dimensional surface embedded in three-dimensional space. The orbital motion will take place on a *torus*, in which case closed orbits are obtained only if ω_2/ω_1 is a rational number. If the frequency ratio is an irrational number, there is no periodicity in the motion but the orbit will approach infinitesimally close to every point in the course of time; in this case one says that the motion is *ergodic* on the torus.

1.4.2. A Perturbed Hamiltonian System

The next case to consider is that in which the integrable system H_0 is perturbed with a term εH_1 let us now try to integrate the perturbed problem in terms of the action and angle variables **J** and **q** and write

$$H(\mathbf{J}, \theta) = H_0(\mathbf{J}) + \varepsilon H_1(\mathbf{J}, \theta) \tag{1.4.1}$$

The standard method is to apply classical perturbation theory and, in lowest order, to obtain the correction to order ε. The details will not be worked out here, but it is found that the perturbation expansion in ε has denominators which may give rise to strong resonances and even divergences. In fact, because of vanishing denominators the expansion diverges whenever

$$\omega_1 n_1 + \omega_2 n_2 = 0 \qquad \text{or} \qquad \omega_1/\omega_2 = \text{a rational number} \tag{1.4.2}$$

This is what was noted in section 1.2, and, in this case, the result can easily be worked out. The conclusion is that the system cannot be integrated by perturbation theory when one has rational frequency ratios. It can possibly be integrated when one has irrational values for ω_1/ω_2, provided the perturbation expansion converges.

A detailed analysis of the properties of nonintegrable systems has been given in fundamental work by Kolmogorov,[2] Arnold,[3] and Moser,[4] and their major result is presented in a famous theorem called the KAM theorem. Incompletely and very simply put, the KAM theorem states that, if certain technical conditions are fulfilled, then those tori for which ω_1/ω_2 is sufficiently irrational will be stable under a perturbation εH_1 (provided the perturbation is not too large). The implication is that, in phase space, even in the presence of a perturbation, there will be regions of regular motion as for an integrable system and other regions in which the system will show a stochastic or *chaotic* behavior.

1.4.3. Poincaré Maps

For systems having several degrees of freedom it is not very practical to discuss the orbital motion in a multidimensional phase space. A more appropriate way is to study the intersections of the orbit with a plane in phase space, e.g., the (p_1, q_1) plane. In principle one can study Poincaré maps where the N-dimensional phase space is cut with a $(N-1)$-dimensional hyperplane, but they are generally far too complicated to be treated analytically. Therefore, one usually deals with simple systems which can be discussed in terms of one-

or two-dimensional maps. Although these maps do not contain the full dynamics of the systems, they are extremely useful for understanding some key properties of chaotic motion, e.g., about the "route to chaos."

1.4.4. Two Coupled Harmonic Oscillators

Let us consider here the case of two simple harmonic oscillators with a nonlinear coupling, described by the Hamiltonian

$$H = \tfrac{1}{2}(p_1^2 + q_1^2 + p_2^2 + q_2^2) + q_1^2 q_2 - q_2^3/3 \qquad (1.4.3)$$

a model first studied in a classical paper by Henon and Heiles.[5] When the total energy is very small, one has essentially the motion of two independent oscillators. In the limit of zero coupling, the energy of each oscillator is a constant of motion, i.e., one has two constants of motion. For small energies, the orbit generates a curve in two dimensions, which indicates that all motion is orderly and that the Hamiltonian is integrable at this energy. For higher energies only the total energy is a constant of motion, so that the motion takes place on a three-dimensional surface in phase space. For moderate energy some of the orbits still lie on two-dimensional surfaces and correspond to order motion, but there are also splattered dots which are generated by a single orbital that wanders around over the three-dimensional energy surface.

For even higher energies there is a complete transition from order to chaos. There is now only a small region in which one still has closed curves and most of the area is covered by dots that correspond to a single orbit, which covers part of the three-dimensional energy surface.

Here, intuitively, this example exhibits a very important property that distinguishes the orderly motion from the chaotic one. If one looks at points which are close to each other in the (q_2, p_2) plane, for orderly motion they will separate only linearly with time while in the chaotic case they will separate exponentially. This characteristic feature is described by the *Liapunov exponent*, to be defined precisely later.

It should be noted that the type of chaos described by this simple model occurs in several Hamiltonian systems. A famous example is the distribution of asteroids in the solar system. The motion of an asteroid around the sun is perturbed by Jupiter, so that one has effectively a three-body problem. The system is characterized by the angular frequency ω of the asteroid and the frequency of Jupiter ω_J. It was noted earlier that the equations are nonintegrable whenever the ratio between the unperturbed and perturbed quantities is a rational number, and this is clearly seen in the distribution of the asteroids as a function of the frequency ratio. It has been verified by some beautiful theoretical and computational work by Wisdom.

1.5. Dissipative Systems with Two Frequencies

We begin this section with, some introductory remarks about the properties of various simple dissipative systems having two competing frequencies ω_1 and ω_2. The system could be, for example, two clocks (pendulums) with a non-linear interaction or a damped pendulum driven by a periodic external field, or a parametrically driven system such as a pendulum wiggled up and down periodically at its support. A general feature of such systems is that of *phase locking*. The Dutch physicist Huyghens noted that two clocks hanging back to back on the wall tend to synchronize their motion. If, for example, one of the frequencies is varied the system will pass through a sequence of resonant regimes which are phase-locked and regions which are not. As long as the nonlinear coupling is weak, the regions with phase-locking will comprise only a small fraction. The motion will either be periodic when ω_1/ω_2 is a rational number or (much more likely) quasi-periodic when it is an irrational number. With increasing nonlinearity the phase-locked portions increase in size and, eventually, chaotic motion will occur in addition to the periodic or quasi-periodic motion. The mechanism leading to chaotic motion is the interaction between the different resonant regions as well as the overlap between these regions, when the coupling exceeds a critical value.

The following subsections are devoted to a few examples of dissipative systems characterized by the competition between two frequencies.

1.5.1. The Driven Damped Oscillator

This is the obvious example and is described by the equation

$$\frac{d^2\theta}{dt^2} + \gamma\frac{d\theta}{dt} + \omega_0^2 \sin\theta t = A + B\cos\omega t \tag{1.5.1}$$

Another version is the parametrically driven pendulum in which, for example, the gravitational force is modulated periodically by wiggling the support of the pendulum up and down:

$$g(t) = g_0 + g_1 \sin\omega t \tag{1.5.2}$$

If one lets the damping $\gamma \to 0$, a well-known model of transition to chaos due to Chirikov[6] (see also Kadanoff[7]) is obtained.

1.5.2. Josephson Junction in a Microwave Field

We next consider a Josephson junction, driven by a constant current with amplitude A and a microwave current with amplitude B, possessing a resistive shunting. The current through the junction is given by $I = I_c \sin\theta$, and the voltage across the junction is given by $V = (h/2e)\, d\theta/dt$, where θ is the phase difference across the junction. The time dependence of the phase is determined by the equation

$$\alpha \frac{d^2\theta}{dt^2} + \beta \frac{d\theta}{dt} + \gamma \sin \theta = A + B \cos \omega t \tag{1.5.3}$$

where γ is the critical current I_c. The differential equation is that of the forced damped pendulum with mass α, damping β, and gravitational field γ; it is experimentally well known that for certain values of the parameters, the junction can be driven into a noisy state and numerical calculations have indicated that the noise arises from chaotic solutions to the differential equation.

1.5.3. A Charge Density Wave in an Electric Field

As the third model system we consider that of a charge density wave (CDW). The CDW tends to move under the influence of a dc and ac electric field $E = A + B \sin \omega t$. In addition, the CDW is influenced by an oscillatory "pinning" potential. Again, this model will be represented by the differential equation for a damped driven oscillator:

$$\alpha \frac{d^2\theta}{dt^2} + \beta \frac{d\theta}{dt} + \gamma \sin \omega t = A + B \cos \omega t \tag{1.5.4}$$

The constants α, β, and γ are phenomenological constants representing the effective mass, damping, and the pinning potential; A is the dc depinning field and B is the amplitude of a radio-frequency field.

1.5.4. Phase Locking. The Circular Map

The phase-locking phenomenon in these three examples shows up in all these systems as a dependency for the average angular velocity $\langle dq/dt \rangle$ to lock into rational multiples of the frequency of the external field,

$$\left\langle \frac{d\theta}{dt} \right\rangle = \frac{N}{M}\, \omega \tag{1.5.5}$$

For the pendulum one knows that application of a small torque A will cause the pendulum to stay near the downward position. When one increases A, it is known that at a critical value the pendulum goes into a rotating mode in which the angular velocity is of the order A/B.

In the case of the Josephson junction, the voltage across the junction is given by the relation

$$V = \frac{h}{2e}\frac{d\theta}{dt} \tag{1.5.6}$$

This means that the phase-locking implies a locking $\langle V \rangle$. Therefore, steps will be seen in the current as a function of the voltage ($I-V$ curve). For $M = 1$, these are the well-known steps basic for the Josephson junctions. However, subharmonic steps with $M > 1$ have often been observed in between them.

In the charge-density wave system the current carried by the sliding charge-density wave is proportional to the velocity $d\theta/dt$, so the average current is

$$I \propto \left\langle \frac{d\theta}{dt} \right\rangle \tag{1.5.7}$$

Therefore a locking of the velocity implies a locking of the current carried by the charge-density wave. The ordinary current by the normal electrons behaves in the usual smooth way, being proportional to the dc electric field. One observes that the roles of voltages and currents are reversed for the Josephson junction and the CDW system.

The most efficient way to study phase-locking in these systems is to study the *return maps*, which are the analogues of the Poincaré maps introduced earlier. The system will be examined at the discrete times $T_n = 2\pi n/\omega$. Since the differential equation is second order in time, the values of the phase θ_n and $(d\theta/dt)_n$ at time T_n contain all the information about the subsequent motion and consequently determine the values at T_{n+1}. This defines a two-dimensional return map for the differential equation. Because the system is dissipative ($\beta > 1$), the area of the map will contract with time. The damping implies that the initial condition will soon be forgotten; one might hope that the motion will be asymptotically described by a unique invariant curve $\theta(t)$,[8] which also means that the velocity, asymptotically, will be a given function of θ,

$$\left(\frac{d\theta}{dt}\right)_n = g(\theta_n) \tag{1.5.8}$$

so that one gets just a smooth invariant curve on which the motion takes place. We now have effectively a reduction from two to one dimension and simply a map of the form

$$\theta_{n+1} = f(\theta_n)$$

where the function $f(\theta_n)$ maps the circle $0 < \theta_n \leq 2\pi$, which means that $f(\theta_n)$ is a *circular map.*

In general, $f(\theta)$ could be any periodic function of θ. However, there are reasons for believing that the specific form of f is not important for the key features of the transition to chaos, and one often uses the so-called "sine" circle maps defined by

$$\theta_{n+1} = f_\Omega(\theta_n) = \theta_n + \Omega - (K/2\pi) \sin 2\pi\theta \tag{1.5.9}$$

where the periodicity has been changed to 1 so that the points Θ and $\theta + n$ can be identified. The mode locking in the circle is examined by studying the iterations of the map, and the frequency of the dynamical system is given by the *winding number* of the mapping,

$$W = \lim (f_\Omega^n - \Theta)/n, \qquad n \to \infty \tag{1.5.10}$$

In the absence of nonlinear coupling, $W = \Omega$. On iterating the equation it may converge to a series in which W is either *periodic* with a rational winding number P/Q, *quasi-periodic* with irrational winding number, or *chaotic*, where the series has an irregular behavior. We shall not treat this topic at present, but the properties of circle maps will be discussed in detail later in the volume. For a further discussion of the topics in this section reference may be made to the review by Bak *et al.*[8]

1.6. Systems Described by Partial Differential Equations

Up to this point, discussion has been limited to a few extremely simple systems having only one or two degrees of freedom that are described by ordinary differential equations. However, a number of the most interesting phenomena for real systems are based on fundamental equations which are partial differential equations, i.e., they are systems with an infinite number of degrees of freedom. Probably the best studied case is that of *fluid dynamics.* The velocity field $\mathbf{v}$ of a fluid is described by the Navier–Stokes equation

$$\frac{\partial \mathbf{v}}{\partial t} + (\mathbf{v} \cdot \nabla)\mathbf{v} = -\frac{\nabla p}{\rho} + \eta \nabla^2 \mathbf{v} \tag{1.6.1}$$

where p is the pressure, ρ is the density, and η is the viscosity. The nonlinearity appears via the term $(\mathbf{v} \cdot \nabla)\mathbf{v}$ and the dissipative nature enters through the viscosity term.

The outstanding problem in fluid dynamics is that of *turbulent motion.* The flow pattern is often extremely complicated and frequently shows a very strong time-dependence, so that local physical measurements exhibit highly irregular (chaotic) behavior.

We now briefly discuss a specific example: the flow around a circular cylinder of diameter D. The parameter characterizing the flow pattern is the dimensionless *Reynolds number* R given by

$$R = \rho / \eta v D \tag{1.6.2}$$

One can distinguish the following phases starting from low speed: (a) $R < 1$: at very small Reynolds number the flow around the cylinder is smooth and regular. (b) $R \approx 20$–30: the pattern changes and one now obtains a pair of *vortices* behind the cylinder. (c) $R > 40$: there is a complete change in the pattern of motion. One of the vortices becomes so long that it breaks away from the cylinder and moves downstream. Then the fluid creates a new vortex behind the cylinder. The vortices peel off alternatively from each side. The velocity at any point now becomes *time-dependent.* (d) $R \approx 10^3$–10^4: there is a partially periodic and partially irregular flow. The irregular flow now turns and twists in all three dimensions. (e) $R \approx 10^5$: the turbulent region now works its way all the way back to the surface of the cylinder. There is probably no periodicity.

The transition toward turbulence can be seen if one looks at the time dependence of the local velocity in the five regimes just discussed. In (a) and (b) one has a steady flow, so that the local velocity will not change with time. In (c) the flow varies in a regular, cyclic fashion, and in (d) one has a time dependence that is partially periodic and partially irregular. In regime (e) one no longer sees any periodic structure and the velocity as a function of time looks completely chaotic.

Let us now turn to other phenomena for which the transition to chaos is much better understood both experimentally and theoretically. Best known is probably *convection or Bénard instability.* We consider a layer of fluid heated from below and kept at a fixed temperature at the top. As long as the temperature is sufficiently low, heat is transported by ordinary heat conduction. When the gradient reaches a critical value, a macroscopic motion starts in the fluid. The heated parts expand and will move upward, being cooled, and then fall back again to the bottom. This motion is extremely well regulated. Typically, one observes rolls, but hexagons (seen from above) have also been observed. When the temperature is further increased, the rolls start a wavy motion along the axis—a motion that varies with time. Ultimately the orderly pattern breaks up and a chaotic motion ensues, with no ordered structure. These patterns play a fundamental role in meteorology as they determine the movement of air and the formation of clouds.

The basic equations governing the Bénard motion of the system are:

(a) The Navier–Stokes equation in which one adds the force to gravitation, $\mathbf{F} = (0, 0, g)$.
(b) The equation for heat conduction

$$\frac{dT}{dt} = \kappa \nabla^2 T$$

(c) The equation of continuity

$$\frac{\partial \rho}{\partial t} + \operatorname{div}(\rho \mathbf{V}) = 0$$

The boundary conditions are that

$$T = T_0 + \Delta T \quad \text{(at the bottom)} \qquad \text{and} \qquad T = T_0 \quad \text{(at the top)}$$

The partial differential equations governing this problem can be transformed into a set of ordinary differential equations for the Fourier components of the various hydrodynamical quantities. The classic work in this field was the famous paper by Lorenz,[9] who truncated the equation taking only the leading Fourier component into account. The corresponding differential equations are

$$\frac{dx}{dt} = \sigma(y - x), \qquad \frac{dy}{dt} = x(r - z) - y, \qquad \text{and} \qquad \frac{dz}{dt} = xy - bz \tag{1.6.3}$$

where σ, b, and r are constant parameters.

Lorenz used his model as an analogue to weather forecasting. He showed that the final state is an extremely sensitive function of the initial state, and as the prediction period becomes longer, both the needs of accuracy in the initial data and the computational power required grow exponentially. This means in practical terms that true long-range detailed weather predictions are impossible.

This type of Bénard convection experiment provides us with ideal and beautiful examples of pattern formation in systems far from equilibrium and they can be extremely well controlled. A second example takes the form of Taylor instabilities, and in these experiments the motion of a fluid between coaxial cylinders is studied. One usually lets the inner cylinder rotate while the outer cylinder is kept fixed, but experiments have also been conducted with both cylinders rotating. At low speeds of rotation, coaxial streamlines are formed. Above a critical speed a pattern occurs in which rolls are formed where the fluid periodically moves outward and inward in horizontal layers. At still higher speeds, above some critical value, the rolls start to oscillate with one basic frequency and at even higher speeds one finds two frequencies. Indeed in some experiments a sequence of frequencies which are $\frac{1}{2}, \frac{1}{4}, \frac{1}{8}, \frac{1}{16}, \ldots$ of the fundamental frequency has been observed. This is an example of *period doubling*, which is a common feature of many non-linear systems. Eventually, at sufficiently high speed chaotic motion sets in. These experiments show several features that are typical in *self-organizing* systems. When an external parameter is changed (such as the speed of rotation) the system can pass through a variety of patterns, which can become more and more complex in their spatial and time structure.

As a second example of a system described by partial differential equations, one may take the reaction-diffusion equations, which describe chemical reactions in an inhomogeneous medium. They can be written in the form

$$\frac{\partial x_i}{\partial t} = f_i(\mathbf{x}) + D_i \nabla^2 x_i \tag{1.6.4}$$

The reaction kinetics is represented by the functions f_i, and the diffusion term pertaining to the inhomogeneity in space is described by the second term, D_i being the diffusion constant of the component i. The most famous example of the formation of ordered patterns in chemical reactions is the *Belousov–Zhabotinsky reaction.* The reaction itself is too complex to be described here, but the chemical patterns formed are of great interest in the present context. In the course of time the color of the liquid changes periodically from red to blue and back again. Indeed a clock could be built on the basis of this reaction. In the original experiment the substances were brought together and thoroughly mixed. In these cases the reaction died out after a few minutes and the system came to a final state of rest. In a modified experiment one continuously supplies fresh substances into the vessel and removes the reaction products. In this case one obtains a permanent reaction of periodic color change.

There are, however, much more spectacular and complex phenomena resulting from the Belousov–Zhabotinsky reaction. For example, in initially random centers, blue dots may form on a red background and grow into blue disks, in which a red dot appears, which quickly grows into a red disk. Another blue dot is produced in this and the cycle repeats itself. Concentric blue rings will travel outward. Under other experimental conditions spirals will be formed and travel through the liquid.

The molecules involved must be able to move in both the waves and the spirals formed in this reaction and, the diffusion term in the equation of motion takes care of this. The processes and patterns mentioned here are based on an interplay between the chemical reactions, on the one hand, and the diffusion mechanism, on the other, i.e., an interaction between the two terms in the reaction–diffusion equation.

1.7. Some Basic Properties of Chaotic Systems

This section in addition to being a summary of some of the work touched upon in the preceding pages will also include some slightly firmer statements, concerning a few of the points mentioned briefly earlier in the chapter. One important aspect to be dealt with is how to characterize chaotic motion and to possibly derive something of an initial level of understanding as to how chaos arises in completely deterministic systems. An equally important aspect is related to the experimental characterization of chaos. What kind of entities does one study in the experiments and what does chaos look like in the experimental curves? Let us discuss these properties in elementary terms in relation to the simple systems discussed in the previous sections.

1.7.1. Chaos in Poincaré Maps

Poincaré maps were first encountered in this chapter in connection with the two coupled oscillators (the Henon–Heiles model), where the orbit in phase space was projected on a plane in phase space. For regular motion, the projection of an orbit with constant energy is a closed curve, and if the energy is changed slightly one obtains a slightly shifted curve. At sufficiently high energies the system becomes chaotic and one gets a splatter of isolated points. These properties are seen also in general systems. In the case of a quasi-periodic system, the intersection with a plane gives rise to points that fall on a closed curve. In the chaotic region there is no predictable relation between one intersection with the plane and the next. Therefore one obtains a set of points that do not fall on any simple curve but show an irregular pattern all over the plane.

In the case of chaotic motion a characteristic feature is that curves which pass through neighboring points in phase space do not stay close as for nonchaotic motion; instead they separate exponentially. This implies an extremely high sensitivity to the initial conditions, as was noted in connection with the Lorenz model.

In order to characterize this sensitivity in a quantitative way, let us turn to the Poincaré map. For simplicity, we consider only the chaotic motion generated by one-dimensional maps. We consider a map

$$x_{n+1} = f(x_n) \tag{1.7.1}$$

that corresponds to chaotic motion, starting from the curve passing through two neighboring points, x_0 and $x_0 + \varepsilon$. The *Liapunov* exponent $\lambda(x_0)$ measures the exponential separation and we write after N iterations of equation (1.7.1)

$$\varepsilon \exp N\lambda(x_0) = |f^N(x_0 + \varepsilon) - f^N(x_0)| \tag{1.7.2}$$

Let us now pass to the limits $\varepsilon \to 0$ and $N \to \infty$ to obtain

$$\begin{aligned}\lambda(x_0) &= \lim \left|[f^N(x_0+\varepsilon) - f^N(x_0)]/\varepsilon\right| \\ &= \lim \frac{1}{N} \log \left|\frac{df^N(x_0)}{dx_0}\right|\end{aligned} \tag{1.7.3}$$

The meaning of the Liapunov exponent is that $\exp \lambda(x_0)$ measures the stretching of the distance between two adjacent points after one iteration. One can obtain an alternative definition by starting from the corresponding nonlinear differential equation in time (still only in one dimension)

$$\frac{dx}{dt} = F(x) \tag{1.7.4}$$

We consider again two neighboring trajectories,

$$x(t) = x_0(t) + \delta x(t) \tag{1.7.5}$$

By comparison with the case of the map one sees that one can find the exponential separation and define the Liapunov exponent by the formula

$$\lambda = \lim \frac{1}{t} \log |\delta x(t)|, \qquad t \to \infty \tag{1.7.6}$$

1.7.2. Chaos as Seen Directly in the Signal

As an example one can take the damped pendulum driven by an external field $A \cos \omega t$ and study the amplitude $q(t)$. As long as the strength is less than the critical strength A_c, one obtains a smooth and periodic signal as a function of time. For $A > A_c$ the signal becomes chaotic and the periodicity is lost. Hence in this and many other cases the chaotic behavior can be read off directly from the signal.

1.7.3. The Power Spectrum

In the preceding sections examples ranging from single or double and up to multiple periods, and the transitions all the way to chaos, have been given. In order to distinguish clearly between a complicated, multiple periodic behavior and chaos, it is often useful to Fourier transform the signal $X(t)$ with the aid of the equation

$$X(\omega) = \int dt \exp(i\omega t) X(t) \tag{1.7.7}$$

and study the *power spectrum*

$$P(\omega) = |X(\omega)|^2 \tag{1.7.8}$$

For multiple periodic motions one obtains a set of discrete lines, one line for each frequency, while a chaotic motion, which is completely aperiodic, has a broad continuous power spectrum at low frequencies. The previous qualitative discussion of turbulence is one example of the transition to chaos. One starts from a static response, passes into a regime with an oscillatory spectrum, at even higher velocity one obtains a superposition of oscillatory and aperiodic behavior, and ends with a broad chaotic spectrum.

Another example is the Bénard experiment. Again one starts out for a small temperature gradient with a static response, then comes a phase of periodic oscillations, followed by a sequence of period doublings, leading finally to a broad chaotic spectrum.

1.7.4. The Autocorrelation Function

As the last example let us consider the case of chemical reactions in a system with several chemical components $\mathbf{c} = (c_1, c_2, \ldots, c_n)$. The general equation of motion is a set of first-order nonlinear equations:

$$\frac{d\mathbf{c}}{dt} = \mathbf{F}(\mathbf{c}, \lambda) \tag{1.7.9}$$

The components of $\mathbf{c}$ are the concentrations of the chemical components and $\mathbf{F}$ is a nonlinear function of the concentrations; λ denotes an external parameter.

We consider here the Belousov-Zhabotinsky reaction. The variable that signals the chaotic behavior is the concentration c of the C^{4+} ions. We denote by $c(t)$ the *deviation* of $c(t)$ from its equilibrium value.

Next let us introduce the *autocorrelation function*, defined by

$$C(t) = \lim \frac{1}{T} \int dt\, c(t) c(t + \tau) \tag{1.7.10}$$

This function is a measure of the correlations between subsequent signals. For regular motions it remains constant in time or oscillates, but in the chaotic regime it exhibits a chaotic behavior with an envelope that decays essentially exponentially to zero.

1.8. Some Different Routes to Chaos

So far, how the transition from regular to chaotic behavior takes place has been treated superficially. A brief discussion will follow which indicates some possible routes to chaos. We first note that there is no universal route to chaos. However, it is well-known that there is a rich hierarchy of different instabilities before the chaotic state is reached. The brief presentation to be given below will serve as a preview and summary of material that will be presented in much more detail in later chapters.

1. When a typical control parameter such as the energy, Reynolds number, or the pumping power of a laser is increased, the number of frequencies $\omega_1, \omega_2, \omega_3, \ldots$ of the system gradually increases. In fluid dynamics this is called the *Landau–Hopf picture.* In this picture the turbulent state of the liquid can be characterized by an infinite number of oscillations with no simple relations (e.g., rational) between the frequencies. In fluids, this picture is no longer accurate since there is evidence that chaos sets in as early as after one observes oscillations at two (or three) frequencies. However, in laser systems situations have been found where more and more frequencies occur.
2. A different picture was proposed by Ruelle and Takens[10] on mathematical grounds, according to which chaos should set in after the system has reached an oscillatory state at two basic frequencies. This corresponds to motion on a two-dimensional torus and seems to be observed in fluid dynamics, although motion on a three-dimensional torus has also been observed. However, this theory is essentially based on some mathematical properties of the system, and may not be immediately applicable to real physical situations.
3. A very popular model has been that of a route based on a sequence of period doublings. In this picture one finds, with increasing value of the control parameter, that the period of oscillations undergoes a doubling at specific values. The universality of this type of transition was discovered and discussed by Feigenbaum.[11] This type of period doubling has been observed in several systems. However, a number of other types of subharmonic generation have also been found. Therefore, the universal character of the route to chaos may be more subtle and complex than what appears from the period-doubling theory.
4. As a final point in this brief review we refer to the phenomenon of *intermittency.* In this situation a typical physical quantity remains static

for a while, then suddenly shows a chaotic outburst for some period of time, then is static again, has another chaotic outburst, and so on.

It should finally be noted that one sometimes finds alternating sequences between periodic and chaotic oscillations on increasing the control parameter(s) of the system. Indeed there is a variety of possibilities and this field opens up opportunities for much further research.

1.9. Some Comments about Attractors

1.9.1. Introductory Remarks

Dynamical systems can be divided into two broad categories—conservative and dissipative—depending on whether or not the energy is conserved. Some examples have already been treated, e.g., the damped oscillator, driven or not, and the Van der Pol oscillator, as examples of systems with one or two degrees of freedom. The Navier-Stokes equations for a fluid have also been mentioned as an important example of a dissipative dynamical system with infinitely many degrees of freedom. In these systems the viscosity converts the energy into heat. Of special interest is the long-time behavior of dissipative systems. This is controlled by various *attractors*, which means that, starting from different initial conditions, the motion evolves in time toward an attractor, and after a sufficiently long time during which the initial transients die out, the motion will reduce to motion *on the attractor.* A couple of examples have already been encountered and the discussion will be extended here by introducing the following three categories from the simple examples introduced earlier:

1. The *fixed point* of a damped pendulum is the simplest case of an attractor. Analogous situations are found in the motion of a fluid at low Reynolds numbers, or in the Bénard experiment at small temperature gradients when steady convection rolls are formed. In this case the liquid is moving, because one has a constant flow pattern, but the attractor is a fixed point.
2. A second type of attractor was found in the discussion of the Van der Pol oscillator. In this case one has regions in phase space with ordinary positive damping and other regions with negative damping. However, irrespective of the starting point, the oscillator moves with time asymptotically into a closed curve in phase space, an attractor called a *limit cycle.* In the Bénard experiment, a limit cycle corresponds to a situation in which the convection rolls oscillate periodically, so that the flow pattern changes periodically in time.
3. The third type of attractor is much more subtle and less familiar. It is called a *strange attractor*, a term first introduced by Ruelle[(12)]—also

sometimes called a chaotic attractor. If one considers a map of the motion, one finds that the strange attractor describes a chaotic motion of the system in the sense that the sequence of successive points is random. The turbulence observed in a fluid can be described by a strange attractor.

It should be pointed out that all of these kinds of attractor can occur in the same physical system when one varies a control parameter (such as the Reynolds number in a fluid).

This introduction is concluded with a few remarks that will be further clarified in later chapters.

One of the most remarkable aspects of this subject is that there is now very strong evidence from a range of experiments that the chaos in a number of real physical systems can be described in terms of strange attractors in just a few dimensions. This implies that out of the infinite number of degrees of freedom in a large dissipative system, only very few will be active. This is a most fundamental aspect of the physics of these systems and will be discussed in several later chapters and, in particular, in the chapter on synergetics. This property is the basis of the fundamental concept *order in chaos.* Since most of the initial conditions lead to a motion that collapses on the attractor, the number of degrees of freedom actively participating in the chaos can be extremely small in comparison with the total. However, the chaos is real since points that are close to each other on the attractor separate initially at an exponential rate (determined by the Liapunov exponent). This causes small errors to increase very rapidly and therefore produces an extreme sensitivity to the initial conditions.

Someone has remarked that strange attractors act like a baker. If one thinks of the whole space of all possible initial conditions as the dough, one can picture the strange attractor grabbing the dough and stretching it, and then folding the whole thing back on itself. Suppose a tiny piece of saffron is quickly mixed in throughout the dough under the baker's hands. This is intuitively what happens when one has a strange attractor creating chaos. Such a situation can be illustrated with a simple "baker's transformation" in two dimensions.

A fundamental aspect of this idea is the reduction of dimensions in a large system when one reaches the chaotic regime, where the system can be described in a very small number of dimensions. Of particular interest is the dimensionality of the strange attractor and its other properties. If we think about the outcome of the folding described above, we arrive at the notion of the degree of complexity. One could just think about the chaotic motion of a particle and compare its orbit as a function of time with a plate of very thoroughly mixed spaghetti (and this system is much too simplified even to serve as a good illustration!). Indeed the complicated structure of the strange attractor does not fit in with our usual concept of dimensionality, and a new geometry developed by Mandelbrot[13] is needed. Known as *fractal geometry* this geometry is playing a central role in many developments in physics. In fact the complicated structure of a strange attractor cannot be described conventionally in terms of integer dimensions, and the fractal dimension, used in Mandelbrot's geometry, provides the proper tools. Later on, some key concepts will be introduced and, in particular, the use of fractal dimensions.

1.9.2. Some Properties of Attractors

The properties of attractors were mentioned in the earlier sections and here this introductory discussion will be extended somewhat. We recall some simple dissipative systems like the Van der Pol oscillator. In such systems the trajectory converges toward a *limit cycle*, wherever one starts in the phase plane.

We remind the reader that the volume in phase space always decreases for a dissipative system and that this effect for a many-dimensional system leads typically to a dramatic reduction in the dimensionality, when one approaches an attractor with a dimension much smaller than that of the dynamical system itself. In the case of a simple system, such as the pendulum, it is easy to see that there is not only a shrinkage in size but also a change in shape, so that one gets gradually a very narrow strip that closes up on the attractor, i.e., one has effectively a reduction in dimension from two to one. We note that when a system has reached the limit cycle, the motion from then on is always periodic (in one dimension).

A trajectory does not always end up on an attractor. It is convenient to define a *basin of attraction*, such that any trajectory starting within the basin of attraction will end up on the attractor.

A characteristic feature of chaotic motion is that periodicity is lost. As discussed earlier, one has an aperiodic behavior in properties such as the power spectrum, the autocorrelation function, and the Poincaré maps. Also characteristic is an extreme sensitivity to initial conditions.

The attractor for a dissipative system now becomes a much more complex quantity and Ruelle introduced the name *strange attractor*. A more detailed account will be given in later chapters and therefore it will suffice here to note only a few key properties of strange attractors:

1. It is a bounded region in phase space to which all trajectories from the basin of attraction will be attracted asymptotically. The basin of attraction itself can have a highly complex structure. The attractor should possess the property that the trajectory passes every point on the attractor in the course of time. That means that it cannot be broken up into disconnected parts, or be a collection of isolated fixed points.
2. A key property that makes an attractor strange is sensitivity to initial conditions. This means that points on the trajectory that are initially

close will separate from each other on the attractor, so that there will be a macroscopic separation between the two trajectories. We note that this happens in spite of the contraction of volume in phase space. This means that lengths will not shrink uniformly in all directions and, as a result, one has this reduction in the number of dimensions. A typical case is a flow that contracts the volume in some direction and stretches it in others. Since the attractor must be confined to a bounded region, the volume element should be folded at the same time. This type of stretching and backfolding is what produces a chaotic motion on the strange attractor.

3. In order to describe a physical system, one requires a kind of structural stability of the attractor and it should be generic. This means that the structure of the attractor changes in a continuous way when changing the parameters of the system.

It should be noted that the concept of strange attractors does not hold only for the trajectories in phase space, but for discrete dissipative maps, such as Poincaré maps, as well.

All strange attractors discussed in the literature correspond to nonintegral dimensions—they are called *fractals.* We therefore leave strange attractors for a while in order to give a brief and elementary introduction to fractals.

1.10. Fractals

1.10.1. Introductory Remarks

One is, by now, quite used to describing physics in 0, 1, 2, 3 and 4 dimensions and, in high-energy physics, one uses many more dimensions. Here the concept of *fractal dimension* will be introduced. As was noted earlier, the name fractal was introduced by Mandelbrot, who also demonstrated how these concepts provided the proper geometry with which to describe a variety of phenomena in many disciplines of natural science. Much of his work up to 1982 is summarized in his most interesting book.[13] In this chapter the fractal properties will only be considered in connection with the geometrical nature of strange attractors. However, it should be pointed out that the fractal concept also plays an important role in the description of a number of complex ordered patterns and applies to a variety of systems from galaxies to human flesh!

The concept of dimension really plays a different role in physics from that in geometry because the science of physics deals with observation of physical phenomena and one often has reason to talk about an *effective dimension.* Mandelbrot takes the example of the many different effective dimensions of a ball of thread as a consequence of the resolution of the observation. An observer who sees the ball from afar sees it as a point, i.e., a zero-dimensional object. When seen from a closer vantage point the ball appears as a three-dimensional figure. Close-up the ball looks like a mass of one-dimensional threads. At an even higher resolution each thread is seen as a column and the system again becomes a three-dimensional object. On increasing the resolution further, the column is resolved into fibers and the ball looks one-dimensional. One can proceed in this way until the individual atoms are seen.

Most objects in the physics of fractals are similar to a ball of thread, in that they exhibit a succession of different effective dimensions. However, there is an important new element added: there are certain ill-defined transitions between regimes of well-defined dimensions, and these zones are interpreted as being fractal zones in which the dimension is larger than the topological dimension.

1.10.2. The Fractal Dimension

Following Mandelbrot[13] a fractal will be defined as an object of *a set for which the Hausdorff dimension strictly exceeds the topological dimension.* This definition asks, in turn, definitions of the term *set*, the *Hausdorff dimension* D, and the *topological dimension* D_T. We will proceed in an intuitive way by employing examples rather than by entering into a mathematical discussion. The fractals to be discussed can be considered as sets of points embedded in space. The topological dimension D_T is always an integer.

The concept of the distance between points in space is central for the definition of the Hausdorff dimension D. We consider a simple example in two dimensions, $D = 2$, and set $N(r)$ to be the number of points inside a circle of radius r. The fractal dimension is determined by the variation of $N(r)$ with r. In the case of a regular lattice one has $N(r) \propto R^2$, so that $D = 2$. In the case of a less dense lattice, $N(r) \propto R^D$ with $D < 2$. If all the points lie on a single smooth curve, then it is clear that in this case $N(r) \propto r$ so that $D = 1$. When, however, the curve is ill-behaved, e.g., so that it folds back close to itself many times for small r, then one has again a case of a fractal dimension with $1 < D < 2$. If instead the density of points on the line is less dense than that corresponding to a uniform density (such as a Cantor set), one will have dimension $D < 1$. In the limiting case, with only one point inside the circle, we have of course $D = 0$.

This kind of reasoning can obviously be extended to spaces of higher dimension, replacing the circle in the example by a hypersphere, where $N(r)$ is the number of points in a hypersphere of radius r. The dimension D defined by these arguments is called the Hausdorff–Besicovich dimension.

Next we make these arguments a bit more precise. A set of points is examined in a p-dimensional space and we try to cover the set with cubes of side ε. If the smallest number of points $N(\varepsilon)$ needed to cover the set increases as

$$N(\varepsilon) \propto \varepsilon^{-D} \quad \text{for } \varepsilon \to 0 \tag{1.10.1}$$

where D is the Hausdorff dimension, then equivalently one can write

$$D = \lim_{\varepsilon \to 0} \frac{N(\varepsilon)}{\ln (1/\varepsilon)} \tag{1.10.2}$$

We just check that this definition makes sense. If the set contains just one point, then $N(\varepsilon) = 1$, so that $D = 0$ in agreement with the usual definition. Similarly, if the set comprises points distributed along a curve of length L, one has $N(\varepsilon) = L\varepsilon^{-1}$ giving $D = 1$. For a set of points covering a surface S uniformly, one obtains $N(\varepsilon) = \varepsilon^2$ giving $D = 2$. What has been shown is just that the definition agrees with the Euclidean notions, as it should.

New insight is only obtained by going beyond the trivial examples. In the following sections a few characteristic examples will be mentioned briefly. The book by Mandelbrot[13] is recommended for a detailed discussion with many illustrations and many more examples.

1.10.3. The Cantor Set ($0 < D < 1$)

Let us here illustrate what has just been introduced. We construct a set with fractal dimension between zero and one, by starting from a straight line and cutting out successively shorter and shorter gaps. This procedure was introduced by Cantor in 1883 and the resulting set is referred to as a *Cantor set.*

The best-known example is the so-called ternary set, obtained as follows. Start from the interval $[0, 1]$, divide it into three equal pieces, and then remove the middle piece. Next remove the open middle part of each of the two remaining thirds and keep on repeating this procedure. This produces extremely short segments after only a few steps. After an infinite number of steps, what remains of the unit interval is just an infinite number of points scattered over the interval.

This set is often called *Cantor dust.* It is clear from the construction that the set has the property of being *self-similar,* in the sense that the structure in each step of the division is always the same (a stringent definition of self-similarity would require more care). Since at each step one is left with two pieces and the interval is divided into three parts, one can simply take $N = 2$ and $e = \frac{1}{3}$ in our formula to obtain

$$D = \ln 2/\ln 3 = 0.6309 \tag{1.10.3}$$

Let us mention briefly a modified meaning of the Cantor set. We replace the unit interval by a bar of some material of unit length and density $r = 1$. The previous operation is now replaced by the following procedure: Cut the bar into two halves of equal mass $p_1 = 0.5$ and then hammer them so that the length of each part becomes $I_1 = \frac{1}{3}$. The density in each part now increases to $r_1 = p_1 I_1 = \frac{3}{2}$. Then repeat the same procedure over and over again. In the nth generation we have $N = 2^n$ small bars, each of length $I_i = 3^{-n}$ and mass $p_i = 2^{-n}$ for $i = 1, 2, \ldots, N$. It follows from the construction that the total mass is conserved, hence $\sum p_i = 1$. This treatment means that the original Cantor bar with a uniform mass distribution is replaced by a large number of tiny slugs in small regions of high density. Mandelbrot calls this process *curdling*.

The mass of a segment of length I_i is given by $p_i = I_i^{\alpha}$, where the *scaling exponent* α is given by $\alpha = \ln 2/\ln 3$. The density of each of the small segments is given by

$$\rho_i = p_i/I_i = I_i^{\alpha-1} \tag{1.10.4}$$

It can be seen that the density diverges in the limit $N \to \infty$ and that the divergency is controlled by the scaling exponent.

What has been shown is just an example, where p is selected as the mass of a segment. One could as well have considered p to be the charge, magnetic moment, or the measure for some physical phenomenon.

This section concludes with a remark about the *Devil's staircase.* Calculate the total mass found in the interval $[0, R]$ as a function of R:

$$M(R) = \int \rho(x)\, dx \tag{1.10.5}$$

We note that the density $r(x)$ is zero in the gaps and infinite on all of the infinite number of points that constitute the Cantor set. The integrated mass $M(R)$ remains constant on the intervals that correspond to the gaps. The total lengths of the gaps add up to the length of the whole bar, so one might be tempted to conclude that $M(R) = 0$. However, the mass increases by infinitesimal amounts at all the points of the Cantor set. The mass as a function of R looks like a staircase—the *Devil's staircase*—being horizontal (almost) everywhere. The self-similar property is obvious from the construction.

1.10.4. The Koch Curve ($1 < D < 2$)

Before presenting the famous example of the Koch curve, a general remark also applicable to the Cantor set will be made, and seems to be a fundamental property of fractal systems—the property of *self-similarity.* This applies to systems in all dimensions and simply means that when each piece of a shape is geometrically similar to the whole, both the shape and the cascade generating it are called self-similar. The property of self-similarity implies a very important property, fundamentally different from what was learned in school about smooth variations that can be described by low-order differential equations. The Koch curve to be presented is an example of obtaining in the limit a closed curve that encloses a finite area but has infinite length. It was first constructed by Koch in 1904.

The construction is as follows: Consider an equilateral triangle with sides of unit length. At each side one pastes onto the middle third another triangle with side of length $\frac{1}{3}$. This new construction ends with a star of David. One continues by adding more and more triangles to the sides. It is clear that one generates the whole pattern (which is similar to a snowflake) by the starting pattern. The self-similarity of this process is evident and in each step $N = 4$ and $e = \frac{1}{3}$, so that the dimension becomes

$$D = \log 4/\log 3 = 1.2618 \tag{1.10.6}$$

It should be noted that each step in the construction increases the total length of the curve by a factor of $\frac{4}{3}$. Therefore the limiting curve will be of infinite length. The Koch curve also serves as an example of a continuous curve having no tangent anywhere—a continuous function having no derivative anywhere. A number of interesting constructions with higher fractal dimensions are discussed in Mandelbrot's book.[13]

1.10.5. Fractal Lattices and Nets: The Menger Sponge ($2 < D < 3$)

In standard geometry a lattice is created by a set of parallel lines forming a regular design. This seems to apply also to regular fractals in which two points can be linked by at least two paths that do not otherwise overlap. When the pattern is not that regular it is more appropriate to talk about *fractal nets* or, sometimes, *fractal lattices.* There are indeed some important differences between regular lattices and fractal systems. Ordinary lattices are invariant under translations but not under scaling. The reverse is true for fractal systems; they scale but generally have no translational invariance.

Mere mention will be made here of the construction by Sierpinsky[14] and referred to as the *Sierpinsky carpet* (see especially chapter 9, Section 9.2). The analogue in three dimensions is the *Menger sponge.*[15] Reference is made to the illustration of the Sierpinski carpet and the Menger sponge in Mandelbrot's book.[13]

1.11. Examples of Strange Attractors

A few illustrations of strange attractors from both theoretical models and experimental studies will now be discussed very briefly, but for a real understanding of these objects, the reader must refer to the books by Cvitanovic or

Hao Bai Lin, or to the original papers. It is appropriate to start with the Lorenz model,[9] described by the differential equations

$$\frac{dx}{dt} = \beta(y - x), \qquad \frac{dy}{dt} = x(r - z) - y, \qquad \text{and} \qquad \frac{dz}{dt} = xy - bz \tag{1.11.1}$$

First consider a part of the trajectory projected on the y-z plane. There are two unstable fix points C and C'. The chaotic behavior is clearly seen from the time dependence of any of the variables.

For small values of β the stable solutions of the model are time-independent. For arbitrary initial values the trajectories are rapidly attracted toward the bounded region. The motion is highly erratic—the trajectory makes one loop to the right, then a few loops to the left, then to the right again, and so on. There is also as we noted before a very sensitive dependence on the initial conditions. If even a slightly different initial condition is used, the solution will soon start to deviate from the original one and even the number of loops will be different.

One should also mention that the system is strongly dissipative, a fact not immediately seen from graphic illustrations of the solution. The volume element contracts exponentially in time:

$$V(t) = V(0)\exp[-(\sigma + b + 1)] \tag{1.11.2}$$

Mention should be made finally that the fractal dimension of the attractor is $D = 2.06$, which means that the attractor is almost planar.

A technique that will be used frequently in this volume is to study the development of a system at discrete times $t_1, t_2, t_3, \ldots$. This means that one can replace a set of differential equations as in the Lorenz model, by a set of recursion relations. A much-studied case is the model introduced by Henon[16] in two dimensions and defined by the relations

$$x_{n+1} = 1 - ax_n^2 + y_n \qquad \text{and} \qquad y_{n+1} = bx \tag{1.11.3}$$

This system is area-contracting, i.e., dissipative for $|b| < 1$. The parameter a controls the nonlinearity and b measures the dissipation. The parametric values mostly used are $a = 1.4$ and $b = 0.3$. Starting from a point x_0, y_0 one can calculate x_1, y_1 and so on. For the parameters mentioned the fractal dimension of the Henon attractor is $D = 1.26$, so the attractor is not so far from one-dimensional.

The motion on the attractor is highly erratic, i.e., the point corresponding to the next iteration is quite some distance away. It turns out that the attractor has the characteristic property of fractals, i.e., *self-similarity*.

The Henon model shows that the dynamics of the system are highly complex and not characterized by the fractal dimension alone; in order to characterize the chaotic motion one often uses the *Kolmogorov entropy.*[17] It is known from thermodynamics that the entropy S measures the disorder in a system and increases with increasing disorder. This increase in disorder is related to an increasing ignorance about the state—with decreasing entropy one has more order and knows more about the system. From statistical mechanics we have

$$S = -k_B \sum \ln p_i \ln p_i \tag{1.11.4}$$

where p_i are the probabilities to find the system in states i. The founder of information theory, Claude Shannon,[18] is of the opinion that this measures the information needed to locate the system in a certain state i. This means, according to Shannon, that S as defined by equation (1.11.4) is a measure of our ignorance of the system.

The thermodynamical entropy is defined only for systems at or near thermodynamical equilibrium and is a function only of the energy of the system. The function defined by equation (1.11.4) can be used in very general situations. In fact disorder is essentially a concept from information theory, and in relation to the problems of chaos and order, information theory has become one of the most powerful tools. It is therefore not surprising that one can define a Kolmogorov entropy K that measures how chaotic a dynamical system is. It can in fact be defined by Shannon's formula in such a way that it comes proportional to the rate at which information about the system is lost with increasing time. The Kolmogorov entropy is defined so that it becomes zero for regular motion, is a constant larger than zero for the kind of chaotic motion we have been discussing, and becomes infinitely large in systems that are completely random. Methods from information theory and other related fields are very important for an understanding of complex systems and such properties as how patterns are formed.

Concepts such as strange attractors and fractal dimensions are not only relevant in theoretical physics, but represent quantities that can be measured. In the last decade there has been a steadily increasing volume of most significant experimental studies concerning these quantities that has provided a great stimulus for the entire field. We will note here just some of this work. The beautiful experiments by Libchaber[19] were crucial in attracting attention to the field. In fluid dynamics mention should be made of work by Ahlers,[20] Gollub[21] and Swinney,[22] in solid state physics the work by Jeffries[23] is notable, and in laser physics and nonlinear optics there have been many studies.

1.12. More about Maps

This section is devoted to a few more comments about some simple maps in order to introduce some additional important concepts. In particular such maps show how even the simplest maps lead to an unexpectedly rich and complex structure and how in appropriate cases chaos emerges out of the very simplest deterministic equations.

1.12.1. A Very Simple Model

Consider the forward iteration of the simple first-order difference equation

$$x_{n+1} = f(x_n) = 2x_n \text{ (Mod 1)} \tag{1.12.1}$$

which is obviously a deterministic one. Mod 1 means that one should drop the integer part, so that the variable is always in the unit interval: one has a mapping of the unit interval onto itself.

This simple difference equation has the simple analytical solution

$$x_n = 2nx_0 \text{ (Mod 1)} \tag{1.12.2}$$

In order to gain further insight into the nature of this solution, the initial value in binary representation is expressed in the form

$$x_0 = Sa_i 2^{-i} = (0, a_1, a_2, \ldots) \tag{1.12.3}$$

where all the coefficients a_i assume values 0 or 1.

In this representation one sees that the effect of multiplying by 2 (Mod 1) just means that one has to delete the first digit and shift the whole sequence one step to the left, i.e., moving the "decimal" point. For example, if one starts from $x_0 = 0.101101\ldots$ one obtains

$$x_1 = 0.01101\ldots, \qquad x_2 = 0.1101\ldots, \qquad x_3 = 0.101\ldots, \qquad x_4 = 0.01\ldots$$

This means that for large i the value of x_i is extremely sensitive to the exact value of x_0. Suppose one has two initial values that differ by a small quantity δ, and this difference grows to $2^n\delta$ after n iterations.

All the iterates lie on the unit interval. We partition the map into two parts: one left cell in which $0 < x < \frac{1}{2}$, called L, and one right cell with $\frac{1}{2} < x < 1$, called R. A given x_n falls in the right cell if the first digit is one, while it falls in the left if the first digit is zero. For example, the number written above for x_0 can equally well be characterized by the cell number, in which case one obtains RLRRLR.... On completing this cell number sequence we find that it is identical to the binary expansion of x_0.

This means that the binary sequence for x_0 is isomorphous to the sequence RL.... Thus there is an interesting analogy with the random process of tossing a coin. Indeed, choosing a sequence R,L... becomes equivalent to tossing a coin (an interesting article by Ford[24] is recommended for further study). This isomorphism has some interesting consequences. The original equation is completely deterministic and allows one to compute the future value as soon as the initial values are known. On the other hand, the sequence R, L... could be obtained as a result of just tossing a coin, and can be considered as a truly random process. The completely deterministic iterates will hop between right and left according to a rule that cannot be distinguished from a random sequence produced by tossing a coin.

This simple map clearly illustrates the mechanism of stretching and back-bending, which is a characteristic feature of general nonlinear maps. If one starts from a small value of $x_0 < \frac{1}{2}$, one sees that each iteration stretches x_0 to a larger value by a factor of two. But after a certain number of iteration x_n would become larger than one, and then x_n is folded back to the unit interval.

This stretching property of nonlinear maps have some strong physical consequences. The initial conditions are only known with a finite precision. Any small error becomes exponentially amplified via the nonlinear equation. This means that the equation acts like a microscope, which is sensitive to the precision in our measurement. This raises the interesting question about the concept of the continuum in connection with physical measurements. In fact equation (1.12.1) is often used as an algorithm to obtain pseudorandom numbers. Already the brief indication given here about the possible relation between deterministic equations and random numbers leads to one of the central problems in probability theory: How does one characterize randomness in a given string of digits?

Let us consider a given sequence of binary digits. Each single digit carries by definition one bit of information. Therefore a sequence of length n could carry n bits of information. Very often, however, the digits are correlated, and the information carried by the whole sequence can be expressed by a much shorter sequence. The shorter sequence could, for example, be a brief computer code that might generate the original n-digit string. This kind of thought led Kolmogorov,[(25)] Chaitin,[(26)] and Solomonov[(27)] independently to the important step of formulating an algorithmic theory of *complexity*. The concept complexity will recur many times throughout this volume and it should be pointed out that the concept is not at all unique. What is mentioned here has to do with the complexity related to sequences of numbers, and in the description of the complexity of physical systems this definition may not be suitable.

Here we limit ourselves to some statements about the algorithmic theory of randomness and define the concept of complexity as follows: *The complexity of a series of digits is the number of bits that must be put into a computer to obtain the original series as an output. This means that the complexity equals the size (in bits) of the minimal program of the series.*

With this concept one can now define randomness in a more rigorous manner: *A random series of digits is one whose complexity is approximately equal to its size in bits.* By this definition of complexity one cannot calculate an n-bit sequence with maximum complexity by any algorithm, if bit lengths are appreciably less than the length of the sequence itself. The simplest way to define the sequence is to provide a copy of it. In fact the sequence of these digits in the sequences of maximum complexity are so difficult to calculate and define, and therefore so unpredictable, that the term "random" seems most natural.

With these remarks let us follow Kolmogorov and others to define a finite string to be *random if it has maximum complexity.* Using this definition, one arrives at the somewhat disturbing conclusion that most finite strings of binary digits are random. Take for example, the 2^n series of n digits. Most of the sequences have a complexity that is within a few bits of n. As the complexity decreases, the number of series decreases in an essentially exponential manner. Really ordered series are very rare indeed.

One can intuitively extrapolate these results to more general situations. For more degrees of freedoms, one can define cells in phase space or in Poincaré maps. Generalizing from the one-dimensional case, a chaotic orbit is one which generates a cell number sequence having maximum complexity. Such an orbit cannot be calculated by any finite algorithm and its information content is essentially infinite.

What has just been discussed seems to have some profound implications. In fact complexity theory and nonlinear dynamics together do establish the important fact that determinism in our usual sense only holds over quite finite domains. Outside these domains one has the vast territory of chaotic behavior.

1.12.2. Mapping a Baker's Work

Let us consider here first the work done by a baker working on an elementary mode, which preserves the area. We start from a square piece of dough. He first stretches the dough in the x direction and then folds it back on the rectangle repeating the action *ad infinitum.* The more interesting example is the work of the baker showing dissipation, i.e., the area of the dough is decreasing in each step. This kind of mapping can be described by the following two-dimensional map:

$$x_{n+1} = 2x_n \bmod 1 \tag{1.12.4}$$

$$y_{n+1} = \begin{cases} ay_n, & 0 < x_n < \frac{1}{2} \\ \frac{1}{2} + ay_n, & \frac{1}{2} < x_n < 1 \end{cases} \tag{1.12.5}$$

The first term is just the one-dimensional map already discussed. It has a Liapunov exponent $\lambda = \log 2 > 0$. This means extreme sensitivity associated with the initial conditions, and makes repeated applications of the map into a strange attractor.

The basin of attraction in this case consists of all points within the unit square. The Liapunov exponent in the x direction is $\lambda = \log 2 > 0$ and leads to a sensitive dependence on the initial conditions. This makes the object that results from repeated applications of this map a strange attractor in the unit square. The Lyapunov exponent for this map in the y direction is $\lambda = \log a < 0$. Therefore lengths are contracted in this direction. The end result is that the stretching in the x direction and the shrinking in the y direction results in a surface contraction, as required from a dissipative map.

The fractal dimension of the strange attractor corresponding to the baker's transformation can be calculated from these considerations and yields the expression

$$D = 1 + \log 2/|\log a| \tag{1.12.6}$$

1.12.3. Some Final Remarks about Fishing

Some brief remarks were made in the introduction about fish populations; we return to these questions here. Suppose N_i is a population of fish, insects, ..., in a closed area in year i. One wishes to calculate the population N_i the following year. The simplest nonlinear equation to describe the system is the quadratic map

$$N_{i+1} = N_i(a - bN_i) \tag{1.12.7}$$

This seems to have been first studied by Verhulst back in 1845 and formed the basis of the early important work by May[28] as well as the seminal work by Feigenbaum[11] and a vast body of later papers. The reader is referred to the original papers (reprinted in books by Cvitanovic[29] and Hao Bai Lin[30]); here we shall limit ourselves to only a few comments.

The equation is rescaled by writing $x = bN/a$ and then obtained in the standard form

$$x_{n+1} = f_z(x_n) = ax_n(1 - x_n) \tag{1.12.8}$$

This is often referred to as the *logistic map.* We wish to study the long-term behavior of the system. In equation (1.12.8) a plays the role of a strength parameter of the system, similar to the Reynolds number in the hydrodynamic example. In order to maintain the value of x between 0 and 1 one limits the range of parameter a to values betwen 0 and 4. For small values, $a < 1$, one has the unfavorable situation in which the population steadily decreases and dies out for all initial values.

A different behavior is found for values between 1 and 3. For any initial value, the population now goes to a constant value. Such equilibrium points—or *fixed points*—are found by putting $x_{j+1} = x_j = x^*$, i.e., one solves the equation

$$x^* = ax^*(1 - x^*)$$

which has two solutions: $x^* = 0$ and $x^* = 1 - 1/a$.

An alternative method, used in more general discussions, is to draw the curve $R(x)$ that maps x_j into x_{j+1} and find the intersection with the 45° line. One can now ask about the stability of the fixed point. This depends on the slope of the $R(x)$ curve at the fixed point x^*. If the slope is between +1 and −1, the fixed point is stable, but for a large slope it becomes unstable. In our model the slope is $2 - a$. Therefore the fixed point is stable for all values of a between 1 and 3.

If the parameter a is increased to values larger than 3, one passes to the regime where the fixed point becomes unstable and one now asks what happens in the regime $a > 3$.

Before discussing this region, let us consider first the special values $a = 4$. This case can be solved exactly and here one encounters a situation of fully developed chaos.

When $a = 4$ we have the dynamical equation

$$x_{j+1} = 4x_j(1 - x_j)$$

It can be solved by introducing the new variable θ through

$$x = (1 - \cos 2\pi\theta)/2$$

to give the new equation

$$\tfrac{1}{2}(1 - \cos 2\pi\theta_{j+1}) = \tfrac{1}{2}(1 - \cos 4\pi\theta_j)$$

with solution

$$\theta_{j+1} = 2\theta_j \qquad \text{or} \qquad \theta_j = 2^j\theta_0$$

It is seen that changing the sign of θ or adding an integer yields the same solution x. This means that for $a = 4$ the map reduces to the sample map discussed earlier in this section. Writing the initial value in binary numbers one sees again the extreme sensitivity to the initial conditions that is the signature of chaotic behavior. If we have two initial values differing by a small value ε, the difference after j steps would have grown to $2j\varepsilon$. This is a parallel to the "butterfly" effect introduced in the famous paper by Lorenz, showing that long-time weather forecasting is in principle impossible.

Let us now turn to the region where a assumes values between 3 and 4. For a smaller than 3, the behavior is completely regular and one ends up at the fixed point x^*. For $a = 4$ one has fully developed chaos. It can be seen that as soon as a becomes larger than 3, the fixed point becomes unstable and one must now further examine the dynamics in that interval.

In order to gain additional insight, it is useful to relate the populations that are two generations apart, i.e., one relates x_{j+2} to x_j. Hence the equation is iterated to obtain

$$x_{j+2} = F[F(x_j)] = F^{(2)}(x_j)$$

Let us now look for solutions where the populations return to the same value *every second generation.* They are given by the fixed points x_2^* of the iterated equation

$$x_2^* = F^{(2)}(x_2^*)$$

With increasing a the fixed points of period 2 become unstable. One then repeats the same argument and iterates the equation to find a stable cycle of period 4. This will in turn give way to a cycle of period 8. Repeating the argument, one obtains a hierarchy of bifurcating stable cycles of period 16, 32, 54, ...

Feigenbaum[11] was the first to prove some unexpected universal properties of this particular "road to chaos" via successive bifurcations. He found through a numerical experiment that the shift between successive bifurcations follows the law

$$\delta = \lim\,(a_{n-1} - a_n)/(a_n - a_{n+1})$$

This holds for any mapping $f(x)$ with $f(0) = f(1)$, f being a smooth function with a single maximum in the interval between 0 and 1.

Feigenbaum also found that the splitting between successive bifurcations is determined by the universal constant

$$\alpha = 2.502907875$$

Feigenbaum's discoveries led to a tremendous interest in the field of chaos. The problem of *universality in chaos* is still one of the most exciting areas and will be discussed in great detail in the chapter by Cvitanovic in this volume.

References

1. A. Einstein, *Verh. Dtsch. Phys. Ges.* **19**, 82 (1917).
2. A. N. Kolmogorov, *Dokl. Akad. Nauk SSSR* **98**, 527 (1954).
3. V. I. Arnold, *Izv. Akad. Nauk SSSR* **25**, 21 (1961).
4. J. Moser, *Nachr. Akad. Wiss. Goettingen, Math.-Phys. Kl.* **IIa** 1 (1962).
5. M. Henon and C. Heiles, *Astron. J.* **69**, 73 (1964).
6. B. Chirikov, *Phys. Rev.* **52**, 265 (1979).
7. L. P. Kadanoff, *Phys. Scr*, **T9**, 5 (1985).

8. P. Bak, T. Bohr and M. H. Jensen, *Phys. Scr.* **T9**, 50 (1985).
9. E. N. Lorenz, *J. Atmos. Sci.* **20**, 130 (1963).
10. D. Ruelle and F. Takens, *Commun. Math. Phys.* **20**, 167 (1971).
11. M. J. Feigenbaum, *J. Stat. Phys.* **21**, 669 (1979); *Los Alamos Science* **1**, 4 (1980); *Commun. Math. Phys.* **77**, 65 (1980).
12. D. Ruelle, *Math. Intell.* **2**, 126 (1980).
13. B. B. Mandelbrot, *The Fractal Geometry of Nature*, Freeman, San Francisco (1982).
14. W. Sierpinsky, *C. R. Akad. Sci. Paris* **162**, 302 (1916).
15. K. Menger, *Am. Math. Mon.* **50**, 2 (1943).
16. M. Henon, *Commun. Math. Phys.* **50**, 69 (1976).
17. A. N. Kolmogorov, *Dokl. Akad. Nauk. SSSR* **98**, 527 (1959).
18. C. E. Shannon and W. Weaver, *The Mathematical Theory of Information*, University of Illinois Press (1949).
19. A. Libchaber and J. Maurer, in: *Non-linear Phenomena at Phase Transitions and Instabilities* (T. Riste, ed.), Plenum Press, New York (1982).
20. V. Steinberg, G. Ahlers, and D. S. Cannell, *Phys. Scr.* **T9**, 97, (1985).
21. J. P. Gollub, *Phys. Scr.* **T9**, 95, (1985).
22. J. Maselko and H. Swinney, *Phys. Scr.* **T9**, 35 (1985).
23. C. D. Jeffries, *Phys. Scr.* **T9**, 11 (1985).
24. J. Ford, *Phys. Today* **36**, 40 (1983).
25. V. M. Alekseev and M. V. Yakobson, *Phys. Rep.* **75**, 287 (1981).
26. G. J. Chaitin, *Sci. Am.* **232**, 47 (1975).
27. R. J. Solomonov, *Inf. Control* **7**, 224 (1964).
28. R. M. May, *Nature* **261**, 459 (1976).
29. P. Cvitanovic, *Universality in Chaos*, Adam Hilger, Bristol (1984).
30. B. L. Hao, *Chaos*, World Scientific, Singapore (1984).

2

An Introduction to the Properties of One-Dimensional Difference Equations

G. Rowlands

2.1. Introduction

Until quite recently most physical systems have been described by continuous equations, either differential or integral ones. Difference equations have been studied, but usually as a finite-difference approximation to be used as a computational algorithm for the continuous system. In this case the approximation is designed such that the difference equation mirrors the corresponding continuous system. However, it is now realized that simple, but nonlinear, difference equations can have very complicated solutions, a complexity not found in what one would consider the analogous differential equation. This was brought to the attention of scientists by May.[1] Although the original application involved problems in ecology, where the difference equations have an immediate interpretation, it was soon realized that they had a much wider range of applicability. In particular, the pioneering work of Feigenbaum showed their relevance to such problems as turbulence. However, it is not the purpose of this chapter to relate the mathematical results to fields of application. Here, some of the mathematical properties of one-dimensional difference equations will be considered. Surprisingly, even for such a restricted class of equations, there is a bewildering range of properties.

In the next section some properties of linear difference equations will be discussed. These are in fact very similar to those of linear differential equations, which is perhaps why little interest had been shown in difference equations. However, the introduction of simple nonlinearities changes this picture entirely.

G. Rowlands • Department of Physics, University of Warwick, Coventry CV4 7AL, England.

The nonlinear difference equations exhibit properties which have no analogy in differential equations. To study such phenomena the logistic equation is studied in detail in Section 2.3. The second major surprise that comes from a study of simple nonlinear difference equations involves certain universal properties shown by a wide class of equations. These universal properties will be discussed in Section 2.4.

A few references are given in the text but further details are provided in the book by Collet and Eckmann[(2)] and the reprints and references cited by Cvitanovic.[(3)] No attempt has been made to relate the equations studied in this chapter with the physical world. A good introduction to this aspect of the subject is given in the book by Lichtenberg and Lieberman[(4)] and Part 2 of Cvitanovic.[(3)]

2.2. Linear Difference Equations

In analogy with second-order differential equations with constant coefficients we consider the linear difference equation

$$a\phi_{n+1} + b\phi_n + c\phi_{n-1} = 0 \tag{2.2.1}$$

where a, b, and c are constants and n takes integer values. Such equations are simple to solve numerically. Using the initial conditions, namely the values of ϕ_0 and ϕ_1 say, equation (2.2.1) can be used to obtain ϕ_2 and subsequently values of ϕ_n for all n. This simplicity of obtaining solutions is very characteristic of difference equations. They are in an ideal form for solution using computers or even calculators.

Though these equations do not necessarily have anything to do with dynamical systems, it is customary to use the jargon of dynamical systems to describe properties of their solution. For example, the iterates of the equation, i.e., the values ϕ_n, are likened to the position of a particle as a function of time (at discrete time intervals) and so one speaks of the trajectory of the solution.

Analytically one looks for solutions of the form

$$\phi_n = Ae^{n\lambda} \equiv \mu^n \tag{2.2.2}$$

where $\mu = e^{\lambda}$. Substitution of this form into equation (2.2.1) gives the condition for a nontrivial solution to exist as

$$a\mu^2 + b\mu + c = 0 \tag{2.2.3}$$

so that the complete solution of equation (2.2.1) is

$$\phi_n = A\mu_+^n + B\mu_-^n \equiv Ae^{n\lambda_+} + Be^{n\lambda_-} \tag{2.2.4}$$

where A and B are constants to be determined from initial conditions and $\mu_\pm$ are the two solutions of equation (2.2.3).

Though n takes integer values only, the solution (2.2.4) gives ϕ_n in the form of a continuous function of n. Thus one writes $\phi_n = \phi(\tau)$, where $\tau = n\Delta$ and Δ is a constant step length. Then using solution (2.2.4) one sees that $\phi(\tau)$ satisfies the differential equation

$$\alpha \frac{d^2\phi}{d\tau^2} + \beta \frac{d\phi}{d\tau} + \gamma\phi = 0 \tag{2.2.5}$$

with α, β, and γ satisfying the equation

$$\alpha\lambda^2 + \beta\lambda + \gamma = 0 \tag{2.2.6}$$

and λ assuming the values λ_+ and λ_-. Hence one can associate with the linear difference equation (2.2.1) the linear differential equation (2.2.5). Both have identical solutions with the correspondence defined by equation (2.2.6), with $\lambda = \ln \mu$ and μ satisfying equation (2.2.3). This leads to the identification

$$c/a = (1/\Delta^2) \exp(-\beta/\alpha)$$

and

$$b/a = -(2/\Delta) \exp(-\beta/2\alpha) \cosh\{[(\beta/2\alpha)^2 - \gamma/\alpha]^{1/2}\} \tag{2.2.7}$$

This exact correspondence between a linear difference equation (2.2.1) and a linear differential equation (2.2.5) also holds for higher-order equations.

If $\lambda_\pm$ is small, then one may approximate $\phi_{n+1} \equiv \phi(n\Delta + \Delta) = \phi(\tau + \Delta)$ by the first two terms in a Taylor series. In this case equation (2.2.1) reduces to a second-order differential equation

$$\frac{(a+c)}{2} \frac{d^2\phi}{d\tau^2} + \frac{(a-c)}{\Delta} \frac{d\phi}{d\tau} + \frac{(a+b+c)}{\Delta^2} \phi = 0$$

For consistency one must have $a + b + c$ small. The relationship between equations (2.2.1) and (2.2.7) lies at the heart of finite-difference schemes used to solve differential equations such as (2.2.7). However, this is only an approximate relationship, and it must be stressed that the interesting behavior found in the next section does not apply to the regime where the above correspondence holds. That is, difference equations must be considered in their own right, not as approximate schemes for solving differential equations.

A linear difference equation which will constantly reappear in subsequent analysis is the simple first-order equation

$$\phi_{n+1} = a\phi_n$$

where a is some constant. It is readily seen that the complete solution is $\phi_n = \phi_0 a^n$, where ϕ_0 is the initial value. In particular, this equation shows quite distinct behavior depending on the value of a and almost independently of ϕ_0. For $|a| < 1$, $\phi_n \to 0$ for $n \to \infty$, while if $|a| > 1$, $\phi_n \to \infty$ for $n \to \infty$. In subsequent chapters the condition $|a| < 1$ will often be referred to as a stability condition.

A first-order equation, even where the coefficient of ϕ_n depends on n, can also be formally solved. If

$$\phi_{n+1} = a_n \phi_n$$

then simple iteration gives $\phi_{n+1} = a_n a_{n-1} \phi_{n-1}$, which suggests the complete solution

$$\phi_{n+1} = \phi_0 \prod_{j=0}^{n} a_j$$

It is sometimes convenient to introduce the quantity S_j defined such that $S_j = \ln(a_j)$, in which case

$$\phi_{n+1} = \phi_0 \exp\left(\sum_{j=0}^{n} S_j\right)$$

This is valid as long as quantities a_j are positive. If some are negative, then we define $S_j = \ln(|a_j|)$ in which case

$$\phi_{n+1} = \hat{\phi}_0 \exp \sum_{j=0}^{n} S_j$$

where now $\hat{\phi}_0 = \phi_0(-1)^q$, q being the number of negative values of a_j. The stability criterion is now

$$\operatorname*{Lt}_{n\to\infty}\left(\frac{1}{n}\sum_{j=0}^{n} S_j\right) < 0$$

2.3. The Logistic Equation

If y_n is the population number or density after n generations, then the simplest ecological model of population dynamics is

$$y_{n+1} = \mu y_n$$

This is of the form discussed in the last section and has the solution ($y_n = A\mu^n$). Thus for $\mu > 1$ the population density increases without bounds. Of course, such a solution will not exist in the real world due to some saturation mechanism. The simplest model equation describing such a mechanism is

$$y_{n+1} = \mu y_n(1 - \alpha y_n)$$

Normalization of y_n such that $\alpha y_n = x_n$ gives

$$x_{n+1} = \mu x_n(1 - x_n) \tag{2.3.1}$$

This is the logistic equation. One imposes the conditions $0 < \mu < 4$ and $0 < x_0 < 1$, where x_0 is the initial value; otherwise, as will be seen later, the solution of equation (2.3.1) diverges for large n.

An entirely equivalent equation is

$$y_{n+1} = 1 - ay_n^2$$

which is related to equation (2.3.1) through the transformations $y = (x - \frac{1}{2})/(\mu/4 - \frac{1}{2})$ and $a = \mu(\mu/4 - \frac{1}{2})$. This is the form of the logistic map that has been studied in detail by Collet and Eckmann.[2]

If one carries out the approximation procedure discussed in the last section, namely, one assumes that x_n is a slowly varying function of τ ($= n\Delta$), then equation (2.3.1) is replaced by

$$\frac{dx}{d\tau} = (\mu - 1)x/\Delta - \mu x^2/\Delta \tag{2.3.2}$$

the solution of which is

$$x = \frac{x_0(\mu - 1)}{\mu x_0\{1 - \exp[-(\mu - 1)\tau/\Delta]\} + (\mu - 1)\exp[-(\mu - 1)\tau/\Delta]}$$

where x_0 is the value at $\tau = 0$. Now if $\mu < 1$, then for $\tau \to +\infty$, $x \to 0$, while if $\mu > 1$ and $\tau \to +\infty$, $x \to (\mu - 1)/\mu$. Thus the exponential growth found for the linear equation when $\mu > 1$ develops into a final asymptotic state given by $x_s = (\mu - 1)/\mu$. It should be noted that this final state is independent of the initial conditions, namely the value of x_0, but depends on the parameter in the differential equation. In fact it is readily obtained from equation (2.3.2) by imposing the condition $dx/d\tau \equiv 0$. In what follows asymptotic states will be considered associated with the difference equation (2.3.1), and though these turn out to be infinitely more complicated they are independent, to all intents and purposes, of the initial conditions. An interesting question is: does the difference equation (2.3.1) exhibit similar properties?

One can define an asymptotic state x_s, independent of n, by demanding that such a solution satisfies equation (2.3.1). That is, $x_s = x_s(1 - x_s)$. There are two solutions, $x_s = 0$ and $x_s = (\mu - 1)/\mu$. The latter is identical to the asymptotic state obtained from the differential equation. This suggests that the differential equation mirrors the behavior of the difference equation. This view is further strengthened if one compares the full time evolution of equation (2.3.2) with the solution of equation (2.3.1). This comparison is illustrated in Figure 2.1 for $\mu = 2.5$, from which it may be concluded that the essential features of the solution of equation (2.3.1) are captured by the differential equation.

These results can be expressed in another way, by focusing attention not on the differential equation but on the asymptotic state; there are two distinct such states. One, $x_s = 0$, exists for all $\mu > 0$, while the other, $x_s = (\mu - 1)/\mu$, only exists for $\mu > 1$ (since $1 > x_n > 0$). The existence of the states is intimately bound up with their linear stability. They, of course, only exist if they are linearly stable. To test for stability one writes $x(\tau) = x_s + \delta x(\tau)$, substitutes into equation (2.3.2), and neglects products of δx. This leads to the linear equation

$$\frac{d\delta_x}{d\tau} = (\mu - 1 - 2\mu x_s)\delta x/\Delta$$

which shows exponential growth and hence instability for $\mu - 1 - 2\mu x_s > 0$. Thus the asymptotic state $x_s = 0$ is unstable for $\mu > 1$ while the asymptotic state $x_s = (\mu - 1)/\mu$ is always stable. These conclusions are illustrated graphically in Figure 2.2, where the values of x_s are plotted as a function of μ, with solid curves corresponding to stable states and dashed ones to unstable states. Thus as μ passes through unity, there is a transfer of stability from one asymptotic state to the other.

These same sort of considerations can now be applied to the difference equation (2.3.1). As stated above, the asymptotic state defined such that $x_n \to x_s$, independent of n, is given by $x_s = (\mu - 1)/\mu$ and so exists for $\mu > 1$. A linear

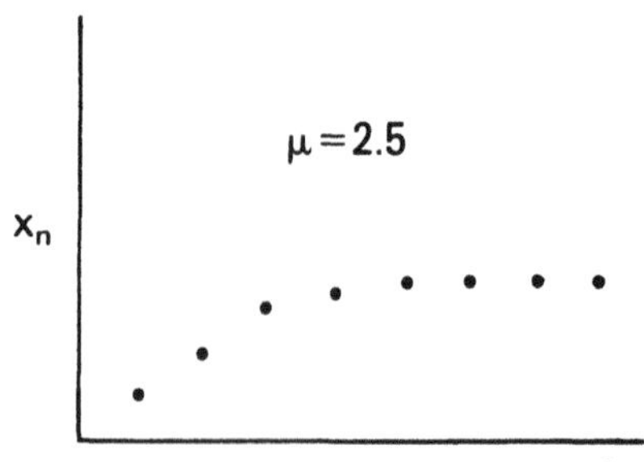

Figure 2.1. Comparison of the solution of the difference equation with that of the corresponding differential equation.

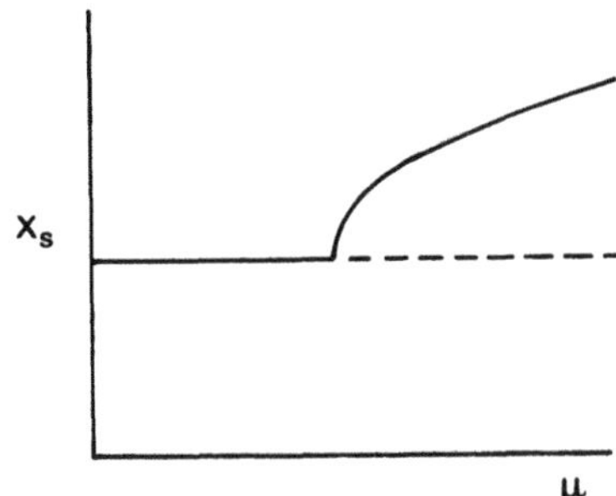

Figure 2.2. Asymptotic states of the logistic equation as a function of the control parameter. Full lines correspond to stable solutions.

analysis of equation (2.3.1) carried out by writing $x_n = x_s + \delta x_n$ and neglecting terms such as δx_n^2 gives rise to the linear difference equation

$$\delta x_{n+1} = (2 - \mu)\delta x_n$$

so that $\delta x_n = A(2 - \mu)^n$. Thus the asymptotic state is stable for $1 < \mu < 3$. Similar analysis shows that the asymptotic state $x_s = 0$ is stable only for $\mu < 1$. The onset of instability of the state $x_s = 0$ ($\mu > 1$) coincides with the existence of the new asymptotic state $x_s = 1 - 1/\mu$.

It was considerations such as those given above that led people to think that one could always find a differential equation equivalent to a difference equation and hence the latter were not worthy of study in their own right. The paper by May,[1] in which he discussed the complexity of the solutions of (2.3.1) for different values of μ, changed this attitude for many people. For example if one considers $\mu = 3.2$ then simple numerical iteration shows that a simple asymptotic constant value is no longer reached as n gets large but rather a final state which hops between two well-defined values. This is illustrated in Figure 2.3. Such behavior is not present in the solution of the differential equation (2.3.2) for any value of the parameters. Such behavior is hinted at by the stability of the asymptotic state x_s of the difference equation which we found to be unstable for $\mu > 3$.

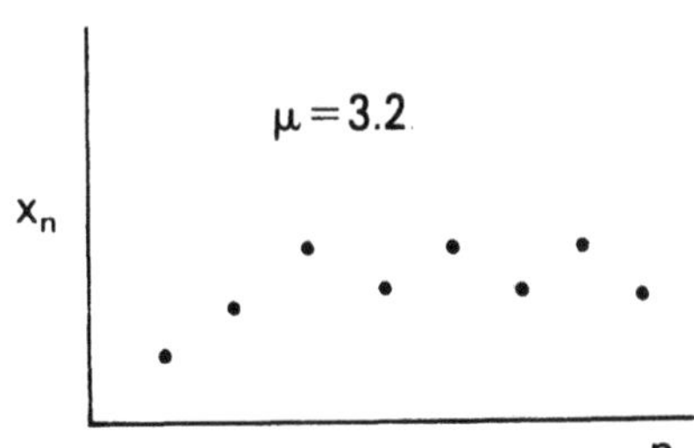

Figure 2.3. Solution of the logistic equation as a function of the number of iterates for $\mu = 2.3$.

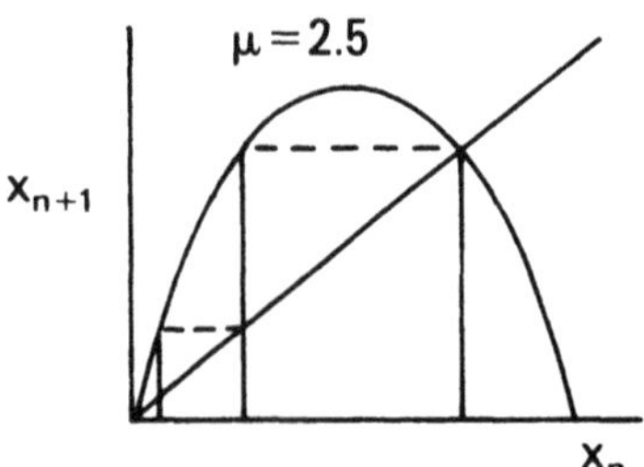

Figure 2.4. Pictorial way of representing the evolution of the solution of a one-dimensional difference equation ($\mu = 2.5$).

A pictorial way of solving equation (2.3.1) is illustrated in Figure 2.4 for $\mu = 2.5$. Here, x_{n+1} as a function of x_n and as given by equation (2.3.1) is shown as a full curve while the identity operation defined by $x_{n+1} = x_n$ is shown as the diagonal line. The iterates of equation (2.3.1) starting at x_0 are shown by a series of alternating vertical lines and horizontal dashed lines. For this value of μ a simple asymptotic state is finally reached. In Figure 2.5, a similar construction for $\mu = 3.2$ is illustrated. Now the final state, represented by the little box, corresponds to two values with iterates changing from one value to the other in perpetuity.

This new type of asymptotic state can be studied analytically in the following manner. Let us assume that for large n, x_n takes values x_+ and x_- alternatively. Thus

$$x_+ = \mu x_-(1 - x_-)$$

and

$$x_- = \mu x_+(1 - x_+)$$

Elimination of either x_+ or x_- leads to a quartic equation. However, the above considerations also apply to the case where a unique asymptotic state is reached, so that $x = 0$ and $x = 1 - 1/\mu$ must be solutions of the quartic equation. If these two solutions are factored out, the quartic equation reduces to the quadratic

$$\mu^2 x_\pm^2 + \mu(1 - \mu)x_\pm + 1 = 0$$

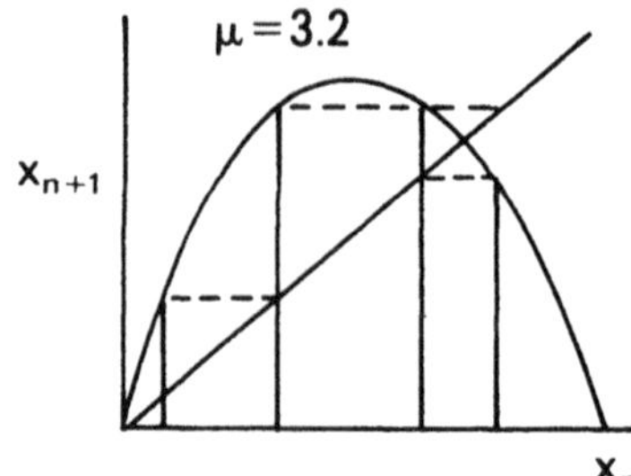

Figure 2.5. Same construction as in Figure 2.4 but now for $\mu = 3.2$. The final state of period two is enclosed by the box.

The two solutions correspond to the two asymptotic states and are given by

$$x_{\pm} = [\mu - 1 \pm (\mu^2 - 2\mu - 3)^{1/2}]/2\mu \tag{2.3.3}$$

Since $x_{\pm}$ must be real, these solutions only exist for $\mu^2 - 2\mu - 3 > 0$, that is, $\mu > 3$.

One can now study the stability of these states using a linearized treatment. Thus one writes $x_n = \bar{x}_n + \delta x_n$ to give

$$\delta x_{n+1} = \mu(1 - 2\bar{x}_n)\delta x_n$$

However, x_n takes the alternative values $x_{\pm}$ and must be considered as a function of n. In the present case this difficulty may be avoided by considering every other iterate, in which case

$$\begin{aligned}\delta x_{n+2} &= \mu^2(1 - 2\bar{x}_{n+1})(1 - 2\bar{x}_n)\delta x_n \\ &= \mu^2(1 - 2x_+)(1 - 2x_-)\delta x_n \\ &= (4 + 2\mu - \mu^2)\delta x_n\end{aligned}$$

This is now a linear equation with constant coefficient, a form treated in Section 2.2. The condition for stability, namely $4 + 2\mu - \mu^2 < 1$, leads to the condition $3 < \mu < 1 + \sqrt{6}$. It should be noted that for $\mu > 1 + \sqrt{6}$, the coefficient of δx_n in the above equation becomes *negative* with magnitude greater than unity, so that δx_n *alternates* in sign with n and also increases in magnitude. The same type of behavior is also apparent about the stability boundary $\mu = 3$.

Let us now reiterate what the above stability analysis has shown. For $\mu > 0$, an asymptotic state $x_s = (\mu - 1)/\mu$ exists and is stable for $1 < \mu < 3$. Then for $\mu > 3$, an asymptotic solution exists and hops between the values x_+ and x_- for alternate values of n. This state is stable for $3 < \mu < 1 + 6$. The whole behavior is illustrated in the form of an amplitude diagram in Figure 2.6. The amplitudes of the asymptotic states are plotted as functions of the parameter μ, solid curves corresponding to stable states and dashed curves to unstable states.

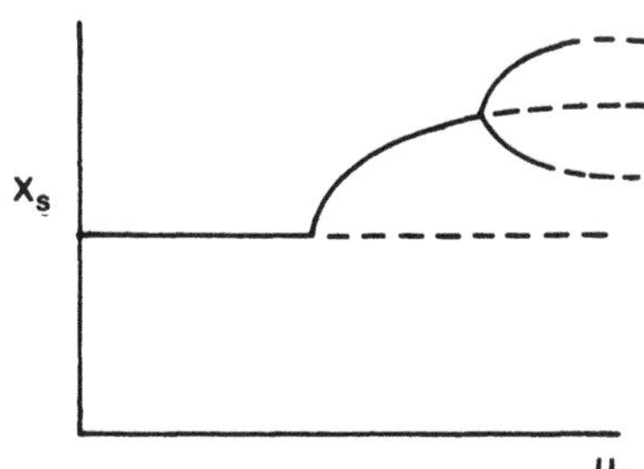

Figure 2.6. Amplitudes of the asymptotic states as a function of the parameter μ.

An interesting feature of Figure 2.6 is that the asymptotic state breaks into two for $\mu > 3$. This phenomenon is called a *bifurcation* for rather obvious reasons, or, more exactly, a pitchfork bifurcation. Comparison of Figures 2.2 and 2.6 shows a weakness of the comparison of the difference and differential equations (2.3.1) and (2.3.2), respectively, namely, the subsequent bifurcation is not present.

However, the phenomenon of bifurcation can be illustrated by reference to a simple nonlinear differential equation, namely

$$\frac{dA}{d\tau} = \alpha A + \beta A^3$$

Asymptotic states, i.e., time-independent ones, exist and are given by the solutions of $\alpha A + \beta A^3 = 0$, namely $A = 0$ and $A = \pm(-\alpha/\beta)^{1/2}$. The latter only exist for α and β having opposite signs, since we demand that A is real. Linear stability analysis shows the state $A = 0$ to be unstable for $\alpha > 0$ while the other state is stable. This is illustrated by a simple bifurcation diagram in Figure 2.7 which, by comparison with Figure 2.6, is seen to give the essential features of the bifurcation found for the case of the difference equation. Again, however, there is an important qualitative difference. For the difference equation, the bifurcated solution itself becomes unstable for large values of ($\mu > 1 + \sqrt{6}$).

What does this instability signify? Numerical iteration of equation (2.3.1) readily reveals that for μ slightly greater than $1 + \sqrt{6}$ the two stable states that existed for $\mu < 1 + \sqrt{6}$ themselves bifurcate. Thus the asymptotic state now consists of four distinct values with x_n taking each value in turn in a periodic manner with n. Further increase in μ reveals that these states themselves bifurcate, giving eight distinct states, which again bifurcate with further increase in μ. This whole process of continuing bifurcation proceeds with increase in μ and is illustrated by an amplitude diagram in Figure 2.8. This was obtained numerically by choosing a value of μ and iterating equation (2.3.1) a large number of times (say one hundred) and then printing the next one hundred or so iterates. In this way *stable* asymptotic states are selected.

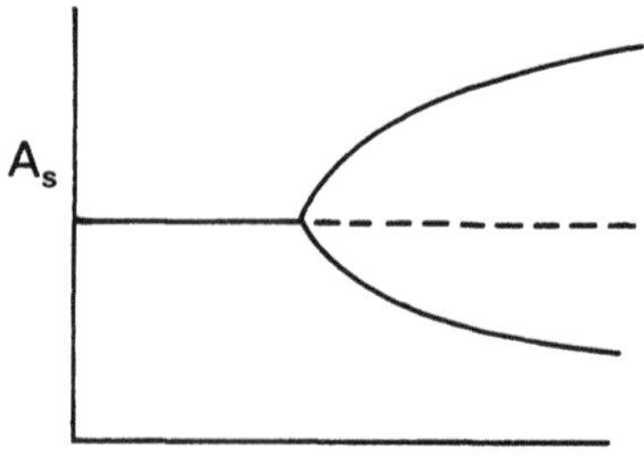

Figure 2.7. A simple bifurcation diagram.

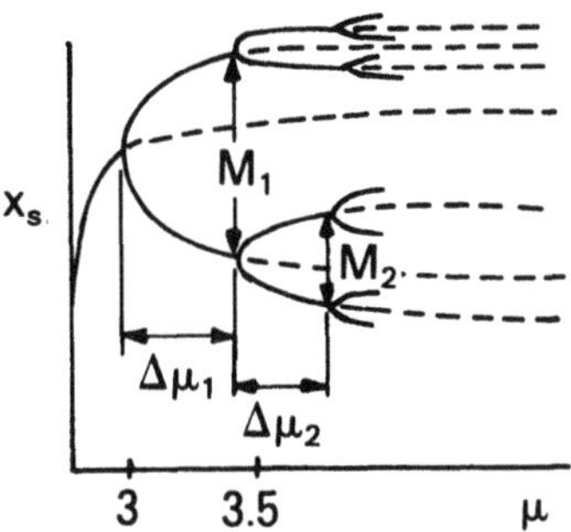

Figure 2.8. The bifurcation diagram for the logistic equation, showing the scaling of the amplitude M_j and width S_j.

The dashed lines shown in Figure 2.6, for example, are not revealed as they correspond to *unstable* solutions. These *stable* asymptotic solutions are independent of the initial value chosen to start the iteration procedure. [This is not strictly true, since if one chooses an initial value *identically* equal to one of the unstable states, then subsequent iterates will remain in this state. However, because of the unstable nature of these states any slight deviation in the initial value will cause subsequent iterations to diverge from the unstable to the stable state. The simplest example is to take $x_0 = 0$. All subsequent iterates give $x_n = 0$. However, an initial state $x_0 = \varepsilon$ gives $x_n = \varepsilon\mu^n$ which, for $\mu > 1$, diverges away from the origin no matter how small ε is, as long as it is not identically zero. (Rounding off errors in any computer will also lead to the final state being the stable one.)]

A study of Figure 2.8 shows that the change in μ between successive bifurcations seems to get smaller, which suggests that the sequence possesses a limit. To this end one considers the values of μ_j, namely, the value of μ for which the solution bifurcates for the jth time. Table 2.1 gives the first few values of μ_j. These values enable one to compute $1/\delta_j$ where

$$\frac{1}{\delta_j} = \frac{\mu_{j+1} - \mu_j}{\mu_j - \mu_{j-1}} \equiv \frac{\Delta\mu_{j+1}}{\Delta\mu_j} \tag{2.3.4a}$$

Table 2.1. Values of Parameter μ_i for the Onset of the jth Bifurcation

μ_j	$1/\delta_j$
3	
3.449490	0.2105
3.544090	0.2148
3.564407	0.2142
3.5687593	0.2144
3.5696915	0.2142
3.5698912	

These are also listed in the table. A plot of $1/\delta_j$ against j suggests that an asymptotic value exists with a value of order $\delta = 4.67$. Thus for sufficiently large values of j, $\Delta\mu_j \approx 1/\delta^j$, that is, the sequence of values of μ_j has an accumulation point with a value μ_c say. The numerical results suggest $\mu_c = 3.5700\ldots$.

One can associate another sequence with the series of bifurcations, namely, the maximum distance between the asymptotic values of x describing the bifurcation. If these are called M_n (they are shown in Figure 2.8), then one may calculate the quantity

$$\frac{1}{\alpha} = \operatorname*{Lt}_{j\to\infty}\left(\frac{M_{j+1} - M_j}{M_j - M_{j-1}}\right) \tag{2.3.4b}$$

and find a value of order 2.50.

This for $\mu < \mu_c$ the system undergoes a series of bifurcations as μ increases, each bifurcation needing a smaller change in μ. What happens for $\mu > \mu_c$?

Again one must resort to numerical iteration of equation (2.3.1). For $\mu > \mu_c$, it is found that the asymptotic periodic state of period 2^m, associated with an m-fold bifurcation, is replaced by a nonperiodic solution that appears random. Such a solution is called *chaotic*. For $\mu > \mu_c$ and for most initial conditions, the iterates of equation (2.3.1) have no apparent order as a function of n even for large n and, unlike the bifurcated solution, depend very sensitively on the initial condition. That is, a small change in the initial value generates an entirely different sequence of values of x_n. (Again, this is only true for most initial conditions, as the now unstable bifurcation solutions still exist.)

How does one quantitatively define chaos? One way is through the concept of a Liapunov number introduced in Chapter 1. First let us linearize the logistic equation, not about an asymptotic state, but about any solution. Thus we write $x_n = \bar{x}_n + \delta x_n$, where $\bar{x}_n$ is assumed known as a function of n. Then the linearized version of equation (2.3.1) is

$$\delta x_{n+1} = \mu(1 - 2\bar{x}_n)\delta x_n$$

which can be solved (see Section 2.2). The solution can be expressed in the form

$$\delta x_{n+1} = \delta\hat{x}_0 \exp\left(\sum_{j=0}^{n} S_j\right)$$

where $S_j = \ln[\mu|(1 - 2x_j)|]$. If the values of x_j do in fact attain an asymptotic state, then for sufficiently large n the above summation is dominated by the contribution from the asymptotic state. For example, if $1 < \mu$, we have $x_s = (\mu - 1)/\mu$ and then $\delta x_{n+1} \approx \delta\tilde{x}_0 \exp[n \ln(|\mu - 2|)]$.

Now we see that δx_n has exponential growth that is linearly unstable if $\mu > 3$. Similar features hold if one considers $\mu > 3$ and examines the first bifurcated solution. Then x_n takes the two values x_+ and x_- defined by equation (2.3.3). This gives $\delta x_{n+1} = \delta\hat{x}_0 \exp[(n/2)\ln|(4+2\mu-\mu^2)|]$ giving instability for $\mu > 1+\sqrt{6}$. Similar considerations apply to the higher-order bifurcations, that is, periodic solutions; in other words, for sufficiently large n we can write $\sum_{s=0}^{n} S_j = n\lambda$, where λ is some constant and is a simple measure of instability.

With these considerations in mind one defines the Liapunov number λ for any solution, periodic or not, by

$$\lambda = \operatorname*{Lt}_{n\to\infty} \frac{1}{n} \sum_{j=0}^{n} \ln[|\mu(1-2\bar{x}_j)|] \tag{2.3.5}$$

For periodic solutions this simply reduces to the constant values calculated above. In general, one interprets the Liapunov number as a measure of the stability of the orbit defined by x_j for all j. (Periodic stable orbits are such that $\lambda < 0$.)

One can now give a more mathematical definition of chaos, namely, a *necessary* condition for the existence of a chaotic solution is that the corresponding Liapunov number is greater than zero.

From its definition one sees that λ depends on $\bar{x}_0$, the initial value used in the iteration. However, because λ is an average over the whole sequence of values $\bar{x}_j$, it is found that λ is independent of $\bar{x}_0$ *almost always.* Thus though the $\bar{x}_n$ values themselves depend sensitively on initial conditions, the value of λ does not. The sensitivity on initial conditions can in fact be appreciated in terms of λ. If we consider two adjacent initial conditions, $\bar{x}_0$ and $\bar{x}_0 + \delta\bar{x}_0$, then the deviation after n iterates (n large) is just $\delta\bar{x}_0 \exp(n\lambda)$ so that for $\lambda > 0$ this deviation increases exponentially (at least according to linear theory).

For the low-order bifurcated solutions the value of λ can be calculated analytically, as seen above. In fact we find

$$\begin{aligned}\lambda &= \ln(|2-\mu|), & 1 < \mu < 3\\ &= \tfrac{1}{2}\ln(|4+2\mu-\mu^2|), & 3 < \mu < 1+\sqrt{6}\end{aligned}$$

The value of λ for other values of μ is difficult to obtain and one must resort to approximate numerical methods. However, before discussing these methods it is necessary to introduce the concepts of a distribution or probability function and an invariant measure.

It seems perfectly reasonable to introduce the idea of a probability function to describe a chaotic solution which, as a function of n, is seemingly random. Thus we define $P(x, n)\,dx$ as the probability of finding the solution of equation (2.3.1), after n iterates, in the interval between x and $x + dx$. From this definition and equation (2.3.1) we may write a Master equation

$$P(x, n+1) = \int_0^1 P(x', n)\delta[x - \mu x'(1-x')]\,dx' \tag{2.3.6}$$

where $\delta(x)$ is the Dirac delta function. This equation merely states that the value x' is converted to the point x after one iterate of equation (2.3.1). The content of this equation is entirely equivalent to that of equation (2.3.1).

Now suppose one starts the iteration process such that $P(x, 0)$ is a smooth function of x, for example $P(x, 0) = 1$, $0 < x < 1$. Then the question may be asked: "does $P(x, n)$ settle down to a value independent of n for large n"? Let us assume that this is so, in which case a function $P(x)$ exists such that

$$P(x) = \int_0^1 P(x')\delta[x - \mu x'(1 - x')]\,dx' \tag{2.3.7}$$

$P(x)$ may be thought of as $\mathrm{Lt}_{n\to\infty} P(x, n)$ and is in some ways equivalent to a thermodynamic equilibrium in conventional statistical physics. This function is usually called the distribution function or density of states while the invariant measure is just $P(x)\,dx$.

The advantage of considering $P(x)$, rather than $P(x, n)$, is that one does not need to worry about transient phenomena associated with particular initial conditions. For the case of $1 < \mu < 3$, one knows that the final state of the system is the asymptotic value $x_s = (\mu - 1)/\mu$. Thus one expects

$$P(x) = \delta[x - (\mu - 1)/\mu] \tag{2.3.8}$$

Substitution of this form into the right-hand side of equation (2.3.7) simply reproduces the same form, showing that this expression is a valid solution of equation (2.3.7). For $3 < \mu < 1 + \sqrt{6}$, the range of μ corresponding to a single stable bifurcation, we have

$$P(x) = \tfrac{1}{2}[\delta(x - x_+) + \delta(x - x_-)] \tag{2.3.9}$$

where $x_\pm$ are given by expression (2.3.3). This is most readily seen if it is remembered that by definition $x_- = \mu x_+(1 - x_+)$ and $x_+ = \mu x_-(1 - x_-)$. By analogy we see that for a periodic solution associated with an m-fold bifurcation, $P(x)$ will consist of 2^m distinct delta functions.

Unfortunately, little is known analytically about $P(x)$ in the chaotic region. Before considering these special cases we re-express equation (2.3.7) in an equivalent form by carrying out the integration using the well-known properties of the delta function. This gives

$$P(x) = \frac{P[(1 + \sqrt{1 - 4x/\mu})/2] + P[(1 - \sqrt{1 - 4x/\mu})/2]}{\mu\sqrt{1 - 4x/\mu}}$$

The two contributions on the right-hand side result from the backward iteration of equation (2.3.1), that is, expressing x_n as a function of x_{n+1} is double-valued. On replacing $(1 + \sqrt{1 - 4x/\mu})/2$ by x, we may also write

$$P(\mu x(1 - x)) = \frac{P(x) + P(1 - x)}{\mu(|1 - 2x|)} \tag{2.3.10}$$

Thus, in general, one has to solve a nasty functional equation and little is known about such equations.

An important exception is when $\mu = 4$. In this case it is convenient to change the variable by setting $x = \sin^2(\theta/2)$ and defining $\bar{P}(\theta)\, d\theta = P(x)\, dx$, in which case the above equation reduces to

$$P(2\theta) = [\bar{P}(\theta) + \bar{P}(\pi - \theta)]/2$$

A solution to this equation is that $\bar{P}(\theta)$ equals a constant, which corresponds to

$$P(x) = 1/[\pi\sqrt{x(1-x)}] \tag{2.3.11}$$

where the factor $1/\pi$ arises from a normalization condition. In terms of the variable θ defined above, the logistic equation for $\mu = 4$ reduces to

$$\theta_{n+1} = [2\theta_n]_{2\pi}$$

where $[\quad]_{2\pi}$ denotes modulo 2π.

The only other analytic solutions of equation (2.3.10) are associated with so-called "island splitting" bifurcations. These exist for certain values of μ, such that the maximum point $x = \frac{1}{2}$ is iterated, after a finite number of times, into an unstable fixed point. The simplest is the case discussed above when $\mu = 4$. Then, after one iterate, the point $x = \frac{1}{2}$ becomes the fixed point $x = 1$, which is unstable. The next simplest is where the third iterate of $x = \frac{1}{2}$ is equal to the fixed point defined by $x = (\mu - 1)/\mu$. This happens for $\mu = 3.67$, and the iterates of $\frac{1}{2}$ are $x_1 = \mu/4$, $x_2 = (\mu^2/4)(1 - \mu/4)$, and $x_3 = (\mu - 1)/\mu$. Any point in the range $x_2 < x < x_3$ is iterated into the range $x_2 < x < x_1$, and vice versa. The distribution function $P(x)$ is then continuous in each of these regions with square-root singularities similar to equation (2.3.11) at the ends of the ranges. Even for this case there is no analytic expression for $P(x)$, though the singularity structure is identified. There are other values of μ such that the point $x = \frac{1}{2}$ is iterated, after a *finite* number of times, to the unstable fixed point. The corresponding distribution function is a piecewise continuous function of x with square-root singularities at the edges of the various regions. There is also an even greater number of values of μ where the iteration process never ends up on an unstable fixed point. Then the distribution function is extremely complicated, consisting of an infinite number of discontinuities of the type of equation (2.3.11) but now occurring at the infinite number of points corresponding to the various iterates of $x = \frac{1}{2}$. The distribution function is then most probably a *fractal* (see Mandelbrot[(5)] and also Chapter 1). Examples of distribution functions obtained numerically are given in Shaw[(6)] and Collett and Eckmann.[(2)]

In the above, the concept of a Liapunov number was used as defined in equation (2.3.5), an average being taken over the iterates of equation (2.3.1). Assuming an invariant measure exists, one can give an alternate definition of the Liapunov number, namely

$$\lambda = \int_0^1 P(x)\, dx \ln(\mu|1-2x|) \tag{2.3.12}$$

Expression (2.3.5) for λ is essentially a time average (n playing the role of time) while the above definition is a space average. Let us assume these two definitions are equivalent, though the proof is the subject of ergodic theory and is beyond the scope of the present discussion; it is discussed by Eckmann and Ruelle.[7] That they are equivalent for the periodic solutions is readily seen by combining equations (2.3.8), (2.3.9), and (2.3.12) to yield expressions for λ identical to those obtained above using expression (2.3.5).

One is now in a position to discuss the value of λ for all values of μ, including the chaotic region. Shaw[6] used equation (2.3.5) to numerically calculate λ as a function of μ. Such results are reproduced in Figure 2.9. For $\mu < 1+\sqrt{6}$, the results agree with the analytic value given above. In particular, $\lambda < 0$, until one enters the chaotic region for $\mu > \mu_c = 3.57\ldots$. One can extend the consideration given above; for a stable periodic solution of period N with the individual iterates taking the values x_i, $i = 1, 2, \ldots, N$, we can set

$$P(x) = \frac{1}{N}\sum_{j=0}^{N} \delta(x - x_j) \tag{2.3.13}$$

Then, using equation using (2.3.12) one may readily obtain the corresponding value of λ, namely

$$\lambda = \frac{1}{N}\sum_{j=0}^{N} \ln(\mu|1-2x_j|)$$

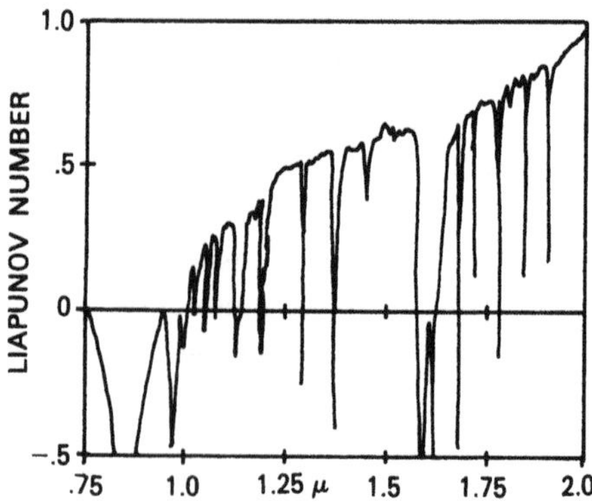

Figure 2.9. Variation of the Liapunov number λ with control parameter μ (numerical).

For the chaotic region, following a suggestion by Kai and Tomita,[8] one writes

$$P(x) = \sum_{j=0}^{N} W_j \delta(x - x_j) \tag{2.3.14}$$

where now x_j are unstable points and W_j a suitably chosen weighting. Kai and Tomita used a variational principle to derive a form for W_i in terms of iterates of equation (2.3.1). This formulation has been used by McCreadie and Rowlands[9] to derive analytic approximations for λ for all values of μ. One of these is shown in Figure 2.10. It is noteworthy that the analytic form captures the main features of Shaw's numerical results, though it is apparent that the numerical results show much more fine structure. An important point to note is that, for $\mu > \mu_c$, there are regions where $\lambda < 0$ which must correspond to stable periodic solutions of equation (2.3.1).

Another point is that though equation (2.3.13) is exact, equation (2.3.14) is not and other approximation forms are possible. Also, we note that for a periodic solution W_j is independent of j, that is, each iterate x_j has the same weight in which case equation (2.3.14) reduces to equation (2.3.13).

The structure of the solution of equation (2.3.1) for all μ is found by simply extending the numerical method used to obtain Figure 2.6 to all $\mu < 4$. The results are shown in Figure 2.11, from which it follows that, for a range of values of μ, a period-three solution exists. Other periodic solutions exist and so do other regions of chaos. However, it is worth considering the period-three solution in a little more detail, as this is intimately connected with the phenomenon of intermittency. To do this it is best to generalize our notation and consider an equation of the form

$$x_{n+1} = F(x_n, \mu) \tag{2.3.15}$$

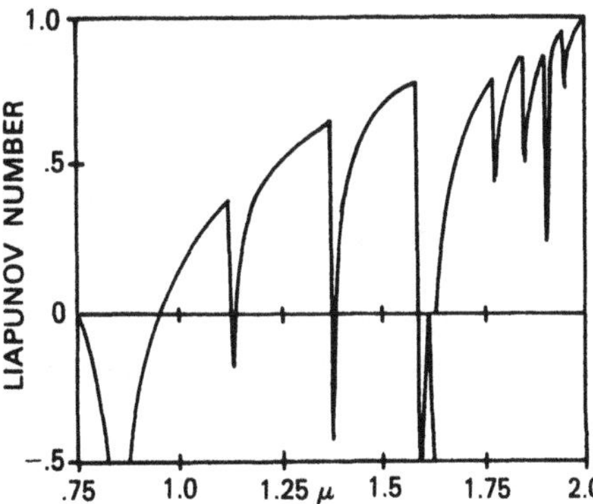

Figure 2.10. Variation of λ with μ (analytic).

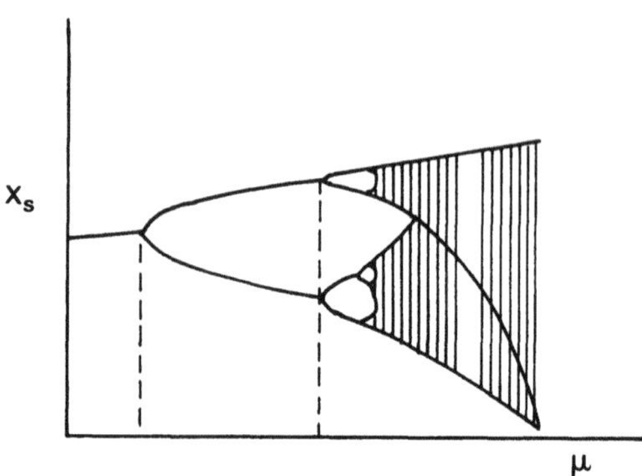

Figure 2.11. The complete bifurcation diagram for the logistic map for all $\mu \leq 4$.

where F is some function of x_n and parameter μ. The logistic map is, of course, of this type. Now suppose an asymptotic state exists such that $x_n = x_s$. Linear stability analysis about this solution ($x_n = x_s + \delta x_n$) gives

$$\delta x_{n+1} = F_x(x_s, \mu)\,\delta x_n \tag{2.3.16}$$

where the subscript denotes differentiation with respect to x. The stability criterion (and hence the existence of x_s) is that $|F_x| < 1$, namely, a condition on μ. One is interested in the solution near the stability boundary and so one writes $x_s = \bar{x} + \Delta x$, $\mu = \bar{\mu} + \Delta\mu$. Then the condition for an asymptotic state $x_s = F(x_s, \mu)$ gives [by definition $\bar{x} = F(\bar{x}, \bar{\mu})$]

$$\Delta x = F_x\,\Delta x + \tfrac{1}{2}F_{xx}\,\Delta^2 x + F_\mu\,\Delta\mu + \cdots$$

Now suppose $F_x = -1$ (stability boundary), then $\Delta x = F_\mu\,\Delta\mu/2$. Expansion of equation (2.3.16) about this critical point yields

$$\delta x_{n+1} = [-1 + (F_{x\mu} + F_{xx}F_\mu/2)\,\Delta\mu]\,\delta x_n$$

A change in sign of $\Delta\mu$ then takes the system from a stable to an unstable asymptotic state, a change characteristic of a pitchfork bifurcation. This was the behavior studied in detail for the logistic equation and which leads to the bifurcation sequences.

We consider $F_x = +1$, again on the stability boundary but now with $\Delta^2 x = -(2F_\mu/F_{xx})\,\Delta\mu$. The linear deviations then satisfy

$$\delta x_{n+1} = (1 \pm \sqrt{-2F_\mu F_{xx}\,\Delta\mu})\,\delta x_n$$

The behavior is now very different. For one sign of $\Delta\mu$ the square root is real and for the negative sign the values of δx_n decrease exponentially with n, i.e., this state is stable. A change in the sign of $\Delta\mu$ makes the square-root term imaginary, so now the state is marginally stable with $x_n = \exp(i\alpha n)$, $\alpha^2 = |2F_\mu F_{xx}\,\Delta\mu|$. We note that, with this sign for $\Delta\mu$, δx is imaginary and hence the asymptotic state is also imaginary. Thus near the stability boundary $F_x = +1$, a change in μ makes a marginally stable state change to a stable one, while for $F_x = -1$, a stable state changes to an unstable one.

The behavior near this type of critical point can be illustrated by considering the special case where $F(x, \mu) = \mu - 1/x$. The condition $F(\bar{x}, \bar{\mu}) = x$ gives

$$2\bar{x} = \mu \pm \sqrt{\mu^2 - 4} \tag{2.3.17}$$

so that for $\mu^2 > 4$ two asymptotic states exist while for $\mu^2 < 4$ these states become complex. The other condition, namely $F_x(\bar{x}, \bar{\mu}) = +1$, gives $x = 1$ so that $\mu = 2$. With the above form for $F(x, \mu)$, equation (2.3.15) can be solved exactly to yield

$$x_n = \cos\lambda - \sin\lambda\,\tan(n\lambda + \phi) \tag{2.3.18}$$

where ϕ is an arbitrary constant and $\cos\lambda = \mu/2$.

For $\mu^2 > 4$, we write $\lambda = i\eta$ in which case

$$x_n = \cosh\eta + \sinh\eta\,\tanh(n\eta + \phi)$$

and for large n one has $x_n \to e^\eta$. Since $\cosh\eta = \mu/2$, this final state is as given by expression (2.3.17). This state is the stable one of the two. For $\mu^2 < 4$, one no longer has real asymptotic states [see equation (2.3.17)] but λ is real and x_n increases with n until finally the linear theory breaks down. These features are illustrated graphically in Figure 2.12. For μ just less than 2, the iterates

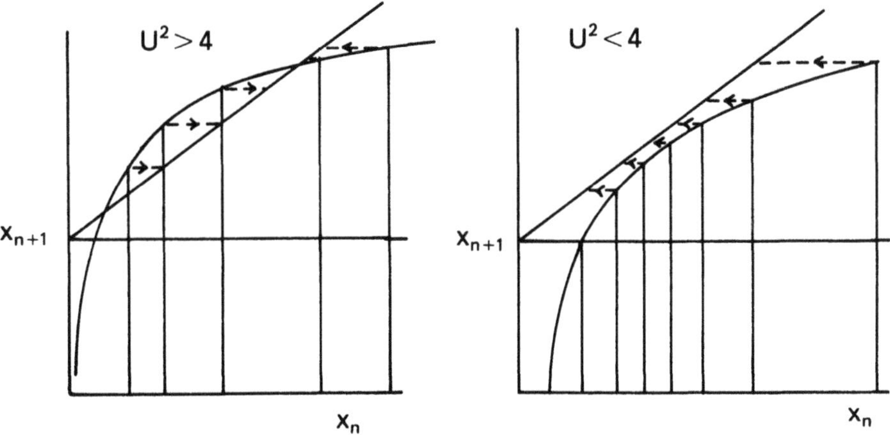

Figure 2.12. Illustration of the phenomenon of intermittency.

of the map spend much time in the region where $F(x, \mu)$ is close to the 45° line, before being whisked away, though returning again as n increases. Though no real asymptotic state exists, this behavior suggests a memory of the states that would exist if μ was slightly greater than 2. This behavior shows up as chaotic in Figure 2.11 with the density of states being enhanced in the region near $x = \bar{x}$. This type of behavior is called intermittency and has been proposed by Pomeau and Manneville[10] as a possible mechanism to explain certain types of turbulence. It will be noted that this behavior is brought about by the function $F(x, \mu)$ becoming tangent to the 45° line and for this reason it is called a tangent bifurcation. All that is required is that the function $F(x, \mu)$ has the general shape of the function shown in Figure 2.12 or its mirror reflection about the 45° line.

If one considers the logistic equation with $F(x, \mu) = \mu x(1 - x)$, one cannot satisfy both conditions $\bar{x} = F(\bar{x}, \bar{\mu})$ and $F_x(\bar{x}, \bar{\mu}) = +1$ in the range $1 < \bar{\mu} < 4$. However, if one considers every third iterate of the logistic equation, i.e., one makes the identification

$$F(x, \mu) = f(f(f(x, \mu)))$$

where $f(x, \quad) = \mu x(1 - x)$, then it is possible to satisfy both these conditions. The critical value of μ separating period-three behavior from intermittency is found to be $\mu = 1 + 2\sqrt{2}$.

More detailed numerical studies of the logistic equation show that as μ increases above the value $\mu = 1 + 2\sqrt{2}$, the period-three state undergoes a bifurcation sequence eventually leading to chaos. What is more, quantities δ and α defined by equations (2.3.4a) and (2.3.4b) take the same values for this bifurcation sequence as for the main sequence. Further, the existence of a tangent bifurcation is not limited to a period-three solution, so that the chaotic region of the logistic equation is broken up into small regions in which periodic solutions, each with their own bifurcation sequence, exist. A really complicated situation.

The above discussion has revealed some of the complicated but fascinating nature of the solution of the logistic equation.

2.4. Universality

The previous section has revealed some fascinating properties of the logistic equation, features far more intricate than seemingly corresponding differential equations. The main features are the bifurcation sequence ending in chaos and intermittent chaos changing to periodic behavior, as the parameter μ changes. A natural question to ask is: how typical is this type of behavior? In other words: do other difference equations exhibit this type of behavior? The short answer is yes, but there is more to the question. In fact, it is found that the quantities δ and α, defined by equations (2.3.4), are universal numbers and are found to be the parameters describing the bifurcation sequence for a whole class of maps. This universality is an unexpected property and perhaps makes up for the added complication of the solution of such maps. It is the purpose of this section to introduce the idea of universality but a detailed discussion will be found in later chapters, in particular that of Cvitanovic (Chapter 12).

First one considers a general map of the form

$$x_{n+1} = F(x_n, \mu)$$

where we restrict attention to functions F such that $F(0, \mu) = F(1, \mu) = 0$, and restrict initial conditions to $0 < x_0 < 1$. Then as long as $F(x_m, \mu) < 1$ where $F_x(x_m, \mu) = 0$, i.e., the maximum of F is mapped back into the interval $0 < x < 1$, all iterates of the map remain in the interval $0 < x < 1$.

The simplest asymptotic state is given by $\bar{x} = F(\bar{x}, \mu)$. In a more mathematical context such a solution is called a fixed point of the map. This state is stable if $|F_x(\bar{x}, \mu)| < 1$, which is a condition on μ. The first bifurcation occurs when a solution exists such that $x_2 = F(x_1, \mu)$, $x_1 = F(x_2, \mu)$. This is equivalent to $x_i = F(F(x_i, \mu), \mu)$ $(i = 1, 2)$, i.e., a fixed point of the map $F(F(x, \mu), \mu)$. Somewhat unfortunately such maps are usually written as $F^{(2)}(x, \mu)$. This is not to be confused with the simple square of the function, in fact a quantity that rarely occurs in this field of research. The condition for stability is now $|F^{(2)}(x, \mu)| < 1$, and by simple differentiation we see that

$$F_x^{(2)}(x, \mu) = F_x(F(x))F_x(x) \equiv F_x(\bar{x}_2)F_x(\bar{x}_1)$$

so the stability criterion can be expressed as the product of derivatives calculated at the fixed points of the bifurcation (x_0, x_1). By analogy, it is readily seen that the nth bifurcation is defined by $F^{(2n)}(\bar{x}, \mu) = \bar{x}$ and the stability condition $|F_x^{(2n)}(\bar{x}, \mu)| < 1$, where $F_x^{(2n)}(x, \mu) = F_x(x_{2n})F_x(x_{2n-1}) \cdots F_x(x_1)$ and $x_1, \ldots, x_{2n}$ are the values of x describing the nth bifurcation. A particular state of interest is where $F_x(x_1) \equiv 0$ (without loss of generality one can always choose the first iterate as starting at this point), that is, $F(x_1)$ assumes a maximum or minimum value. For such a situation, the stability factor $F_x^{(2n)}(\bar{x}, \mu) \equiv 0$ and such states are said to be superstable. In the case of the logistic map, $F(\bar{x}, \mu) = \bar{x}$ and $F_x(\bar{x}, \mu) = 0$ are satisfied by $x = \frac{1}{2}$ and $\mu = 2$, while $F^{(2)}(\bar{x}, \mu) = x$ and $F_x^{(2)}(\bar{x}, \mu) = 0$ are satisfied by $x = \frac{1}{2}$, $(1 + \sqrt{5})/4$ and $\mu = 1 + \sqrt{5}$. The maps at these critical values of μ are shown in Figure 2.13. The important point to note is that those parts of the map shown in the shaded areas are similar, namely, of general parabolic form with a fixed point at the extreme (one map, of course, is simply upside down). The other fixed points in the $F^{(2)}(x, \mu)$ map are uninteresting as they are unstable, just as the fixed point $x = 0$ is for the map $F(x, \mu)$.

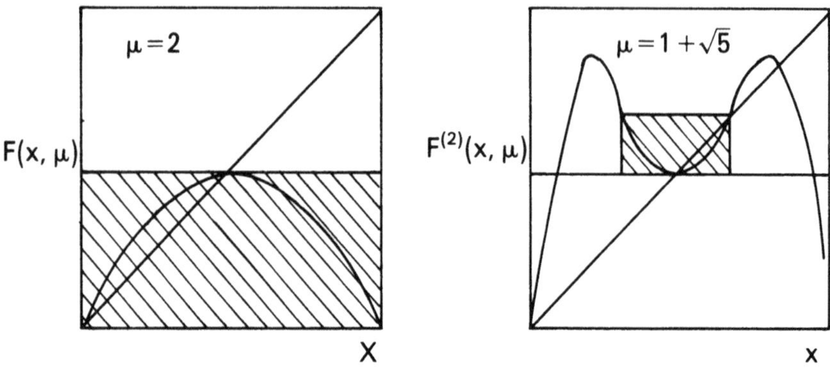

Figure 2.13. Illustration of the universality of the maps in the vicinity of a maximum.

This process of considering the higher maps $F^{(2n)}(x, \mu)$ can be continued, and at each stage the fixed point is at an extreme and the map in the region of this point is of the general shape as in the shaded area in Figure 2.13. The shaded area gets smaller and smaller as the value of n is increased, which means that the detailed shape of $F^{(2n)}(x, \mu)$ in this region depends on the form of $F(x, \mu)$ for values of x nearer and nearer the maximum at $x = \frac{1}{2}$. Thus as long as the original map has an isolated maximum, then for large enough values of n the form of $F^{(2n)}(x, \mu)$ in the shaded area will be the same for *all* such maps. This is the basic reason for the universality nature of the bifurcation sequence.

Let us now assume that a universal function $g(x, \mu)$ exists such that after a bifurcation, brought about by a change in a parameter μ, a new function is formed and is simply a scaled version of the original function. We write the critical value of μ corresponding to the state of superstability for the nth bifurcation as μ_n, so near this state we may write $\mu = \mu_n + P\Delta_n$ where $\mu_{n+1} - \mu_n = \Delta_n$, and now the parameter P describes the map in the vicinity of the superstable state $(0 < P < 1)$. The whole universality of the bifurcation sequence is now represented by the functional equation

$$g(x, p) = -\alpha g^{(2)}(-x/\alpha, a + p/\delta) \tag{2.4.1}$$

The scaling parameter α simply changes the size of the shaded areas, at each bifurcation. The minus signs simply arise owing to the inversion of the map, as is evident from Figure 2.13. The second scaling parameter δ is introduced to take care of the fact that, as the bifurcation sequence progresses, the range of μ between adjacent bifurcations gets smaller, as is evident from Table 2.1. The whole theory of universality rests on the assumption that the function $g(x, p)$ exists.

Unfortunately it is not possible to solve equation (2.4.1) analytically. The nature of the solution and the values of the scaling parameters α and δ can only be obtained from a numerical solution.

Equation (2.4.1) has a fixed point in p space such that $\bar{p} = 1 + \bar{p}/\delta$ and

$$\bar{g}(\bar{x}, \bar{p}) = -\alpha \bar{g}^{(2)}(-x/\alpha, \bar{p}) \tag{2.4.2}$$

This equation has been solved numerically and found to have a solution for $\alpha = 2.50290787$. [The solution is subjected to the normalization condition $\bar{g}(x = 0, \bar{p}) = 0$.] This one identifies with the scaling of the amplitudes of the bifurcation solutions, M_n, as shown in Figure 2.8, and with α defined by equation (2.3.4b).

This solution can be employed to find the value of the other scaling parameter δ. To do this one looks for solutions of equation (4.1) near the fixed point $p = \bar{p}$ by linearizing (2.4.1) about solution (2.4.2). Thus one writes $g(x, p) = \bar{g}(x, \bar{p}) + (p - \bar{p})h(x)$, substitutes into equation (2.4.1), and only retains terms proportional to $(p - \bar{p})$. This gives the following equation for $h(x)$:

$$h(\bar{g}(x)) + \bar{g}_x(\bar{g}(x)) = -(\delta/\alpha)g(-\alpha x)$$

where the subscript denotes differentiation of $g(x)$ with respect to x. This is an equivalent equation for $h(x)$ and has a solution (obtained numerically) for $\delta = 4.6692016$. This is identified with the scaling of the parameter μ between bifurcations, illustrated in Figure 2.8, and defined by expression (2.3.4a) in the limit $j \to \infty$.

Thus the universal nature of the bifurcation sequence is described by the two universal numbers α and δ. These are called the Feigenbaum numbers after the person who first appreciated the universal nature of one-dimensional maps.

It will now be appreciated why it was found that the values of α and δ corresponding to the bifurcation sequence following the period-three state were the same as those describing the major sequence.

2.5. Summary

The logistic map has been used to illustrate the wealth of structure found in the solution of one-dimensional difference equations. Periodic solutions have been found and their structure related to the sequence of bifurcations found for this map. Such solutions are independent of initial conditions. Nonperiodic-type solutions, called chaotic, have also been studied and their nature elucidated by considering the statistical nature of such solutions. These are very sensitive to initial conditions. The phenomenon of intermittency, intimately related to tangent bifurcations, has also been considered. Finally, the universal aspects of these various solutions have been discussed and, in particular, it has been shown how the bifurcation sequences are described in terms of the Feigenbaum numbers α and δ.

It could be said that the logistic map plays the same role to difference equations as simple harmonic oscillator equations play to general periodic motion.

Two major problems remain. The first is associated with the complicated nature of the solution that naturally arises for quite simple one-dimensional difference equations. Such solutions can only be understood by developing appropriate statistical formulations. A start has been made in this direction by considering a Master equation of the form (2.3.6) and such parameters as the Liapunov number defined by expression (2.3.12). However, much remains to be done before one can really say such equations are understood.

The second problem arises when one considers these equations to be physically relevant such that μ is an experimentally controlled parameter, for instance, temperature or rotation velocity. For good experimental reasons such quantities can only be controlled to a finite degree of accuracy. By contrast, the higher-order bifurcations only exist for extremely small ranges of μ. In practice such bifurcations would be washed out and not observed. Thus it is necessary to consider the effect on maps, such as the logistic map, of external noise. This can be done by either taking μ to vary in a manner appropriate to this noise (for example, to be random valued), or to add to the equation a term incorporating such effects, or in fact both together. Then new ways must be developed of analyzing such equations. Some results have already been obtained (see Part 4 of Cvitanovic[3]) but much remains to be done.

References

1. R. M. May, *Nature* **261**, 459 (1976).
2. P. Collet and J. P. Eckmann, *Iterated Maps on the Interval as Dynamical Systems*, Birkhauser, Boston (1980).
3. R. Cvitanovic, *Universality in Chaos*, Adam Hilger, Bristol (1984).
4. A. J. Lichtenberg and M. A. Lieberman, *Regular and Stochastic Motion*, Springer-Verlag, New York (1983).
5. B. B. Mandelbrot, *The Fractal Geometry of Nature*, Freeman, San Francisco (1982).
6. R. S. Shaw, *Z. Naturforsch.* **36A**, 80 (1981).
7. J. P. Eckmann and D. Ruelle, *Rev. Mod. Phys.* **57**, 617 (1985).
8. T. Kai and K. Tomita, *Prog. Theor. Phys.* **64**, 1532 (1980).
9. G. A. McCreadie and G. Rowlands, *Phys. Lett.* **91A**, xxx (1982).
10. Y. Pomeau and P. Manneville, *Commun. Math. Phys.* **74**, 189 (1980).

3

Spectral Transform and Solitons: How to Solve and Investigate Nonlinear Evolution Equations

F. Calogero and A. Degasperis

3.1. Introduction

The soliton was discovered (and named) in 1965 by Zabusky and Kruskal,[1] who were experimenting with the numerical solution by computer of the Korteweg–de Vries (KdV) equation. This nonlinear partial differential equation had been introduced at the end of the last century to describe wave motion in shallow canals.[2] Zabusky and Kruskal studied the equation because of its relevance to plasma physics, as well as to the Fermi–Pasta–Ulam puzzle[3] (for a fascinating account of the motivations that led to the "birth of the soliton," see Kruskal.[4]) The first scientific description of the soliton as a natural phenomenon, however, goes back to the first half of the nineteenth century, and was reported by J. Scott-Russell in the following prose:[5]

> I was observing the motion of a boat which was rapidly drawn along a narrow channel by a pair of horses, when the boat suddenly stopped—not so the mass of water in the channel which it had put in motion; it accumulated round the prow of the vessel in a state of violent agitation, then suddenly leaving it behind, rolled forward with great velocity, assuming the form of a large solitary elevation, a rounded, smooth and well defined heap of water, which continued its course along the channel apparently without change of form or diminution of speed. I followed

---

F. Calogero and A. Degasperis • Dipartimento di Fisica, Università degli Studi di Roma, "La Sapienza," 00185 Rome, Italy. This chapter has been reprinted, with minor modifications, from F. Calogero and A. Degasperis, *Spectral Transform and Solitons: Tools to Solve and Investigate Nonlinear Evolution Equations*, Vol. 1, North-Holland Publishing Company, Amsterdam (1982), pp. 1-67, with permission of the authors and publisher.

> it on horseback, and overtook it still rolling on at a rate of some eight or nine miles an hour, preserving its original figure some thirty feet long and a foot to a foot and a half in height. Its height gradually diminished, and after a chase of one or two miles I lost it in the windings of the channel. Such, in the month of August 1834, was my first chance interview with that singular and beautiful phenomenon....

But the real breakthrough occurred in 1967, when the idea of the spectral transform technique was introduced by Gardner, Greene, Kruskal, and Miura as a means to solve the Cauchy problem for the KdV equation.[6] Soon afterwards Lax put the method into a framework that provided a clear indication of its generality and greatly influenced future developments;[7] and a few years later Zakharov and Shabat, by a nontrivial extension of the approaches of Gardner, Greene, Kruskal, Miura, and Lax, were able to solve the Cauchy problem for another important nonlinear evolution equation, the so-called nonlinear Schrödinger equation.[8] The way was thereby opened for the search and discovery of several other nonlinear evolution equations, or rather classes of such equations, solvable by these techniques, a process that continues unabated to the present day. The subject has moreover branched out into other areas of mathematics (algebraic and differential geometry, functional and numerical analysis), and its applications are percolating through the whole of physics (from nonlinear optics to hydrodynamics, from plasma to elementary particle physics, from lattice dynamics to electrical networks, to superconductivity and to cosmology), and are indeed appearing also in other scientific disciplines (epidemiology, neurodynamics, etc.). This is of course related to the central rôle played by nonlinear evolution equations in mathematical physics, and more generally in applied mathematics, and to the fact that the spectral transform approach constitutes in some sense (which will be made more precise below) an extension to a nonlinear context of the Fourier transform technique, whose all-pervading rôle for solving and investigating linear phenomena is of course well known.

The broadness of scope, in both pure and applied mathematics, that has been outlined here, as well as the dynamical stage of development of this field of enquiry, excludes the possibility of providing a systematic and complete coverage of the theory and/or its applications. This chapter focuses on one approach, and leaves out any specific treatment of applications. The guiding thread is the analogy of the spectral transform technique for solving (certain classes of) nonlinear evolution equations, to the Fourier transform method for solving linear partial differential equations.

The material covered in this chapter is delineated clearly enough by the list of subheadings in the Table of Contents not to require an additional review here. Let us rather mention the main topics we have omitted. These include: the relationship to Hamiltonian dynamics; discretized problems (finite-difference equations, mappings, dynamical systems with a finite number of degrees of freedom); problems on a finite interval, and in particular problems with periodic boundary conditions; the "inverse" problem of ascertaining, given a nonlinear evolution equation, whether it belongs to some class of equations tractable by spectral transform techniques and, if so, what is then the appropriate spectral problem that provides the basis to introduce the spectral transform (this is still largely an open problem, whose investigation is now actively pursued mainly in the framework of differential geometry); the study of equations "close" to those solvable by these techniques, or equivalently the use, as the point of departure of a perturbative approach, of some nonlinear evolution equation solvable via the spectral transform (rather than a "brutally" linearized equation); the many other approaches that overlap and complement that treated in this chapter; and of course the open-ended range of applications.

Notes to Section 3.1

In addition to those referred to in the text, the main contributions to the early history of "soliton" theory can be found in the literature.(9–18)

It is perhaps of interest to note that the keyword "soliton" became an entry of the Subject Index of Physics Abstracts in 1973 (January–June issue).

A list of "solvable" equations is given in Section 3.13. Among the more important papers that have enlarged the class of solvable equations are those of Zakharov and Shabat(19) and Ablowitz *et al.*;(20) the main papers that have extended the class of solvable equations using the approach described in this book can be found elsewhere.(21–27)

There are many publications on developments toward other branches of mathematics.(29–31)

The literature on applications is enormous, and the many review papers and books provide useful guidance.(32–43)

There are numerous books and survey papers devoted entirely to these topics or including some treatment of them which should prove useful, especially to readers interested in an approach different from that followed here.(28–33,35,38,39,42–72)

A large number of papers(21–27,73–143) closely related to the approach followed here should prove particularly useful to the reader who wishes to pursue some detail insufficiently covered in this chapter, or to test whether some development, suggested by the results given here, has already been accomplished.

The main original papers on the relationship with Hamiltonian dynamics are those of Gardner(12) and Zakharov and Faddeev.(16) The subsequent literature is vast.(31,144–157)

Discretized equations are dealt with in several review papers.(53,97,158,159)

The recent literature on dynamical systems is very large,(160–166) and numerous review papers have been published.(56,61,108,167–172)

The treatment here is confined to problems on the whole line, generally with vanishing boundary conditions at infinity. The mathematical techniques relevant to analogous problems, but in a finite interval, bear, in the case with periodic boundary conditions, the same sort of relationship to the problems treated here as the Fourier series to the Fourier integral. In fact, however, such problems generally involve a higher dose of mathematics than is used here, for instance, familiarity with elliptic functions and some algebraic geometry. The relevant literature is fairly ample.[28,173-176]

The research on prolongation structures (the geometricodifferential technique to investigate whether a given nonlinear evolution equation is integrable) is surveyed by Estabrook and Wahlquist[177] and also by Hermann[29,30] and Pirani *et al.*[178] An elementary recent introduction requiring hardly any geometricodifferential background is provided by Kaup.[179]

The construction of a "perturbation theory" based on integrable nonlinear evolution equations is an important task that is far from completed.

It is impossible to outline here the many other approaches that now exist to solve and investigate nonlinear evolution equations and to detail their applications; but the books and survey papers cited above should be sufficient to provide adequate guidance.

The implicit notion that there exist different degrees of mathematical rigor may appear disturbing to some readers: a theoretical physicist friend of ours, having once claimed to have proved something "rigorously," was promptly asked by a mathematician whether it would not have sufficed to prove it "normally," the implication being of course that a proof is a proof and needs no adjective to qualify it. Not so: for even within "pure" mathematics the standards of rigor vary immensely. This is shown by the popularized analysis given on pp. 37-39 and 104-109 of the book by Nagel and Newman[186] of the hidden assumptions—that ought to be brought out in order to make the proof "rigorous"—contained, in the context of formalized mathematical logic, in the proof of even the most elementary theorem of arithmetic, such as that negating the existence of a largest prime number.

3.2. The Main Idea and Results: An Overview

In this section we introduce, in the very simplest context, the basic idea of the technique to solve (certain classes of) nonlinear evolution equations via the spectral transform, and we outline its main implications. The results surveyed here are then taken up and treated in more detail in subsequent sections.

3.2.1. Solution of Linear Evolution Equations by Fourier Transform

Consider the linear partial differential equation

$$u_t(x, t) = -i\omega\left(-i\frac{\partial}{\partial x}\right)u(x, t) \tag{3.2.1}$$

where $\omega(z)$ is, say, a polynomial. The study of many natural phenomena can be reduced to the investigation of the solution $u(x, t)$ of equation (3.2.1) characterized by the initial condition

$$u(x, 0) = u_0(x), \qquad -\infty < x < \infty \tag{3.2.2}$$

where $u_0(x)$ is regular (for all real values of x) and vanishes asymptotically, say

$$\lim_{x\to\pm\infty} [|x|^{1+\varepsilon} u_0(x)] = 0, \qquad \varepsilon > 0 \tag{3.2.3}$$

Note that, if the odd part of $\omega(z)$ is real and the even part imaginary {i.e., if $[\omega(z^*)]^* = -\omega(-z)$}, equation (3.2.1) is a real equation; then $u(x, t)$ is also real for $t > 0$ if $u_0(x)$ is real. Moreover, if $\omega(z)$ is real {$[\omega(z^*)]^* = \omega(z)$}, then equation (3.2.1) is purely dispersive (i.e., nondissipative); for instance, it is then easily seen [see equations (3.2.4) and (3.2.7) below] that the integral over all values of x of $|u(x, t)|^2$ is time-independent.

The *initial-value problem* or *Cauchy problem* characterized by a boundary condition of type (3.2.2) and (3.2.3) is the typical kind of problem we will be investigating throughout this chapter, although our main focus below will be on *nonlinear* evolution equations rather than on *linear* evolution equations like (3.2.1).

A central role in the solution of equation (3.2.1) is played by the (direct and inverse) Fourier transform equations for (the dependence on the variable x of) $u(x, t)$:

$$u(x, t) = (2\pi)^{-1} \int_{-\infty}^{+\infty} dk \exp(ikx)\hat{u}(k, t) \tag{3.2.4}$$

$$\hat{u}(k, t) = \int_{-\infty}^{+\infty} dx \exp(-ikx) u(x, t) \tag{3.2.5}$$

This is due to the fact that, if $u(x, t)$ evolves according to the *partial* differential equation (3.2.1), the Fourier transform $\hat{u}(k, t)$ evolves according to the *ordinary* differential equation

$$\hat{u}_t(k, t) = -i\omega(k)\hat{u}(k, t) \tag{3.2.6}$$

that can be immediately integrated to yield

$$\hat{u}(k, t) = \hat{u}(k, 0) \exp[-i\omega(k)t] \tag{3.2.7}$$

Thus the solution of equations (3.2.1) and (3.2.2) [with equation 3.2.3] is accomplished in three steps. First, at the initial time $t = 0$, the Fourier transform

$$\hat{u}(k, 0) = \hat{u}_0(k) = \int_{-\infty}^{+\infty} dx \exp(-ikx) u_0(x) \tag{3.2.8}$$

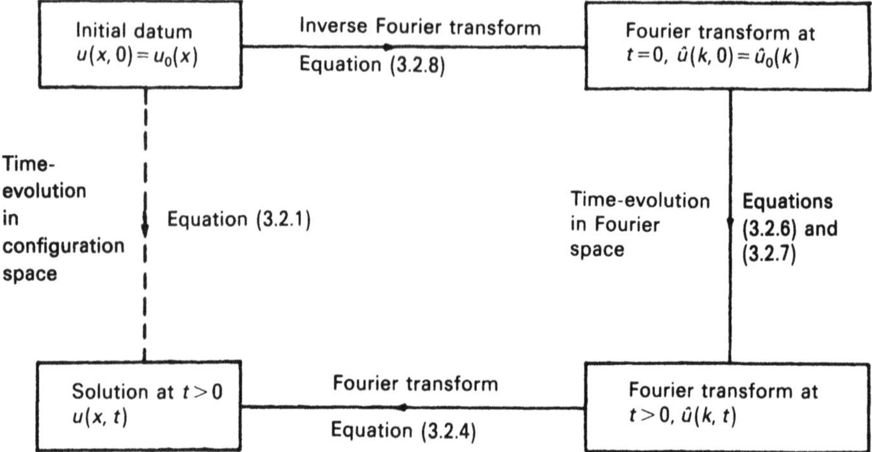

Figure 3.1. Schematic of solution technique.

is evaluated [see equations (3.2.5) and (3.2.2)]; then, the Fourier transform $\hat{u}(k, t)$ is obtained from equation (3.2.7); and finally, at time t, the function $u(x, t)$ is recovered from $\hat{u}(k, t)$ using the Fourier transform equation (3.2.4).

The technique of solution that we have illustrated here is conveniently summarized by the schematic in Figure 3.1, where the broken line indicates the (difficult) problem that is generally directly related to applications, while the three continuous lines indicate the three (easier) steps whose sequence yields the solution.

We submit that the main reason why the Fourier transform is such an important tool in mathematical physics and in applied mathematics is because it provides, as we have indicated above, the appropriate technique to solve the problem characterized by equations (3.2.1) and (3.2.2); for indeed this mathematical problem, as well as its generalizations that are also analogously solvable by Fourier methods, constitute the prototypical (often, of course, approximate) schematization of many natural phenomena.

In fact what is really important is not the possibility of exhibiting explicitly the solution of equations (3.2.1) and (3.2.2) [indeed only in rare instances can the integrals in equations (3.2.4) and (3.2.5) be analytically calculated], but rather the insight into the behavior of $u(x, t)$ implied by this technique of solution. Clearly the time evolution is completely determined by the "dispersion function" $\omega(k)$; and the (standard) analysis of the long-time behavior of $u(x, t)$, as given by equations (3.2.7) and (3.2.4), establishes that a solution $u(x, t)$, characterized by initial data $u_0(x)$ that are localized around a position x_0 and moreover have a Fourier transform $\hat{u}_0(k)$ that is localized around a value k_0, behaves generally as a "wave packet" moving with the *group velocity*

$$v_g = \left.\frac{d\omega(k)}{dk}\right|_{k=k_0} \tag{3.2.9}$$

and dispersing asymptotically (the peak amplitude of its envelope is localized around $x = x_0 + v_g t$ and decreases proportionally to $t^{-1/2}$ as $t \to \infty$). We note that this behavior, though of course well understood, is by no means evident from the stucture of the partial differential equation (3.2.1). Contrast this with the extremely simple time evolution (3.2.7) of the Fourier transform $\hat{u}(k, t)$. It is thus seen that *the dynamics is much simpler in k-space than in x-space.* The implications of this message, that synthesizes the lesson to be drawn from the solvability of equation (3.2.1) via Fourier transform, are pervasive: not only do they suggest the proper language to theorize about many natural phenomena [all those whose mathematical description can be, one way or another, patterned on equation (3.2.1)], but indeed they dictate how best to experiment with them (for instance, the motivation for experimenting, in optics, with monochromatic beams, can be traced precisely to this origin; and the same applies, in a way, even to the drive to build particle accelerators of higher and higher energy!). It is reasonable to expect an analogous situation to develop for those *nonlinear* evolution equations that are solvable by a technique—based on the spectral transform—that, as we will see, presents a close similarity, but also some significant novelties, relative to the Fourier transform approach discussed thus far.

Here we also mention some other properties of the solutions of equation (3.2.1) that will also be shown to have analogues in the nonlinear cases treated in the following.

First, note that, if $u^{(1)}(x, t)$ satisfies equation (3.2.1) and $u^{(2)}(x, t)$ is related to $u^{(1)}(x, t)$ by the formula

$$g\left(-i\frac{\partial}{\partial x}\right)u^{(2)}(x, t) + h\left(-i\frac{\partial}{\partial x}\right)u^{(1)}(x, t) = 0 \tag{3.2.10}$$

then $u^{(2)}(x, t)$ also satisfies equation (3.2.1). Indeed, the relation in Fourier space corresponding to equation (3.2.1) reads of course

$$g(k)\hat{u}^{(2)}(k, t) + h(k)\hat{u}^{(1)}(k, t) = 0 \tag{3.2.11a}$$

or, equivalently,

$$\hat{u}^{(2)}(k, t) = -[h(k)/g(k)]\hat{u}^{(1)}(k, t) \tag{3.2.11b}$$

clearly implying that, if $u^{(1)}(k, t)$ satisfies equation (3.2.6), so does $u^{(2)}(k, t)$. Although expression (3.2.10), for given $u^{(1)}(x, t)$, reads as an ordinary differential equation for $u^{(2)}(x, t)$ only if $g(z)$ and $h(z)$ are polynomials, or rational functions, the functions $g(z)$ and $h(z)$ here are largely arbitrary, except for those conditions which guarantee that $u^{(2)}(x, t)$ is regular and vanishes asymptotically according to equation (3.2.3), if also $u^{(1)}(x, t)$ does [for instance, $g(z)$ should not have any real zero].

Second, we note that, if $u^{(1)}(x, t)$ and $u^{(2)}(x, t)$ are solutions of equation (3.2.1), this equation is also satisfied by any linear combination of them (with constant coefficients),

$$u(x, t) = c_1 u^{(1)}(x, t) + c_2 u^{(2)}(x, t) \tag{3.2.12}$$

This is of course the *superposition principle*, which corresponds to the linear structure of equation (3.2.1).

Third, we remark that the solution of expressions (3.2.1) and (3.2.2) can be formally written through the *resolvent equation*

$$u(x, t) = \exp.\left[-it\omega\left(-i\frac{\partial}{\partial x}\right)\right] u_0(x) \tag{3.2.13}$$

corresponding to a straightforward integration of equation (3.2.1) [from 0 to t, using equation (3.2.2)]. In the particular case $\omega(z) = -z$, when equation (3.2.1) reads

$$u_t(x, t) = u_x(x, t) \tag{3.2.14}$$

and has the general solution

$$u(x, t) = f(x + t) \tag{3.2.15}$$

equation (3.2.13) corresponds therefore to the well-known *operator identity*

$$f(x + a) = \exp\left(a\frac{d}{dx}\right) f(x) \tag{3.2.16}$$

where we have written a in place of t to emphasize that the validity of this expression has nothing to do with the time-evolution problem; indeed equation (3.2.16) expresses a property of the "translation operator" $\exp(ad/dx)$, and it holds for any function $f(x)$ [strictly speaking, for any function $f(z)$ holomorphic in the disc $|z - x| \leq |a|$].

While for the sake of simplicity we have introduced no explicit time-dependence in equation (3.2.1), it should be emphasized that the technique of solution described in this section works (with trivial changes) also if this partial differential equation is explicitly time-dependent, i.e.,

$$u_t(x, t) = -i\omega\left(-i\frac{\partial}{\partial x}, t\right) u(x, t) \tag{3.2.17}$$

Then of course in place of equation (3.2.6) one has

$$\hat{u}_t(k, t) = -i\omega(k, t)\hat{u}(k, t) \tag{3.2.18}$$

and equation (3.2.7) is replaced by

$$\hat{u}(k, t) = \hat{u}(k, 0) \exp\left[-i \int_0^t dt' \,\omega(k, t')\right] \tag{3.2.19}$$

If instead the partial differential equation (3.2.1) contains an explicit x-dependence, the applicability of the Fourier transform approach becomes less simple. A case in which it continues to work (in the sense of reducing the solution to quadratures, and to the solution of a, generally nonlinear, *ordinary* differential equation; see below) is when the x-dependence is linear, so that in place of equation (3.2.1) one has

$$u_t(x, t) = -i\left[\omega\left(-i\frac{\partial}{\partial x}, t\right) + x\omega'\left(-i\frac{\partial}{\partial x}, t\right)\right] u(x, t) \tag{3.2.20}$$

This equation is now characterized by the two functions $\omega(z, t)$ and $\omega'(z, t)$ (for the sake of generality we are also introducing a t-dependence). Its counterpart in Fourier space reads

$$\hat{u}_t(k, t) = [-i\omega(k, t) + \omega'_k(k, t)]\hat{u}(k, t) + \omega'(k, t)\hat{u}_k(k, t) \tag{3.2.21}$$

and it is explicitly solvable (by the method of characteristics), yielding

$$\hat{u}(k, t) = \hat{u}_0[k_0(t, k)] \exp\left\{\int_0^t dt' \,[-i\omega(\chi, t') + \omega'_k(\chi, t')]\right\} \tag{3.2.22}$$

where

$$\chi \equiv \chi[t', k_0(t, k)] \tag{3.2.23}$$

the function $\chi(t, k_0)$ being defined by the (ordinary) differential equation

$$\chi_t(t, k_0) = -\omega'[\chi(t, k_0), t] \tag{3.2.24}$$

and by the boundary condition

$$\chi(0, k_0) = k_0 \tag{3.2.25}$$

while the function $k_0(t, k)$ is defined by χ through the (implicit) equation

$$\chi(t, k_0) = k \tag{3.2.26}$$

Thus the solution of the (Cauchy) problem characterized by equations (3.2.20) and (3.2.2) is given by equations (3.2.4) and (3.2.22), $\hat{u}_0(k)$ being given by equation (3.2.8). Of course we are here assuming that all integrals converge, and that the integrations by parts necessary to derive equation (3.2.21) from equation (3.2.20) are permissible. That this is not always the case is clear from equation (3.2.22), since this expression, together with equation (3.2.23)-(3.2.26), need not imply that $\hat{u}(k, t)$, for $t > 0$, vanish asymptotically in k, even though $\hat{u}(k, 0) \equiv \hat{u}_0(k)$ does.

Our motivation for reviewing in this section some well-known facts concerning the solution of *linear* evolution equations by Fourier transform is because of the close similarity of this approach to the method of solution, via the spectral transform, of (certain classes of) *nonlinear* evolution equations that constitutes our main interest. The correspondence applies also to the extensions that we have just mentioned (equations with t-dependent and linearly x-dependent coefficients). We end this section by mentioning two directions instead in which such a close correspondence does not yet seem to exist.

First and most important is the extension of the approach to more (space) variables. This can be done rather trivially in the linear case by introducing the multidimensional Fourier transform; a comparably straightforward extension to problems with more space variables does not exist in the nonlinear case (although there are some ways to introduce extra variables).

Returning to the simplest case of one space and one time variable, there is another kind of extension that can be done very simply in the linear case but still has no simple counterpart in the nonlinear context: the inclusion of certain classes of integrodifferential equations. Consider, for instance, in place of equation (3.2.1), the evolution equation

$$u_t(x, t) = \int_{-\infty}^{+\infty} dy\, K(x - y, t)u(y, t) \tag{3.2.27}$$

Then the treatment described above is again applicable, with equation (3.2.6) [or rather equation (3.2.18)] replaced by

$$\hat{u}_t(k, t) = \hat{K}(k, t)\hat{u}(k, t) \tag{3.2.28}$$

where of course $\hat{K}(k, t)$ is the Fourier transform of $K(x, t)$:

$$K(x, t) = (2\pi)^{-1} \int_{-\infty}^{+\infty} dk\, \hat{K}(k, t) \exp(ikx) \tag{3.2.29a}$$

$$\hat{K}(k, t) = \int_{-\infty}^{+\infty} dx\, K(x, t) \exp(-ikx) \tag{3.2.29b}$$

No spectral transform technique is as yet available for solving simple nonlinear evolution equations of integral type, although recently there appears to have been progress also in that direction (see Section 3.13).

3.3. A Class of Solvable Nonlinear Evolution Equations

A class of *nonlinear* evolution equations solvable via the spectral transform technique can be written in compact form as follows:

$$u_t(x, t) = \alpha(\boldsymbol{L})u_x(x, t) \tag{3.3.1}$$

Here $\alpha(z)$ is, say, a polynomial, and $\boldsymbol{L}$ is the integrodifferential operator defined by the following formula that specifies its action on a generic function $f(x)$:

$$\boldsymbol{L}f(x) = f_{xx}(x) - 4u(x, t)f(x) + 2u_x(x, t)\int_x^{+\infty} dy\, f(y) \tag{3.3.2}$$

Here, and generally below, consideration is restricted to functions such that all integrals are convergent. Of course f may also depend on other variables, for instance on t.

Note that the operator $\boldsymbol{L}$ depends on u; this causes the right-hand side of equation (3.3.1) to be nonlinear in u. Since $\boldsymbol{L}$ is integrodifferential, it might appear that equation (3.3.1) is generally an integrodifferential equation. But this is not the case: as long as $\alpha(z)$ is a polynomial in z, equation (3.3.1) is a (nonlinear) *partial differential* evolution equation. This is due to an important and nontrivial property of the operator $\boldsymbol{L}$, that can be synthetically formulated as follows:

$$\boldsymbol{L}^n u_x(x, t) = g_x^{(n)}, \qquad n = 0, 1, 2, \ldots \tag{3.3.3}$$

where $g^{(n)}$ is a polynomial (of degree $n + 1$) in u and its x-derivatives (up to the derivative of order $2n$). For instance:

$$\boldsymbol{L}u_x = u_{xxx} - 6uu_x = (u_{xx} - 3u^2)_x \tag{3.3.3a}$$

$$\begin{aligned}\boldsymbol{L}^2 u_x &= u_{xxxxx} - 10uu_{xxx} - 20u_x u_{xx} + 30u^2 u_x \\ &= (u_{xxxx} - 10uu_{xx} - 5u_x^2 + 10u^3)_x\end{aligned} \tag{3.3.3b}$$

Thus the simplest nonlinear evolution equation contained in the class (3.3.1) is the Korteweg-de Vries (KdV) equation

$$u_t + u_{xxx} - 6uu_x = 0 \tag{3.3.4}$$

corresponding to $\alpha(z) = -z$.

As in the case discussed in the preceding section, our main interest will be in the Cauchy problem associated with equation (3.3.1), characterized by a given initial condition

$$u(x, 0) = u_0(x), \qquad -\infty < x < +\infty \tag{3.3.5}$$

with $u_0(x)$ regular for all real values of x and vanishing asymptotically.

We have written out the class of evolution equations (3.3.1) at this early stage to provide a motivation for subsequent developments. We shall return to this class several times in the following. The technique of solution of equation (3.3.1) hinges on the spectral transform of (the x-dependence of) $u(x, t)$, that plays an analogous role to that of the Fourier transform for the solution of equation (3.2.1). Thus the following section is devoted to a terse outline of the spectral transform.

But before closing this section we would like to point out that, in the limit in which $u(x, t)$, and its derivatives, are so small that it is justified to neglect all powers of them, the *nonlinear* evolution equation (3.3.1) goes over into the *linear* equation

$$u_t(x, t) = \alpha\left(\frac{\partial^2}{\partial x^2}\right) u_x(x, t) \tag{3.3.6}$$

or equivalently

$$u_t(x, t) = -\omega\left(-i\frac{\partial}{\partial x}\right) u(x, t) \tag{3.3.6a}$$

with

$$\omega(z) = -z\alpha(-z^2). \tag{3.3.6b}$$

Note that, for real $\alpha(z)$, $\omega(z)$ turns out automatically to be real and odd, yielding therefore not only, via equation (3.3.6a), a real equation (as of course it must), but indeed a purely dispersing (i.e., nondissipative) one.

3.4. The Spectral Transform

This topic is divided into three parts. In the first the Schrödinger spectral problem is tersely described, and the spectral transform is introduced. In Section 3.5, the solution of the corresponding *inverse* spectral problem is outlined. In the third, the spectral transform is discussed.

The spectral problem that we discuss in this section is that appropriate to the solution, via the spectral transform technique, of the class of nonlinear evolution equations introduced in the preceding section. Other spectral problems are appropriate for the solution of other classes of nonlinear evolution equations.

We emphasize that the topic treated in this section has no reference to the time-evolution problem; just as the study of the Fourier transform has nothing to do with the eventual use of it to solve (linear) evolution equations.

3.4.1. Direct Spectral Problem

The spectral problem that we discuss here is familiar to students of quantum mechanics, being based on the stationary Schrödinger equation

$$-\psi_{xx}(x, k) + u(x)\psi(x, k) = k^2\psi(x, k), \qquad -\infty < x < +\infty \tag{3.4.1}$$

This is a second-order linear ordinary differential equation. In the context of this section $u(x)$ is a given real function, that we assume to be regular for all (real) values of x and to vanish (sufficiently fast) asymptotically, say

$$\lim_{x\to\pm\infty} [|x|^{1+\varepsilon}u(x)] = 0, \qquad \varepsilon > 0 \tag{3.4.2}$$

In the quantum-mechanical context the stationary Schrödinger equation (3.4.1) obtains from a time-dependent Schrödinger equation. It is perhaps useful to emphasize that this has nothing to do with the time evolution we are discussing in (other sections of) this chapter. Indeed it is probably advisable to forget altogether the physical interpretation of equation (3.4.1) in terms of one-dimensional quantal scattering or bound states, but rather to consider the spectral problem defined by equation (3.4.1) as a typical Sturm–Liouville eigenvalue problem; in fact, a *singular* Sturm–Liouville problem, since one is considering the differential equation (3.4.1) on the whole real line, namely on an infinite interval. In the following we shall take this point of view, although we keep some of the terminology (such as "reflection" and "transmission" coefficients; see below) that clearly has originated from the quantal scattering problem.

The spectrum of the Sturm–Liouville problem characterized by equation (3.4.1) has two components: a continuum, including all positive values of the eigenvalue k^2, and a number of discrete negative eigenvalues, $k^2 = -p_n^2, p_n > 0$, $n = 1, 2, \ldots, N$.

It is convenient, in order to characterize the continuum part of the spectrum (corresponding to real values of k, so that $k^2 > 0$), to introduce the solution of equation (3.4.1) characterized by the asymptotic boundary conditions

$$\psi(x, k) \to T(k) \exp(-ikx), \qquad x \to -\infty \tag{3.4.3a}$$

$$\psi(x,k) \to \exp(-ikx) + R(k)\exp(ikx), \qquad x \to +\infty \tag{3.4.3b}$$

This asymptotic behavior is clearly consistent with equation (3.4.2), and it identifies uniquely the eigenfunction $\psi(x, k)$, as well as the *transmission coefficient* $T(k)$ and the *reflection coefficient* $R(k)$.

Let N be the number of (discrete) negative eigenvalues,

$$k^2 = -p_n^2, \qquad p_n > 0, \qquad n = 1, 2, \ldots, N \tag{3.4.4}$$

To each of these eigenvalues there corresponds a solution $f_n(x)$ of equation (3.4.1), uniquely identified by the asymptotic boundary condition

$$\lim_{x\to+\infty} [\exp(p_n x) f_n(x)] = 1, \qquad n = 1, 2, \ldots, N \tag{3.4.5}$$

This solution vanishes also as $x \to -\infty$, proportionally to $\exp(p_n x)$, and is therefore normalizable; indeed it is precisely this condition that identifies the (discrete negative) eigenvalue $-p_n^2$. It is convenient to define the *normalization coefficient* ρ_n by the formula

$$\rho_n = \left[\int_{-\infty}^{+\infty} dx\, f_n^2(x)\right]^{-1}, \qquad n = 1, 2, \ldots, N \tag{3.4.6}$$

We assume hereafter that the number of discrete eigenvalues N is finite; a sufficient condition is that equation (3.4.2) holds for some $\varepsilon > 1$. Of course N might vanish, namely, there might be no discrete eigenvalue; a sufficient condition for this is that $u(x)$ be nowhere negative, $u(x) \geq 0$ for $-\infty < x < +\infty$; a sufficient condition to exclude this is that $u(x)$ be nowhere positive, $u(x) \leq 0$ for $-\infty < x < +\infty$. Indeed N may be viewed as a (global) measure of the negativeness of the function $u(x)$.

The *spectral transform* S of the function $u(x)$ is, by definition, the collection of data

$$S[u] = \{R(k), \ -\infty < k < +\infty; \ p_n, \rho_n, \ n = 1, 2, \ldots, N\} \tag{3.4.7}$$

The motivation for such a definition is that there is a one-to-one correspondence between functions $u(x)$ (in an appropriate functional class, as indicated above), and the spectral transform equation (3.4.7). The analysis of this section indicates how S is determined (clearly uniquely) by u; this is the *direct* spectral problem. In the following section the *inverse* spectral problem is discussed, namely the determination of u from a given S.

3.5. Inverse Spectral Problem

Given the spectral transform

$$S = \{R(k), -\infty < k < +\infty; p_n, \rho_n, n = 1, 2, \ldots, N\} \tag{3.5.1}$$

the following procedure yields the corresponding function $u(x)$.

Define first of all the function

$$M(x) = (2\pi)^{-1} \int_{-\infty}^{+\infty} dk \exp(ikx)R(k) + \sum_{n=1}^{N} \rho_n \exp(-p_n x) \tag{3.5.2}$$

We note that, except for the contribution from the discrete spectrum, $M(x)$ is precisely the inverse Fourier transform of $R(k)$.

Consider next the *Gel'fand-Levitan-Marchenko* (GLM) integral equation

$$K(x, y) + M(x + y) + \int_{x}^{+\infty} dz\, K(x, z)M(z + y) = 0, \qquad y > x \tag{3.5.3}$$

This is a Fredholm integral equation, and it determines uniquely the function $K(x, y)$. Note that the integral equation (3.5.3) refers to the dependence of $K(x, y)$ on its second argument, y; as for the dependence of $K(x, y)$ on its first argument, x, this occurs, as it were, parametrically, being caused by the appearance of x both in the argument of the inhomogeneous term $M(x + y)$ and as lower limit of integration.

Once $K(x, y)$ is determined, the function $u(x)$ follows from the simple equations

$$w(x) = 2K(x, x + 0) \tag{3.5.4}$$

$$w(x) = \int_{x}^{+\infty} dy\, u(y) \tag{3.5.5a}$$

$$u(x) = -w_x(x) \tag{3.5.5b}$$

Of course, the function $u(x)$ yielded by this procedure is real and regular for all x, and vanishes asymptotically, only provided the spectral transform S satisfies certain properties; for instance, the reflection coefficient $R(k)$ must obviously be Fourier transformable.

We shall now treat some examples. Consider the very special case of a spectral transform having a vanishing contribution from the continuum, and containing a single discrete eigenvalue:

$$S = \{R(k) = 0, -\infty < k < +\infty;\ N = 1, p_1 = p, \rho_1 = \rho\} \tag{3.5.6}$$

As we shall see, this case is going to play a very important rôle. In the present context, it provides a simple explicit illustration of the way the inverse spectral problem works. For in this case, $M(x)$ becomes simply an exponential,

$$M(x) = \rho \exp(-px) \tag{3.5.7}$$

and therefore the GLM equation (3.5.3) becomes separable and is easily solved, yielding

$$K(x, y) = -p \exp[p(\xi - y)]/\cosh[p(x - \xi)] \tag{3.5.8}$$

$$w(x) = -4p/\{1 + \exp[2p(x - \xi)]\} \tag{3.5.9}$$

$$u(x) = -2p^2/\cosh^2[p(x - \xi)] \tag{3.5.10}$$

In these equations

$$\xi = (2p)^{-1} \ln(\rho/2p) \tag{3.5.11}$$

Actually this procedure can also be carried out explicitly in the more general case of a spectral transform, having again no contribution from the continuum, but with N discrete eigenvalues:

$$S = \{R(k) = 0, \ -\infty < k < +\infty; \ p_n, \rho_n, \ n = 1, 2, \ldots, N\} \tag{3.5.12}$$

Indeed in such a case the GLM Fredholm equation (3.5.3) is still separable, although now of rank N. The function $u(x)$ corresponding to equation (3.5.12) can be written in the compact form

$$u(x) = -2\frac{d^2}{dx^2}\{\ln \det[\boldsymbol{I} + \boldsymbol{C}(x)]\} \tag{3.5.13}$$

where $\boldsymbol{I}$ is the unit matrix of order N and $\boldsymbol{C}(x)$ is the symmetrical matrix of order N with elements

$$C_{mn}(x) = (\rho_m\rho_n)^{1/2}(p_m + p_n)^{-1} \exp[-(p_m + p_n)x] \tag{3.5.14}$$

Consider finally the case when there are no discrete eigenvalues and moreover $R(k)$ is very small (in modulus), so that $u(x)$ is also very small (see below). Note the consistency of this last assumption with the absence of discrete eigenvalues; indeed if $u(x)$ is very small, there can be at most one discrete eigenvalue.

Assume then that the smallness of R, and therefore of M [see equation (3.5.2)], justifies the neglect of the last term in the left-hand side of equation (3.5.3), yielding

$$K(x, y) \approx -M(x + y) \tag{3.5.15}$$

and therefore

$$u(x) \approx (2\pi)^{-1} \int_{-\infty}^{+\infty} dk \exp(ikx) ikR(\tfrac{1}{2}k) \tag{3.5.16}$$

It is therefore seen that, in this approximation, there is a simple relationship between the reflection coefficient $R(k)$ and the Fourier transform $\hat{u}(k)$ of $u(x)$, namely

$$\hat{u}(k) \approx ikR(\tfrac{1}{2}k) \tag{3.5.17a}$$

Indeed this formula, or rather the completely equivalent version

$$R(k) \approx (2ik)^{-1}\hat{u}(2k) \tag{3.5.17b}$$

corresponds, in the context of the quantum-mechanical scattering problem, just to the familiar "Born approximation" formula. As is clear from equation (3.5.17b), this approximation generally breaks down in the neighborhood of $k \approx 0$.

3.6. Discussion of the Spectral Transform

The spectral transform has been introduced at the end of subsection 3.4.1. The results of that subsection, and of Section 3.5, imply that there is a one-to-one correspondence between the function $u(x)$ and its spectral transform $S[u]$:

$$u(x) \Leftrightarrow S[u] \tag{3.6.1}$$

There are moreover constructive procedures to go from a function u to its spectral transform S, and from a spectral transform S to the corresponding function u. Both these procedures involve the solution of *linear* problems: the Schrödinger differential equation (3.4.1) for the direct spectral problem ($u \Rightarrow S$), and the GLM Fredholm integral equation (3.5.3) for the inverse spectral problem ($S \Rightarrow u$). The relation (3.6.1), between a function and its spectral transform S is, on the other hand, clearly *nonlinear*, except in the approximate case of small $u(x)$ when, as discussed at the end of the preceding subsection (and see also below), the spectral transform coincides essentially with the Fourier transform.

The analogy with the Fourier transform is moreover apparent from the exact equations

$$u(x) = \pi^{-1}\int_{-\infty}^{+\infty} dk[f^{(+)}(x,k)]^2 2ikR(k) - 4\sum_{n=1}^{N} p_n\rho_n[f_n(x)]^2 \tag{3.6.2}$$

$$w(x) = -\pi^{-1}\int_{-\infty}^{+\infty} dk f^{(+)}(x,k)\exp(ikx)R(k)$$

$$-2\sum_{n=1}^{N} \rho_n f_n(x)\exp(-p_n x) \tag{3.6.3}$$

where $f^{(+)}(x, k)$ is the solution of the Schrödinger equation (3.4.1) characterized by the asymptotic boundary condition

$$\lim_{x\to+\infty} [\exp(-ikx)f^{(+)}(x, k)] = 1 \tag{3.6.4}$$

while $f_n(x)$ is analogously defined, but for $k = ip_n$, $p_n > 0$. Note that equation (3.6.3) provides an expression for the integral of $u(x)$,

$$w(x) = \int_x^{+\infty} dy\, u(y) \tag{3.6.5}$$

rather than for $u(x)$ itself. The equivalence of equation (3.6.3) to (3.6.2), as well as the validity of these representations of $u(x)$ and $w(x)$, are nontrivial results.

We also display here three equations that may be viewed as the analogs of the inverse Fourier transform:

$$R(k) = (2ik)^{-2} \int_{-\infty}^{+\infty} dx\, [\psi(x, k)]^2 u_x(x) \tag{3.6.6}$$

$$R(k) = (2ik)^{-1} \int_{-\infty}^{+\infty} dx\, \psi(x, k) \exp(-ikx) u(x) \tag{3.6.7}$$

$$R(k) = (2ik)^{-2} \int_{-\infty}^{+\infty} dx\, \psi(x, k) \exp(-ikx)[u_x(x) + u(x)w(x)] \tag{3.6.8}$$

Here the function $\psi(x, k)$ is the solution of the Schrödinger equation (3.4.1) characterized by the boundary conditions (3.4.3); again the equivalence of these three expressions of $R(k)$ is nontrivial. Note that, in writing equation (3.6.8), we have used the definition (3.6.5).

The nonlinear character of the relation between a function $u(x)$ and its spectral transform is well displayed by these equations; it appears from the fact that the functions $f^{(+)}(x, k)$ and $f_n(x)$ in equations (3.6.2) and (3.6.3) and the function $\psi(x, k)$ in equations (3.6.6), (3.6.7) and (3.6.8) depend on $u(x)$. In the approximation in which this dependence is ignorable because $u(x)$ is negligibly small, so that [as implied by equations (3.4.1), (3.6.4) and (3.4.3)]

$$f^{(+)}(x, k) \approx \exp(ikx) \tag{3.6.9}$$

$$\psi(x, k) \approx \exp(-ikx) \tag{3.6.10}$$

and [see equation (3.5.17)]

$$R(\tfrac{1}{2}k) \approx (ik)^{-1}\hat{u}(k) \tag{3.6.11}$$

then equations (3.6.2) and (3.6.3) go over into the Fourier transform formula (3.2.4) (provided the contribution of the discrete part of the spectrum is neglected; we have commented in the previous subsection on the consistency of this assumption), while equations (3.6.6), (3.6.7) and (3.6.8) go over into the inverse Fourier transform formula (3.2.5).

3.7. Solution of Nonlinear Evolution Equations via the Spectral Transform

In the preceding section we have introduced the spectral transform S of a function $u(x)$ (regular for all real values of x and vanishing at infinity). Imagine now that u depends also on another variable, call it t ("time"): $u \equiv u(x, t)$. Then, of course, in one-to-one correspondence to $u(x, t)$, there is a spectral transform that is also time-dependent:

$$u(x, t) \Leftrightarrow S(t) \tag{3.7.1}$$

Thus, if $u(x, t)$ evolves in time, so does $S(t)$.

The all-important, and highly nontrivial, discovery that has opened up this field of scientific enquiry is that there exists a class of *interesting* time-evolutions of $u(x, t)$ to which there corresponds *simple* time-evolutions of $S(t)$. These time-evolutions of $u(x, t)$ can therefore be investigated by following the evolution in the spectral space, namely, by following the evolution of $S(t)$ rather than, directly, the evolution of $u(x, t)$, taking then advantage, to gain information on the (*interesting*) evolution of $u(x, t)$, of the possibility to go from u to S (say, at the initial time) and from S to u (say, at any later time), via the direct and inverse spectral problems described in the preceding section.

This is, of course, a closely analogous procedure to the technique of solution via Fourier transform of the class of linear partial differential equations discussed in Section 3.2. This analogy is further discussed below, in this section and in the following one, and in several other places as well, since it constitutes the main idea on which this chapter is based.

As the reader should have guessed, the simplest class of *nonlinear* evolution equations that are solvable in this way are those that were introduced in Section 3.3, namely, the class of *nonlinear partial differential equations*

$$u_t(x, t) = \alpha(\mathbf{L})u_x(x, t) \tag{3.7.2}$$

where $\alpha(z)$ is a polynomial and L is the integrodifferential operator defined by the formula

$$Lf(x) = f_{xx}(x) - 4u(x, t)f(x) + 2u_x(x, t) \int_x^{+\infty} dy\, f(y) \qquad (3.7.3)$$

that specifies its action on a generic function $f(x)$ (vanishing at infinity). (The reader is advised at this stage to review the basic properties of this class of evolution equations, as described in Section 3.3.)

The crucial property of the class of evolution equations (3.7.2) is that the corresponding time-evolution of the spectral transform of $u(x, t)$ is given by *linear ordinary differential equations*, namely

$$R_t(k, t) = 2ik\alpha(-4k^2)R(k, t) \qquad (3.7.4)$$

$$\dot{p}_n(t) = 0, \qquad n = 1, 2, \ldots, N \qquad (3.7.5)$$

$$\dot{\rho}_n(t) = -2p_n\alpha(4p_n^2)\rho_n(t), \qquad n = 1, 2, \ldots, N \qquad (3.7.6)$$

(a subscripted t, or a dot on top, are equivalent, both denoting differentiation with respect to t). These equations are explicitly solved according to the equations

$$R(k, t) = R(k, 0) \exp[2ik\alpha(-4k^2)t] \qquad (3.7.7)$$

$$p_n(t) = p_n(0) = p_n, \qquad n = 1, 2, \ldots, N \qquad (3.7.8)$$

$$\rho_n(t) = \rho_n(0) \exp[-2p_n\alpha(4p_n^2)t], \qquad n = 1, 2, \ldots, N \qquad (3.7.9)$$

It is noteworthy that equation (3.7.5), or equivalently equation (3.7.8), implies that the (discrete) eigenvalues of the Schrödinger differential operator

$$-\frac{d^2}{dx^2} + u(x, t) \qquad (3.7.10)$$

do not change, when $u(x, t)$ evolves in time according to equation (3.7.2). This property of the Schrödinger operator (3.7.10), to experience an *isospectral evolution* when $u(x, t)$ evolves according to (any one of the evolution equations of the class) (3.7.2), plays a crucial rôle in other approaches to these results, more operator-theoretically oriented than the (rather elementary) point of view basically adopted throughout this chapter.

We now consider the solution of the Cauchy problem for the class of nonlinear evolution equations (3.7.2). It clearly proceeds through three steps. First, at the initial time $t = 0$, from the given datum

$$u(x, 0) = u_0(x) \qquad (3.7.11)$$

the spectral transform

$$S(0) = \{R(k, 0), -\infty < k < +\infty; p_n, \rho_n(0), n = 1, 2, \ldots, N\} \quad (3.7.12)$$

is evaluated (solving the *direct* spectral problem; see subsection 3.4.1). Then, the spectral transform at time t,

$$S(t) = \{R(k, t), -\infty < k < +\infty; p_n, \rho_n(t), n = 1, 2, \ldots, N\} \quad (3.7.13)$$

is obtained, from the explicit expressions (3.7.7), (3.7.8), and (3.7.9) [we note that the isospectrality property, (3.7.5) or (3.7.8), also implies that the number of discrete eigenvalues N does not change as time evolves]. Finally, at time t, the function $u(x, t)$ is recovered from its spectral transform $S(t)$ (solving the *inverse* spectral problem; see Section 3.5). This technique of solution may be summarized by the schematic in Figure 3.2, closely analogous to that displayed in Figure 3.1. There the broken line indicates the (difficult and interesting) problem of evaluating the time-evolution phenomenon described by expression (3.7.2); the three continuous lines indicate the steps whose sequence yields the solution. Note that these three steps are easier than the direct solution of equation (3.7.2); in particular, they require only the solution of *linear* problems.

Of course, only in special cases (some of which are discussed below) the operations described above can actually be carried out, yielding in analytic explicit form the solution $u(x, t)$ of (3.7.2). The message of general validity that obtains from the technique of solution we have just described is that for this class of evolution equations *the time evolution is much simpler in the spectral space than in configuration space.* This message, as we will see, has many

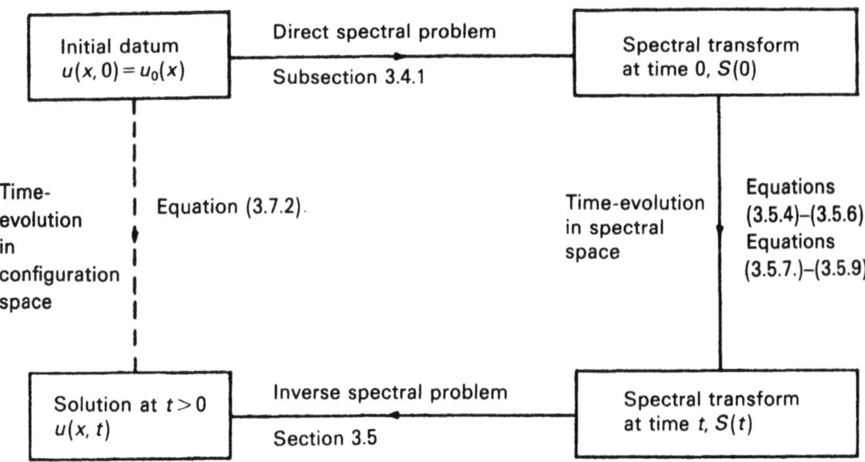

Figure 3.2. Schematic of solution technique.

theoretical implications; it is eventually going to impact also on the experimental techniques used to investigate phenomena that are described, through appropriate, possibly approximate, schematizations, by equations belonging to the class (3.7.2) (or to other classes solvable by analogous techniques, see below). The similarity of this situation to that discussed in Section 3.2, in the context of the linear evolution equations solvable by Fourier transform, should of course be emphasized.

It is possible moreover to utilize the technique of solution we have just described to evince a qualitative understanding of the behavior of the solutions of equation (3.7.2), especially of their long-time behavior, again in analogy to the situation prevailing in the context of the linear evolution equations solvable by Fourier transform, as tersely discussed in Section 3.2. This we do in Section 3.9, after having treated, in the following section, the relationship between the solution via the spectral transform of the nonlinear evolution equations (3.7.2) and the solution via Fourier transform of the linear evolution equations that obtain from expression (3.7.2) if all nonlinear contributions are neglected.

We end this section by noting that the class of nonlinear evolution equations (3.7.2) is not the most general one that can be solved with the help of the spectral transform of Section 3.4.

3.8. Relation to the Fourier Transform Technique to Solve Linear Evolution Equations

In this section we clarify in what sense the spectral transform technique to solve (certain classes of) nonlinear evolution equations constitutes a natural extension of the Fourier transform method to solve linear evolution equations. The analogy between these techniques has already been noted. Here we show that, in the limit of "brutal" linearization corresponding to the neglect of all nonlinear contributions, just as the nonlinear evolution equation becomes a linear equation, so the technique of solution via the spectral transform goes over into the technique of solution via the Fourier transform.

Indeed all the elements leading to this conclusion have been provided in the preceding sections. All we need here is to collect the relevant relations, which are reported below even if this entails some repetition.

The class of nonlinear evolution equations whose solvability via the spectral transform has been described above reads

$$u_t(x, t) = \alpha(\boldsymbol{L})u_x(x, t) \tag{3.8.1}$$

with the integrodifferential operator $\boldsymbol{L}$ defined by expression (3.3.2). In the approximation in which all nonlinear effects are neglected it goes over into the linear partial differential equation

$$u_t(x, t) = -i\omega\left(-i\frac{\partial}{\partial x}\right)u(x, t) \tag{3.8.2}$$

with

$$\omega(z) = -z\alpha(-z^2) \tag{3.8.3}$$

(see end of Section 3.3).

The solution of equation (3.8.2), as discussed in Section 3.2.1, results from the fact that the Fourier transform $\hat{u}(k, t)$ of $u(x, t)$ evolves in time according to the simple formula

$$\hat{u}(k, t) = \hat{u}(k, 0)\exp[-i\omega(k)t] \tag{3.8.4}$$

On the other hand, as $u(x, t)$ evolves according to relation (3.8.1), the corresponding reflection coefficient $R(k, t)$ evolves according to the equation

$$R(k, t) = R(k, 0)\exp[2ik\alpha(-4k^2)t] \tag{3.8.5}$$

as discussed in the preceding Section 3.7. This equation, using expression (3.8.3), can be rewritten in the form

$$R(\tfrac{1}{2}k, t) = R(\tfrac{1}{2}k, 0)\exp[-i\omega(k)t] \tag{3.8.6}$$

which indicates that $R(\frac{1}{2}k, t)$ evolves, when $u(x, t)$ obeys equation (3.8.1), exactly as $\hat{u}(k, t)$ does, when $u(x, t)$ obeys the linearized version (3.8.2) of (3.8.1). And this is precisely consistent with the remark reported at the end of Sections 3.5 and 3.6; in the approximation in which all nonlinear effects are neglected, the reflection coefficient $R(\frac{1}{2}k)$ is simply proportional to the Fourier transform $\hat{u}(k)$:

$$R(\tfrac{1}{2}k, t) = (ik)^{-1}\hat{u}(k, t) \tag{3.8.7}$$

This clinches the argument, albeit with two qualifications. First, we have ignored here the contribution of the discrete eigenvalues; the justification for doing this, in the limit of small u, has already been mentioned above (Section 3.5). Second, it is clear from equation (3.8.7) that the above argument breaks down in the neighborhood of $k = 0$, namely, for the more slowly varying component of $u(x, t)$.

3.9. Qualitative Behavior of the Solutions: Solitons and Background

In Section 3.7 a class of *nonlinear* partial differential equations of evolution type has been identified, that is solvable by the spectral transform technique, in close analogy to the *linear* partial differential equations of evolution type whose solution via Fourier transform has been discussed in Section 3.2. As emphasized there, the availability of a constructive technique of solution is important not so much to obtain explicitly the solution—a task that is only rarely feasible—but rather to provide a qualitative understanding of its behavior. Indeed the main lesson in the *linear* case, solvable via the Fourier transform, is, as we have already emphasized, that the time evolution is much simpler in "k-space" than in "x-space"; this observation constitutes the main guiding principle to investigate these equations, and thereby to understand the behavior of the natural phenomena whose time evolution is, in the framework of mathematical physics or applied mathematics, described by them; indeed the influence of this guiding principle largely determines the appropriate approach to the experimental study of these natural phenomena.

The situation is completely analogous for the class of nonlinear evolution equations identified in the preceding section. Here again the time evolution is much simpler in "k-space" than in "x-space"; however the structure of "k-space" is now richer, due to the presence, in general, of two different components, namely, the continuous ($k^2 > 0$, k real) and discrete ($k^2 = -p_n^2$, p_n real, $n = 1, 2, \ldots, N$) parts of the spectrum (see Section 3.4). Note that the results of Section 3.7 imply that the time evolution of each of these components occurs separately, with no mixing.

It is therefore convenient, in order to arrive at a qualitative understanding of the behavior of the solutions of the class of nonlinear evolution equations (3.3.1), to begin by analyzing separately the solutions that correspond to the two components, discrete and continuous, of the spectrum. We therefore consider first the (rather special) case characterized by the absence of any contribution from the continuum, then the case characterized by the absence of any discrete eigenvalue, and finally we tersely describe the behavior of a generic solution, whose spectral transform contains both contributions, from the discrete spectrum and from the continuum. As in the case of the linear evolution equations discussed tersely in Section 3.2.1, the qualitative analysis focuses largely on the asymptotic, long-time, behavior of the solutions.

3.9.1. Solitons

The simplest solution with no contribution from the continuum corresponds to an initial datum $u_0(x)$ whose spectral transform S_0 has a vanishing reflection coefficient and only one discrete eigenvalue:

$$S_0 = \{R(k) = 0, -\infty < k < \infty; p, \rho_0\} \tag{3.9.1}$$

$$u_0(x) = -2p^2/\cosh^2[p(x-\xi_0)] \tag{3.9.2}$$

with

$$\xi_0 = (2p)^{-1}\ln(\rho_0/2p) \tag{3.9.3}$$

Note the consistency of these, and the following, expressions, with equations (3.5.6), (3.5.10), and (3.5.11). The time evolution of this function is then given, very directly, by the results of Section 3.7:

$$S(t) = \{R(k, t) = 0, -\infty < k < \infty; p, \rho(t) = \rho_0 \exp[-2p\alpha(4p^2)t]\} \tag{3.9.4}$$

$$u(x, t) = -2p^2/\cosh^2\{p[x-\xi(t)]\} \tag{3.9.5}$$

$$\xi(t) = (2p)^{-1}\ln[\rho(t)/2p] \tag{3.9.6a}$$

$$\xi(t) = \xi_0 + vt \tag{3.9.6b}$$

$$v = -\alpha(4p^2) \tag{3.9.7}$$

The solution (3.9.5) [with equations (3.9.6b) and (3.9.7)] describes a wave of constant shape moving with constant speed [in a graphical picture where $-u(x, t)$ marks the profile of the surface of the water in a canal as a function of the distance x, this wave has indeed "the form of a large solitary elevation, a rounded, smooth and well-defined heap of water," moving "without change of form or diminution of speed"]. This is of course the famous *soliton* (indeed the sentences quoted above have been lifted from Scott-Russell,[(5)] whose relevance here is highlighted by the fact that the KdV equation was indeed introduced as the appropriate mathematical schematization to describe "long waves advancing in a rectangular canal."[(2)]

The special *single-soliton solution* (3.9.5) [with equations (3.9.6b) and (3.9.7)] of the nonlinear evolution equation (3.3.1) contains two (real) parameters, ξ_0 and p. The first characterizes the initial localization of the soliton; its arbitrariness corresponds of course to the translation-invariant character of expression (3.3.1), and its relation to the spectral transform is given by expression (3.9.3). The second parameter, p, whose spectral significance is directly related to the value of the discrete eigenvalue (see Section 3.4), determines the shape of the soliton (both its height and its width) and moreover its speed; note that the shape is the same for all equations of the class (3.3.1), while the speed depends on the function $\alpha(z)$, see equation (3.9.7), namely, it depends on which specific equation of the class (3.3.1) one is considering. For instance, for the KdV equation (3.3.4) corresponding to $\alpha(z) = -z$, the speed of the soliton is

$$v = 4p^2 \tag{3.9.8}$$

Thus all solitons of the KdV equation move to the right [the fact that this does not correspond to the behavior of the waves in a canal, that should clearly be parity invariant, need not worry the reader; the KdV equation in the simple form (3.3.4) is only appropriate to describe waves traveling in one direction; moreover it describes the behavior of long waves in shallow canals as seen in a reference frame moving with an appropriately chosen constant speed].

We now proceed and consider the *N-soliton solution*, namely, the function $u(x, t)$, being again characterized by the absence of the continuum contribution in the corresponding spectral transform $[R(k, t) = 0, -\infty < k < \infty]$, but now with N discrete eigenvalues $k^2 = -p_n^2$, $n = 1, 2, \ldots, N$. A closed-form expression for this solution [see equation (3.5.13)] reads

$$u(x, t) = -2\,\partial^2\{\ln \det [\boldsymbol{I} + \boldsymbol{C}(x, t)]\}/\partial x^2 \tag{3.9.9}$$

where $\boldsymbol{I}$ is the unit matrix (of order N) and $\boldsymbol{C}(x, t)$ is the symmetrical matrix of order N defined by

$$C_{mn}(x, t) = [\rho_m(t)\rho_n(t)]^{1/2}(p_m + p_n)^{-1} \exp[-(p_m + p_n)x] \tag{3.9.10}$$

where of course the time-dependence of the normalization coefficients, $\rho_n(t)$, is given by expression (3.7.9).

To discuss this solution further, we hereafter restrict consideration, for simplicity, to the case of the KdV equation, so that the velocity of the soliton is given by the simple expression (3.9.8); but the qualitative essence of the findings discussed below applies equally to all equations of the class (3.3.1).

For definiteness, we order the N positive numbers p_n in increasing order,

$$p_1 < p_2 < \cdots < p_n \tag{3.9.11}$$

and introduce the corresponding soliton velocities

$$v_n = 4p_n^2, \qquad n = 1, 2, \ldots, N \tag{3.9.12}$$

that are then clearly also ordered in increasing order,

$$v_1 < v_2 < \cdots < v_N \tag{3.9.13}$$

The solution (3.9.9) is not simple, especially if N is large; but it is not too difficult to show that its asymptotic behavior in t is quite simple:

$$u(x, t) \approx -2 \sum_{n=1}^{N} p_n^2/\cosh^2[p_n(x - \xi_n^{(\pm)} - v_n t)], \qquad t \to \pm\infty \tag{3.9.14}$$

where the $\xi_n^{(\pm)}$ are $2N$ real constants (related to one another, see below). Thus both in the remote past and future the N-soliton solution (3.9.9) divides up into N separated solitons; in the remote past the solitons are ordered, from left to right, in order of decreasing amplitude and they move toward the right with speeds ordered in decreasing magnitude; then the taller and faster solitons, that are more to the left, gradually catch up and eventually "overtake" the fatter and slower solitons (the quotation marks underscore the fact that, whenever two, or more, solitons get together, their individuality is in fact lost; indeed the "overtaking"—as seen, for instance, in a film—may appear as an exchange of identity, with the taller soliton becoming fatter, and vice versa, as they get close together, until they separate again because the one that has become taller speeds up, while the one that has become fatter slows down; and finally in the remote future the ordering of the solitons gets altogether reversed, with the taller and faster heading the escape to the right. The most important implication of equation (3.9.14) is that precisely the same solitons that existed at the beginning are found at the end, the only effect of their "interaction" having been to shift the position of the nth soliton, relative to what it would have been had it been moving in isolation, by the amount

$$\Delta_n = \xi_n^{(+)} - \xi_n^{(-)} \tag{3.9.15}$$

These shifts are, moreover, determined [while either the N quantities $\xi_n^{(-)}$ or the N constants $\xi_n^{(+)}$ can be chosen arbitrarily; this corresponds to the possibility of choosing arbitrarily the N quantities $\rho_n(0)$ in equation (3.7.9)], being given by the equation

$$\Delta_n = \sum_{m=1}^{n-1} \Delta(p_n, p_m) - \sum_{m=n+1}^{N} \Delta(p_n, p_m) \tag{3.9.16}$$

with

$$\Delta(p, p') = p^{-1} \ln [(p + p')/|p - p'|] \tag{3.9.17}$$

Of course, in writing equation (3.9.16), we employ the usual convention that sets to zero a sum if the lower limit exceeds the upper limit.

The expression (3.9.16) has a simple "physical" interpretation. We will first consider the case of two solitons only, $N = 2$; it is then seen that the second soliton (the faster and taller one) gets *advanced* precisely by the amount $\Delta(p_2, p_1)$, while the first soliton (the slower and fatter one) gets *delayed* by the amount $\Delta(p_1, p_2)$. Thus the shift (3.9.16) experienced by the nth soliton appears as the sum of the $n - 1$ positive shifts derived from its "overtaking" the $n - 1$ slower solitons and the $N - n$ negative shifts derived from its being overtaken by the $N - n$ faster solitons. Note that this result makes perfect sense in the case in which each collision occurs separately; it is highly nontrivial in the present context, since it might well happen (depending on the initial conditions) that, at some intermediate time, several solitons all coalesce together.

This behavior strongly suggests ascribing to each soliton an individuality, even though it does, strictly speaking, show up as a separate entity only in the remote past and future; indeed it was the observation, in numerical experiments with the KdV equation, of precisely such persistence after collisions that motivated Zabusky and Kruskal to introduce the soliton concept.(1) In the context of the spectral transform technique, the individuality of the soliton has an obvious interpretation; and its permanence is clearly related to the fact that the discrete eigenvalues do not vary as $u(x, t)$ evolves according to equation (3.3.1) [thus it will be a feature also of the generic solution of equation (3.3.1); see below].

3.9.2. Background

In contrast to the solutions whose spectral transform contains only contributions from the discrete spectrum (see preceding subsection), solutions corresponding to the continuum part of the spectral transform cannot generally be exhibited in explicit form. Thus the study of these solutions is more difficult.

On the other hand, while the soliton solutions discussed in the preceding subsection reflect in an essential way the nonlinear character of the class of evolution equations (3.3.1), the solutions to be discussed here include those "weak field" situations in which the drastic linearization of equation (3.3.1), consisting in the elimination of all nonlinear terms, provides a reasonable approximation. Indeed, if $u(x, t)$ is very small, its spectral transform contains generally no discrete eigenvalues (or at most only one), so that such solutions generally belong to the class being discussed here.

For such solutions, the standard analysis, based on the Fourier transform technique of solution, becomes applicable. Thus their behavior—in particular, their long-time behavior—is essentially characterized by the phenomenon of dispersion, and by the group velocity associated through expression (3.2.9) with the dispersion function $\omega(k)$, which is itself related by equation (3.8.3) to the function $\alpha(z)$ that determines the right-hand side of (3.3.1), namely, the form of the particular nonlinear evolution equation under consideration.

It might appear that this analysis is only relevant to those solutions $u(x, t)$ that are everywhere sufficiently small to justify the approximation consisting in the drastic elimination of all nonlinear contributions in the right-hand side of expression (3.9.1). But if the spectral transform of $u(x, t)$ has no discrete eigenvalues, namely, if $u(x, t)$ contains no solitons, there is no mechanism available to maintain the solution $u(x, t)$ (or any part of it) localized in the long-time limit. Thus, in the asymptotic long-time limit, the solution becomes locally small everywhere. This indicates that the analysis given above is in fact applicable to any solution of the type being considered in this subsection.

We conclude therefore that the long-time behavior of those solutions of the nonlinear evolution equations (3.3.1) whose spectral transform contains no discrete spectrum contribution may be qualitatively understood by considering the corresponding linearized equation, namely, equation (3.2.1) with equation (3.8.3). They are therefore characterized by dispersion [with the dispersion function defined by expression (3.8.3)]; in contrast to the solitons, these solutions do not remain localized. They are often referred to as *background*; sometimes as *radiation*, to underscore the difference from the particle-like nature of the solitons; and sometimes (perhaps a bit disparagingly—motivated by their lack of persistence?) as *hash.*(187)

3.9.3. Generic Solution

The N-soliton solution of the nonlinear evolution equation (3.3.1) is clearly not generic, since it has the specific form (3.9.9); it is, however, fairly general, since it contains the $2N$ arbitrary constants p_n and $\rho_n(0)$, $n = 1, 2, \ldots, N$, and N itself is arbitrary. That this solution is indeed far from generic is particularly evident if one considers its spectral transform, that is of course characterized by the total absence of the continuum contribution $[R(k, t) = 0]$.

The pure background solution, characterized by the complete absence of solitons, is also not quite generic, although it represents a broader class than the pure multisoliton solutions; the condition on a potential, in the context of the Schrödinger spectral problem, to have no associated discrete eigenvalues (in the language of physics, no bound states) appears indeed less restrictive than the requirement to be completely reflectionless.

The generic solution of the nonlinear evolution equation (3.3.1) contains a soliton part and a background; the separation into these two components is of course not generally visible in x-space (although it may occur, as it were, automatically, in the long-time limit; see below); this separation is on the other hand quite evident, throughout the time evolution, from the point of view of the spectral transform.

The qualitative behavior of the generic solution in x-space as time evolves is a combination of those discussed in the preceding two subsections; of course, only when the two phenomena get disentangled also in x-space can a simple qualitative picture emerge. This generally happens in the asymptotic long-time limit. Thus, in the remote future the solution evolves generally into a finite number of separated well-localized solitons, superimposed on a background that tends locally to zero everywhere according to the standard behavior of solutions of (non-dissipative) dispersive linear evolution equations.

For instance, for the KdV equation (3.3.4) $[\alpha(z) = -z$ in equation (3.3.1)], the group velocity v_g turns out to be negative $[\omega(k) = -k^3$, $v_g = -3k^2$; see expressions (3.8.3) and (3.2.9)], in contrast to the soliton speeds $v_n = 4p_n^2$ [see solution (3.9.12)], which are all positive; thus in this case a complete separation occurs asymptotically between the background part of the solution, that disperses away toward the left, and the solitons, that travel to the right as described in subsection 3.9.1.

Clearly this behavior constitutes a distinctive and remarkable feature of this class of nonlinear evolution equations (and of other classes solvable by analogous technique, based on different spectral problems; see below). The physical, or more generally natural, interpretation of these results—for instance, the emergence of the solitons as localized individual entities—does of course depend on the particular physical, or more generally natural, phenomenon that the nonlinear evolution equation under consideration is supposed, perhaps approximately, to represent, be it in fluid dynamics or in demography, in solid state physics or in epidemiology, in the investigation of signal transmission through nervous fibers or of models of elementary particles or of plasma disturbances.

3.10. Additional Properties of the Solutions

The class of nonlinear evolution equations (3.3.1) has a number of additional remarkable features, all of them originating from the basic property of this class of nonlinear evolution equations, i.e., the fact that the time evolution, although nonlinear in configuration space (x-space), is related (via the spectral transform) to a simple linear evolution in the spectral space (see Section 3.7). In this section we review tersely some of these properties.

3.10.1. Bäcklund Transformations

Consider an evolution equation, say

$$u_t(x, t) = F[u, u_x, u_{xx}, \dots] \tag{3.10.1}$$

Let $u^{(0)}(x, t)$ be a generic solution of equation (3.10.1), and let $u^{(1)}(x, t)$ be related to $u^{(0)}(x, t)$ by some explicit relation (involving the differential operator $\partial/\partial x$ as well as the integral operator $\int_x^\infty dy$) which, for the moment, we indicate by

$$G[u^{(0)}, u^{(1)}] = 0 \tag{3.10.2}$$

If $u^{(1)}(x, t)$ is then also found to satisfy expression (3.10.1), then equation (3.10.2) is, by definition, a *Bäcklund transformation.*

Thus we define here a *Bäcklund transformation* to be any relation between two functions, say $u^{(0)}(x, t)$ and $u^{(1)}(x, t)$, in such a way as to insure that, if one of them satisfies a (generally nonlinear) evolution equation, then the other does so too. Such transformations were introduced long ago in differential geometry and their name goes back to this origin. Here we tersely outline how, in the context of the spectral transform technique to solve the nonlinear evolution equations (3.3.1), a class of Bäcklund transformations for the solutions of this equation naturally emerges. Our purpose and scope here is merely to convey the basic idea.

We begin by reporting here once more the class of evolution equations (3.3.1), reading

$$u_t = \alpha(L)u_x, \qquad u \equiv u(x, t) \tag{3.10.3}$$

where $\alpha(z)$ is a polynomial and L is the integrodifferential operator defined by

$$Lf(x) = f_{xx} - 4uf + 2u_x \int_x^\infty dy f(y) \tag{3.10.4}$$

And we recall that, as $u(x, t)$ evolves according to expression (3.10.1), the reflection coefficient $R(k, t)$ corresponding to $u(x, t)$ via the spectral (or scattering) problem of Section 3.4 evolves in time according to the equation

$$R_t(k, t) = 2ik\alpha(-4k^2)R(k, t) \tag{3.10.5a}$$

implying of course

$$R(k, t) = R(k, 0) \exp[2ik\alpha(-4k^2)t] \tag{3.10.5b}$$

For the sake of simplicity, in the following we shall limit consideration to the reflection coefficient, assuming that there is a one-to-one correspondence between u and R (which is actually the case only if the spectral transform of u has no discrete eigenvalues; see Section 3.4).

We then report a property of the Schrödinger spectral problem of Section 3.4. Let $u^{(0)}(x)$ and $u^{(1)}(x)$ be two functions (both regular for real x and vanishing "sufficiently fast" as $x \to \pm\infty$), related to each other by the (nonlinear integrodifferential) expression

$$g(\mathbf{\Lambda})[u^{(0)}(x) - u^{(1)}(x)] + h(\mathbf{\Lambda}) \cdot \mathbf{\Gamma} \cdot 1 = 0 \tag{3.10.6}$$

Here $g(z)$ and $h(z)$ are two arbitrary entire functions (say, two polynomials) and the integrodifferential operators $\mathbf{\Gamma}$ and $\mathbf{\Lambda}$ are defined by the following expression that specify their action on a generic function $f(x)$ (vanishing as $x \to \infty$):

$$\mathbf{\Gamma} f(x) = [u_x^{(0)}(x) + u_x^{(1)}(x)]f(x) + [u^{(0)}(x) - u^{(1)}(x)] \int_x^\infty dy\, [u^{(0)}(y) - u^{(1)}(y)]f(y) \tag{3.10.7}$$

$$\Lambda f(x) = f_{xx}(x) - 2[u^{(0)}(x) + u^{(1)}(x)]f(x) + \mathbf{\Gamma} \cdot \int_x^\infty dy f(y) \qquad (3.10.8)$$

Note that $\mathbf{\Gamma}$ and Λ depend (in a symmetrical way) on $u^{(0)}(x)$ and $u^{(1)}(x)$; this causes equation (3.10.6) to be nonlinear in $u^{(0)}$ and $u^{(1)}$. Incidentally, for $u^{(0)}(x) = u^{(1)}(x) = u(x)$, Λ goes over into the operator L; see equation (3.10.4).

Then the reflection coefficients $R^{(0)}(k)$ and $R^{(1)}(k)$, corresponding respectively to $u^{(0)}(x)$ and $u^{(1)}(x)$, are related to each other by the simple linear equation

$$g(-4k^2)[R^{(0)}(k) - R^{(1)}(k)] + 2ikh(-4k^2)[R^{(0)}(k) + R^{(1)}(k)] = 0 \quad (3.10.9)$$

implying

$$R^{(1)}(k) = R^{(0)}(k)\frac{[g(-4k^2) + 2ikh(-4k^2)]}{[g(-4k^2) - 2ikh(-4k^2)]} \qquad (3.10.10)$$

Let us emphasize that this result constitutes a property of the spectral problem; it has nothing to do with time-evolution. But if $u^{(0)}$ and $u^{(1)}$ evolve in time (i.e., if they depend on another variable t in addition to x), all the formulas given above remain valid, of course with $R^{(0)}$ and $R^{(1)}$ also depending on t. Indeed also the two arbitrary functions g and h in equations (3.10.6), (3.10.9), and (3.10.10) could well depend on t; but the following development hinges on the assumption that they be time-independent.

Suppose then that $u^{(0)}(x, t)$ evolves in time according to equation (3.10.3); then the corresponding reflection coefficient, $R^{(0)}(k, t)$, evolves in time according to (3.10.5). But then $R^{(1)}(k, t)$, being related to $R^{(0)}(k, t)$ by equation (3.10.10) (with time-independent g and h), also evolves according to equation (3.10.5); hence $u^{(1)}(x, t)$ also evolves according to equation (3.10.3). Conclusion: if $u^{(0)}(x, t)$ satisfies equation (3.10.3) and $u^{(1)}(x, t)$ is related to $u^{(0)}(x, t)$ by equation (3.10.6), then $u^{(1)}(x, t)$ also satisfies (3.10.3).

It is thus seen that equation (3.10.6) is a Bäcklund transformation for the (class of) evolution equations (3.10.3). Indeed, due to the arbitrariness of the polynomials g and h, this formula yields a large class of Bäcklund transformations. The simplest of the transformations corresponds to g and h being just constants; setting for notational convenience $g/h = -2p$, it reads

$$u_x^{(0)}(x, t) + u_x^{(1)}(x, t) = [u^{(0)}(x, t) - u^{(1)}(x, t)] \cdot \left\{2p - \int_x^\infty dy\, [u^{(0)}(y, t) - u^{(1)}(y, t)]\right\} \qquad (3.10.11)$$

An equivalent, but generally more convenient, formula obtains for the integrals $w^{(m)}(x, t)$ of the $u^{(m)}(x, t)$,

$$w^{(m)}(x, t) = \int_x^{\infty} dy\, u^{(m)}(y, t), \qquad m = 0, 1 \tag{3.10.12a}$$

$$w_x^{(m)}(x, t) = -u^{(m)}(x, t), \qquad m = 0, 1 \tag{3.10.12b}$$

reading

$$w_x^{(0)}(x, t) + w_x^{(1)}(x, t) = \tfrac{1}{2}[w^{(0)}(x, t) - w^{(1)}(x, t)] \cdot \{4p - [w^{(0)}(x, t) - w^{(1)}(x, t)]\} \tag{3.10.13}$$

To get equation (3.10.13) from equation (3.10.11) an integration has been performed; the vanishing of $w^{(m)}(x, t)$ as $x \to \infty$,

$$w^{(m)}(\infty, t) = 0, \qquad m = 0, 1 \tag{3.10.14}$$

clearly implied by the definition (3.10.12a), as well as the asymptotic vanishing of $w_x^{(m)}(x, t)$ implied by expression (3.10.12b), have moreover been invoked to set to zero the integration constant. Note that equation (3.10.13) is, if one assumes $w^{(0)}$ to be known, a Riccati equation for $w^{(1)}$, and vice versa.

The relationship (3.10.10) corresponding to the simplest Bäcklund transformation (3.10.11) or (3.10.13) reads simply

$$R^{(1)}(k, t) = -R^{(0)}(k, t)[(k + ip)/(k - ip)] \tag{3.10.15}$$

An elementary application of the Bäcklund transformation (3.10.11) or (3.10.13) is to use it in order to discover some (special) solutions of equation (3.10.3). The simpler instance obtains setting

$$w^{(0)}(x, t) = 0 \tag{3.10.16a}$$

in equation (3.10.13), since this expression, or equivalently the equation

$$u^{(0)}(x, t) = 0 \tag{3.10.16b}$$

implied by it, clearly provides a solution of equation (3.10.3)—indeed, a very trivial solution. The solution of equation (3.10.13) with equation (3.10.16a) [and equation (3.10.14)] is an elementary task, and it yields

$$w(x, t) = -2p\Big\{1 - \mathrm{tgh}\,\{p[x - \xi(t)]\}\Big\} \tag{3.10.17a}$$

namely

$$u(x, t) = -2p^2/\cosh^2\{p[x - \xi(t)]\} \qquad (3.10.17b)$$

In the last two equations we have omitted for notational simplicity the superscript 1; what we are displaying is merely a solution of equation (3.10.3). It should be noted that this result obtains for positive p [although equation (3.10.17b) is eventually independent of the sign of p]; for negative p there is no solution of equation (3.10.13) [with equation (3.10.16a)] consistent with equation (3.10.14). The quantity $\xi(t)$ in the right-hand side of expressions (3.10.17a) and (3.10.17b) originates as an integration constant, which means, in the present context, that it does not depend on the variable x [equation (3.10.13) is a differential equation in this variable]; this however does not prevent it from depending on t, as we have explicitly indicated.

Obviously equation (3.10.17b) is just the single-soliton solution, see equation (3.9.5); indeed the time dependence of $\xi(t)$ can be determined by inserting equation (3.10.17b) back into equation (3.10.3), and by using the expression

$$\boldsymbol{L}u_x(x, t) = 4p^2u_x(x, t) \qquad (3.10.18)$$

that is easily verified from equation (3.10.4) and (3.10.17b). There obtains

$$\dot{\xi}(t) = -\alpha(4p^2) \qquad (3.10.19)$$

consistently with equations (3.9.6) and (3.9.7).

This derivation of the single-soliton solution from the Bäcklund transformation (3.10.13) is however rather objectionable, for clearly the simplification adopted in this subsection, namely, to ignore the discrete part of the spectral transform, is not applicable to the soliton solution. Indeed, while the crucial rôle in the derivation of the Bäcklund transformation (3.10.6) resp. (3.10.11) or (3.10.13), has been played by the corresponding expressions for the reflection coefficient, (3.10.9) or (3.10.10) resp. (3.10.15), in the special case that we have now considered [and that has yielded the single-soliton solution (3.10.17b)] the latter equations are merely instances of the identity "zero equal zero."

Another important property of the Bäcklund transformations that have been introduced above is their *commutativity*. In the (simplified) context to which consideration has been restricted in this subsection this property is rather obvious. Consider indeed two sets of two polynomials, $g^{(m)}(z)$ and $h^{(m)}(z)$, $m = 1, 2$, and the two Bäcklund transformations (3.10.6) generated by them, say BT1 and BT2. Take as a starting point some given function $u^{(0)}(x)$, and associate with it two functions, say $u^{(1)}(x)$ and $u^{(2)}(x)$, obtained respectively from $u^{(0)}(x)$ via these two Bäcklund transformations, BT1 and BT2. Then associate with $u^{(1)}(x)$ a new function, say $u^{(12)}(x)$, obtained applying the Bäcklund transformation BT2, and with $u^{(2)}(x)$ a function $u^{(21)}(x)$ obtained applying the Bäcklund transformation BT1. The property of commutativity is expressed by the equality

$$u^{(12)}(x) = u^{(21)}(x) \tag{3.10.20}$$

It is a highly nontrivial property when viewed, as we have just done, in configuration space; its validity is implied by the corresponding expression in the spectral space,

$$R^{(12)}(k) = R^{(21)}(k) \tag{3.10.21}$$

which is instead trivially valid, since equation (3.10.10) implies that

$$R^{(12)}(k) = R^{(0)}(k)B_1(k)B_2(k) \tag{3.10.22a}$$

$$R^{(21)}(k) = R^{(0)}(k)B_2(k)B_1(k) \tag{3.10.22b}$$

with

$$B_m(k) = \frac{[g_m(-4k^2) + 2ikh_m(-4k^2)]}{[g_m(-4k^2) - 2ikh_m(-4k^2)]}, \qquad m = 1, 2 \tag{3.10.23}$$

Thus the commutativity of Bäcklund transformations, in the spectral space, is merely the commutativity of ordinary multiplication. Here we see once more the advantage, in terms of simplicity, to work in the spectral space rather than in configuration space.

The property of commutativity that we have just described is conveniently synthesized by Figure 3.3.

We end this subsection noting that in the limit in which both $u^{(0)}(x)$ and $u^{(1)}(x)$ are so small that all nonlinear terms in equation (3.10.6) can be neglected, this Bäcklund transformation reads

$$g(\partial^2/\partial x^2)[u^{(0)}(x) - u^{(1)}(x)] + h(\partial^2/\partial x^2)[u_x^{(0)}(x) + u_x^{(1)}(x)] = 0 \tag{3.10.24}$$

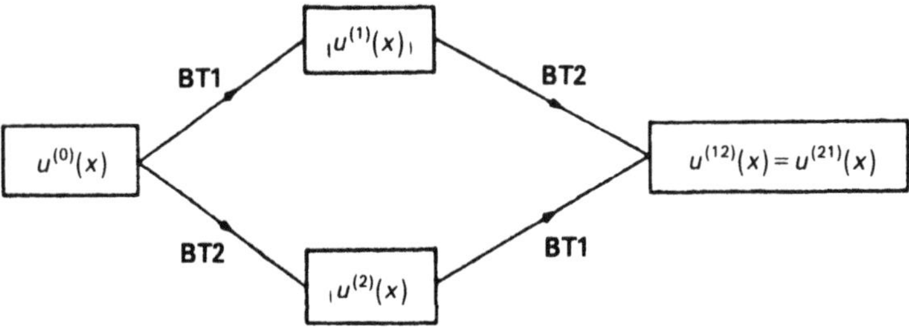

Figure 3.3. Commutativity property.

namely, up to a change of notation it becomes identical to expression (3.2.10), displaying once again the fact that all properties derived in the context of the spectral transform approach have counterparts in the Fourier transform context, to which they go over whenever it is justified to ignore all nonlinear effects.

3.11. Nonlinear Superposition Principle

Another remarkable property of the class of evolution equations (3.3.1) obtains as a straightforward consequence of the commutativity property of Bäcklund transformations described in the preceding subsection, applied to the simplest Bäcklund transformation (3.10.11) or rather (3.10.13) [we work hereafter, in this subsection, with the integrals $w(x, t)$ of the solutions of equation (3.3.1), rather than with the functions $u(x, t)$ themselves, see equation (3.10.12); and, for simplicity of language, we refer directly to these functions w, that are in one-to-one correspondence with their derivatives u, see equation (3.10.12), as the solutions of equation (3.3.1), hoping to cause no misunderstanding].

Let $w^{(0)}(x, t)$ be some solution of equation (3.3.1), $w^{(1)}(x, t)$ resp. $w^{(2)}(x, t)$ the solutions of equation (3.3.1) related to $w^{(0)}(x, t)$ by a Bäcklund transformation (3.10.13) with the parameter p replaced by p_1 resp. p_2, and $w^{(12)}(x, t) = w^{(21)}(x, t)$ [see equation (3.10.20)] the solution of equation (3.3.1) obtained from $w^{(0)}(x, t)$ via two Bäcklund transformations (3.10.13), with the parameter p replaced once by p_1 and once by p_2:

$$w_x^{(0)} + w_x^{(1)} = -\tfrac{1}{2}(w^{(0)} - w^{(1)})(-4p_1 + w^{(0)} - w^{(1)}) \tag{3.11.1}$$

$$w_x^{(0)} + w_x^{(2)} = -\tfrac{1}{2}(w^{(0)} - w^{(2)})(-4p_2 + w^{(0)} - w^{(2)}) \tag{3.11.2}$$

$$w_x^{(1)} + w_x^{(12)} = -\tfrac{1}{2}(w^{(1)} - w^{(12)})(-4p_2 + w^{(1)} - w^{(12)}) \tag{3.11.3}$$

$$w_x^{(2)} + w_x^{(21)} = -\tfrac{1}{2}(w^{(2)} - w^{(21)})(-4p_1 + w^{(2)} - w^{(21)}) \tag{3.11.4}$$

Now subtract equation (3.11.2) from equation (3.11.1), and equation (3.11.4) from equation (3.11.3) [using equation (3.10.20)], and subtract again from each other the two resulting equations. Then all differentiated terms in the left-hand side cancel out, and there results an algebraic relation that can be solved for $w^{(12)} = w^{(21)}$:

$$\begin{aligned} w^{(12)} &= w^{(21)} \\ &= w^{(0)} - (p_1 + p_2)(w^{(1)} - w^{(2)})/[p_1 - p_2 + \tfrac{1}{2}(w^{(1)} - w^{(2)})] \end{aligned} \tag{3.11.5}$$

It should be emphasized that this equation provides a completely explicit expression of a novel solution of equation (3.3.1), in terms of three other solutions of the same equation: an (arbitrary) solution $w^{(0)}$, and the two solutions $w^{(1)}$ and $w^{(2)}$ related to $w^{(0)}$ by the Bäcklund transformations (3.11.1) and (3.11.2).

As an example of application of the *nonlinear superposition principle* (3.11.5), we again consider the case with

$$u^{(0)}(x, t) = w^{(0)}(x, t) = 0 \qquad (3.11.6)$$

that clearly provides a (trivial) solution of equation (3.3.1). Then [see equation (3.10.17)]

$$u^{(m)}(x, t) = -2p_m^2/\cosh^2\{p_m[x - \xi_0^{(m)} + \alpha(4p_m^2)t]\}, \qquad m = 1, 2 \qquad (3.11.7a)$$

$$w^{(m)}(x, t) = -2p_m\{1 - \operatorname{tgh}\{p_m[x - \xi_0^{(m)} + \alpha(4p_m^2)t]\}\}, \qquad m = 1, 2 \quad (3.11.7b)$$

where $\xi_0^{(m)}$ has to be either real,

$$\operatorname{Im} \xi_0^{(m)} = 0 \qquad (3.11.8a)$$

or $\xi_0^{(m)} = \xi_0^{(m)\prime} + i\pi/(2p_m)$ with $\xi_0^{(m)\prime}$ real,

$$\operatorname{Im} \xi_0^{(m)} = \tfrac{1}{2}\pi/p_m \qquad (3.11.8b)$$

in order that $u^{(m)}$ and $w^{(m)}$, $m = 1, 2$, be real. Insertion of equations (3.11.6) and (3.11.7b) in equation (3.11.5) yields the two-soliton solution, provided $0 < p_1 < p_2$ and $\xi_0^{(1)}$ satisfies expression (3.11.8a) while $\xi_0^{(2)}$ satisfies equation (3.11.8b) [otherwise the solution produced by equation (3.11.5) is complex or singular].

Having thus obtained the two-soliton solution, it is possible to apply the nonlinear superposition principle (3.11.5) to get the three-soliton solution, by inserting in place of $w^{(0)}$ the single-soliton expression (with parameter, say, p_1) and in place of $w^{(1)}$ resp. $w^{(2)}$ the two-soliton expression (with parameters p_1 and p_2 resp. p_1 and p_3); and the process can then be continued (*soliton ladder*). In this manner it is in principle possible to construct all multisoliton solutions by a purely algebraic procedure; and very simple rules can be given, detailing the restrictions on the soliton parameters p_m and the reality properties of the $\xi_0^{(m)}$ [equations (3.11.8a) or (3.11.8b)], to insure that the solution so arrived at is indeed real and nonsingular, and thus coincides with equation (3.9.9).

3.12. Conservation Laws

Another important property of the class of nonlinear evolution equations (3.3.1) is the existence of an infinite sequence of local conservation laws. Each of these conservation laws yields, for the class of asymptotically ($x \to \pm\infty$) vanishing solutions to which our consideration is confined, a conserved quantity expressed as the integral over all space of an appropriate nonlinear (polynomial) combination of $u(x, t)$ and its derivatives.

In this section we review tersely these conserved quantities. It should be noted that, in the framework of the approach developed in this chapter the existence of an infinite sequence of conservation laws, remarkable as it is, appears as a secondary feature of the class of evolution equations (3.3.1). In other approaches to this subject the existence of an infinite number of conservation laws plays a more fundamental rôle; indeed the discovery of several independent conservation laws for the KdV equation provided the first hint of its peculiarity, and it was actually in the process of investigating this property that the spectral transform technique was invented.

All the results given in the preceding sections hold for, or are trivially extended to, the class of evolution equations

$$u_t(x, t) = \alpha(\boldsymbol{L}, t)u_x(x, t) \tag{3.12.1}$$

where $\alpha(z, t)$ is an (arbitrary) polynomial (with t-dependent coefficients) in z, and $\boldsymbol{L}$ is the integrodifferential operator defined by equation (3.3.2). This is a more general class than equation (3.3.1), since an explicit (and largely arbitrary) t-dependence has been introduced. In the preceding sections this generalization was omitted, although it could have been easily accommodated, since we are interested throughout this chapter in outlining the basic ideas rather than in presenting the results in full generality. Here, however, we refer to the more general class (3.12.1), since the existence of an infinite sequence of conserved quantities, in spite of the largely arbitrary explicit time dependence appearing in the right-hand side of equation (3.12.1), is sufficiently remarkable to deserve special mention already at this stage.

We already noted (see Section 3.3) that the class of nonlinear evolution equations (3.12.1) has itself the structure of a local conservation law,

$$u_t(x, t) = \gamma_x^{(0)}(x, t) \tag{3.12.2}$$

with $\gamma^{(0)}(x, t)$ a polynomial (now, with t-dependent coefficients) in $u(x, t)$ and its x-derivatives. (The notation employed for the right-hand side of this equation is motivated by future developments; see below.) This implies the asymptotic vanishing of $\gamma^{(0)}(x, t)$,

$$\gamma^{(0)}(\pm\infty, t) = 0 \tag{3.12.3}$$

Integration of equation (3.12.2) over x from $-\infty$ to $+\infty$ therefore shows that the quantity

$$C_0 = c_0 = \int_{-\infty}^{+\infty} dx\, u(x, t) \tag{3.12.4}$$

is a constant of the motion for the flow defined by equation (3.12.1).

It is remarkable that this quantity is merely the first of an infinite sequence. There are in fact two equivalent sequences of constants of the motion, whose expression may be given in explicit and compact form. The first reads

$$C_m = (-1)^m (2m+1)^{-1} \int_{-\infty}^{+\infty} dx\, \boldsymbol{L}^m [x u_x(x, t) + 2u(x, t)], \qquad m = 0, 1, 2, \ldots \tag{3.12.5}$$

where the appearance of the integrodifferential operator $\boldsymbol{L}$, defined by equation (3.3.1), should be noted. An alternative expression for the quantity (3.12.5) is

$$C_m = (-1)^m (2m+1)^{-1} \int_{-\infty}^{+\infty} dx\, \tilde{\boldsymbol{L}}^m u(x, t), \qquad m = 0, 1, 2, \ldots \tag{3.12.6}$$

where $\tilde{\boldsymbol{L}}$ is the integrodifferential operator defined by the equation

$$\tilde{\boldsymbol{L}} f(x) = f_{xx}(x) - 4u(x, t) f(x) + 2 \int_x^{+\infty} dy\, u(y, t) f_y(y, t) \tag{3.12.7}$$

that shows its action on a generic function $f(x)$. As will be seen in the following, also this operator plays an important rôle, comparable to that played by the operator $\boldsymbol{L}$; indeed, in some sense, $\tilde{\boldsymbol{L}}$ is just the adjoint of $\boldsymbol{L}$.

The second sequence of conserved quantities reads

$$c_m = (-1)^m \int_{-\infty}^{+\infty} dx\, u(x, t) \boldsymbol{M}^{2m} \cdot 1, \qquad m = 0, 1, 2, \ldots \tag{3.12.8}$$

with the integrodifferential operator $\boldsymbol{M}$ defined by the equation

$$\boldsymbol{M} f(x) = f_x(x) - \int_{-\infty}^{x} dy\, u(y, t) f(y) \tag{3.12.9}$$

that specifies its action on a generic function $f(x)$ [of course here, as well as in equation (3.12.7), f may depend on other variables besides x, for instance on t]. Also the quantities (3.12.8) possess an alternative expression,

$$c_m = (-1)^m \int_{-\infty}^{+\infty} dx\, \boldsymbol{\Lambda}^{(0)^m} u(x, t), \qquad m = 0, 1, 2, \ldots \tag{3.12.10}$$

where $\Lambda^{(0)}$ is the integrodifferential operator defined by the equation

$$\Lambda^{(0)}f(x) = f_{xx}(x) - 2u(x,t)f(x) + u_x(x,t)\int_x^{+\infty} dy\, f(y)$$

$$+ u(x,t)\int_x^{+\infty} dy\, u(y,t)\int_y^{+\infty} dz\, f(z) \qquad (3.12.11)$$

The first three constants of the motion given by equation (3.12.5) read

$$C_0 = \int_{-\infty}^{+\infty} dx\, u(x,t) \qquad (3.12.12a)$$

$$C_1 = \int_{-\infty}^{+\infty} dx\, u^2(x,t) \qquad (3.12.12b)$$

$$C_2 = \int_{-\infty}^{+\infty} dx\, [2u^3(x,t) + u_x^2(x,t)] \qquad (3.12.12c)$$

Note that the explicit x-dependence in the right-hand side of equation (3.12.5) has disappeared from these expressions; this is indeed the case for all of them, as implied by the alternative equation (3.12.6) [with equation (3.12.7)], or by the relationship of the C_m to the c_m; see below. It is also obvious that all the C_m are independent (none of them can be expressed in terms of the others); and their general structure is apparent from the explicit expressions (3.12.12) (an elementary dimensional argument implies that each of the monomials appearing in the integrand in the definition of C_m must contain q powers of u and r x-derivatives, with the restriction $2q + r = 2m + 2$).

The constants c_m are also all independent of each other. There is instead a relationship between the constants of the two series, which is expressed in compact form by the equation

$$\sum_{m=0} c_m z^{2m+1} = \sin\left[\sum_{m=0} C_m z^{2m+1}\right] \qquad (3.12.13)$$

which is to be understood by expanding in powers of z on the right-hand side and then equating the coefficients of equal powers of z:

$$c_0 = C_0 \qquad (3.12.14a)$$

$$c_1 = C_1 - \tfrac{1}{6}C_0^3 \qquad (3.12.14b)$$

$$c_2 = C_2 - \tfrac{1}{2}C_0^2C_1 + \tfrac{1}{120}C_0^5 \qquad (3.12.14c)$$

and so on.

We emphasize that all these conservation laws hold for the whole class of evolution equations (3.12.1); and note that the definition of the conserved quantities does not depend on $\alpha(z, t)$, namely, on which particular equation of the class (3.12.1) is being considered.

The conservation laws expressed by the time independence of the integrals (3.12.5) and (3.12.6) can be also rewritten in the differential form of a "continuity equation." In fact, to the "densities" [see equation (3.12.5)]

$$f^{(m)}(x, t) = L^m[xu_x(x, t) + 2u(x, t)], \qquad m = 0, 1, 2, \ldots \quad (3.12.15)$$

respectively [see equation (3.12.6)]

$$g^{(m)}(x, t) = \tilde{L}^m u(x, t), \qquad m = 0, 1, 2, \ldots \quad (3.12.16)$$

are associated the "currents" $\varphi^{(m)}(x, t)$, respectively $\gamma^{(m)}(x, t)$, satisfying the equations

$$f_t^{(m)}(x, t) = \varphi_x^{(m)}(x, t), \qquad m = 0, 1, 2, \ldots \quad (3.12.17)$$

$$g_t^{(m)}(x, t) = \gamma_x^{(m)}(x, t), \qquad m = 0, 1, 2, \ldots \quad (3.12.18)$$

together with the conditions

$$\varphi^{(m)}(\pm\infty, t) = 0, \qquad \gamma^{(m)}(\pm\infty, t) = 0, \qquad m = 0, 1, 2, \ldots \quad (3.12.19)$$

that guarantee the time independence of equations (3.12.5) and (3.12.6). Note that, while the densities (3.12.15) and (3.12.16) do not depend on $\alpha(z, t)$ [they are common to all evolution equations of the class (3.12.1)], this is not the case for the currents $\varphi^{(m)}$ and $\gamma^{(m)}$. It should also be mentioned that in the conservation law (3.12.18) both the density $g^{(m)}$ and the current $\gamma^{(m)}$ have the remarkable property of being polynomial expressions in u and its x-derivatives [this implies in particular that all the integrations that appear in the right-hand side of equation (3.12.16) as a consequence of the definition of $\tilde{L}$, see equation (3.12.7), can be performed in closed form]. For instance, for the (trivial) case $\alpha(z, t) = 1$, the current $\gamma^{(m)}$ coincides with its corresponding density, $\gamma^{(m)} = g^{(m)}$, while for the KdV equation

$$u_t + u_{xxx} - 6u_x u = 0, \qquad u \equiv u(x, t) \quad (3.12.20)$$

[the particular equation of the class (3.12.1) corresponding to $\alpha(z, t) = -z$] the first two currents are

$$\gamma^{(0)} = -u_{xx} + 3u^2 \quad (3.12.21a)$$

$$\gamma^{(1)} = -u_{xxxx} + 3u_x^2 + 12uu_{xx} - 12u^3 \tag{3.12.21b}$$

It should be finally pointed out that there exists an additional conserved quantity, for the evolution equation (3.12.1). For instance, for the KdV equation (3.12.20) the quantity

$$C = \int_{-\infty}^{+\infty} dx\,[xu(x, t) + 3tu^2(x, t)] \tag{3.12.22}$$

is another constant of the motion, independent of all those written above; its time-independence is readily verified differentiating equation (3.12.22) with respect to t and using equation (3.12.20). Note that in the definition (3.12.22) the time variable t, as well as the space coordinate x, appear explicitly.

By using the constancy of C, together with that of C_0 and C_1 [see equation (3.12.12)], it is immediately seen that, for the KdV equation (3.12.20), *the center of mass*

$$X(t) = \int_{-\infty}^{+\infty} dx\, xu(x, t) \Big/ \int_{-\infty}^{+\infty} dx\, u(x, t) \tag{3.12.23}$$

undergoes a *uniform motion*:

$$X(t) = X_0 + Vt \tag{3.12.24}$$

the constants X_0 and V being given by

$$X_0 = C/C_0, \qquad V = -3C_1/C_0 \tag{3.12.25}$$

We emphasize that this property of the center of mass $X(t)$, to move with constant speed (as a free particle), holds for a generic solution of the KdV equation (3.12.20).

This result in fact generalizes to the whole class (3.12.1), the additional conserved quantity being

$$C = \int_{-\infty}^{+\infty} dx \left[xu(x, t) + \int_0^t dt'\, \alpha(\tilde{\boldsymbol{L}}, t')u(x, t) \right] \tag{3.12.26}$$

with $\tilde{\boldsymbol{L}}$ defined by equation (3.12.7). This implies that, for the generic solution $u(x, t)$ of equation (3.12.1), the center of mass $X(t)$, see equation (3.12.23), moves according to the expression

$$X(t) = X_0 + \sum_{m=0}^{M} (-1)^{m+1}(2m+1)(C_m/C_0) \int_0^t dt'\, \alpha_m(t') \tag{3.12.27}$$

where

$$X_0 = C/C_0 \tag{3.12.28}$$

and the functions $\alpha_m(t)$ are the coefficients of the polynomial $\alpha(z, t)$ in equation (3.12.1):

$$\alpha(z, t) = \sum_{m=0}^{M} \alpha_m(t) z^m \tag{3.12.29}$$

Thus, for all "autonomous" evolution equations of the class (3.12.1) [characterized by the lack of an explicit time-dependence: $\alpha(z, t) \equiv \alpha(z)$], the center of mass of the generic solution moves uniformly, see equation (3.12.24), with the (constant) speed

$$V = \sum_{m=0}^{M} (-1)^{m+1}(2m+1)\alpha_m C_m / C_0 \tag{3.12.30}$$

3.13. A List of Solvable Equations

To complete this chapter we list some nonlinear evolution equations of applicative and/or theoretical relevance. This list is not meant to be complete, but merely representative; it includes equations that are treated in this volume, and equations that do not fit into the classes that are actually discussed here, but are nevertheless solvable by techniques that are (nontrivial) extensions of those treated here. And some examples that are *not* solvable by the spectral transform technique are also mentioned.

The purpose of this section is to provide an idea of the scope of the spectral transform technique to deal with nonlinear evolution equations, not to discuss in any detail the properties of the equations being listed or their applicative relevance.

The *Korteweg-de Vries* (KdV) *equation*

$$u_t + u_{xxx} - 6u_x u = 0 \tag{3.13.1}$$

was introduced in 1895 in the context of fluid dynamics[2] and has great applicative and theoretical relevance: the former, especially in plasma physics and fluid dynamics; the latter, inasmuch as it is, in some sense, the simplest truly nonlinear novel equation beyond the class of second-order partial differential equations for which a fairly general theory exists (consisting in the classification, at least locally, of the three possible behaviors, hyperbolic, parabolic, or elliptic). The technique of solution of this equation has been described in the preceding sections and is treated in more detail in the following chapters.

It should be emphasized that, in writing the KdV equation (3.13.1), as well as the equations that follow, we are generally adopting the choice of variables that yields the neater notation [or the more conventional one; see for instance the factor 6 in equation (3.13.1)]. The flexibility implied by the possibility to rescale (dependent or independent) variables should however be kept in mind; for instance, this allows the introduction of arbitrary constants multiplying the three terms in equation (3.13.1).

The regularized long-wave (RLW) *equation*, or *PBBM* (Peregrine, Benjamin, Bona, and Mahoney) *equation*, is

$$u_t + u_x - u_{txx} - 6u_x u = 0 \tag{3.13.2}$$

This is a competitor of the KdV equation (3.13.1) to describe waves in a rectangular canal, in the limit when the wavelength is large relative to the depth of the canal. The RLW equation (3.13.2) differs from the KdV equation (3.13.1) in two respects. In the first place, due to the presence of the term u_x; but this difference is marginal, since the introduction of such a term in the KdV equation (3.13.1) corresponds to a trivial change of variables, equivalent to measuring the spatial coordinates in a frame of reference moving with unit speed. The second difference, namely the appearance of the term $-u_{txx}$ in equation (3.13.2) in place of u_{xxx} in equation (3.13.1) is, from an analytical point of view, more substantial; it implies indeed that equation (3.13.2) is not an evolution equation at all, at least in the most straightforward and strict sense. On the other hand, to the extent that dispersive and nonlinear effects are negligible in equation (3.13.2), there holds the approximate equation $-u_t \approx u_x$, and this implies $-u_{txx} \approx u_{xxx}$, a relation that eliminates the difference between equations (3.13.2) and (3.13.1) [with u_x appropriately added in equation (3.13.1)]. This remark is at the root of the physical "equivalence" of the KdV and RLW equations, in the sense that both have been used to describe the same physical phenomenon.

The RLW equation is *not* known to be solvable by the spectral transform technique, and there are reasons to suspect this is an intrinsic property, namely, that any search for a spectral transform appropriate to solve equation (3.13.2) is doomed to failure. [It is, however, worth noting that the equation

$$u_t + u_x - u_{txx} + 4u_t u - 2u_x \int_x^{+\infty} dy\, u_t(y, t) = 0 \tag{3.13.2a}$$

whose linear part coincides with the linear part of equation (3.13.2), belongs to the (solvable) class (3.3.1), with the (nonpolynomial) choice of $\alpha(z) = (z-1)^{-1}$].

The *transitional Korteweg–de Vries equation* is

$$u_t + u_{xxx} - 6f(t)u_x u = 0 \tag{3.13.3}$$

The name originates from the assumption that $f(t)$ is a monotonically increasing function, with $f(\pm\infty) = \pm 1$, so that equation (3.13.3) described the transition from a wave propagation regime modeled by the KdV equation (3.13.1) with a positive coefficient of the nonlinear term to a regime where the evolution equation is still the KdV but with the nonlinear term having the opposite sign. This equation has three constants of the motion. The transitional KdV equation, for any choice of $f(t)$ that has the properties stated above, is *not* solvable by the spectral transform technique; indeed the exceptional case $f(t) = t^{-1/2}$, whereby equation (3.13.3) can be easily transformed into the cylindrical KdV equation (see below), is not of the type we have described above as "transitional." When $f(t)$ is sufficiently smooth compared to the wavelength (3.13.3) it describes wave propagation on the surface separating two shallow layers of fluid, the function $f(t)$ accounting for a (smooth) variation of the depth of the bottom layer. In particular, the zero of $f(t)$ corresponds to the point where the two layers have equal depth, and the propagation across this turning point can be approximately modeled by the equation

$$u_t + u_{xxx} - 6(1 - \varepsilon t)u_x u = 0 \tag{3.13.3a}$$

that has been investigated by perturbation and numerical methods in the neighborhood of $t = \varepsilon^{-1}$.

The (so-called) *cylindrical Korteweg-de Vries* (cKdV) *equation* is

$$u_t + u_{xxx} - 6u_x u + (2t)^{-1}u = 0 \tag{3.13.4}$$

Note the explicit time-dependence and the singularity at $t = 0$. The Cauchy problem for this equation (from an initial condition given at $t = t_0 \neq 0$) can be solved, within the (physically interesting) class of functions $u(x, t)$ that vanish asymptotically, $u(\pm\infty, t) = 0$, by a spectral transform technique based on a novel spectral problem, although still related to the one-dimensional linear Schrödinger equation. The cKdV equation (3.13.4) had been introduced, and investigated numerically, in plasma physics, before its solvability was demonstrated. It obtains by performing an analogous schematization to that leading to the (standard) KdV equation (3.13.1), but in a cylindrical rather than a rectangular geometry. [The variable x in (3.13.4) does not have, however, the significance of a radial coordinate; indeed (3.13.4) is considered, as all other equations discussed in this chapter, on the whole line, $-\infty < x < +\infty$.]

As implied by its solvability via the spectral transform technique, the cKdV equation (3.13.4) possesses an infinite sequence of conserved quantities, in the form of integrals over all space of polynomials in u and its x-derivatives.

The (so-called) *spherical Korteweg-de Vries equation* is

$$u_t + u_{xxx} - 6u_x u + t^{-1}u = 0 \tag{3.13.5}$$

This equation is *not* solvable (so far at least) by any spectral transform technique. It arises, and it has therefore been numerically investigated, in the same context as the standard and cylindrical KdV equations, (3.13.1) and (3.13.4), with the modifications obviously implied by its name. Only three conserved quantities are known for this equation (this is the standard endowment for an evolution equation of this general form).

Note that it is not possible to change the numerical factor multiplying the last term in equations (3.13.4) and (3.13.5) by rescaling variables without changing the first term as well.

The modified Korteweg-de Vries (mKdV) *equation* is

$$u_t + u_{xxx} \pm 6u_x u^2 = 0 \tag{3.13.6}$$

Note that this equation differs from the KdV equation (3.13.1) only because of its cubic (rather than quadratic) nonlinearity. It has applicative relevance; for instance it has been used to describe acoustic waves in anharmonic lattices and Alfvén waves in a collisionless plasma. The Cauchy problem for this equation is solvable by an appropriate spectral transform, based on a spectral problem that is analogous to that discussed in the preceding sections, except for the replacement of the linear Schrödinger equation by the spinor Pauli equation. We refer to this spectral problem as the (generalized) Zakharov-Shabat spectral problem.

As it is typical of all equations solvable via the spectral transform, the mKdV equation (3.13.6) possesses an infinite number of conservation laws, its solutions may be related by explicit Bäcklund transformations, and it gives rise to the soliton phenomenology [note however that for real u, solitons may occur in equation (3.13.6) only if a positive sign appears in front of the last term; in contrast to the KdV case (3.13.1), this sign cannot be modified, relative to that of the second term, by any real rescaling of variables]. The mKdV equation is related to the KdV equation via a transformation known as the Miura transformation; this transformation has played an important rôle in the early history of the spectral transform method.

It is generally believed that the generalized KdV equation $u_t + u_{xxx} \pm 6u_x u^m = 0$ is solvable by spectral transform techniques only if $m = 1$ [ordinary KdV; see equation (3.13.1)] or $m = 2$ [mKdV; see equation (3.13.6)]; and of course by Fourier transform if $m = 0$ (linear case). Indeed only in these cases there exist an infinite number of conservation laws (of polynomial type).

As an example of a highly nonlinear evolution equation that differs from the KdV equation (3.13.1) only for the nonlinear terms [and includes the mKdV equation (3.13.6) as a special case] we report the equation

$$u_t + u_{xxx} - 6u_x\{u^2 - (u + u_{xx} - 2u^3)^2/[a^2 - 4(u^2 + u_x^2 - u^4)]\} = 0 \tag{3.13.7}$$

whose solvability can be demonstrated. Clearly equation (3.13.7) reduces to equation (3.13.6) if the constant a^2 is sufficiently large.

Another highly nonlinear equation whose solvability can be displayed reads

$$u_t + u_{xxx} - \tfrac{1}{8}u_x^3 + u_x[a \exp(u) + b \exp(-u) + c] = 0 \qquad (3.13.8)$$

This includes, as a limiting case, the equation

$$u_t + u_{xxx} + u_x(\alpha u^2 + \beta u + \gamma) = 0 \qquad (3.13.9)$$

that obtains from (3.13.8) by replacing u with εu, by setting $a = \alpha\varepsilon^{-2} + \frac{1}{2}\beta\varepsilon^{-1} + \gamma$, $b = \alpha\varepsilon^{-2} - \frac{1}{2}\beta\varepsilon^{-1}$, $c = -2\alpha\varepsilon^{-2}$, and then letting $\varepsilon \to 0$. The constants α, β and γ in equation (3.13.9) [as well as the constants a, b and c in equation (3.13.8)] are arbitrary. For $\alpha = \gamma = 0$, $\beta = -6$, equation (3.13.9) becomes the KdV equation (3.13.1); for $\beta = \gamma = 0$, $\alpha = \pm 6$, it becomes the mKdV (3.13.6).

Two other highly nonlinear evolution equations that are also tractable by the spectral transform method are the *Harry Dym* (HD) *equation*

$$v_t = v_{xxx}v^3 \qquad (3.13.10a)$$

$$u_t + 2(u^{-1/2})_{xxx} = 0, \qquad u = v^{-2} \qquad (3.13.10b)$$

and the *Wadati-Konno-Ichikawa-Schimizu* (WKIS) *equation* (for the *complex* field u)

$$iu_t + [(1 + |u|^2)^{-1/2}u]_{xx} = 0 \qquad (3.13.11)$$

The corresponding spectral problems are, however, somewhat unconventional. A related way to investigate these equations is by reducing them to other equations by a change of variable (that depends however on the solution itself). For instance, it can be shown that the HD equation (3.13.10) can be transformed into the KdV equation (3.13.1).

The last few examples we have reported are notable because of their theoretical significance rather than their applicative relevance (so far at least). We now return to some other nonlinear evolution equations solvable via the spectral transform technique that are directly related to applications.

The *Benjamin-Ono* (BO) *equation*

$$u_t + \boldsymbol{H}u_{xx} - 6u_x u = 0 \qquad (3.13.12)$$

resembles the KdV equation (3.13.1), except for the replacement, in the second term, of one space derivative by the Hilbert operator defined by the equation

$$\boldsymbol{H}f(x) = \pi^{-1}\mathrm{P}\int_{-\infty}^{+\infty} dy\, f(y)/(y - x) \qquad (3.13.12a)$$

[Incidentally, this implies that the dispersion function for the linear part of equation (3.13.12) reads $\omega(k) = -k^2 \operatorname{sign} k$, which can be compared with the corresponding dispersion functions for the linear part of the KdV equation (3.13.1), $\omega(k) = -k^3$, and of the RLW equation (3.13.2), $\omega(k) = k/(1+k^2)$.] The BO equation (3.13.12) is of course an integrodifferential, rather than a partial differential, equation; it can however be solved by a method analogous to the spectral transform technique. As a consequence, it possesses all the attributes of the solvable equations discussed above: an infinite number of conservation laws, Bäcklund transformations, and the soliton phenomenology.

The BO equation (3.13.12) was introduced to describe internal waves in stratified fluids (linear geometry, scale of disturbance small relative to the depth of the fluid). There is actually a more general nonlinear evolution equation, that is appropriate to describe a stratified fluid of finite depth. It reads

$$u_t + Tu_{xx} - 6u_x u = 0 \tag{3.13.13}$$

with the integrodifferential operator $\boldsymbol{T}$ defined by the equation

$$Tf(x) = \mathrm{P}\int_{-\infty}^{+\infty} dy\, f(y)\{\operatorname{cotgh}[\pi(y-x)/(2d)] - \operatorname{sign}(y-x)\}/(2d) \tag{3.13.13a}$$

The (constant) parameter d represents the distance between the bottom and the internal wave layer. It is easily seen, by appropriate rescaling, that in the limit $d \to 0$ (shallow fluid), equation (3.13.13) goes over into the KdV equation (3.13.1), while in the limit $d \to \infty$ (deep fluid), equation (3.13.13) goes over into the BO equation (3.13.12). Most remarkably, the general equation (3.13.13) is itself solvable, and it does possess all the attributes of a solvable equation.

The *Boussinesq* (BSQ) *equation*

$$u_{tt} - u_{xx} - u_{xxxx} + 3(u^2)_{xx} = 0 \tag{3.13.14}$$

was introduced by Boussinesq in 1871 to describe the propagation of long waves in shallow water in rectilinear geometry; it is more general than the KdV equation (3.13.1), since it accounts for motion in both directions. Indeed the KdV equation (3.13.1) (with u_x appropriately added) derives from equation (3.13.14) in the approximation, $u_t \approx -u_x$, that corresponds to restricting consideration to the motion around a wave traveling only in one direction. The BSQ equation (3.13.14) has many other applications: long waves in a one-dimensional nonlinear lattice, vibrations in a nonlinear string, and ion sound waves in a plasma. Remarkably, the BSQ equation (3.13.14) is itself solvable by the spectral transform technique. It is therefore graced by all the paraphernalia that come with such solvability: infinite number of conservation laws, Bäcklund transformations, and multisoliton solutions.

The *Kadomtsev–Petviashvili* (KP) *equation*

$$(u_t + u_{xxx} - 6u_x u)_x \pm u_{yy} = 0 \qquad (3.13.15)$$

is also called the *two-dimensional KdV equation*; it describes slow variations in the y-direction of wave propagation with dispersive and nonlinear effects in the x-direction. Thus, it models the "transverse" perturbation of solutions of the KdV equation (3.13.1); and, in particular, the single soliton solution of expression (3.13.1) is stable with respect to this perturbation if equation (3.13.15) is written with the "plus" sign in the last term, and unstable in the other case. This equation can be treated by techniques that are similar to those discussed here, though the extension required to introduce the additional spatial coordinate y is far from trivial. The KP equation (3.13.15) possesses multisoliton solutions that are however localized only in one direction (not in the other); explicit solutions that are localized in both directions are also known [they depend rationally on the spatial coordinates and are nonsingular only if the "plus" sign prevails in expression (3.13.15)]. Other particular solutions of equation (3.13.15) with the "plus" sign can be obtained from solutions of the cKdV equation (3.13.4) by using the fact that, if $\bar{u}(x, t)$ is a solution of equation (3.13.4), then

$$u(x, y, t) = \bar{u}(x + y^2/4t, t) \qquad (3.13.15a)$$

satisfies equation (3.13.15).[188]

The *Burgers equation*

$$u_t = u_{xx} + 2uu_x \qquad (3.13.16)$$

is the simplest physically interesting nonlinear evolution equation (it may be considered as a one-dimensional reduction of the Navier–Stokes equations). Its solvability, discovered in 1950, before the introduction of the spectral transform, is easily achieved via the Hopf–Cole transformation

$$u = (\ln v)_x \qquad (3.13.16a)$$

that maps the Burgers equation into the linear ("diffusion") equation

$$v_t = v_{xx} \qquad (3.13.16b)$$

The *nonlinear Schrödinger* (NLS) *equation*

$$iu_t + u_{xx} \pm 2|u|^2 u = 0 \qquad (3.13.17)$$

appears in many applicative contexts; indeed, it is generally appropriate to describe the time evolution of the envelope of an almost monochromatic wave of moderate amplitude in a weakly nonlinear, dispersive system (note that u is now complex). It is solvable by the spectral transform technique, belonging to the same class as the mKdV equation (3.13.6) (the underlying spectral problem is the generalized Zakharov–Shabat spectral problem). It possesses all the features of solvable equations: infinite number of conservation laws, Bäcklund transformations, and the soliton phenomenology [note however that solitons may appear only if the "plus" sign occurs in front of the last term in equation (3.13.17)]. A remarkable property of the NLS equation (3.13.17) is that its solutions can be mapped into the solutions of the one-dimensional *Heisenberg ferromagnet equation*

$$\mathbf{S}_t = \mathbf{S} \wedge \mathbf{S}_{xx}, \qquad \mathbf{S} \cdot \mathbf{S} = 1 \tag{3.13.18}$$

whose solution [the three-dimensional unit-vector $\mathbf{S} \equiv \mathbf{S}(x, t)$] describes a linear (continuous) spin density. This of course entails the solvability of expression (3.13.18) as well.

The *derivative nonlinear Schrödinger equation*

$$iu_t + u_{xx} \pm i(|u|^2 u)_x = 0 \tag{3.13.19}$$

also has a number of applications, for instance, in plasma physics (propagation of circularly polarized nonlinear Alfvén waves or of radio-frequency waves). This equation is also solvable by an appropriate spectral transform; the spectral problem is a variation of the Zakharov–Shabat problem (the dependence on the eigenvalue is quadratic rather than linear). It has all the properties characteristic of the equations solvable by the spectral transform technique.

The *Hirota equation* is

$$u_t + iau + ib(u_{xx} - 2\eta|u|^2 u) + cu_x + d(u_{xxx} - 6\eta|u|^2 u_x) = 0 \tag{3.13.20}$$

We have written this equation in a form intended to emphasize its generality; indeed, this equation is solvable by the spectral transform technique based on the generalized Zakharov–Shabat spectral problem: even if a, b, c, and d are (arbitrarily) time-dependent, while η is an (arbitrary) constant. For $a = c = d = 0$, $b = -1$, $\eta = \pm 1$, equation (3.13.20) becomes the NLS equation (3.13.17); for $a = b = c = 0$, $d = 1$, $\eta = \pm 1$ and u real, it becomes the mKdV (3.13.6).

An interesting extension of the NLS equation (3.13.17) is the system of two coupled equations

$$iu_t + u_{xx} - 2\eta|u|^2 u + iuv_x + uv^2 = 0 \tag{3.13.21a}$$

$$v_t - 2\eta(|u|^2)_x = 0 \tag{3.13.21b}$$

that is also solvable by the spectral transform technique; it has been investigated as a model for the interaction between long waves and short envelope waves, respectively described by the fields v and u.

A different model for interacting waves is described by the *3-dimensional 3-wave resonant interaction* (3D3WRI) *equation*

$$(\partial_t + \mathbf{v}_j \cdot \nabla) u_j + \eta_j u_k^* u_n^* = 0, \qquad jkn = 123 \text{ cyclic} \tag{3.13.22}$$

Here the three complex fields u_j are functions of the three-dimensional position vector $\mathbf{r}$, and of time; the three 3-vectors $\mathbf{v}_j$ are given constants, the dot indicates the usual scalar product, ∇ is the gradient operator, and $\eta_j = \pm 1$. Note that this example, in contrast to all those reported above and below, is imbedded in three-dimensional space (although the first term in each of the three-coupled equations (3.13.22) actually describes wave propagation in one direction only, that of the velocity $\mathbf{v}_j$). The system of nonlinear evolution equations (3.13.22) is important in many applications describing the resonant coupling between different plane waves, that in many contexts provides the first nonlinear correction to a basically linear dispersionless wave-propagation phenomenon. The solution of expression (3.13.22) involves a clever interplay of three spectral problems of the Zakharov–Shabat type, referring to three different space coordinates, each of them collinear to one of the vectors $\mathbf{v}_j$. Also the 3-wave interaction equations, that read as equation (3.13.22) but with only one or two spatial coordinates, are solvable by the spectral transform method.

The *Davey–Stewartson equations* are

$$iu_t - u_{xx} + u_{yy} + u|u|^2 - 2uv = 0 \tag{3.13.23a}$$

$$v_{xx} + v_{yy} - (|u|^2)_{yy} = 0 \tag{3.13.23b}$$

This system of two coupled differential equations is relevant to the description of two-dimensional long surface waves on water of finite depth. It is solvable by techniques that are analogous to those developed for the KP equation (3.13.15).

The *boomeron equation* reads

$$u_t = \mathbf{b} \cdot \mathbf{v}_x \tag{3.13.24a}$$

$$\mathbf{v}_{xt} = u_{xx}\mathbf{b} + \mathbf{a} \wedge \mathbf{v}_x - 2\mathbf{v}_x \wedge [\mathbf{v} \wedge \mathbf{b}] \tag{3.13.24b}$$

Here u and the three components of the 3-vector $\mathbf{v}$ are the dependent variables, x and t are the independent variables, $\mathbf{a}$ and $\mathbf{b}$ are two given 3-vectors, and the scalar and vector products are defined in the standard way [equation (3.13.24) is actually solvable even if $\mathbf{a}$ and $\mathbf{b}$ are time-dependent, although for simplicity we assume here they are constant]. This system of four coupled equations is equivalently described by the single matrix equation of order 2,

$$2\boldsymbol{W}_{xt} = i[\boldsymbol{W}_x, \boldsymbol{A}] + \{\boldsymbol{W}_{xx}, \boldsymbol{B}\} + [\boldsymbol{W}_x, [\boldsymbol{W}, \boldsymbol{B}]] \tag{3.13.24c}$$

Here the matrices $\boldsymbol{W}$, $\boldsymbol{A}$ and $\boldsymbol{B}$ are related to the scalar u and to the 3-vectors $\mathbf{v}$, $\mathbf{a}$ and $\mathbf{b}$ by the equations $\boldsymbol{W} = u + \mathbf{v} \cdot \boldsymbol{\sigma}$, $\boldsymbol{A} = \mathbf{a} \cdot \boldsymbol{\sigma}$, $\boldsymbol{B} = \mathbf{b} \cdot \boldsymbol{\sigma}$, $\boldsymbol{\sigma}$ being the 3-vector having as components the three Pauli matrices, and the square and curly brackets indicate commutation and anticommutation ($[\boldsymbol{A}, \boldsymbol{B}] \equiv \boldsymbol{AB} - \boldsymbol{BA}$, $\{\boldsymbol{A}, \boldsymbol{B}\} \equiv \boldsymbol{AB} + \boldsymbol{BA}$). The evolution equation (3.13.24) is solvable by the spectral transform technique; the spectral problem is based on the *matrix* Schrödinger equation. The class of equations solvable in this manner is treated elsewhere; all these evolution equations possess the properties characteristic of solvable equations, including an infinite number of conservation laws, explicit Bäcklund transformations, and multisoliton solutions. The basic property of solitons, to interact with each other, as it were, "elastically," is also preserved; but a novel phenomenon appears, namely, the fact that each soliton generally moves with variable speed, i.e., as a particle acted upon by an external force rather than as a free particle. Indeed the interest (and name) of the evolution equation (3.13.24) originates from the fact that it provides the simplest instance of this novel phenomenon; generally its solitons boomerang back, in the remote future, to where they came from in the remote past (in the case in which the vectors $\mathbf{a}$ and $\mathbf{b}$ are mutually orthogonal another generic behavior of solitons may occur, namely, they may oscillate without ever escaping to infinity, i.e., behave as *trappons* rather than *boomerons*; both behaviors may coexist in a multisoliton solution).

Another equation, that is essentially a subcase of the boomeron equation (3.13.24) (in the particularly interesting case with $\mathbf{a} \cdot \mathbf{b} = 0$ mentioned above), but has the advantage to describe the evolution of a single scalar field (rather than four coupled fields), is the *zoomeron equation*

$$(\partial^2/\partial t^2 - \partial^2/\partial x^2)(Z_{xt}/Z) + 2(Z^2)_{xt} = 0 \qquad (3.13.25)$$

This highly nonlinear equation is a convenient one to display the novel phenomenology associated with boomerons and trappons; moreover its solitons ("*zoomerons*") have an amplitude that changes with time along with their speed.

The *reduced Maxwell-Bloch equations* are

$$E_t - v = 0 \qquad (3.13.26a)$$

$$v_x - \omega r - Eq = 0 \qquad (3.13.26b)$$

$$q_x + Ev = 0 \qquad (3.13.26c)$$

$$r_x + \omega v = 0 \qquad (3.13.26d)$$

Here E is the electric field, r and v are real combinations of the density matrix related to the atomic polarization, and q is a measure of the atomic polarization; these four functions depend on the variables x and t that are appropriate linear combinations of the physical space and time coordinates, while ω is constant, proportional to the atomic density. All these quantities have been put in dimensionless form by appropriate rescaling. The physical phenomenon (approximately) described by this system of nonlinear evolution equations is the interaction, with a medium of two-level atoms, of a short intense beam of light traveling to the right.

The *Sine-Gordon* (SG) *equation* comes in two forms,

$$u_{xt} \pm \sin u = 0, \qquad u \equiv u(x, t) \tag{3.13.27a}$$

$$\varphi_{\xi\xi} - \varphi_{\tau\tau} \pm \sin \varphi = 0, \qquad \varphi \equiv \varphi(\xi, \tau) \tag{3.13.27b}$$

the transition from one form to the other corresponding to the change from "light-cone" (x, t) to "laboratory" (ξ, τ) coordinates, i.e.,

$$\varphi(\xi, \tau) = u(x, t), \qquad x = \tfrac{1}{2}(\xi + \tau), \qquad t = \tfrac{1}{2}(\xi - \tau) \tag{3.13.27c}$$

The initial value problem for equation (3.13.27a) is the Goursat problem (the initial datum is given on the characteristic $t = 0$), while for equation (3.13.27b) it is the Cauchy problem (initial datum given at $\tau = 0$). Both these problems are solvable by the spectral transform; in fact expression (3.13.27a) belongs to the same class (associated with the Zakharov-Shabat spectral problem) as the mKdV equation (3.13.6) and the NLS equation (3.13.17), or the Hirota equation (3.13.20).

The SG equation (3.13.27), in one or the other avatar, appears in many applications, from differential geometry (its relation to two-dimensional pseudospheres was actually known a century ago) to nonlinear optics, where solitons of (3.13.27a) account for the *self-induced transparency* phenomenon [in fact, the reduced Maxwell-Bloch equations (3.13.26) become the SG equations (3.13.27) when $\omega = 0$, $r = 0$, $E = u_x$, $v = \mp\sin u$, $q = \mp\cos u$], to superconductivity. Moreover it has been extensively investigated as a model of relativistic field theory (both classical and quantal; this is due to its relativistically invariant nature). The SG equation (3.13.27) possesses of course all the attributes of a solvable equation (infinite number of conservation laws, Bäcklund transformations, and solitons, which however appear as "kinks," namely, not u is localized, but its derivative u_x).

The *double Sine-Gordon equation* is

$$u_{xt} \pm [\sin u + \eta \sin \tfrac{1}{2}u] = 0 \tag{3.13.28}$$

η being a constant. The relevance of this equation is motivated by its physical applications in nonlinear optics (propagation of ultrashort pulses in a resonant 5-fold degenerate medium) and in low-temperature physics (propagation of spin waves in anisotropic magnetic liquids). There is strong evidence that this equation is not solvable via the spectral transform method.

The (so-called) *phi-four equation* is

$$\varphi_{tt} - \varphi_{xx} - \varphi + \varphi^3 = 0 \tag{3.13.29}$$

its name being due to the expression $\mathcal{H} = \frac{1}{2}(\varphi_t^2 + \varphi_x^2) - \frac{1}{2}\varphi^2 + \frac{1}{4}\varphi^4$ of the corresponding Hamiltonian density. This equation has been the subject of intensive investigation in the context of classical and quantized field theory. It has kink-like solutions (similar to the SG kinks) but they are not solitons (they are solitrons); in fact the phi-four equation is not solvable by means of the spectral transform.

The *Liouville equation* is

$$u_{xt} = \exp(\eta u) \tag{3.13.30}$$

η being an arbitrary constant. The general solution of this equation was given by Liouville in 1853, and reads

$$u(x, t) = f(x) - g(t) - (2/\eta) \ln\left\{p \int_{x_0}^{x} dy \exp[\eta f(y)] + (\eta/2p) \int_{t_0}^{t} ds \exp[-\eta g(s)]\right\} \tag{3.13.30a}$$

where p is an arbitrary parameter, and $f(x)$ and $g(t)$ are arbitrary functions. Indeed the Liouville equation can be transformed into the wave equation $\varphi_{xt} = 0$, whose general solution is precisely $\varphi(x, t) = f(x) + g(t)$. The Liouville equation can be imbedded in the same class of solvable equations as the NLS, mKdV, and SG equations, and it has some relevance as a field-theoretical model.

Another scalar equation that, like the SG and Liouville equations, has the property to be relativistically invariant and that is solvable via the spectral transform technique reads

$$u_{xt} + \exp(-u) - \exp(2u) = 0 \tag{3.13.31}$$

The spectral problem that provides the basis for the spectral transform used to solve equation (3.13.31) involves matrices of order 3.

There are several other relativistically invariant equations that are now known to be solvable by the spectral transform technique (though in several cases this technique requires a substantial extension of the formulation described in this chapter), but they generally involve two, or more, coupled fields. For instance, the SG equation (3.13.27b) is the reduced case, with $v = 0$ and $\varphi = 2u$, of the *Pohlmeyer–Lund–Regge-model*

$$u_{\xi\xi} - u_{\tau\tau} \pm \sin u \cos u + (\cos u/\sin^3 u)(v_\xi^2 - v_\tau^2) = 0 \quad (3.13.32a)$$

$$(v_\xi \operatorname{cotg}^2 u)_\xi = (v_\tau \operatorname{cotg}^2 u)_\tau \quad (3.13.32b)$$

which is also solvable, and the equation (3.13.31) is the reduced case, with $v = 0$, of the solvable system

$$u_{xt} + \cosh(3v) \exp(-u) - \exp(2u) = 0 \quad (3.13.33a)$$

$$v_{xt} - \sin^3 v \exp(-u) = 0 \quad (3.13.33b)$$

Among other interesting field equations, we mention below (without further comment) a few more that stand out as particularly representative of the present stock of nonlinear partial differential equations that are treatable by the spectral transform method.

The O(n) *sigma-model* is

$$\mathbf{v}_{xt} + (\mathbf{v}_x \cdot \mathbf{v}_t)\mathbf{v} = 0 \quad (3.13.34)$$

here $\mathbf{v}$ is an n-dimensional real unit vector, $(\mathbf{v} \cdot \mathbf{v}) = 1$.

The principal SU(n) *chiral field equation* is

$$(U^\dagger U_x)_t + (U^\dagger U_t)_x = 0 \quad (3.13.35)$$

where $U(x, t)$ is a unitary matrix, $U^\dagger U = \mathbf{1}$, of SU(n).

The (reduced) *Einstein equation* is

$$(g^{1/2} G_x G^{-1})_t + (g^{1/2} G_t G^{-1})_x = 0 \quad (3.13.36)$$

$G(x, t)$ being a 2×2 symmetrical matrix, and $g(x, t)$ being its determinant, $g = \det G$.

The *Ernst equation* is

$$(\text{Re } E)(E_{\rho\rho} + \rho^{-1}E_\rho + E_{zz}) = E_\rho^2 + E_z^2 \tag{3.13.37}$$

where E is a complex function of the radial and axial coordinates ρ, z, this equation being another reduced case of the Einstein equation, corresponding to a static axially symmetric vacuum field.

The (reduced) *Yang-Mills equations* is

$$(U^\dagger U_t)_t - (U^\dagger U_z)_{\bar{z}} = 0 \tag{3.13.38}$$

where $U(x, y, t)$ is a matrix of SU(n), $\det U = 1$, and $\partial_z \equiv \partial_x - i\partial_y$, $\partial_{\bar{z}} = \partial_x + i\partial_y$. Note that this equation involves two spatial coordinates.

The *massive Thirring model* reads

$$iu_x + v + u|v|^2 = 0 \tag{3.13.39a}$$

$$iv_t + u + v|u|^2 = 0 \tag{3.13.39b}$$

The *Nambu-Jona Lasinio-Vaks-Larkin model* reads

$$iu_x^{(n)} = v^{(n)} \sum_{m=1}^{N} v^{(m)*}u^{(m)} \tag{3.13.40a}$$

$$iv_t^{(n)} = u^{(n)} \sum_{m=1}^{N} u^{(m)*}v^{(m)} \tag{3.13.40b}$$

The *Gross-Neveu model* reads

$$iu_x^{(n)} = v^{(n)} \sum_{m=1}^{N} (v^{(m)*}u^{(m)} + u^{(m)*}v^{(m)} \tag{3.13.41a}$$

$$iv_t^{(n)} = u^{(n)} \sum_{m=1}^{N} (u^{(m)*}v^{(m)} + v^{(m)*}u^{(m)}) \tag{3.13.41b}$$

It is remarkable that these relativistic invariant equations have been recently applied in solid state physics.

We terminate here this list, that has included 37 "solvable" equations and five that are, at least so far, not solvable by the spectral transform method. We reiterate that this is by no means a complete list; for one thing, novel solvable equations are being discovered at a sustained rate. Moreover, we have confined our list to partial differential or integrodifferential equations, omitting any mention of finite difference equations; yet the techniques based on the spectral transform can be extended to discretized contexts (either the space variable or the time variable or both can be discretized).

Finally, our attention to solvable equations involving more than two variables has been limited to the mere mention of the KP equation (3.13.15), the 3D3WRI equation (3.13.22), the DS equations (3.13.23), and the reduced YM equations (3.13.28).

Notes to Chapter 3

3.2.1. The solution of linear PDEs via the Fourier transform is standard textbook material.(46) The progress mentioned at the end of this section refers to the recent demonstration that certain nonlinear *integrodifferential* evolution equations have been recently shown to be solvable; see Section 3.13.

3.8. Among the first researchers to emphasize the analogy between the spectral transform technique to solve *nonlinear* evolution equations and the Fourier transform method to solve *linear* partial differential equations were Tanaka(189) and Ablowitz *et al.*(20) The analogy between the direct and inverse spectral (or scattering) problems and the Fourier transform technique has on the other hand always been well understood by researchers in this field, including the pioneers (see the standard reference texts on the inverse problem).(51,52,190–194)

3.9.1. The pure multisoliton solution (3.9.9) is most easily derived (once the spectral transform formalism is available) by the solution of the inverse spectral problem; see Section 3.5. There also exist alternative techniques, among which stands out as particularly effective the so-called "direct method" (for the KdV equation see Hirota;(15) a review of this technique is given by Hirota(195)); this method, mainly due to Hirota, is largely manipulative, but has proved its worth in many instances, by discovering multisoliton solutions for novel equations before other techniques to deal with these equations were developed.

3.13. *The KdV equation*, being the prototype of the nonlinear evolution equations whose integrability has been recently discovered, has been investigated in an enormous number of papers in the last decade. The argument that singles it out as worthy of study from an *a priori* mathematical point of view is given by Kruskal.[196] For a short readable introduction to the KdV equation, and to its properties, see, for instance, Miura.[197] The fact that the KdV equation is just the first of a sequence of nonlinear evolution equations, the so-called *higher KdV equations* [set $\alpha(z) = z^n$ in equation (3.7.2)],

$$u_t = L^n u_x, \qquad n = 2, 3, 4, \ldots$$

was pointed out by Lax.[7] The static version of equation (3.7.2), the so-called Novikov equation, $\alpha(L)u_x = 0$, has been thoroughly investigated in Novikov[198] (see also Gel'fand and Dikii[145,146]). The wide applicative relevance of the KdV equation is due to its being the natural equation to describe the long-time behavior of any wave-like phenomenon that is nondissipative, weakly nonlinear and weakly dispersive.[11,199-202]

The theory of the *RLW equation* has been thoroughly investigated by Benjamin *et al.*,[203] and for this reason it is also known as the BBM equation; numerical solutions of this equation were first given by Peregrine[204] (hence the name PBBM is also largely used). It is a matter of dispute[187] whether this equation or KdV is more appropriate for applications, and in particular to describe long waves.[205]

For a description of the fluid dynamical phenomena modeled by the *transitional KdV equation*, see Knickerbocker and Newell[206] and the references cited by them.

The applicative relevance of the *cylindrical and spherical KdVs* (and some numerical results) are discussed by Maxon,[207] and in the literature cited by this author. The solvability of the cKdV equation via the spectral transform method has been first proved by Calogero and Degasperis.[24]

The *mKdV equation* has been extensively studied, due to its applicative relevance, but especially because of its close relation to the KdV equation. Indeed the *Miura transformation* that relates the solutions of these two equations has played an important rôle in bringing about the discovery of the spectral transform technique.[197,208] Its solvability (via the Zakharov–Shabat spectral transform) was first proved by Wadati.[209]

The generalized KdV equation $u_t + u_{xxx} \pm 6u_x u^m = 0$ is tersely discussed by Scott *et al.*;[32] see also Miura.[199,210]

The evolution equations (3.13.7) and (3.13.8) were introduced (and shown to be solvable) by Calogero and Degasperis.[27]

The *WKIS equation* (3.13.11) is treated by Wadati *et al.*[211] and Shimizu and Wadati.[212]

There are numerous references for the *BO equation* (3.13.12).[213-227] For a related *modified BO equation*, see Nakamura,[228] while for the "*higher*" *BO equations* see Matsuno.[229,230]

The *finite depth fluid equation* (3.13.8–3.13.13) has been treated extensively.[231-235] For the "*higher*" *finite depth fluid equations* see

Matsuno,[236] and for a *modified finite depth fluid equation* see Nakamura[237] and Gibbons and Kupershmidt.[235]

The solution of the *BSQ equation* (3.13.11) (introduced by Boussinesq[238,239]) via the spectral transform technique is based on the spectral problem associated with the linear third-order equation $\psi_{xxx} + v(x)\psi_x + u(x)\psi = -ik^3\psi$; this problem has been solved, with various degrees of generality, by Kaup[240] and Caudrey;[241] the solvability of the BSQ equation has been dealt with widely.[242-247]

The *KP equation* (3.13.15) has been discussed extensively.[64,126,148-253]

For the treatment of the *Burgers equation* (3.13.16) see Burgers;[254] the linearizing transformation (3.13.16a) has been independently given by Hopf[255] and by Cole.[256] For a symmetry approach to a class of evolution equations including the Burgers equation see Fokas and Yortos,[257] and for its formulation in the Hamiltonian formalism see Taflin.[258]

For the applicative relevance of the *NLS equation* (3.13.17) see, for instance, Taniuti and Yajima,[259] Segur,[200] Lamb and MacLaughlin,[260] Newell,[261] and the references cited by them. The discovery that the NLS equation is solvable via (an appropriate) spectral transform is due to Zakharov and Shabat.[8] For its relation to the *one-dimensional Heisenberg ferromagnet equation* (3.13.18) see Takhtajan[262] and Lakshmanan.[263]

The *derivative nonlinear Schrödinger* (*DNLS*) equation (3.13.19) has been solved by Kaup and Newell.[264] A complete treatment of the spectral transform method to solve this equation is given by Gerdjikov *et al.*[265] An extended version of the DNLS equation, that also has applicative relevance, has also been shown to be solvable, see Morris and Dodd.[266]

The *Hirota equation* (3.13.20) was investigated by the "direct method" by Hirota,[267] and so named by Scott *et al.*[32]

The wave interaction modeled by the equations (3.13.21) was investigated via the spectral transform technique by Newell.[268]

The *3D3WRI equation* (3.13.22) in the one-dimensional case has been solved by Zakharov and Manakov[269,270] and Zakharov and Shabat.[19] The most extensive investigation of this problem has been carried out in a series of papers by Kaup.[271-276] The last four papers are based on an advance due to Cornille.[277]

The *Davey-Stewartson equations* (3.13.23) were introduced by Davey and Stewartson,[278] and their solvability has been demonstrated by Anker and Freeman.[279]

The *boomeron and zoomeron equations* were first investigated by Calogero and Degasperis;[78,79] a review on the corresponding soliton phenomenology is given by Degasperis.[280] There exist two films that display the behavior of the one- and two-soliton solutions of these equations.[80,81]

The *reduced Maxwell-Bloch* (rMB) *equations* (3.13.26) provide a model to describe the phenomenon of Self-Induced Transparency (SIT).[281] They are treated, for instance, by Gibbon *et al.*;[282] see also Bullough and Bullough *et al.*[283] The rMB equations approximate the Maxwell-Bloch (MB) equations through the restriction to waves traveling only in one direction. The MB equations themselves are presumably not integrable; they seem to possess "solitrons," not solitons.[284]

The literature on the *SG equation* is very extensive, due to its ubiquitous presence in mathematical and physical problems, and also to its providing the first instance of a relativistically invariant solvable equation possessing multi-soliton solutions (in contrast with the Liouville equation that has no soliton solutions). For the early physical literature see Skryme[285,286] and Barone;[287] more recently, its applications have been discussed, for instance, by Bullough,[33,34] Cercignani,[35] Bishop,[288] and Timonen and Bullough.[41] The early mathematical literature is much older, see for instance Hilbert,[289] Bianchi,[290] and Eisenhart.[291] The solvability of the SG equation was first demonstrated by Ablowitz *et al.*[292] in the form (3.13.27a) (for the solvability of the SG equation (3.13.27b) in "laboratory coordinates," see Kaup[293] and Kaup and Newell[294] and by Zakharov *et al.*[295] (see also Tahtadžjan and Faddeev[296,297])).

For the *double SG equation* see Bullough *et al.*,[298] and the references quoted there.

The *phi-four equation* has been mainly investigated as the simplest non-trivial relativistic invariant field theoretical model (see, for instance, Dashen *et al.*[299] and Goldstone and Jackiw.[300] Numerical investigations are reported by Aubry[301] and Getmanov.[302]

The general solution of the *Liouville equation* was first found by Liouville.[303] For its treatment in the context of the spectral transform method, see, for instance, Andreev,[304] Chaichian and Kulish,[305] and Tsutsumi.[306]

The relativistic invariant equation (3.13.31) is a reduced case of the system (3.13.33), that was introduced by Mikhailov[307] and Fordy and Gibbons[308] (see also Zhiber and Shabat,[309] Kupershmidt and Wilson,[310] and Mikhailov *et al.*[311])

For the *Pohlmeyer-Lund-Regge model* (3.13.32), see Pohlmeyer,[312] Lund and Regge,[313] and Lund.[314]

The *sigma-model* (3.13.34) has been considered by Weinberg,[315] and then shown to be solvable by Pohlmeyer;[312] for other sigma-models see Orfanidis.[316]

For the solvability of the *principal chiral field equation*, see Zakharov and Mikhailov,[317] and the references quoted there.

The integrability of the *Einstein equation* (3.13.36) was first pointed out by Maison.[318] A general scheme for its solution has been given by Belinsky and Zakharov.[319]

The *Ernst equation* (3.13.37) was introduced by Ernst;[320] for the more recent findings concerning its solvability see Harrison,[321,326] Maison,[322] Belinsky and Zakharov,[323] Neugebauer,[324] Cosgrove,[325] Hauser and Ernst,[327] Morris and Dodd,[266] and Omote and Wadati.[329,330]

References for the solvability of the *Yang-Mills equations* (3.13.38) are Belavin and Zakharov[331] and Manakov and Zakharov;[332] see also Atiyah *et al.*[333]

The *massive Thirring model* was introduced by Thirring,[334] and shown to be solvable by Mikhailov[335] (see also Kuznetsov and Mikhailov[336] and Gerdjikov *et al.*[265]).

For the *Nambu-Jona Lasinio-Vaks-Larkin model* (3.13.40) and the *Gross-Neveu model* (3.13.41) see the general treatment given in Zakharov and Mikhailov;[337] for the applications to solid state physics, see Campbell and Bishop[338] and the literature quoted there.

References

1. N. J. Zabusky and M. D. Kruskal, Interaction of "solitons" in a collisionless plasma and the recurrence of initial states, *Phys. Rev. Lett.* **15**, 240-243 (1965).
2. D. J. Korteweg and G. de Vries, On the change of form of long waves advancing in a rectangular canal, and on a new type of long stationary waves, *Phil. Mag.* **39**, 422-443 (1895).
3. E. Fermi, J. R. Pasta, and S. M. Ulam, Studies of nonlinear problems. Los Alamos Sci. Lab. Rep. LA-1940 (1955). Reprinted in *Collected Works of Enrico Fermi*, Vol. II, 1978, University of Chicago Press, Chicago (1965); and also in: *Nonlinear Wave Motion* (A. C. Newell, ed.), pp. 143-156, Lect. Appl. Math. **15**, American Mathematical Society, Providence, RI (1974).
4. M. D. Kruskal, The birth of the soliton, in: *Nonlinear Evolution Equations Solvable by the Spectral Transform* (F. Calogero, ed.), pp. 1-8, Research Notes in Mathematics **26**, Pitman, London (1978).
5. J. Scott-Russell, Report on waves, in: *Report of the Fourteenth Meeting of the British Association for the Advancement of Science*, pp. 311-390, John Murray, London (1845).
6. C. S. Gardner, J. M. Greene, M. D. Kruskal, and R. M. Miura, Method for solving the Korteweg-de Vries equation, *Phys. Rev. Lett.* **19**, 1095-1097 (1967).
7. P. D. Lax, Integrals of nonlinear equations of evolution and solitary waves, *Comm. Pure Appl. Math.* **21**, 467-490 (1968).
8. V. E. Zakharov and A. B. Shabat, Exact theory of two-dimensional self-focusing and one-dimensional self-modulation of waves in nonlinear media, *Soviet Phys. JETP* **34**, 62-69 (1972) [*Zh. Eksp. Teor. Fiz.* **61**, 118-134 (1971)].
9. R. M. Miura, Korteweg-de Vries equation and generalizations. I: A remarkable explicit nonlinear transformation, *J. Math. Phys.* **9**, 1202-1204 (1968).
10. R. M. Miura, C. S. Gardner, and M. D. Kruskal, Korteweg-de Vries equation and generalizations. II: Existence of conservation laws and constants of motion, *J. Math. Phys.* **9**, 1204-1209 (1968).
11. C. H. Su and C. S. Gardner, Korteweg-de Vries equation and generalizations. III: Derivation of the Korteweg-de Vries equation and Burgers' equation, *J. Math. Phys.* **10**, 536-539 (1969).
12. C. S. Gardner, Korteweg-de Vries equation and generalizations. IV: The Korteweg-de Vries equation as a Hamiltonian system, *J. Math. Phys.* **12**, 1548-1551 (1971).
13. M. D. Kruskal, R. M. Miura, C. S. Gardner, and N. J. Zabusky, Korteweg-de Vries equation and generalizations. V: Uniqueness and nonexistence of polynomial conservation laws, *J. Math. Phys.* **11**, 952-960 (1970).
14. C. S. Gardner, J. M. Greene, M. D. Kruskal, and R. M. Miura, Korteweg-de Vries and generalizations. VI: Methods for exact solution, *Comm. Pure Appl. Math.* **27**, 97-133 (1974).
15. R. Hirota, Exact solution of the Korteweg-de Vries equation for multiple collisions of solitons, *Phys. Rev. Lett.* **27**, 1192-1194 (1971).

16. V. E. Zakharov and L. D. Faddeev, Korteweg-de Vries equation, a completely integrable Hamiltonian system, *Funct. Anal. Appl.* **5**, 280-287 (1971) [*Funk. Anal. Pril.* **5**, 18-27 (1971)].
17. N. J. Zabusky, G. S. Deem, and M. D. Kruskal, Formation, propagation and interaction of solitons, Computer-produced film, Bell Laboratories (1968).
18. F. D. Tappert, Nonlinear wave propagation as described by the Korteweg-de Vries equation and its generalizations, Computer-produced film, Bell Laboratories (1971).
19. V. E. Zakharov and A. B. Shabat, A scheme for integrating the nonlinear equations of mathematical physics by the method of the inverse scattering problem. I, *Funct. Anal. Appl.* **8**, 226-235 (1974) [*Funk. Anal. Pril.* **8**, 43-53 (1974)].
20. M. J. Ablowitz, D. J. Kaup, A. C. Newell, and H. Segur, The inverse scattering transform—Fourier analysis for nonlinear problems, *Stud. Appl. Math.* **53**, 249-315 (1974).
21. F. Calogero, A method to generate solvable nonlinear evolution equations, *Lett. Nuovo Cimento* **14**, 443-448 (1975).
22. F. Calogero and A. Degasperis, Nonlinear evolution equations solvable by the inverse spectral transform. I, *Nuovo Cimento* **32B**, 201-242 (1976).
23. F. Calogero and A. Degasperis, Nonlinear evolution equations solvable by the inverse spectral transform. II, *Nuovo Cimento* **39B**, 1-54 (1977).
24. F. Calogero and A. Degasperis, Extension of the spectral transform method for solving nonlinear evolution equations, *Lett. Nuovo Cimento* **22**, 131-137 (1978).
25. F. Calogero and A. Degasperis, Extension of the spectral transform method for solving nonlinear evolution equations. II, *Lett. Nuovo Cimento* **22**, 263-269 (1978).
26. F. Calogero and A. Degasperis, Solution by the spectral transform method of a nonlinear evolution equation including as a special case the cylindrical KdV equation, *Lett. Nuovo Cimento* **23**, 150-154 (1978).
27. F. Calogero and A. Degasperis, Reduction technique for matrix nonlinear evolution equations solvable by the spectral transform, *J. Math. Phys.* **22**, 23-31 (1981).
28. B. A. Dubrovin, V. B. Matveev, and S. P. Novikov, Nonlinear equations of Korteweg-de Vries type, finite-zone linear operators and abelian varieties, *Russian Math. Surveys* **31**(1), 59-146 (1976) [*Usp. Mat. Nauk* **31**(1), 55-136 (1976)].
29. R. Hermann, *The Geometry of Nonlinear Differential Equations, Bäcklund Transformations, and Solitons. Part A*, Math. Sci. Press, Brookline, MA (1976).
30. R. Hermann, *Geometric Theory of Nonlinear Differential Equations, Bäcklund Transformations and Solitons. Part B*, Math. Sci. Press, Brookline, MA (1977).
31. Yu. I. Manin, Algebraic aspects of nonlinear differential equations, *J. Sov. Math.* **12**, 1-122 (1979) [*Sovrem. Probl. Matem.* **11**, 5-152 (1978)].
32. A. C. Scott, F. Y. F. Chu, and D. McLaughlin, The soliton: a new concept in applied science, *Proc. IEEE* **61**, 1443-1483 (1973).
33. R. K. Bullough, Solitons, in: *Interaction of Radiation with Condensed Matter*, Vol. I, pp. 381-469, IAEA, Vienna (1977).
34. R. K. Bullough, Solitons in physics, in: *Nonlinear Equations in Physics and Mathematics* (A. O. Barut, ed.), pp. 99-141, Reidel, Dordrecht (1978).
35. C. Cercignani, Solitons, theory and applications, *Riv. Nuovo Cimento* **7**, 429-469 (1977).
36. A. R. Bishop and T. Schneider (eds.), *Solitons and Condensed Matter Physics*, Springer Series in Solid-State Sciences **8**, Springer-Verlag, Berlin (1978).
37. R. K. Bullough, Solitons, *Phys. Bull.*, 78-82 (February 1978).
38. H. Flaschka and D. W. McLaughlin (eds.), Proceedings of a conference on the theory and applications of solitons, *Rocky Mountain J. Math.* **8**, 1, 2 (1978).
39. K. Lonngren and A. Scott, *Solitons in Action*, Academic Press, New York (1978).
40. W. Guttinger and H. Eikemeier (eds.), *Structural Stability in Physics*, Springer Series in Synergetics **4**, Springer-Verlag, Berlin (1979).
41. J. Timonen and R. K. Bullough, Solitons in physics: an application to spin waves in the one-dimensional ferromagnet $CsNiF_3$, in: *Problèmes Inverses Evolution Non Linéaire* (P. C. Sabatier, ed.), pp. 133-187, CNRS, Paris (1980).

42. H. Wilhelmsson (ed.), *Solitons in Physics*, Topical issue of Physica Scripta, Royal Swedish Academy of Sciences, *Phys. Scr.* **20**, 291-680 (1979).
43. R. K. Bullough and P. J. Caudrey (eds.), *Solitons*, Topics in Current Physics **17**, Springer, Berlin (1980).
44. V. Arnold, *Mathematical Methods of Classical Mechanics*, Nauka, Moscow (1974) (in Russian). [French translation: *Les Méthodes Mathématiques de la Mécanique Classique*, Mir, Moscow (1976); English translation: *Mathematical Methods of Classical Mechanics*, Graduate Texts in Mathematics **60**, Springer-Verlag, New York (1978)].
45. A. C. Newell (ed.), *Nonlinear Wave Motion*, Lect. Appl. Math. **15**, American Mathematical Society, Providence, RI (1974).
46. G. B. Whithan, *Linear and Nonlinear Waves*, Wiley, New York (1974).
47. I. A. Kunin, *Theory of Elastic Media with Microstructures*, Nauka, Moscow (1975) (in Russian). Chapter 5 (The inverse scattering method) is due to V. E. Zakharov.
48. J. Moser (ed.), *Dynamical Systems, Theory and Applications*, Lecture Notes in Physics **38**, Springer-Verlag, Berlin (1975).
49. V. B. Matveev, *Abelian Functions and Solitons*, Institute of Theoretical Physics, Wroclaw University preprint 373 (1976).
50. R. M. Miura (ed.), *Bäcklund Transformations*, Lecture Notes in Mathematics **515**, Springer-Verlag, Berlin (1976).
51. K. Chadan and P. C. Sabatier, *Inverse Problems in Quantum Scattering Theory*, Springer-Verlag, Berlin (1977).
52. V. A. Marchenko, *Sturm-Liouville Operators and their Applications*, Naukova Dumka, Kiev (1977) (in Russian).
53. M. J. Ablowitz, Lectures on the inverse scattering transform, *Stud. Appl. Math.* **58**, 17-94 (1978).
54. A. O. Barut (ed.), *Nonlinear Equations in Physics and Mathematics*, Reidel, Dordrecht (1978).
55. F. Calogero (ed.), *Nonlinear Evolution Equations Solvable by the Spectral Transform*, Research Notes in Mathematics **26**, Pitman, London (1978).
56. D. V. Chudnovsky and G. V. Chudnovsky, Seminaire sur les equations non linaires. I, Mimeographed lecture notes, Centre de mathématiques, Ecole polytechnique, Paris (1977-78).
57. M. G. Crandall (ed.), *Nonlinear Evolution Equations*, Proceedings of a Symposium at the Mathematics Research Center, University of Wisconsin-Madison, October 17-19, 1977, Academic Press, New York (1978).
58. G. Dell'Antonio, S. Doplicher, and G. Jona-Lasinio, *Mathematical Problems in Theoretical Physics*, Lecture Notes in Physics **80**, Springer-Verlag, Berlin (1978).
59. V. G. Makhankov, Dynamics of classical solitons (in nonintegrable systems), *Phys. Rep.* **35C**, 1-128 (1978).
60. P. C. Sabatier (ed.), *Applied Inverse Problems*, Lecture Notes in Physics **85**, Springer-Verlag, Berlin (1978).
61. D. V. Chudnovsky, One and multidimensional completely integrable systems arising from the isospectral deformation, in: *Complex Analysis, Microlocal Calculus and Relativistic Quantum Theory* (D. Iagolnitzer, ed.), pp. 352-416, Lecture Notes in Physics **126**, Springer-Verlag, Berlin (1980).
62. S. V. Manakov and V. E. Zakharov, *Soliton Theory*, Proceedings of the Soviet-American Symposium on Soliton Theory, Kiev, September 1979, *Physica* **3D**, 1, 2 (July 1981).
63. A. F. Rañada (ed.), *Nonlinear Problems in Theoretical Physics*, Lecture Notes in Physics **98**, Springer-Verlag, Berlin (1979).
64. V. E. Zakharov and S. V. Manakov, Soliton theory, *Soviet Sci. Rev.* **A1**, 133-190 (1979).
65. C. Bardos and D. Bessis (eds.), *Bifurcation Phenomena in Mathematical Physics and Related Topics*, Reidel, Dordrecht (1980).
66. M. Boiti, F. Pempinelli, and G. Soliani (eds.), *Nonlinear Evolution Equations and Dynamical Systems*, Proceedings of a meeting in Lecce, June 1979. Lecture Notes in Physics **120**, Springer, Berlin (1980).

67. G. L. Lamb, Jr., *Elements of Soliton Theory*, Wiley, New York (1980).
68. V. E. Zakharov, S. V. Manakov, S. P. Novikov, and L. P. Pitaievski, *Theory of Solitons. The Inverse Problem Method*, Nauka, Moscow (1980) (in Russian) [English edition: S. Novikov, S. V. Manakov, L. P. Pitaevskii, and V. E. Zakharov, *Theory of Solitons*, Consultants Bureau, New York (1984)].
69. M. J. Ablowitz and H. Segur, *Solitons and the Inverse Scattering Transform*, SIAM, Philadelphia (1981).
70. G. Eilenberger, *Solitons. Mathematical Method for Physicists*, Springer Series in Solid-State Science **19**, Springer-Verlag, Berlin (1981).
71. W. Eckhaus and A. Van Harten, *The Inverse Scattering Transformation and the Theory of Solitons. An Introduction*, North-Holland Mathematics Studies **50**, North-Holland, Amsterdam (1981).
72. R. K. Dodd, J. C. Eilbeck, J. D. Gibbon, and H. C. Morris, *Solitons and Nonlinear Waves*, Academic Press, New York (1982).
73. F. Calogero, Generalized wronskian relations: a novel approach to Bargmann-equivalent and phase-equivalent potentials, in: *Studies in Mathematical Physics* (*Essays in honor of Valentine Bargmann*) (E. H. Lieb, B. Simon, and A. S. Wightman, eds.), Princeton University Press, Princeton, NJ (1976).
74. F. Calogero, Generalized wronskian relations, one-dimensional Schrödinger equation and nonlinear partial differential equations solvable by the inverse-scattering method, *Nuovo Cimento* **31B**, 229–249 (1976).
75. F. Calogero, Bäcklund transformations and functional relation for solutions of nonlinear partial differential equations solvable via the inverse-scattering method, *Lett. Nuovo Cimento* **14**, 537–543 (1975).
76. F. Calogero and A. Degasperis, Transformations between solutions of different nonlinear evolution equations solvable via the same inverse spectral transform, generalized resolvent formulas and nonlinear operator identities, *Lett. Nuovo Cimento* **16**, 181–186 (1976).
77. F. Calogero and A. Degasperis, Nonlinear evolution equations solvable by the inverse spectral transform associated with the multichannel Schrödinger problem, and properties of their solutions, *Lett. Nuovo Cimento* **15**, 65–69 (1976).
78. F. Calogero and A. Degasperis, Coupled nonlinear evolution equations solvable via the inverse spectral transform and solitons that come back: the boomeron, *Lett. Nuovo Cimento* **16**, 425–433 (1976).
79. F. Calogero and A. Degasperis, Bäcklund transformations, nonlinear super-position principle, multisoliton solutions and conserved quantities for the "boomeron" nonlinear evolution equation, *Lett. Nuovo Cimento* **16**, 434–438 (1976).
80. J. C. Eilbeck, Boomerons. A computer-produced film, Mathematics Dept., Heriot-Watt University, Edinburgh (1977).
81. J. C. Eilbeck, Zoomerons. A computer-produced film, Mathematics Dept., Heriot-Watt University, Edinburgh (1977).
82. K. M. Case and S. C. Chiu, Some remarks on the wronskian technique and the inverse scattering transform, *J. Math. Phys.* **18**, 2044–2052 (1977).
83. F. Calogero and A. Degasperis, Special solutions of coupled nonlinear evolution equations with bumps that behave as interacting particles, *Lett. Nuovo Cimento* **19**, 525–533 (1977).
84. F. Calogero and A. Degasperis, Exact solution via the spectral transform of a nonlinear evolution equation with linearly x-dependent coefficients, *Lett. Nuovo Cimento* **22**, 138–141 (1978).
85. F. Calogero and A. Degasperis, Exact solution via the spectral transform of a generalization with linearly x-dependent coefficients of the modified Korteweg–de Vries equation, *Lett. Nuovo Cimento* **22**, 270–273 (1978).
86. F. Calogero and A. Degasperis, Exact solution via the spectral transform of a generalization with linearly x-dependent coefficients of the nonlinear Schrödinger equation, *Lett. Nuovo Cimento* **22**, 420–424 (1978).

87. F. Calogero and A. Degasperis, Conservation laws for classes of nonlinear evolution equations solvable by the spectral transform, *Commun. Math. Phys.* **63**, 155-176 (1978).
88. F. Calogero and A. Degasperis, Inverse spectral problem for the one-dimensional Schrödinger equation with an additional linear potential, *Lett. Nuovo Cimento* **23**, 143-149 (1978).
89. F. Calogero and A. Degasperis, Conservation laws for a nonlinear evolution equation that includes as a special case the cylindrical KdV equation, *Lett. Nuovo Cimento* **23**, 155-160 (1978).
90. F. Calogero, Spectral transform and nonlinear evolution equations, in: *Nonlinear Problems in Theoretical Physics* (A. F. Reñada, ed.), pp. 29-34, Lecture Notes in Physics **98**, Springer-Verlag, Berlin (1979).
91. F. Calogero, M. A. Olshanetsky, and A. M. Perelomov, Rational solutions of the KdV equation with damping, *Lett. Nuovo Cimento* **24**, 97-100 (1979).
92. M. Bruschi, D. Levi, and O. Ragnisco, Evolution equations associated to the triangular-matrix Schrödinger problem solvable by the inverse spectral transform, *Nuovo Cimento* **45A**, 225-237 (1978).
93. M. Bruschi, D. Levi, and O. Ragnisco, Nonlinear evolution equations solvable by the inverse spectral transform associated to the matrix Schrödinger equation of rank 4, *Nuovo Cimento* **43B**, 251-270 (1978).
94. F. Calogero and A. Degasperis, Nonlinear evolution equations solvable by the inverse spectral transform associated with the matrix Schrödinger equation, in: *Solitons* (R. K. Bullough and P. J. Caudrey, eds.), pp. 301-323, Topics in Current Physics **17**, Springer-Verlag, Berlin (1980).
95. F. Calogero, Motion of poles and zeros of special solutions of nonlinear and linear partial differential equations, and related "solvable" many-body problems, *Nuovo Cimento* **43B**, 177-241 (1978).
96. S. C. Chiu and J. F. Ladik, Generating exactly soluble nonlinear discrete evolution equations by a generalized Wronskian technique, *J. Math. Phys.* **18**, 690-700 (1977).
97. D. Levi, The spectral transform as a tool for solving nonlinear discrete evolution equations, in: *Nonlinear Problems in Theoretical Physics* (A. F. Rañada, ed.), pp. 91-106, Lecture Notes in Physics **98**, Springer-Verlag, Berlin (1979).
98. D. Levi and O. Ragnisco, Extension of the spectral transform method for solving nonlinear differential-difference equations, *Lett. Nuovo Cimento* **22**, 691-696 (1978).
99. D. Levi and O. Ragnisco, Nonlinear differential-difference equations with n-dependent coefficients. I, *J. Phys. A* **12**, L157-L162 (1979).
100. D. Levi and O. Ragnisco, Nonlinear differential-difference equations with n-dependent coefficients. II, *J. Phys. A* **12**, L163-L167 (1979).
101. F. Kako and N. Mugibayashi, Complete integrability of general nonlinear differential-difference equations solvable by the inverse method. II, *Prog. Theor. Phys.* **61**, 776-790 (1979).
102. P. M. Santini, Asymptotic behaviour (in t) of solutions of the boomeron equation, *Nuovo Cimento* **47B**, 228-243 (1978).
103. P. M. Santini, Asymptotic behavior (in t) of solutions of the cylindrical KdV equation. I, *Nuovo Cimento* **54A**, 241-258 (1979).
104. L. Pilloni, Extension of the spectral transform method for solving nonlinear evolution equations, and related conservation laws, *Nuovo Cimento* **56B**, 87-109 (1980).
105. V. S. Gerdjikov and E. Kh. Khristov, On the evolution equations solvable through the inverse scattering method. I: Spectral Theory, *Bulg. J. Phys.* **7**, 28-41 (1980) (in Bulgarian).
106. V. S. Gerdjikov and E. Kh. Khristov, On the evolution equations solvable through the inverse scattering method. II: Hamiltonian structure and Bäcklund transformations, *Bulg. J. Phys.* **7**, 119-133 (1980) (in Bulgarian).
107. F. Calogero, Nonlinear evolution equations solvable by the spectral transform: some recent results, in: *Nonlinear Evolution Equations and Dynamical Systems* (M. Boiti, F. Pempinelli, and G. Soliani, eds.), pp. 1-14, Proceedings of a meeting in Lecce, June 1979, Lecture Notes in Physics **120**, Springer-Verlag, Berlin (1980).

108. F. Calogero, Solvable many-body problems and related mathematical findings (and conjectures), in: *Bifurcation Phenomena in Mathematical Physics and Related Topics* (C. Bardos and D. Bessis, eds.), pp. 371-384, Reidel, Dordrecht (1980).
109. F. Calogero, Spectral transform and solitons: an introduction to a novel technique to solve (certain classes of) nonlinear evolution equations, in: *Fundamental Problems in Statistical Mechanics V.* (E. G. D. Cohen, ed.), pp. 143-150, Proceedings of the 1980 Enschede Summer School, North-Holland, Amsterdam (1980).
110. F. Calogero and A. Degasperis, Conserved quantities for generalized KdV equations, *Lett. Nuovo Cimento* **28**, 12-14 (1980).
111. M. Bruschi, D. Levi, and O. Ragnisco, Discrete version of the modified Korteweg-de Vries equation with x-dependent coefficients, *Nuovo Cimento* **48A**, 213-226 (1978).
112. M. Bruschi, D. Levi, and O. Ragnisco, Discrete version of the nonlinear Schrödinger equation with x-dependent coefficients, *Nuovo Cimento* **53A**, 21-30 (1979).
113. A. Degasperis, Reduction technique for matrix nonlinear evolution equations, in: *Nonlinear Evolution Equations and Dynamical Systems* (M. Boiti, F. Pempinelli, and G. Soliani, eds.), pp. 16-34, Proceedings of a meeting in Lecce, June 1979, Lecture Notes in Physics **120**, Springer-Verlag, Berlin (1980).
114. A Degasperis, Solutions of the Korteweg-de Vries equation and their spectral transform, in: *Problèmes Inverses, Evolution Non Linéaire* (P. C. Sabatier, ed.), pp. 189-222, CNRS, Paris (1980).
115. M. Bruschi, D. Levi, and O. Ragnisco, Toda lattice and generalized Wronskian technique, *J. Phys. A* **13**, 2531-2533 (1980).
116. M. Bruschi, S. V. Manakov, O. Ragnisco, and D. Levi, The non-abelian Toda lattice (discrete analogue of the matrix Schrödinger spectral problem), *J. Math. Phys.* **21**, 2749-2753 (1980).
117. M. Bruschi and O. Ragnisco, Existence of a Lax pair for any member of the class of nonlinear evolution equations associated to the matrix Schrödinger spectral problem, *Lett. Nuovo Cimento* **29**, 321-326 (1980).
118. M. Bruschi and O. Ragnisco, Extension of the Lax method to solve the class of nonlinear evolution equations with x-dependent coefficients associated to the matrix Schrödinger spectral problem, *Lett. Nuovo Cimento* **29**, 327-330 (1980).
119. M. Bruschi and O. Ragnisco, Bäcklund transformations and Lax technique, *Lett. Nuovo Cimento* **29**, 331-334 (1980).
120. A. Degasperis, The spectral transform method to integrate nonlinear dynamical systems, in: *Proceedings of the 3rd International School on Modern Trends in Solid State Theory* (U. Lindner, ed.), pp. 5-27, Reinhardsbrunn, November 1980, Karl-Marx-Universität, Leipzig (1980).
121. A Degasperis, M. A. Olshanetsky, and A. M. Perelomov, Group-theoretical approach to a class of Lax equations, including those solvable by the spectral transform, *Nuovo Cimento* **59A**, 245-262 (1980).
122. D. Levi and R. Benguria, Bäcklund transformations and nonlinear differential-difference equations, *Proc. Natl. Acad. Sci. USA* **77**, 5025-5027 (1980).
123. D. Levi, M. A. Olshanetsky, A. M. Perelomov, and O. Ragnisco, Group-theoretical approach to nonlinear evolution equations of Lax type. III: The Boussinesq equation, *Phys. Lett.* **77A**, 307-311 (1980).
124. D. Levi, O. Ragnisco, and M. Bruschi, Extension of the generalized Zakharov-Shabat inverse method for solving differential-difference and difference-difference equations, *Nuovo Cimento* **58A**, 56-66 (1980).
125. L. Martina and P. M. Santini, Propagation of ion-acoustic waves in cold inhomogeneous plasmas, *Lett. Nuovo Cimento* **29**, 513-516 (1980).
126. S. V. Manakov, P. M. Santini, and L. A. Takhtajan, Asymptotic behaviour of the solutions of the Kadomtsev-Pyatviashvili equation (two-dimensional Korteweg-de Vries equation), *Phys. Lett.* **75A**, 451-454 (1980).
127. P. M. Santini, Asymptotic behaviour (in t) of solutions of the cylindrical KdV equation. II, *Nuovo Cimento* **57A**, 387-396 (1980).

128. M. Bruschi and O. Ragnisco, Nonlinear differential-difference equations, associated Bäcklund transformations and Lax technique, *J. Phys. A* **14**, 1075-1081 (1981).
129. M. Bruschi and O. Ragnisco, Nonlinear differential-difference matrix equations with n-dependent coefficients, *Lett. Nuovo Cimento* **31**, 492-496 (1981).
130. M. Bruschi, O. Ragnisco, and D. Levi, Evolution equations associated with the discrete analogue of the matrix Schrödinger spectral problem solvable by the inverse spectral transform, *J. Math. Phys.* **22**, 2463-2471 (1981).
131. D. Levi, Nonlinear differential-difference equations as Bäcklund transformations, *J. Phys. A* **14**, 1083-1098 (1981).
132. D. Levi, L. Pilloni, and P. M. Santini, Integrable three-dimensional lattices, *J. Phys. A* **14**, 1567-1575 (1981).
133. D. Levi, L. Pilloni, and P. M. Santini, Bäcklund transformations for (2+1)-dimensional integrable systems, *Phys. Lett.* **A81**, 419-423 (1981).
134. O. Ragnisco, Conservation laws for the whole class of nonlinear evolution equations associated to the matrix Schrödinger spectral problem, *Lett. Nuovo Cimento* **31**, 651-656 (1981).
135. P. M. Santini, On the evolution of two-dimensional packets of water waves over an uneven bottom, *Lett. Nuovo Cimento* **30**, 236-240 (1981).
136. M. Bruschi, D. Levi, M. A. Olshanetsky, A. M. Perelomov, and O. Ragnisco, The quantum Toda lattice, *Phys. Lett.* **88A**, 7-12 (1982).
137. M. Bruschi, D. Levi, and O. Ragnisco, The chiral field hierarchy, *Phys. Lett.* **88A**, 379-382, (1982).
138. M. Bruschi, D. Levi, and O. Ragnisco, Nonlinear partial differential equations and Bäcklund transformations related to the four-dimensional self-dual Yang-Mills equations, *Lett. Nuovo Cimento* **33**, 263-266 (1982).
139. M. Bruschi, D. Levi, and O. Ragnisco, The discrete chiral field hierarchy, *Lett. Nuovo Cimento* **33**, 284-288 (1982).
140. M. Bruschi and O. Ragnisco, Nonlinear evolution equations associated to the 3rd order scalar differential operator, Rome preprint, No. 254 (1981).
141. A. Degasperis, On the conservation laws associated with Lax equations, *Lett. Nuovo Cimento* **33**, 425-432 (1982).
142. D. Levi and O. Ragnisco, Bäcklund transformations for chiral field equations, *Phys. Lett.* **87A**, 381-384 (1982).
143. D. Levi, O. Ragnisco and A. Sym, Bäcklund transformation vs. the dressing method, *Lett. Nuovo Cimento* **33**, 401-406 (1982).
144. H. Flaschka and A. C. Newell, Integrable systems of nonlinear evolution equations, in: *Dynamical Systems, Theory and Applications* (J. Moser, ed.), pp. 355-440, Lecture Notes in Physics **38**, Springer-Verlag, Berlin (1975).
145. I. M. Gel'fand and L. A. Dikii, Asymptotic behaviour of the resolvent of Sturm-Liouville equations and the algebra of the Korteweg-de Vries equations, *Russian Math. Surveys* **30**(5), 77-113 (1975) [*Usp. Mat. Nauk.* **30**(5), 67-100 (1975)].
146. I. M. Gel'fand and L. A. Dikii, Fractional powers of operators and Hamiltonian systems, *Funt. Anal. Appl.* **10**, 259-273 (1976) [*Funk. Anal. Priloz.* **10**(4), 13-29 (1976)].
147. I. M. Gel'fand and L. A. Dikii, The resolvent and Hamiltonian systems, *Funct. Anal. Appl.* **11**, 93-105 (1977) [*Funk. Anal. Priloz.* **11**(2), 11-27 (1977)].
148. P. D. Laz, A Hamiltonian approach to the KdV and other equations, in: *Nonlinear Evolution Equations* (M. G. Crandall, ed.), pp. 207-224, Proceedings of a Symposium at the Mathematics Research Center, University of Wisconsin-Madison, Oct. 17-19, 1977, Academic Press, New York (1978).
149. F. A. Berezin, Models of Gross-Neveu type are quantization of a classical mechanics with nonlinear phase space, *Commun. Math. Phys.* **63**, 131-153 (1978).
150. I. M. Gel'fand and L. A. Dikii, A family of Hamiltonian structures connected with integrable, nonlinear differential equations, Preprint Inst. Prikl. Mat. Akad. Nauk SSSR. No. 136 (1978).

151. I. M. Gel'fand and L. A. Dikii, The calculus of jets and nonlinear Hamiltonian systems, *Funct. Anal. Appl.* **12**, 81-94 (1978) [*Funk. Anal. Priloz.* **12**(2), 8-23 (1978)].
152. I. M. Gel'fand and L. A. Dikii, Integrable nonlinear equations and the Liouville theorem, *Funct. Anal. Appl.* **13**, 6-15 (1979) [*Funk. Anal. Priloz.* **13**(1), 8-20 (1979)].
153. I. M. Gel'fand and I. Ya. Dorfman, Hamiltonian operators and algebraic structures related to them, *Funct. Anal. Appl.* **13**, 248-262 (1979) [*Funk. Anal. Priloz.* **13**, 13-30 (1979)].
154. M. Adler, Completely integrable systems and symplectic actions, *J. Math. Phys.* **20**, 60-67 (1979).
155. M. Adler, On a trace functional for formal pseudodifferential operators and symplectic structure of the Korteweg-de Vries type equations, *Invent. Math.* **50**, 219-248 (1979).
156. F. A. Berezin and A. M. Perelomov, Group-theoretical interpretation of the Korteweg-de Vries type equations, *Commun. Math. Phys.* **74**, 129-140 (1980).
157. L. Martinez Alonso, Gel'fand-Dikii method and nonlinear equations associated to Schrödinger operators with energy-dependent potentials, *Lett. Math. Phys.* **4**, 215-222 (1980).
158. M. J. Ablowitz, Nonlinear evolution equations—continuous and discrete, *SIAM Rev.* **19**, 663-684 (1977).
159. M. J. Ablowitz, The inverse scattering transform—continuous and discrete, and its relationship with Painlevé transcendents, in: *Nonlinear Evolution Equations Solvable by the Spectral Transform* (F. Calogero, ed.), pp. 9-32, Research Notes in Mathematics **36**, Pitman, London (1978).
160. M. Toda, Vibration of a chain with nonlinear interaction, *J. Phys. Soc. Jpn.* **22**, 431-436 (1967).
161. F. Calogero, Solution of the one-dimensional *N*-body problems with quadratic and/or inversely quadratic pair potentials, *J. Math. Phys.* **12**, 419-436 (1971).
162. M. Hénon, Integrals of the Toda lattice, *Phys. Rev. B* **9**, 1921-1923 (1974).
163. H. Flaschka, The Toda lattice. I: Existence of integrals, *Phys. Rev. B* **9**, 1924-1925 (1974).
164. H. Flaschka, On the Toda lattice. II: Inverse scattering solution, *Prog. Theor. Phys.* **51**, 703-716 (1974).
165. S. V. Manakov, Complete integrability and stochastization of discrete dynamical systems, *Soviet Phys. JETP* **40**, 269-274 (1975) [*Zh. Eksp. Teor. Fiz.* **67**, 543-555 (1974)].
166. J. Moser, Three integrable Hamiltonian systems connected with isospectral deformations, *Adv. in Math.* **16**, 197-220 (1975).
167. M. Toda, Studies of a nonlinear lattice, *Phys. Rep.* **18C**, 1-124 (1975).
168. F. Calogero, Integrable many-body problems, in: *Nonlinear Equation in Physics and Mathematics* (A. O. Barut, ed.), pp. 3-53, Reidel, Dordrecht (1978).
169. F. Calogero, Integrable many-body problems and related mathematical results, in: *Fundamental Problems in Statistical Mechanics V.* (E. G. D. Cohen, ed.), pp. 151-164 Proceedings of the 1980 Enschede Summer School, North-Holland, Amsterdam (1980).
170. M. Toda, On a nonlinear lattice (the Toda lattice), in: *Solitons* (R. K. Bullough and P. J. Caudrey, eds.), pp. 143-155, Topics in Current Physics, Springer-Verlag, Berlin (1980).
171. M. A. Olshanetsky and A. M. Perelomov, Classical integrable finite-dimensional systems related to Lie algebras, *Phys. Rep.* **71**, 313-400 (1981).
172. M. Toda, *Theory of Nonlinear Lattices*, Springer Series in Solid-State Sciences **20**, Springer-Verlag, Berlin (1981).
173. P. D. Lax, Periodic solutions of the KdV equation, *Comm. Pure Appl. Math.* **28**, 141-188 (1975).
174. H. P. McKean and P. van Moerbeke, The spectrum of Hill's equation, *Invent. Math.* **30**, 217-274 (1975).
175. H. P. McKean and E. Trubowitz, Hill's operator and hyperelliptic function theory in the presence of infinitely many branch points, *Comm. Pure Appl. Math.* **29**, 143-226 (1976).
176. I. M. Krichever, Methods of algebraic geometry in the theory of nonlinear equations, *Russian Math. Surveys* **32**(6), 185-213 (1977) [*Usp. Math. Nauk* **32**(6), 183-208 (1977)].
177. F. B. Estabrook and H. D. Wahlquist, Prolongation structures, connection theory and Bäcklund transformation, in: *Nonlinear Evolution Equations Solvable by the Spectral Trans-*

form (F. Calogero, ed.), pp. 64-83, Research Notes in Mathematics **26**, Pitman, London (1978).
178. F. A. E. Pirani, D. C. Robinson, and W. F. Shadwick, Local jet bundle formulation of Bäcklund transformations, *Mathematical Physics Studies* **1**, 1-132 (1979).
179. D. J. Kaup, The Estabrook-Wahlquist method with examples of applications, *Physica* **1D**, 391-411 (1980).
180. A. C. Newell, Near integrable systems, nonlinear tunnelling and solitons in slowly changing media, in: *Nonlinear Evolution Equations Solvable by the Spectral Transform* (F. Calogero, ed.), pp. 127-179, Research Notes in Mathematics **26**, Pitman, London (1978).
181. J. P. Keener and D. W. McLaughlin, Solitons under perturbations, *Phys. Rev. A* **16**, 777-790 (1977).
182. D. W. McLaughlin and A. C. Scott, Soliton perturbation theory, in: *Nonlinear Evolution Equations Solvable by the Spectral Transform* (F. Calogero, ed.), pp. 225-243, Research Notes in Mathematics **26**, Pitman, London (1978).
183. V. I. Karpman and E. M. Maslov, Structure of tails produced under the action of perturbations on solitons, *Sov. Phys. JETP* **48**, 252-259 (1978) [*Zh. Eksp. Teor. Fiz.* **75**, 504-517 (1978)].
184. D. J. Kaup and A. C. Newell, Solitons as particles, oscillators, and in slowly-changing media: a singular perturbation theory, *Proc. R. Soc. London, Ser. A* **361**, 413-446 (1978).
185. Yu. Kodama and M. J. Ablowitz, Perturbations of solitons and solitary waves, *Stud. Appl. Math.* **64**, 225-245 (1981).
186. E. Nagel and J. R. Newman, *Gödel's Proof*, New York University Press, New York (1958).
187. M. D. Kruskal, Nonlinear wave equations, in: *Dynamical Systems, Theory and Applications* (J. Moser, ed.), pp. 310-354, Lecture Notes in Physics **38**, Springer-Verlag, Berlin (1975).
188. R. S. Johnson, On the inverse scattering transform, the cylindrical Korteweg-de Vries equation and similarity solutions, *Phys. Lett.* **72A**, 197-199 (1979).
189. S. Tanaka, Analogue of Fourier's method for Korteweg-de Vries equation, *Proc. Jpn. Acad.* **48**, 647-650 (1972).
190. L. D. Faddeev, On the relation between S-matrix and potential for the one-dimensional Schrödinger operator, *Dokl. Akad. Nauk SSSR* **121**, 63-66 (1958) (in Russian).
191. L. D. Faddeev, The inverse problem in the quantum theory of scattering, *J. Math. Phys.* **4**, 72-104 (1963) [*Uspekhi Matem. Nauk* **14**, 57 (1959)].
192. L. D. Faddeev, On the relation between the S-matrix and potential for the one-dimensional Schrödinger operator, *Trudi Mat. Inst. Steklov* **73**, 314-336 (1964) (in Russian).
193. L. D. Faddeev, The inverse problem of quantum scattering theory. II, *Soviet J. Math.* **5** (1976) [*Sovrem. Probl. Mat.* **3**, VINITI, Moscow, 93-181 (1974)].
194. R. G. Newton, *Scattering Theory of Waves and Particles*, Chapter 20, McGraw-Hill, New York (1966).
195. R. Hirota, Direct methods in soliton theory, in: *Solitons* (R. K. Bullough and P. J. Caudrey, eds.), pp. 157-176, Topics in Current Physics **17**, Springer, Berlin (1980).
196. M. D. Kruskal, The Korteweg-de Vries equation and related evolution equations, in: *Nonlinear Wave Motion* (A. C. Newell, ed.), pp. 61-83, Lect. Appl. Math. **15**, American Mathematical Society, Providence, RI (1974).
197. R. M. Miura, An introduction to solitons and the inverse scattering method via the Korteweg-de Vries equation, in: *Solitons in Action* (K. Lonngren and A. Scott), pp. 1-19, Academic Press, New York (1978).
198. S. P. Novikov, The periodic problem of the Korteweg-de Vries equation. I, *Funct. Anal. Appl.* **8**, 236-246 (1974) [*Funk. Anal. Priloz.* **8**, 54-66 (1974)].
199. R. M. Miura, The Korteweg-de Vries equation: a model equation for nonlinear dispersive waves, in: *Nonlinear Waves* (S. Leibovich and A. R. Seebass, eds.), pp. 212-234, Cornell University Press, Ithaca (1974).
200. H. Segur, Solitons as approximate descriptions of physical phenomena, *Rocky Mountain J. Math.* **8**, 15-24 (1978).
201. A. R. Osborne and T. L. Burch, Internal solitons in the Andaman sea, *Science* **208**, 451-460 (1980).

202. J. W. Miles, The Korteweg-de Vries equation: a historical essay, *J. Fluid Mech.* **106**, 131-147 (1981).
203. T. B. Benjamin, J. L. Bona, and J. J. Mahony, Model equations for long waves in nonlinear dispersive systems, *Phil. Trans. R. Soc. London, Ser. A* **272**, 47-78 (1972).
204. D. H. Peregrine, Calculations of the development of an undular bore, *J. Fluid Mech.* **25**, 321-330 (1966).
205. T. B. Benjamin, Lectures on nonlinear wave motion, in: *Nonlinear Wave Motion* (A. C. Newell, ed.), pp. 3-47, Lect. Appl. Math. **15**, American Mathematical Society, Providence, RI (1974).
206. C. J. Knickerbocker and A. C. Newell, Internal solitary waves near a turning point, *Phys. Lett.* **75A**, 326-330 (1980).
207. S. Maxon, Cylindrical and spherical solitons, *Rocky Mountain J. Math.* **8**, 269-281 (1978).
208. M. J. Ablowitz, M. Kruskal, and H. Segur, A note on Miura's transformation, *J. Math. Phys.* **20**, 999-1003 (1979).
209. M. Wadati, The exact solution of the modified Korteweg-de Vries equation, *J. Phys. Soc. Jpn.* **32**, 1681 (1972).
210. R. M. Miura, The Korteweg-de Vries equation: a survey of results, *SIAM Rev.* **18**, 412-459 (1976).
211. M. Wadati, K. Konno, and Y. H. Ichikawa, New integrable nonlinear evolution equations, *J. Phys. Soc. Jpn.* **47**, 1698-1700 (1979).
212. T. Shimizu and M. Wadati, A new integrable nonlinear evolution equation, *Prog. Theor. Phys.* **63**, 808-820 (1980).
213. T. Benjamin, Internal waves of permanent form in fluids of great depth, *J. Fluid Mech.* **29**, 559-592 (1967).
214. H. Ono, Algebraic solitary waves in stratified fluids, *J. Phys. Soc. Jpn.* **39**, 1082-1091 (1975).
215. R. I. Joseph, Solitary waves in a finite depth fluid, *J. Phys. A* **10**, L225-L227 (1977).
216. R. I. Joseph and R. Egri, Multisoliton solutions in a finite depth fluid, *J. Phys. A* **11**, L97-L102 (1978).
217. H.-H. Chen, Y. C. Lee, and N. R. Pereira, Algebraic internal wave solitons and the integrable Calogero-Moser-Sutherland *N*-body problem, *Phys. Fluids* **22**, 187-188 (1979).
218. K. M. Case, The *N*-soliton solution of the Benjamin-Ono equation, *Proc. Natl. Acad. Sci. USA* **75**, 3562-3563 (1978).
219. T. L. Bock and M. D. Kruskal, A two-parameter Miura transformation of the Benjamin-Ono equation, *Phys. Lett.* **74A**, 173-176 (1979).
220. K. M. Case, Properties of the Benjamin-Ono equation, *J. Math. Phys.* **20**, 972-977 (1979).
221. K. M. Case, Benjamin-Ono-related equations and their solutions, *Proc. Natl. Acad. Sci. USA* **76**, 1-3 (1979).
222. K. M. Case, Meromorphic solutions of the Benjamin-Ono equation, *Physica* **96A**, 173-182 (1979).
223. Y. Matsuno, Exact multisoliton solution of the Benjamin-Ono equation, *J. Phys. A* **12**, 619-621 (1979).
224. A. Nakamura, Bäcklund transform and conservation laws of the Benjamin-Ono equation, *J. Phys. Soc. Jpn.* **47**, 1335-1340 (1979).
225. J. Satsuma and Y. Ishimori, Periodic wave and rational soliton solutions of the Benjamin-Ono equation, *J. Phys. Soc. Jpn.* **46**, 681-687 (1979).
226. H.-H. Chen and D. J. Kaup, Conservation laws of the Benjamin-Ono equation, *J. Math. Phys.* **21**, 19-20 (1980).
227. Y. Matsuno, Interaction of the Benjamin-Ono solitons, *J. Phys. A* **13**, 1519-1536 (1980).
228. A. Nakamura, *N*-periodic wave and *N*-soliton solutions of the modified Benjamin-Ono equation, *J. Phys. Soc. Jpn.* **47**, 2045-2046 (1979).
229. Y. Matsuno, *N*-soliton and *N*-periodic wave solutions of the higher order Benjamin-Ono equation, *J. Phys. Soc. Jpn.* **47**, 1745-1746 (1979).
230. Y. Matsuno, Solutions of the higher order Benjamin-Ono equation, *J. Phys. Soc. Jpn.* **48**, 1024-1028 (1980).

231. H.-H. Chen and Y. C. Lee, Internal wave solitons of fluids with finite depth, *Phys. Rev. Lett.* **43**, 264–266 (1979).
232. Y. Matsuno, Exact multi-soliton solution for nonlinear waves in a stratified fluid of finite depth, *Phys. Lett.* **74A**, 233–235 (1979).
233. J. Satsuma, M. J. Ablowitz, and Y. Kodama, On an internal wave equation describing a stratified fluid with finite depth, *Phys. Lett.* **73A**, 283–286 (1979).
234. F. S. Henyey, Finite-depth and infinite-depth internal-wave solitons, *Phys. Rev. A* **21**, 1054–1056 (1980).
235. J. Gibbons and B. Kupershmidt, A linear scattering problem for the finite depth equation, *Phys. Lett.* **79A**, 31–32 (1980).
236. Y. Matsuno, N-soliton solution of the higher order wave equation for a fluid of finite depth, *J. Phys. Soc. Jpn.* **48**, 663–668 (1980).
237. A. Nakamura, Exact N-soliton solution of the modified finite depth fluid equation, *J. Phys. Soc. Jpn.* **47**, 2043–2044 (1979).
238. M. J. Boussinesq, Théorie de l'intumescence liquide appellé onde solitaire ou de translation se propageant dans un canal rectangulaire, *C. R. Acad. Sci. Paris* **72**, 755–759 (1871).
239. M. J. Boussinesq, Théorie des ondes et des remous qui se propagent le long d'un canal horizontal, en communiquant au liquide contenu dans ce canal des vitesses sensiblement pareilles de la surface au fond, *J. Math. Pures Appl.* **7**, 55–108 (1872).
240. D. J. Kaup, On the inverse scattering problem for cubic eigenvalue problems of the class $\psi_{xxx} + 6Q\psi_x + 6R\psi = \lambda\psi$, *Stud. Appl. Math.* **62**, 189-216 (1980).
241. P. J. Caudrey, The inverse problem for the third order equation $u_{xxx} + q(x)u_x + r(x)u = -i\zeta^3 u$, *Phys. Lett.* **79A**, 264-268 (1980).
242. V. E. Zakharov, On stochastization of one-dimensional chains of nonlinear oscillators, *Sov. Phys. JETP* **38**, 108–110 (1974) [*Zh. Eksp. Teor. Fiz.* **65**, 219–225 (1973)].
243. R. Hirota, Exact N-soliton solutions of the wave equation of long waves in shallow-water and in nonlinear lattices, *J. Math. Phys.* **14**, 810–814 (1973).
244. H.-H. Chen, Relation between Bäcklund transformations and inverse scattering problems, in: *Bäcklund Transformations* (R. M. Miura, ed.), pp. 241–252, Lecture Notes in Mathematics **515**, Springer-Verlag, Berlin (1976).
245. H. P. McKean, Boussinesq's equation as a Hamiltonian system, *Topics in Functional Analysis*, Advances in Mathematics Supplementary Studies, Vol. 3, Academic Press, New York (1978) 217-226.
246. M. Boiti and F. Pempinelli, Similarity solutions and Bäcklund transformations of the Boussinesq equation, *Nuovo Cimento* **56B**, 148–156 (1980).
247. P. J. Caudrey, The inverse problem for a general $N \times N$ spectral equation, *Physica* **6D**, 51–66 (1982).
248. B. B. Kadomtsev and V. I. Petviashvili, On the stability of solitary waves in weakly dispersing media, *Sov. Phys. Dokl.* **15**, 539-541 (1970) [*Dokl. Akad. Nauk SSSR* **192**, 753-756 (1970)].
249. V. S. Dryuma, Analytic solution of the two-dimensional Korteweg-de Vries (KdV) equation, *Soviet JETP Lett.* **19**, 387-388 (1974) [*Zh. ETF Pis. Red.* **19**, 753-755 (1974)].
250. S. V. Manakov, V. E. Zakharov, L. A. Bordag, A. R. Its, and V. B. Matveev, Two-dimensional solitons of the Kadomtsev-Petviashvili equation and their interaction, *Phys. Lett.* **63A**, 205-206 (1977).
251. J. D. Gibbon, N. C. Freeman, and R. S. Johnson, Correspondence between the classical $\lambda\phi^4$, double and single sine-Gordon equations for three-dimensional solitons, *Phys. Lett.* **65A**, 380-382 (1978).
252. V. E. Zakharov, The inverse scattering method, in: *Solitons* (R. K. Bullough and P. J. Caudrey, eds.), pp. 243-285, Topics in Current Physics **17**, Springer-Verlag, Berlin (1980).
253. S. V. Manakov, The inverse scattering transform for the time-dependent Schrödinger equation and Kadomtsev-Petviashvili equation, in: *Soliton Theory* (S. V. Manakov and V. E. Zakharov, eds.), pp. 420-427, Proceedings of the Soviet-American Symposium on Soliton Theory, Kiev, Sept. 1979, *Physica* **3D**, 1, 2 (July 1981).
254. J. M. Burgers, *The Nonlinear Diffusion Equation*, Reidel, Dordrecht (1974).

255. E. Hopf, The partial differential equation $u_t + uu_x = \mu u_{xx}$, *Comm. Pure Appl. Math.* **3**, 201-230 (1950).
256. J. D. Cole, On a quasilinear parabolic equation occurring in aerodynamics, *Q. Appl. Math.* **9**, 225-236 (1950).
257. A. S. Fokas and Y. C. Yortos, On the exactly solvable equation $S_t = [(\beta S + \gamma)^{-2} S_x]_x + a(\beta S + \gamma)^{-2} S_x$, occurring in two-phase flow in porous media, *SIAM J. Appl. Math.* **42**, 318-332 (1982).
258. E. Taflin, Analytic linearization, Hamiltonian formalism and infinite sequences of constants of motion for Burgers' equation, *Phys. Rev. Lett.* **47**, 1425-1428 (1981).
259. T. Taniuti and N. Yajima, Perturbation method for a nonlinear wave modulation, I, *J. Math. Phys.* **10**, 1369-1372 (1969).
260. G. L. Lamb, jr. and D. W. MacLaughlin, Aspects of soliton physics, in: *Solitons* (R. K. Bullough and P. J. Caudrey, eds.), pp. 65-106, Topics in Current Physics **17**, Springer-Verlag, Berlin (1980).
261. A. C. Newell, The inverse scattering transform, in: *Solitons* (R. K. Bullough and P. J. Caudrey, eds.), pp. 177-242, Topics in Current Physics **17**, Springer-Verlag, Berlin (1980).
262. L. A. Takhtajan, Integration of the continuous Heisenberg spin chain through the inverse scattering method, *Phys. Lett.* **64A**, 235-237 (1977).
263. M. Lakshmanan, Continuum spin system as an exactly solvable dynamical system, *Phys. Lett.* **61A**, 53-54 (1977).
264. D. J. Kaup and A. C. Newell, An exact solution for a derivative nonlinear Schrödinger equation, *J. Math. Phys.* **19**, 798-801 (1978).
265. V. S. Gerdjikov, M. I. Ivanov, and P. P. Kulish, Quadratic bundle and nonlinear equations, *Theor. Math. Phys.* **44**, 372 (1980) [*Teor. Mat. Fiz.* **44**, 342-357 (1980)].
266. H. C. Morris and R. K. Dodd, The two-component derivative nonlinear Schrödinger equation, in: *Solitons in Physics* (H. Wilhelmsson, ed.), pp. 505-508, Topical issue of Physica Scripta, Royal Swedish Academy of Sciences, Phys. Scr. **20** (1979).
267. R. Hirota, Exact envelope-soliton solutions of a nonlinear wave equation, *J. Math. Phys.* **14**, 805-809 (1973).
268. A. C. Newell, Long waves—short waves; a solvable model, *SIAM J. Appl. Math.* **35**, 650-664 (1978).
269. V. E. Zakharov and S. V. Manakov, Resonant interaction of wave packets in nonlinear media, *JETP. Lett.* **18**, 243-245 (1973).
270. V. E. Zakharov and S. V. Manakov, The theory of resonant interaction of wave packets in nonlinear media, *Soviet Phys. JETP* **42**, 842-850 (1976) [*Zh. Eksp. Teor. Fiz.* **69**, 1654-1673 (1975)].
271. D. J. Kaup, The three-wave interaction: a nondispersive phenomenon, *Stud. Appl. Math.* **55**, 9-44 (1976).
272. D. J. Kaup, Applications of the inverse scattering transform. II: The three-wave resonant interaction, *Rocky Mountain J. Math.* **8**, 283-308 (1978).
273. D. J. Kaup, A method for solving the separable initial-value problem of the full three-dimensional three-wave interaction, *Stud. Appl. Math.* **62**, 75-83 (1980).
274. D. J. Kaup, Determining the final profiles from the initial profiles for the full three-dimensional three-wave resonant interaction, *Lecture Notes in Physics* **130**, 247-252 (1980).
275. D. J. Kaup, The inverse scattering solutions for the full three-dimensional three-wave resonant interaction, *Physica* **1D**, 45-67 (1980).
276. D. J. Kaup, The solution of the general initial value problem for the full three-dimensional three-wave resonant interaction, *Physica* **3D**, 374-395 (1981).
277. H. Cornille, Solutions of the nonlinear three-wave equations in three spatial dimensions, *J. Math. Phys.* **20**, 1653-1666 (1979).
278. A. Davey and K. Stewartson, On three-dimensional packets of surface waves, *Proc. R. Soc. London, Ser. A* **338**, 101-110 (1974).

279. D. Anker and N. C. Freeman, On the soliton solutions of the Davey-Stewartson equation for long waves, *Proc. R. Soc. London, Ser. A* **360**, 529-540 (1978).
280. A. Degasperis, Solitons, boomerons, trappons, in: *Nonlinear Evolution Equations Solvable by the Spectral Transform* (F. Calogero, ed.), pp. 97-126, Research Notes in Mathematics **26**, Pitman, London (1978).
281. J. C. Eilbeck, J. D. Gibbon, P. J. Caudrey, and R. K. Bullough, Solitons in nonlinear optics. I: A more accurate description of the 2π pulse in self-induced transparency, *J. Phys. A* **6**, 1337-1347 (1973).
282. J. D. Gibbon, P. J. Caudrey, R. K. Bullough, and J. C. Eilbeck, An N-soliton solution of a nonlinear optics equation derived by a general inverse method, *Lett. Nuovo Cimento* **8**, 775-779 (1973).
283. R. K. Bullough, P. M. Jack, P. W. Kitchenside, and R. Saunders, Solitons in laser physics, in: *Solitons in Physics* (H. Wilhelmsson, ed.), pp. 364-381, Topical issue of Physica Scripta, Royal Swedish Academy of Sciences, Phys. Scr. **20** (1979).
284. P. J. Caudrey and J. C. Eilbeck, Numerical evidence for breakdown of soliton behaviour in solutions of the Maxwell-Bloch equations, *Phys. Lett.* **62A**, 65-66 (1977).
285. T. H. R. Skyrme, A nonlinear theory of strong interactions, *Proc. R. Soc. London, Ser. A* **247**, 260-278 (1958).
286. T. H. R. Skyrme, Particle states of a quantized field, *Proc. R. Soc. London, Ser. A* **262**, 237-245 (1961).
287. A. Barone, F. Esposito, C. J. Magee, and A. C. Scott, Theory and applications of the Sine-Gordon equation, *Riv. Nuovo Cimento* **1**, 227-267 (1971).
288. A. R. Bishop, Solitons and physical perturbations, in: *Solitons in Action* (K. Lonngren, ed.), pp. 61-87, Academic Press, New York (1978).
289. D. Hilbert, *Grundlagen der Geometrie*, Teubner, Leipzig (1899).
290. L. Bianchi, *Lezioni di Geometria Differenziale*, Vol. 2, 2nd edn., Section 352, Spoerri, Pisa (1903).
291. L. P. Eisenhart, *A Treatise on the Differential Geometry of Curves and Surfaces*, p. 280, Dover, New York (1960).
292. M. J. Ablowitz, D. J. Kaup, A. C. Newell, and H. Segur, Method for solving the Sine-Gordon equation, *Phys. Rev. Lett.* **30**, 1262-1264 (1973).
293. D. J. Kaup, Method for solving the Sine-Gordon equation in laboratory coordinates, *Stud. Appl. Math.* **54**, 165-179 (1975).
294. D. J. Kaup and A. C. Newell, The Goursat and Cauchy problems for the Sine-Gordon equation, *SIAM J. Appl. Math.* **34**, 37-54 (1977).
295. V. E. Zakharov, L. A. Tahtadžjan, and L. D. Faddeev, A complete description of the solution of the Sine-Gordon equation, *Soviet Phys. Dokl.* **19**, 824-826 (1975) [*Dokl. Akad. Nauk SSSR* **219**, 1334-1337 (1974)].
296. L. A. Tahtadžjan and L. D. Faddeev, Essentially nonlinear one-dimensional model of classical field theory, *Theor. Math. Phys.* **21**, 1046-1057 (1974) [*Teoret. Mat. Fiz.* **21**, 160-174 (1974)].
297. L. A. Tahtadžjan and L. D. Faddeev, The Hamiltonian system connected with the equation $u_{\xi\eta} + \sin u = 0$, *Proc. Steklov Inst. Math.* **3**, 277-289 (1979) [*Trudi Mat. Inst. Steklov* **142**, (1976)].
298. R. K. Bullough, P. J. Caudrey, and H. M. Gibbs, The double Sine-Gordon equations: a physically applicable system of equations, in: *Solitons* (R. K. Bullough and P. J. Caudrey, eds.), pp. 107-144, Topics in Current Physics **17**, Springer-Verlag, Berlin (1980).
299. R. F. Dashen, B. Hasslacher, and A. Neveu, Nonperturbative methods and extended-hadron models in field theory. II: Two-dimensional models and extended hadrons, *Phys. Rev. D* **10**, 4130-4138 (1974).
300. J. Goldstone and R. Jackiw, Quantization of nonlinear waves, *Phys. Rev. D* **11**, 1486-1498 (1975).
301. S. Aubry, A unified approach to the interpretation of displacive and order-disorder systems. II: Displacive systems, *J. Chem. Phys.* **64**, 3392-3402 (1976).

302. B. S. Getmanov, Bound states of soliton in the ϕ_2^4 field-theory model, *JETP Lett.* **24**, 291-294 (1976) [*Pis'ma Zh. Eksp. Teor. Fiz.* **24**, 323-327 (1976)].
303. J. Liouville, Sur l'equation aux differences partielles $d^2 \log \lambda / du\, dv \pm \lambda/2a^2 = 0$, *J. Math. Pures Appl.* **18**, 71-72 (1853).
304. V. A. Andreev, Application of the inverse scattering method to the equation $\sigma_{xt} = \exp(\sigma)$, *Theor. Math. Phys.* **29**, 1027-1032 (1976) [*Teor. Mat. Fiz.* **29**, 213-220 (1976)].
305. M. Chaichian and P. P. Kulish, On the method of inverse scattering problem and Bäcklund transformations for supersymmetric equations, *Phys. Lett.* **78B**, 413-416 (1978).
306. M. Tsutsumi, On solutions of Liouville's equation, *J. Math. Anal. Appl.* **76**, 116-123 (1980).
307. A. V. Mikhailov, Integrability of a two-dimensional generalization of the Toda chain, *JETP Letters* **30**(7), 414-418 (1979) [*Pis'ma Zh. Eksp. Teor. Fiz.* **30**(7), 443-448 (1979)].
308. A. P. Fordy and J. Gibbons, A class of integrable nonlinear Klein-Gordon equations in many dependent variables, *Commun. Math. Phys.* **77**, 21-30 (1980).
309. A. V. Zhiber and A. B. Shabat, Klein-Gordon equations with a non-trivial group, *Sov. Phys. Dokl.* **24**(8), 607-609 (1979) [*Dokl. Akad. Nauk SSSR* **247**, 1103-1107 (1979)].
310. B. A. Kupershmidt and G. Wilson, Conservation laws and symmetries of generalized Sine-Gordon equations, *Commun. Math. Phys.* **81**, 189-202 (1981).
311. A. V. Mikhailov, M. A. Olshanetsky, and A. M. Perelomov, Two-dimensional generalized Toda lattice, *Commun. Math. Phys.* **79**, 473-488 (1981).
312. K. Pohlmeyer, Integrable Hamiltonian systems and interactions through quadratic constraints, *Commun. Math. Phys.* **46**, 207-221 (1976).
313. F. Lund and T. Regge, Unified approach to strings and vortices with soliton solutions, *Phys. Rev. D* **14**, 1524-1535 (1976).
314. F. Lund, Solitons and geometry, in: *Nonlinear Equations in Physics and Mathematics* (A. O. Barut, ed.), pp. 143-175, Reidel, Dordrecht (1978).
315. S. Weinberg, Nonlinear realizations of chiral symmetry, *Phys. Rev.* **166**, 1568-1577 (1968).
316. S. J. Orfanidis, σ models of nonlinear evolution equations, *Phys. Rev. D* **21**, 1513-1522 (1980).
317. V. E. Zakharov and A. V. Mikhailov, Relativistically invariant two-dimensional models of field theory which are integrable by means of the inverse scattering problem method, *Sov. Phys. JETP* **47**, 1017-1027 (1978) [*Zh. Eksp. Teor. Fiz.* **74**, 1953-1973 (1978)].
318. D. Maison, On the complete integrability of the stationary, axially symmetric Einstein equations, *J. Math. Phys.* **20**, 871-877 (1979).
319. V. A. Belinsky and V. E. Zakharov, Integration of the Einstein equations by means of the inverse scattering problem technique and construction of exact soliton solutions, *Sov. Phys. JETP* **48**, 985-994 (1978) [*Zh. Eksp. Teor. Fiz.* **75**, 1955-1971 (1978)].
320. F. J. Ernst, New formulation of the axially symmetric gravitational field problem, *Phys. Rev.* **167**, 1175-1178 (1968).
321. B. K. Harrison, Bäcklund transformation for the Ernst equation of general relativity, *Phys. Rev. Lett.* **41**, 1197-1200 (1978).
322. D. Maison, Are the stationary, axially symmetric Einstein equations completely integrable?, *Phys. Rev. Lett.* **41**, 521-522 (1978).
323. V. A. Belinsky and V. E. Zakharov, Stationary gravitational solitons with axial symmetry, *Sov. Phys. JETP* **50**, 1-9 (1979) [*Zh. Eksp. Teor. Fiz.* **77**, 3-19 (1979)].
324. G. Neugebauer, Bäcklund transformations of axially symmetric stationary gravitational fields, *J. Phys. A* **12**, L67-L70 (1979).
325. C. M. Cosgrove, Relationships between the group-theoretic and soliton-theoretic techniques for generating stationary axisymmetric gravitational solutions, *J. Math. Phys.* **21**, 2417-2447 (1980).
326. B. K. Harrison, New large family of vacuum solutions of the equations of general relativity, *Phys. Rev. D* **21**, 1695-1697 (1980).
327. I. Hauser and F. J. Ernst, A homogeneous Hilbert problem for the Kinnersley-Chitre transformations, *J. Math. Phys.* **21**, 1126-1140 (1980).
328. H. C. Morris and R. K. Dodd, A two-connection and operator bundles for the Ernst equation for axially symmetric gravitational fields, *Phys. Lett.* **75A**, 20-22 (1979).

329. M. Omote and M. Wadati, Bäcklund transformations for the Ernst equation, *J. Math. Phys.* **22**, 961–964 (1981).
330. M. Omote and M. Wadati, The Bäcklund transformations and the inverse scattering method of the Ernst equation, *Prog. Theor. Phys.* **65**, 1621–1631 (1981).
331. A. A. Belavin and V. E. Zakharov, Yang–Mills equations as inverse scattering problem, *Phys. Lett.* **73B**, 53–57 (1978).
332. S. V. Manakov and V. E. Zakharov, Three-dimensional model of relativistic-invariant field theory, integrable by the inverse scattering transform, *Lett. Math. Phys.* **5**, 247–253 (1981).
333. M. F. Atiyah, V. G. Drinfeld, N. J. Hitchin, and Yu. I. Manin, Construction of instantons, *Phys. Lett.* **65A**, 185–187 (1978).
334. W. E. Thirring, A soluble relativistic field theory, *Ann. Phys.* **3**, 91–112 (1958).
335. A. V. Mikhailov, Integrability of the two-dimensional Thirring model, *JETP Lett.* **23**, 320–323 (1976) [*Pis'ma Zh. Eksp. Teor. Fiz.* **23**, 356–358 (1976)].
336. E. A. Kuznetsov and A. V. Mikhailov, On the complete integrability of the two-dimensional classical Thirring model, *Theor. Math. Phys.* **30**, 193–200 (1977) [*Teor. Mat. Fiz.* **30**, 303–314 (1977)].
337. V. E. Zakharov and A. V. Mikhailov, On the integrability of classical spinor models in two-dimensional space-time, *Commun. Math. Phys.* **74**, 21–40 (1980).
338. D. K. Campbell and A. R. Bishop, Soliton excitations in polyacetylene and relativistic field theory models. Preprint LA-UR-81-2114, Los Alamos Scientific Lab.

4

Homogeneous Isothermal Oscillations and Spatiotemporal Organization in Chemical Reactions

P. De Kepper

4.1. Introduction

After the discussion of mathematical methods to treat nonlinear problems, this chapter will focus on chemical problems, already referred to briefly in Chapter 1. Physicochemical systems exhibiting periodic phenomena, in time and/or space, are numerous.[1-3] Some of them have been known for more than a century.[4] The most spectacular include periodic dissolution of metals in acid, periodic adsorption of gas, catalytic decomposition of hydrogen peroxide on mercury, periodic chemical luminescence of phosphorus, Liesegang rings[5,6] due to space periodic precipitation, electrode oscillation, and the fascinating electromechanical oscillator known as the "mercury heart."[7]

All the above-mentioned systems are heterogeneous and their mechanisms have not all been elucidated.[3] Periodic phenomena in homogeneous systems are not so common though, as we shall see; their number has increased tremendously within the last decade.[8] In fact, for a time they were even considered impossible on the base of a misinterpretation of the second law of thermodynamics. The development of thermodynamics of nonlinear irreversible processes by Glansdorff and Prigogine,[9,10] demonstrating that homogeneous systems could exhibit spontaneous organization provided they are maintained far enough from equilibrium, gave homogeneous periodic systems their letters patent of nobility.

P. De Kepper • Centre de Recherche Paul Pascal/CNRS, Domaine Universitaire, 33405 Talence Cédex, France.

Depending on the physical nature of the destabilizing mechanism, homogeneous chemical oscillating systems can, in principle, be divided in two categories: thermokinetic oscillators and isothermal oscillators. In the former, instabilities are temperature-driven. They are a consequence of the heat release of the reaction and the heat losses of the reactor. This type of instability is thoroughly studied by chemical engineers concerned with the operational stability of industrial plant reactors.(11,12) Owing to the highly nonlinear dependence of the rate constant on temperature, even simple first-order reactions

$$\text{Reactant} \rightarrow \text{Product} + \text{heat}$$

can exhibit large temperature and composition oscillations in flow reactors.(13) Thermokinetic oscillations occur most commonly in gas-phase reactions.(14) The small heat capacity of gases favors the rise of temperature during the reaction.

Thermal instabilities, however, are not the core of this paper, which is essentially devoted to isothermal systems. The latter instabilities incorporate the result of the sole kinetic mechanisms. Typically, these reactions take place in diluted aqueous solutions and generate only small temperature changes, of the order of 0.1 °C, during oscillatory behavior.(15)

This chapter, which aims at giving the reader a general picture of the experimental techniques used and the experimental results obtained in the field of homogeneous isothermal liquid-phase reactions, is divided into four parts. Section 4.2 presents a brief historical survey and classification of the major chemical oscillating systems presently known. Biochemical systems are not included in this presentation; these are well reviewed elsewhere.(16) Section 4.3 is devoted to the temporal behavior and includes a presentation of the experimental tools. Both periodic and nonperiodic oscillatory dynamics are described in diverse systems. A special section (4.4) describes the method we have developed to design fundamentally new chemical oscillators. Section 4.5 deals with spatiotemporal patterns. Appendix 4.1 contains simple procedures for lecture demonstrations of temporal and spatial organization.

4.2. Brief Historical Survey of Homogeneous Liquid-Phase Oscillating Reactions

The first truly homogeneous liquid-phase oscillating reaction was discovered by Bray(17) while studying the iodate-ion-catalyzed decomposition of hydrogen peroxide to water and oxygen. He recognized the kinetic nature of his observations, but most chemists at that time dismissed his discovery. The common opinion was that oscillations resulted from extraneous effects due to dust particles or oxygen supersaturation phenomena. There was a widespread feeling that chemical oscillations were in all cases forbidden by the second law of thermodynamics.

Some thirty years later the Russian biochemist Belousov[18] accidently discovered another liquid-phase oscillating reaction. He noticed that the cerium-catalyzed bromate oxidation reaction of citric acid would undergo, in appropriate mixtures, visible periodic changes from colorless to pale yellow during a long period of time. It took Belousov about seven years to have his work recognized and published in 1958.[18] Zhabotinsky[19] continued and extended Belousov's work during the 1960s to give the now so-called Belousov-Zhabotinsky (B–Z) reaction its present form (see Appendix 4.1). At that time a number of physical chemists were aware of theoretical developments in thermodynamics of irreversible processes[9,10] and took good notice of the reaction.

The real takeoff in the field of experimental studies of oscillating reactions in Western countries followed the first Prague meeting in 1967,[20a] where chemists and biochemists of both Western and Eastern countries met and realized the similitude of behavior between the B–Z reaction and some oscillatory phenomena observed in several biological systems.[21,23] The development by Field, Körös, and Noyes, in the early 1970s of a chemically consistent detailed mechanism[24] for the B–Z reaction gradually turned the majority opinion in favor of homogeneous oscillating kinetics.

Prior to 1980, the only expansions in chemical oscillators were produced by variants and hybrids of the two previously cited reactions. Among the most noticeable extensions one must mention the so-called "uncatalyzed B–Z reactions"[25] and the iodate-based Briggs–Rauscher (B–R) reaction.[26] In the former, the usual enol-type organic substrate (citric acid, malonic acid, etc.) is substituted by phenol or aniline compounds[25,27] allowing the metal-ion catalyst (Ce^{3+}, Mn^{2+}, etc.) to be suppressed. The latter reaction can be considered as a hybrid of Bray and B–Z reactions; it involves hydrogen peroxide, iodate, malonic acid, manganous ions, and starch in acidic solution and is the most visually impressive oscillating reaction, changing from colorless to gold to blue, and back again (recipe in Appendix 4.1).

At that time I had developed, with my colleagues at Brandeis University, USA, a systematic method of designing new oscillating reactions. The approach, based on a theoretical study[28] analyzing the relationship between the bistability phenomenon and relaxation oscillations, turned out to be very fruitful. It gave rise to a whole family of chlorite-based oscillators in combination with a number of iodine- or sulfur-containing inorganic reducing substrates.[29,30] It also significantly expanded the diversity of bromate oscillators,[31] some of them being quite different from the early B–Z reaction. Later, it also gave rise to a number of sulfide-based oscillators.[32,33] Some of the foundations of our method, which uses the topological properties of phase diagrams, is developed in Section 4.4. The achievements of this technique, now widely used, do not impede further serendipitous discoveries, such as the first simple organic oscillator due to Jensen in 1983.[34] The reaction corresponds to the air oxidation of benzaldehyde catalyzed by cobalt and bromide ions.

Figure 4.1 contains a summary of the main well-characterized homogeneous oscillators. The sketch gives an overall idea of their diversity without entering into any detailed considerations. Reactions are represented by pairs of boxes connected by full curves; species on the line are considered

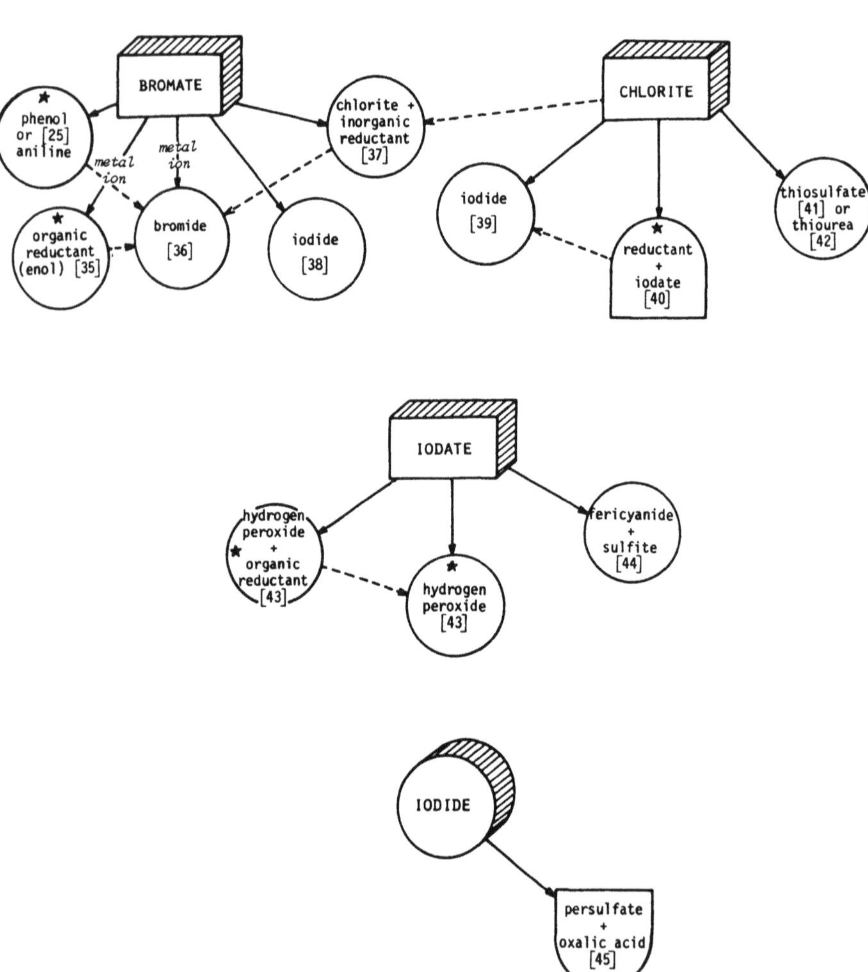

Figure 4.1. Schematic summary of the different homogeneous oscillating reactions. The numbers in square brackets refer to literature references where more information can be found.

SULFUR OSCILLATORS

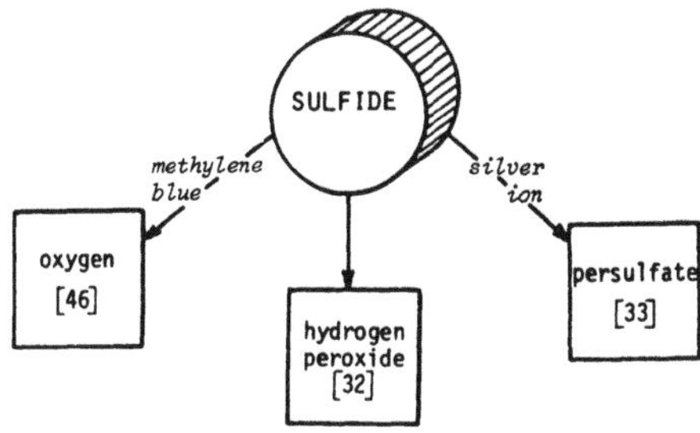

ORGANIC OSCILLATORS (AUTOXIDATION)

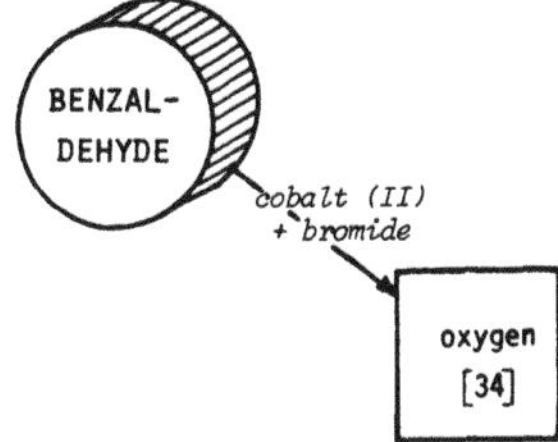

Figure 4.1. (*Continued*)

as catalysts. Reactants in square boxes are oxidants, while those in circular boxes are reductants. When the role of the species in a box is ambivalent, the box exhibits a double character. The species in the central double boxes are regarded as the basic reactant of a family which bear their name, e.g., the bromate family contains the B-Z reaction represented by the bromate-organic reductant (enol) couple. Dashed lines indicate that there are direct simple connections between different oscillators; e.g., the chlorite-reductant-iodate reaction is directly linked to the chlorite-iodide reaction, since the reduction of iodate produces iodide. In some cases the present classification may appear a little arbitrary, for example, the iodide-persulfate-oxalic acid reaction[45] which is classified among the halogen oscillators could also be thought of as the persulfate oxidation of oxalic acid catalyzed by iodide. It is noteworthy that most of these systems only oscillate when they are processed in flow reactors. The starts in Figure 4.1 indicate the few systems which may exhibit batch oscillations. Many other variants, hybrid or compound systems not presented in the figure, have also been studied. Some of these variants were developed for specific characters: to enlarge the domain of oscillation or to produce complex or chaotic dynamics, batch oscillations, spatial structures. For more information the reader is referred to the literature recorded in Figure 4.1.

4.3. Temporal Behavior

4.3.1. Batch Experiments

Early experiments were performed under batch conditions (closed systems) and only transient oscillatory phenomena could be observed. Figure 4.2 shows a typical time trace of an experiment carried out in such conditions. One distinguishes an induction period t_i during which some amount of intermediate species built up, followed by the onset of a transient oscillatory dynamic (t_o) of variable period and amplitude which in turn ceases after some time as the system evolves to thermodynamic equilibrium. Usually the number of oscillations is small; exceptionally some B–Z-type systems can exhibit more than 500 oscillations, in batch.[47] The large majority of the new chemical oscillators only exhibit, after an induction period, a single sudden switch phenomenon when carried out batchwise. Systems that exhibit batch oscillations are kinetically more complicated (Section 4.4). Moreover, in closed vessels dynamical studies are limited, due to unavoidable drifts.

4.3.2. Flow Reactor Experiments

The systematic use of continuous stirred tank reactors (CSTR) elegantly solves the problem of drift to equilibrium, by allowing the constant supply of fresh reactants and the removal of products from a well-mixed reaction vessel.[48] Figure 4.3 shows a schematic of such an experimental arrangement.

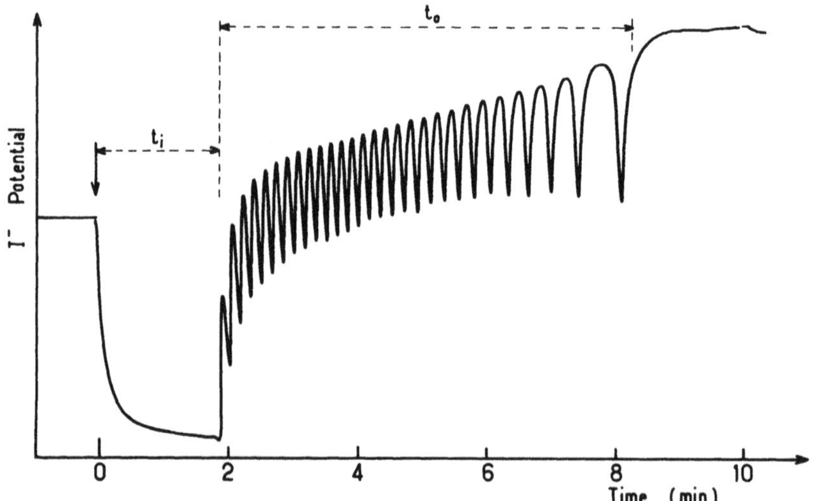

Figure 4.2. Iodide specific electrode time trace for the Briggs–Rauscher reaction. Initial composition given in Appendix 4.1, Experiment 2.

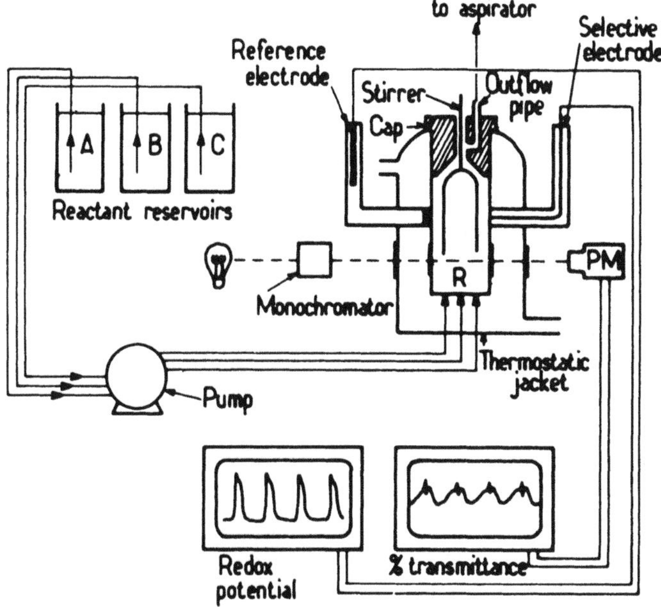

Figure 4.3. Schematic of a continuous-flow stirred tank reactor (CSTR) setup. R denotes reactor tank and PM photomultiplier.

The reactants, held in reservoirs, are continuously pumped into the base of the reaction vessel, where they are mixed. An outlet port at the top allows one to maintain the reacting volume constant. A cooling jacket controls the temperature. The concentration evolutions in the tank are monitored by optical means or by redox electrodes. These signals can be recorded on strip charts or digitalized and filed in a computer for further numerical treatments.[49,50]

If the mixing time is short compared to the reaction time,* then in the above-described device the only relevant control parameters or constraints are: (1) the inlet fluxes of reactants A_i, often expressed (as we do) by $[A_i]$, the concentration it would attain in the reactor if no reaction took place; (2) the residence time θ, which is the ratio of the outlet flow to the reactor volume, or the flow rate $k_0 = 1/\theta$; (3) the temperature of the reactor (heat release during the reaction is small, so the heat loss constant of the reactor is not relevant); and (4) the pressure of the environment, only important when gaseous reactants are involved.

In a CSTR, controlled steady operating conditions can easily be achieved for long periods of time. The system commonly attains, after an initial transient period, a stable stationary state resulting from dynamical equilibrium between

* We do not consider phenomena linked to nonideal mixing. For information on this topic see elsewhere.[51]

the rate of chemical transformation and the transfer rate of reactants through the reactor.[11,12,52] Stable stationary states are, for our purpose, uninteresting on their own and most of the early[53,54] and subsequent[30,54] CSTR experiments are devoted to the determination of conditions where stationary states lose their stability. The sustained oscillations, eventually observed, can usually be maintained within 1% relative stability. This excellent experimental control led to the recognition of a great number of dynamical behaviors, otherwise difficult to fully characterize; examples are simple and complex periodic oscillations, nonperiodic oscillation (see illustrations in Figure 4.4), and several multistability phenomena.[52,55]

In typical experiments one of the constraints is varied stepwise, while all the others are maintained constant, and the long-time behavior of the system is analyzed after each change. Transitions between qualitatively different dynamical or stationary states are noted and their reversibility probed by reversing the changes in the constraint. The results are gathered in response-constraint plots, as exemplified by Figures 4.5 and 4.6 for two quite different reactions. Point symbols represent stationary states while vertical segments denote the relative value and amplitude of oscillating states. On increasing the flow rate k_0 from low initial values, in Figure 4.5, one observes first, at (a), a reversible transition from a stationary state to an oscillating state of growing amplitude. Then, on increasing k_0 beyond (b), the oscillatory state is suddenly replaced by another stationary state characterized by a low potential value. This latter transition is not reversible and the low-potential stationary state persists for k_0 values as low as (c), where a discontinuous transition brings the system back to the high-potential stationary state. The hysteresis phenomenon defines a region of bistability either between the two stationary states or between the low-potential stationary state and the oscillating state. In the multiplicity regions, the stability of states is often tested by appropriate temporary perturbation, which can trigger transitions from one branch of a state to another constant constraint value. In fact this is the only way some isolated stable branches can be reached. This is the case, in Figure 4.6, for the stationary-state branch B_3. On gradually increasing $[I^-]_0$ from low values, one can obtain a spontaneous transition from branch B_1 to branch B_2 and back by reversing the $[I^-]_0$ evolution, while B_3 can only be attained from B_1 by a sudden injection of sulfite or by appropriately initializing the experiment.[30]

When the interplay of more than one constraint is studied, the results are usually summarized in two-dimensional constraint–constraint diagrams. These sections of parameter space are divided into different regions according to the asymptotic behavior exhibited by the system in response to the constraint values and are referred to as phase diagrams. Figure 4.7 and 4.8 present two examples of phase diagrams for two different bromate-based oscillating systems. The first is taken from the classic B-Z reaction[56] and shows in the (malonic acid, bromide ion) phase plane a region of sustained oscillations bordered by two stationary states (SSI and SSII), which have overlapping

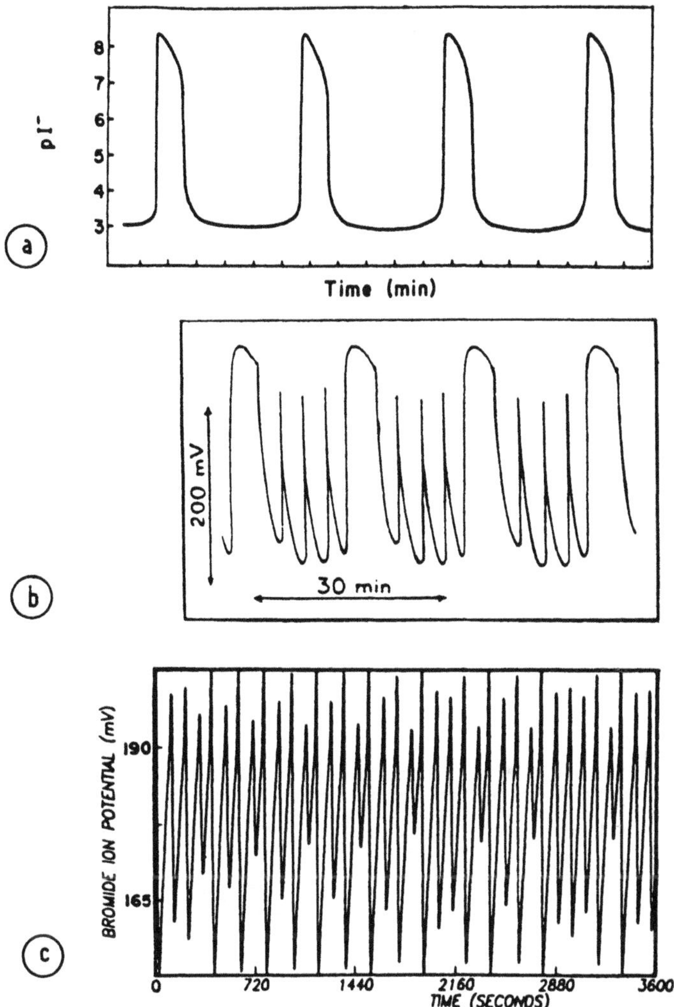

Figure 4.4. Typical oscillation trace for chemical reactions performed in CSTR. (a) Periodic simple oscillation in the chlorite-iodide reaction, with $[ClO_2^-)_0 = 2 \times 10^{-3}$ mol l^{-1}, $[I^-]_0 = 6 \times 10^{-3}$ mol l^{-1}, pH = 2.04; $k_0 = 1.1 \times 10^{-3}$ s^{-1}, and $T = 25$ °C. (b) Complex period oscillations in the chlorite-thiosulfate reaction, with $[ClO_2^-]_0 = 5.10^{-4}$ mol l^{-1}, $[S_2O_3^{2-}]_0 = 3 \times 10^{-4}$ mol l^{-1}, pH = 4, $\theta = 83$ s, and $T = 25$ °C. (c) Nonperiodic oscillations in the B-Z reaction, with $[BrO_3^-]_0 =$ 0.14, $[CH_2(COOH)_2]_0 = 0.25$ mol l^{-1}, $[Ce^{3+}]_0 = 1.66 \times 10^{-3}$ mol l^{-1}, $[H_2SO_4]_0 = 0.2$ mol l^{-1}, $\theta =$ 0.90 h, and $T = 28.3$ °C.

regions of stability i.e., bistability. The other phase diagram (Figure 4.8) results from a compound chemical system,[57] which can be regarded as the combination of several oscillating subsystems, namely, the bromate-iodide oscillating system[39] and the bromate-bromide-manganeous ion oscillator,[41] since bromide is produced during the reaction. As in the previous diagram one can

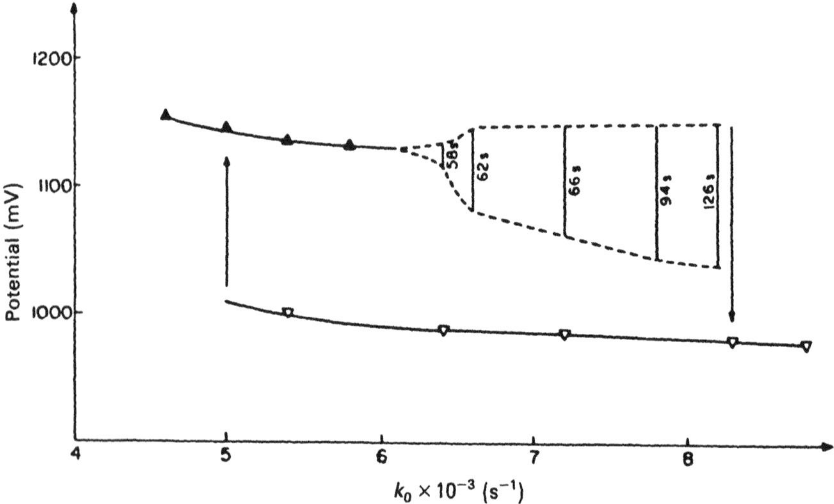

Figure 4.5. Response-constraint plot in the bromate-iodide reaction. Platinum redox potential vs. flow rate k_0. Symbol ▲ denotes high-potential stationary state and ▽ low-potential stationary state; vertical segments correspond to the amplitude of oscillating states. $[BrO_3^-]_0 = 5 \times 10^{-3}$ mol l^{-1}, $[I^-] = 5 \times 10^{-3}$ mol l^{-1}, $[H_2SO_4]_0 = 1.5$ mol l^{-1}, and $T = 25$ °C.

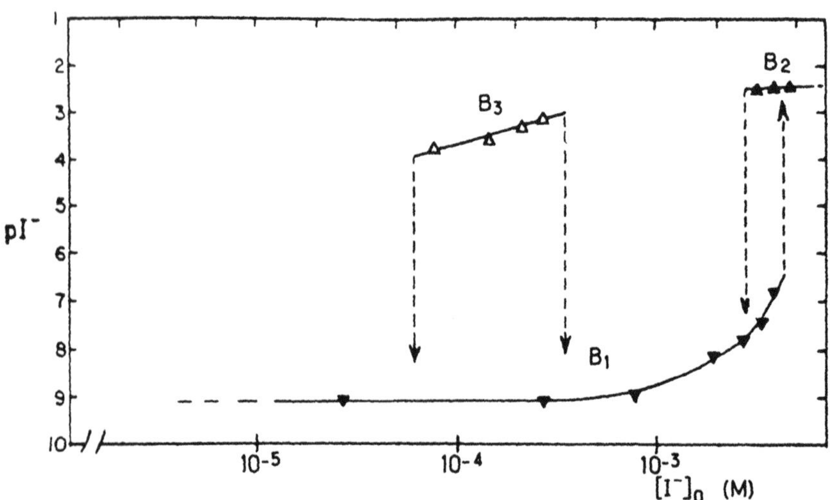

Figure 4.6. Response-constraint plot in the chlorite-iodate-arsenite-iodide reaction. Iodide specific electrode potential vs. iodide input concentration $[I^-]_0$. Symbol ▼ denotes low I^- stationary state, ▲ high I^- stationary state, △ intermediate I^- stationary state. Fixed constraints: $[ClO_2^-]_0 = 2.5 \times 10^{-3}$ mol l^{-1}, $[IO_3^-]_0 = 2.5 \times 10^{-2}$ mol l^{-1}, $[H_3AsO_3]_0 = 10^{-3}$ mol l^{-1}, $k_0 = 5.35$ s^{-1}, pH = 3.35, and $T = 25$ °C.

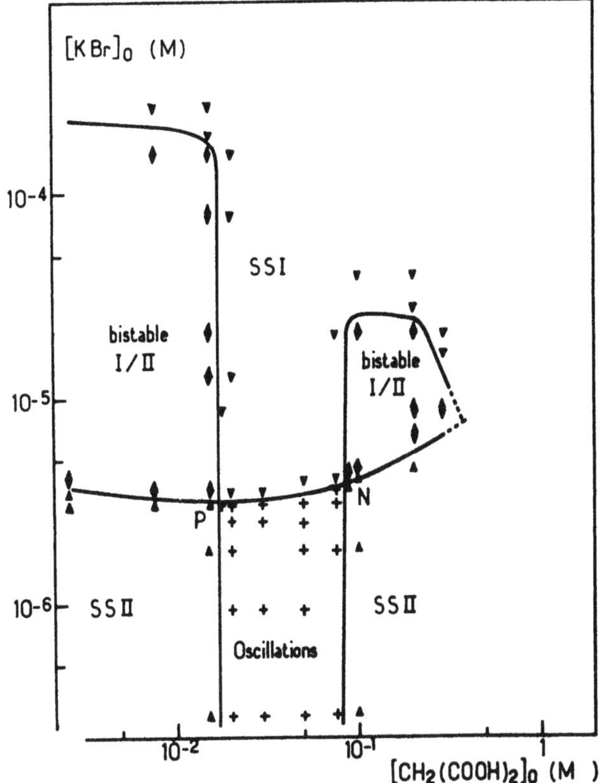

Figure 4.7. B–Z reaction phase-diagram section in the ($[Br^-]_0$, $[CH_2(COOH)_2]_0$) plane. Symbol ▲ denotes high [Ce(IV)] stationary state SSI, ▼ low [Ce(IV)] stationary state SS II, ◆ bistability SSI/SSII, + oscillating state. $[BrO_3^-]_0 = 2.7 \times 10^{-3}$ mol l^{-1}, $[H_2SO_4]_0 = 1.5$ mol l^{-1}, $[Ce^{3+}] = 2.9 \times 10^{-3}$ mol l^{-1}, $\theta = 3.7$ min, and T = 40 °C.

observe two regions of bistability, but in this case the bistability domains overlap at low $[I^-]_0$ thus generating a region of tristability (three different stable states). At high $[I^-]_0$ each region of bistability is terminated by a domain of simple sustained oscillations. It is noteworthy that in both phase diagrams (Figures 4.7 and 4.8) the domain of sustained oscillation exchanges with the region of bistability. This characteristic topology which seems, roughly, to result from the crossover of the stability limits of the two adjacent stationary states (Section 4.4) is widespread in chemical systems.[28,52]

Figures 4.7 and 4.8 are relatively simple examples of nonequilibrium chemical phase diagrams. Some of them can be very complicated.[55,58] Bistability phenomenon is a common feature of all chemical oscillators operating in flow reactors. In fact, stationary bistability is a more widespread phenomenon than oscillation.[52] Bistable systems are known in manganese[59,60] and nitrogen chemistry,[61,62] where no oscillatory conditions have yet been found.

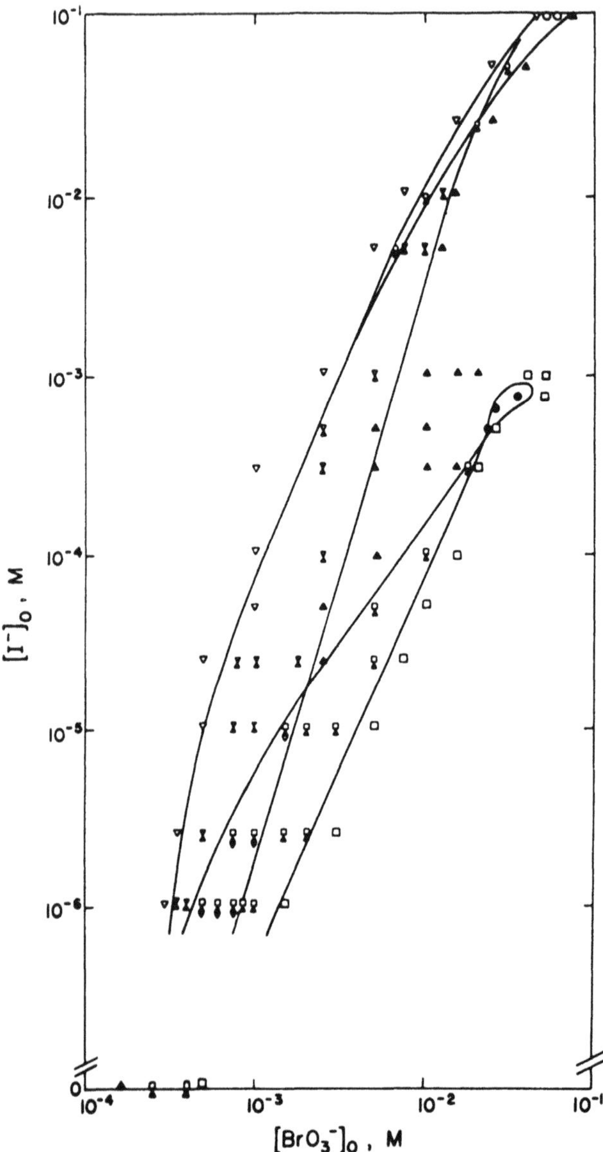

Figure 4.8. Phase-diagram section in the ($[BrO_3^-]_0$, $[I^-]_0$) plane for the bromate, iodide, manganous ion compound reaction. $[Mn^{2+}]_0 = 1 \times 10^{-3}$ mol l^{-1}, $[H_2SO_4]_0 = 1.5$ mol l^{-1}, and $T = 25$ °C. Symbol □ denotes high-potential, low $[I_2]$ stationary state, △ intermediate-potential, $[I_2]$ stationary state, ▽ low-potential, high $[I_2]$ stationary state, ○ and ● oscillatory states.

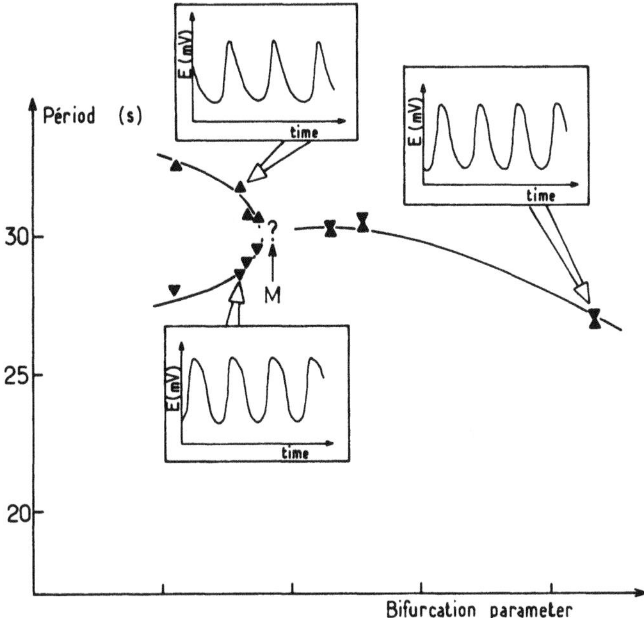

Figure 4.9. Pitchfork bifurcation response-constraint plot in the bromate-iodide reaction. Period of oscillations as a function of the bifurcation parameter (a combination of flow rate and bromate concentration).

Birhythmicity, which is the existence of two different stable oscillatory states under the same set of constraints, is a more exotic aspect. It may result from two independent oscillatory branches as suggested in the bromate-chlorite-iodide compound system[55,58] or from oscillatory states associated with the same singularity. This is the case in the bromate-iodide system, where the birhythmicity phenomenon is shown to originate from a pitchfork bifurcation of periodic orbits. Figure 4.9 shows the resulting bifurcation in the period and the change in the shape of oscillations when one follows the experimentally determined bifurcation direction.[63] Point M, in the immediate vicinity of the critical point, is very sensitive to external fluctuations and slow drifts from one oscillatory behavior to the other was observed due to small temperature fluctuations of the bath.

4.3.3. Aperiodic Oscillations or "Chemical Chaos"

Another salient feature of far-from-equilibrium chemical dynamics is their ability to produce very complicated or even nonperiodic oscillatory behaviors. This property, predicted by the French mathematician Ruelle[64] in the early 1970s, is receiving ever-increasing attention from mathematicians and physicists dealing with nonlinear dynamical systems. The advantage that chemical systems have over most other physical systems, in this type of study, is their flexibility and the numerous control parameters at one's disposal, such as different input chemical concentrations, flow rate, temperature, etc. It is one of the fastest growing areas in chemical dynamics,[65,66] though the deterministic nature of nonperiodic regimes were, for a time, regarded with skepticism in chemical systems. This is to be imputed to the misleading aspect of the most commonly observed nonperiodic behavior arising in alternating sequences of periodic and nonperiodic oscillations. In these sequences the periodic regime is typically formed of one large amplitude oscillation (L) followed by q small amplitude oscillations (S), as illustrated by Figure 4.4b with $q = 3$, thus producing an $(L + qS)$ peak complex oscillation denoted in short by P_q^1. The associated nonperiodic regimes appear in narrow regions of control parameters between two consecutive P_q^1 and P_{q+1}^1 periodic regimes and seem to result from random mixtures of $L + qS$ and $L + (q + 1)S$ cycles. This was thought to result either from ramdom fluctuations in the control parameters (such as temperature or flow rate) or from external forcing due to pump pulses. Progress in experimental setups (high-performance piston pumps or accurately controlled gravity smooth flow) and in data analysis techniques, such as the reconstruction of phase-space trajectories from a single time series,[67] have allowed one to distinguish unambiguously between deterministic chaos and external noise.[66] Alternating sequences were more particularly observed in the chlorite-thiosulfate,[68] chlorite-bromate-iodide,[58] and B–Z reaction.[66]

The most relevant studies of nonperiodic or chaotic behaviors were performed on the B–Z reaction in a CSTR, and in the following only results stemming from this reaction will be considered. Figure 4.10a shows a projection of the reconstructed attractor, by the time delay method[67] in the $X(t)$, $X(t + T)$, $X(t + 2T)$ phase space, obtained from the time series there in Figure 4.4c. This nonperiodic regime is observed between a simple large amplitude periodic regime P_0^1 and P_1^1 complex periodic regime.[69] Figure 4.10a shows that the trajectories remain in a very well defined finite region of space. This restriction in space is even more spectacular from the Poincaré section (Figure 4.10b), obtained by the intersection with more than 300 oscillations on this attractor. They lie on a smooth curve, which infers that the attractor is nearly a two-dimensional manifold (however, owing to the fractal nature of the attractor its actual dimension must be slightly greater than 2[70]). A first return map constructed from the dynamics on the Poincaré section is presented in Figure 4.10c. The data thus obtained provide a well-behaved single-valued curve, with practically no scatter; this indicates that the underlying dynamics obeys a deterministic law described by the iteration of a one-dimensional map with a single extremum.[71] Such a type of map is also consistent with more detailed experiments in this region of parameter space which show that the transition from the large-amplitude relaxation oscillations P_0^1 to the chaotic regime proceeds via a period-doubling sequence;[72] see also Chapter 1.

In addition to the periodic-chaotic alternating sequences and the period-doubling sequence mentioned above, over routes to chaos were characterized

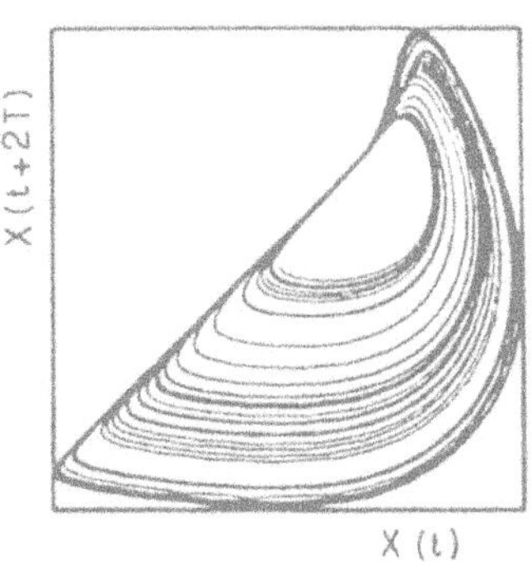

Figure 4.10a. Reconstructed attractor projection in the $X(t)$, $X(t+T)$ plane obtained from the time series in Figure 5c of Takens.[67]

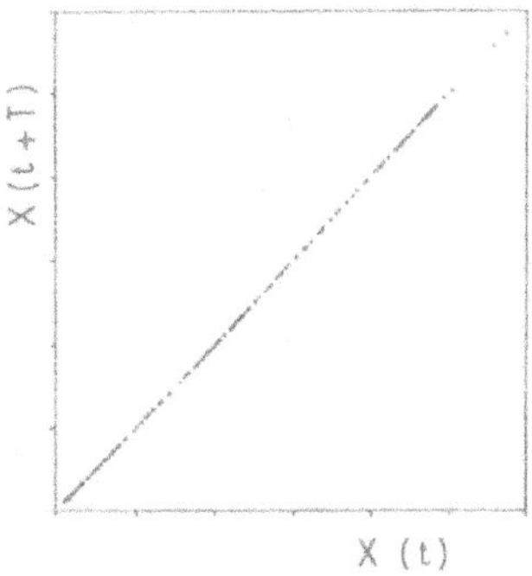

Figure 4.10b. Poincaré section of the previous attractor in the $X(t)$, $X(t+2T)$ plane.

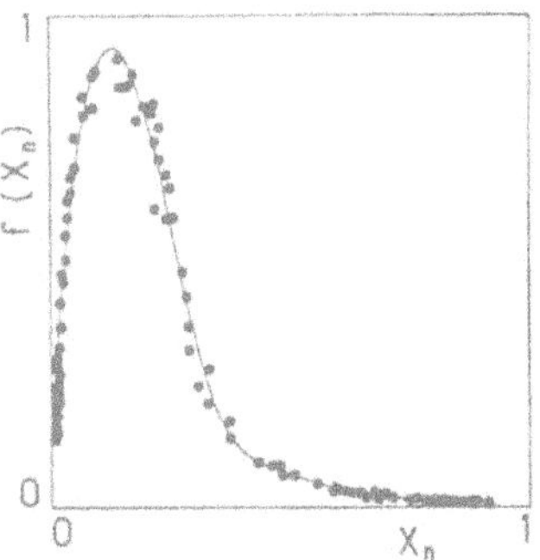

Figure 4.10c. One-dimensional first-return map obtained from the Poincaré section.

in the B-Z reaction. Type I intermittency, which results from a tangent bifurcation of a one-dimensional map, was among the first route to chaos evidenced in chemical systems.[73]

Quasi-periodic regimes, with two incommensurate frequencies, have recently been observed.[74] Figure 4.11a shows a time series of the optical density of such a regime and Figure 4.11b presents, in a pseudoperspective view, the associated reconstructed attractor. For other values of the control

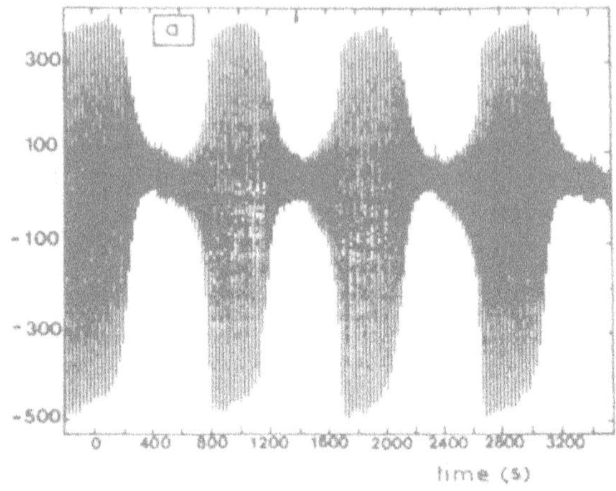

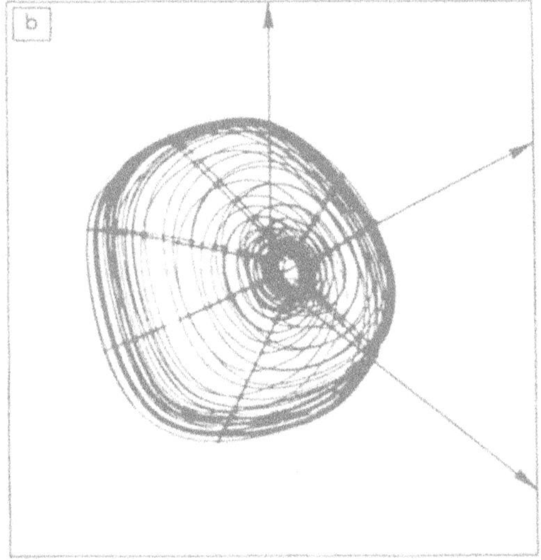

Figure 4.11. (a) Time series of a quasi-periodic regime in the B–Z reaction. Experimental conditions: $[BrO_3^-]_0 = 0.012$ mol l^{-1}, $[CH_2(COOH)_2]_0 = 0.167$ mol l^{-1}, $[Ce^{3+}]_0 = 1.66 \times 10^{-4}$ mol l^{-1}, $H_2SO_4 = 1$ mol l^{-1}, and $T = 39$ °C. (b) The associated reconstructed attractor. Experimental conditions: $[BrO_3^-]_0 = 0.012$ mol l^{-1}, $[CH_2(COOH)_2]_0 = 0.167$ mol l^{-1}, $[Ce^{3+}]_0 = 1.66 \times 10^{-4}$ mol l^{-1}, $H_2SO_4 = 1$ mol l^{-1}, and $T = 39$ °C.

parameter the central hole is shown to shrink, and wrinkles appear on the surface of the torus before the central hole disappears and gives way to a stationary state located in this center.[75,76]

All the above-mentioned results serve as illustrations of the dynamical richness of chemical systems. Other more complex regimes have been observed in the B–Z reaction, some of which have not yet been possible to characterize; these may correspond to dynamics on higher-dimensional attractors.

4.4. Design of Chemical Oscillators

Though there is no set of necessary and sufficient conditions to produce oscillatory dynamics, some helpful conditions are known in chemical systems. One obvious necessary condition for oscillation is that the system evolves far from thermodynamical equilibrium. It is thus recommended to work with open systems to avoid the consequences of reaction reversibility. The other equally important condition is that the reaction mechanism should involve nonlinear feedback loops.

In the following we outline a workable search procedure, proposed in 1981,[29] to design new chemical oscillators. There are two reasons for this; first, the technique produced considerable progress in the development of chemically different oscillators and, second, the simple theoretical concepts on which it is based provide some insight in the mechanism of chemical oscillators without entering into any detailed kinetic considerations.

The basic theoretical concept is as follows. Consider a bistable system controlled by a parameter U, as presented in Figure 4.12a. This bistable is coupled with a feedback dynamic, which modifies the actual value of the control parameter according to the composition on each branch of the bistable. Furthermore, and this is our fundamental hypothesis, the feedback effect settles on a time scale τ_{ri} greater than τ_{bi}, the characteristic relaxation times to each branch i of the bistable. Since composition on the lower and upper branches is often very different, the feedback will shift differently the stability limits U_1 and U_2 of the original bistable. Quantities U_1 and U_2 now depend on the feedback mechanism. Suppose the amplitude or effectiveness of this feedback is controlled by a second independent parameter K; the U_i are then functions

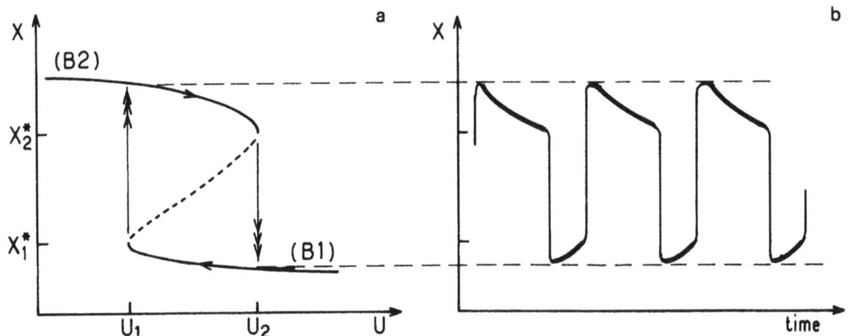

Figure 4.12. (a) Schematic bistability plot and the associated X hysteresis loop as a function of U. (b) Schematic of oscillations obtained when the bistable system is coupled to an appropriate feedback system.

of K, $U_1(K)$ and $U_2(K)$, say. These functions define in the (U, K) space two curves, which delimit the domain of stability of the two stationary states, SS1 and SS2. Generally, these curves intersect at a point P. Beyond this point, a region opens up where neither SS1 or SS2 is stable and, if our fundamental time-scale hypothesis ($\tau_{ri} \gg \tau_{bi}$) holds, relaxation oscillations are observed. They are the result of cyclic switching between the branches, as shown by the arrows in Figure 4.12a; Figure 4.12b presents a typical time trace. The resulting phase diagram is divided into four domains, as depicted in Figure 4.13. Such phase diagrams have been called "cross-shaped phase diagrams."[28,52,56] Figure 4.7 and 4.8 are experimental examples of this topology. Calculations on explicit dynamical models based on the above description show that, in fact, more complicated dynamics can occur in the vicinity of the cross point P, but the gross overall topology is preserved.[28] If the time-scale hypothesis does not hold curves $U_1(K)$ and $U_2(K)$ become tangential at point P and then correspond to a simple cusp point with no oscillatory behavior beyond.

The above schematic view of relaxation oscillations has suggested the following approach to the design of new chemical oscillators: (1) find bistable systems; (2) invent an appropriate feedback reaction; (3) study the resulting phase diagram, focus on the region of vanishing bistability, and examine the dynamical behavior beyond the cross point. It happens that bistability can easily be produced by operating autocatalytic reactions in a CSTR (autocatalytic reactions are reactions in which a product increases its own rate of production). Autocatalysis is not unusual in chemistry. Examples are readily found in the halogen, sulfur, nitrogen, manganese, or carbon chemical

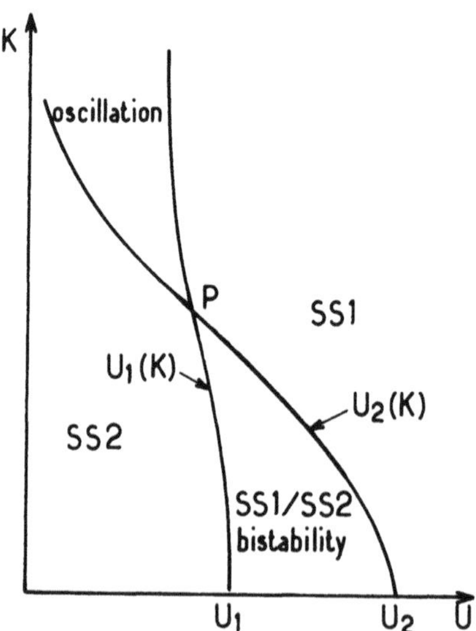

Figure 4.13. Schematic "cross-shaped phase diagram" resulting from the coupling of a bistable system and an appropriate feedback reaction.

literature. As shown in Figure 4.1, oscillatory systems have been observed in several of these areas of chemistry. Bistability has been described in nitrogen[61, 62] and manganese[60] systems, but no oscillatory behavior has yet been produced. Studies in these areas are currently under progress. The most difficult problem is to find feedback reactions with suitable time-scale ratios.

All the new oscillators discovered since 1981, except one,[34] are more or less direct results of the cross-shaped phase-diagram technique. This latter is probably not the only possible method to produce oscillating reactions but it has, so far, been the most efficient.

In conclusion, it is emphasized that our method is directly linked to the use of open reactors; most new chemical systems only oscillate in such reactors. Batch oscillatory reactions are much fewer and call for a higher degree of kinetical complexity. In this case some primary reactions must play roles analogous to that of the flow in an open system, such as the supply in appropriate amounts of active intermediate species and the scavenging of inhibitory end products over a time long compared to the period of oscillations. For example, in some variants of the chlorite-iodide oscillator containing malonic acid, it was possible to have the system roughly mimic, for some period of time, an iodide flow, making use of the slow hydrolysis of organic iodo compounds, and thereby a series of batch oscillations were produced. The same solution poured in an unstirred thin layer can also produce traveling trigger waves,[77] just as in the B-Z reaction.[78] These spatiotemporal phenomena are described in the next section.

4.5. Spatial Organization

4.5.1. Introduction

Turing,[79] in the early 1950s, was the first to suggest that organized concentration profiles could spontaneously emerge in initially homogeneous chemical systems when reaction processes are allowed to couple with molecular diffusion. Following this pioneering approach, many formal chemical models have been shown to produce a variety of standing concentration patterns (stationary spatial structures) and other inhomogeneous time-dependent structures.[9,10,80] As for the temporal instabilities treated before, the spatial organization is well understood in the general framework of nonlinear irreversible processes.[9,10]

Experimentally speaking, chemical spatial structure studies are much less advanced than their temporal counterpart.[81] One can find major reasons for this situation; some impediments are purely practical, others are more fundamental. On the experimental side, the studies of chemical spatial structures suffer from the present lack of an appropriate workable tool, i.e., a device that could play a role analogous to that of the CSTR in studies of homogeneous temporal structures. The requirements of such a device are difficult to meet, since it must allow for spatially controlled chemical fluxes over any period of time without destroying any eventual spatial pattern. At this stage of development, no usable method to produce spatial structures, other than some wave phenomena, has been invented. On the theoretical side, mathematical methods to analyze and predict the formation of spatial patterns are just beginning to be elaborated. Computational studies are tedious and lengthy.

A variety of different spatial organizations has nonetheless been reported in oscillating and nonoscillating chemical reactions over two decades. These structures may be of a very different nature; some are trivial phenomena while others do not seem to be chemically driven. Most of the well-characterized spatial behavior falls in two main categories: the propagating waves and the so-called unmoving mosaic patterns.

4.5.2. Propagating Waves

Although not always easy to distinguish experimentally, this term covers two fundamentally different type of waves.

Pseudowaves. If in an oscillating reaction the phase and/or the period were not uniform over space, a seemingly propagating wave would be observed. They are illusions resulting from preset inhomogeneities. They are not stopped by impermeable barriers, thus no coupling between the different parts of the solution is necessary to produce these pseudowaves. Their velocity is inversely proportional to the phase gradients. This velocity has no upper limit and tends to infinity as the system approaches uniformity. Pseudowaves are of no interest on their own, but they are difficult to avoid in experimental systems and, furthermore, in a continuous medium they may trigger real chemical waves when they are sufficiently slow.

Propagating reaction–diffusion fronts. This class of chemical wave is also known in the literature as "trigger waves" or "autowaves." They are characterized by propagating fronts, which depend both on diffusion and on the dynamics of the reacting medium. They are impeded by impermeable boundaries and propagate in so-called "excitable" media. By excitable media we mean, loosely, chemical systems in which a small but finite perturbation will trigger a dramatic change in composition. This change in turn contaminates the surroundings by diffusion and spreads the perturbation with no loss of strength away from the point of origin. The change in composition may be temporary or definitive, depending on the chemical dynamics. In the former case, the system can return to its initial state and be reexcitable, as in the B–Z reaction, the most popular nonbiological system to produce spatial patterns. In a thin layer of B–Z reagent at rest, repetitive expanding blue (oxidized) rings of chemical activity spontaneously develop in a reddish (reduced) background state from centers or pacemakers, thus producing target patterns. Usually, as shown in Figure 4.14b, the wavelength varies from one pattern to

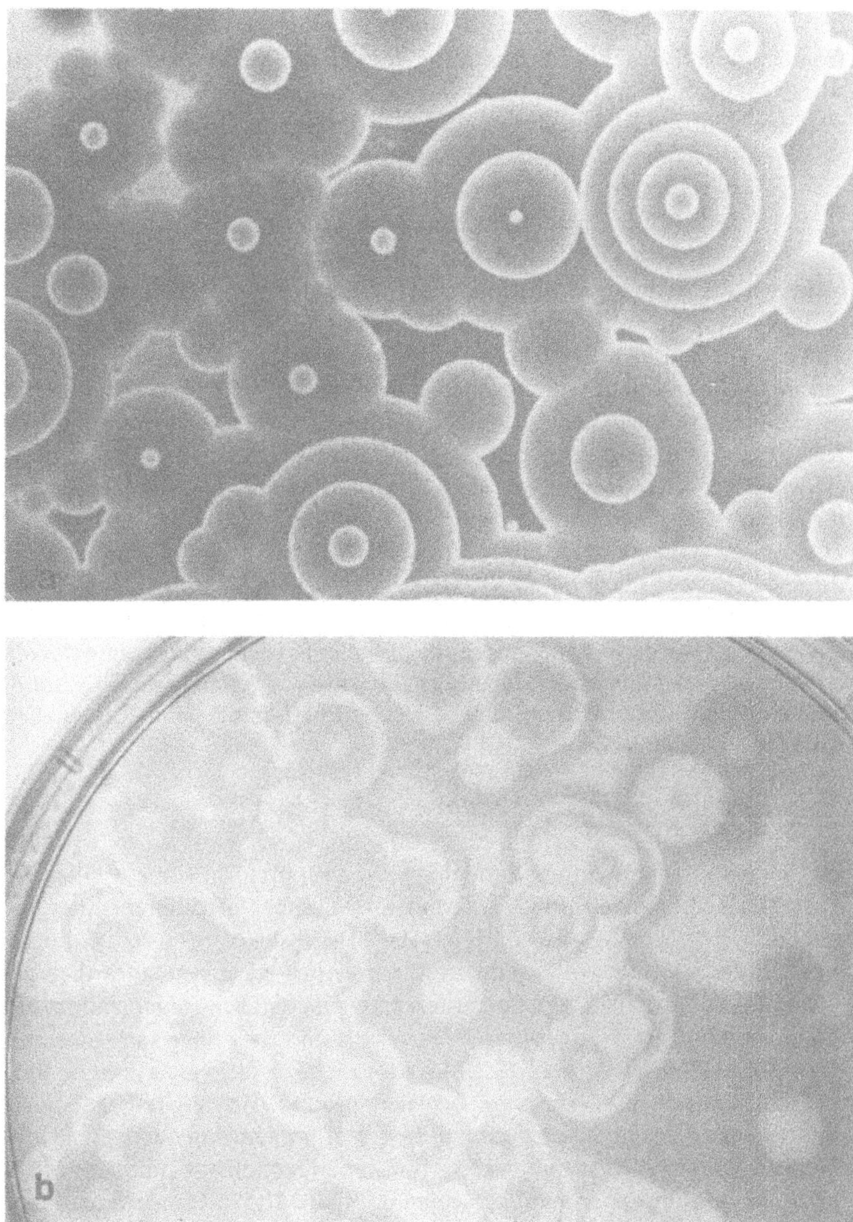

Figure 4.14. Traveling wave spatial structures: (a) in the B–Z reaction (see Appendix, Experiment 3), (b) in a chlorite-type oscillator.

another and wave fronts are annihilated during head-on collisions. In some more restricted composition domain it is also possible to produce reversed color patterns, a red wave on a blue background.[82] Breaking a circular wave front by slightly shaking the solution can produce pairs of counterrotating spirals of shorter wavelength.[78,83] More elaborate perturbations can also produce multiarmed spirals.[84] Target patterns and spirals have also been observed subsequently in some chlorite oscillators[77] (Figure 4.14b). In simpler autocatalytic reactions, single propagating fronts have been studied.[85,86]

The wave-front propagation phenomena in chemical systems is well understood and kinetic model calculations give results in good quantitative agreement with experimental observations.[85-87] The fundamental question that remains experimentally unsolved in these studies is the origin of the leading centers. The source may be endogenous and result from a reaction–diffusion instability in an initially homogeneous medium.[88,89] However, they can also be induced by local heterogeneities, which locally catalyze or modify the reaction to produce leading centers of different frequencies.[87] There is clear evidence for heterogeneous pacemakers in ordinary B–Z experiments. Nonetheless, in experiments in clean dishes, some pacemakers remain even after filtering out nuclei bigger than 0.50 μm.[83] The two types of leading centers do not exclude each other. In principle one can distinguish between them. Homogeneous centers should depend only on the actual dynamic of the reaction e.g., the period and amplitude of the bulk oscillations, while the other types of center should rather depend on the nature of the heterogeneity. Studies are being performed with the aim of reconciling homogeneous center theory with experimental observations.

4.5.3. Unmoving Mosaic Structures

Seemingly unmoving patterns have been reported in some oscillating or excitable B–Z-type reactions[90-92] and in a number of different monotonic reactions,[93-95] when spread in thin layers. These structures usually appear at a gas–liquid interface. In contrast to the previous wave structures they grow and eventually vanish at nearly the same place. Their organization often reminds one of fuzzy convective Bénard cells and they have sometimes been described as "mosaic structures." Mosaic structures seem always to be linked to a surface reaction. They are revealed/produced by light irradiation in photochemical reactions or by the presence of a chemically active gas at the interface, i.e., oxygen in the case of B–Z-type reactions or other vapors like bromine, as in the experiment presented in Figure 4.15.[94] Sometimes thought to be examples of standing spatial structures resulting from a reaction–diffusion instability, they are now rather considered as the mere result of convection cells induced by adverse density gradients. Convective movements have actually been observed in a few cases.[96] The convective motion can be induced by surface evaporative cooling of the solvent and/or from the composition

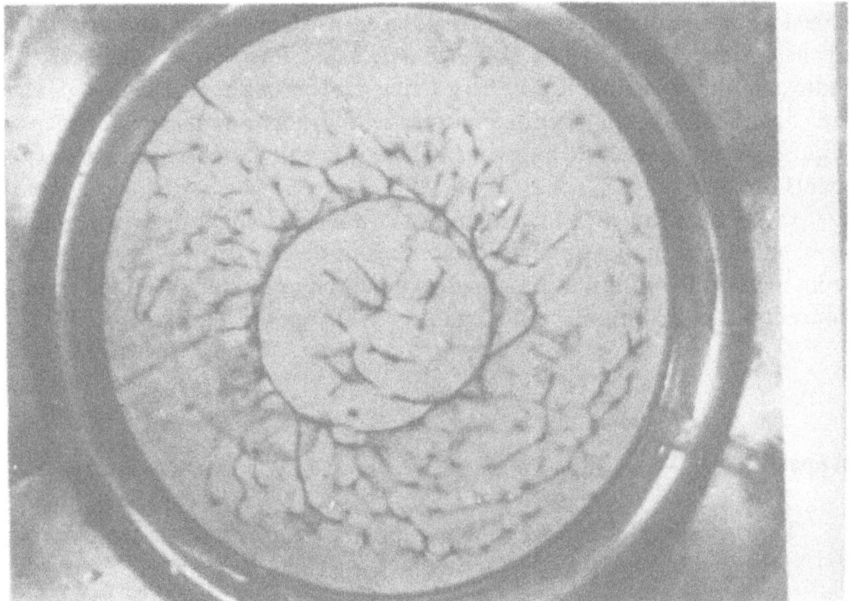

Figure 4.15. Convective mosaic structure in an iodide-bromine system. Patterns of iodine-starch complex resulting from the diffusion of bromine gas into an iodide-starch solution (after Avnir and Kagan[94]).

change resulting from the surface reaction.[97] The colored species formed at the interface are then drawn by the convective streamlines, thus revealing the hydrodynamic structure.

4.6. Conclusion

The above presentation only provides a sampling of the diverse experimental work performed in the field of homogeneous chemical oscillating reactions and does not give a complete picture of the present situation in this field. Before concluding with future developments one should also add to the above studies carried out under periodic[98-100] or random[101] constraint values, and the delicate work on mutually coupled reactors.[102,103] Another important domain of activity linked to oscillating reactions, barely mentioned in this chapter, is the tedious task of unraveling the detailed kinetic mechanism of these reactions. Only a few of them are known satisfactorily; here again the case of the B-Z reaction can be regarded as exemplar.[35]

Kinetic studies in chemical instabilities will remain a major activity and probably expand still further. Since the goal of showing the ubiquity of oscillatory behavior has, in practice, now been attained, the quest for new chemical oscillators will slow down at the expense of a better characterization of the dynamical possibility of the already known oscillators. The search for more complex dynamics is now very active. It seems, though, that to characterize chaotic attractors embedded in a more than three-dimensional space will require yet more accurate experimental devices and extremely lengthy experiments.

In the near future the most important novelties might probably originate from studies of spatiotemporal instabilities in diverse open systems deviating from the CSTR conditions, the final goal being to produce sustained standing concentration profiles as predicted thirty years ago by Turing.[79]

Appendix

The procedures for three attractive illustrative experiments are given here.

Experiment 1

The Beating Mercury Heart. The system is a chemically driven electromechanical oscillator. It is a heterogeneous chemical system. In a watch glass of diameter 6-12 cm, deposit about 1 ml of clean mercury, cover it with a 20% sulfuric acid solution (≈ 4 mol l^{-1}) in which 2-4 small crystals of permanganate have been dissolved. Now gently touch the side of the mercury drop with a clean soft iron wire. The mercury should suddenly contract, break contact with the iron wire, and then immediately relax back to its initial shape to renew the contact; if the iron wire is held in position, the cycle repeats. Different fascinating oscillatory modes may be observed, depending on the size of the mercury drop and the contact position with the iron wire.

Experiment 2

The Briggs-Rauscher reaction is a homogeneous oscillating reaction exhibiting dramatic changes in color. Prepare in distilled water the following starting solutions in three different flasks: (1) 0.14 mol l^{-1} potassium iodate, (2) 0.15 mol l^{-1} malonic acid with 0.024 mol l^{-1} manganeous sulfate and starch (about 20 g/l), (3) 3.2 mol l^{-1} hydrogen peroxide plus 0.17 mol l^{-1} perchloric acid. Mix rapidly equal volumes of each of the above three solutions. After a short induction period the system oscillates with deepening orange-blue-colorless alternations.

Experiment 3

The Belousov–Zhabotinskii reaction phase waves and trigger waves in an initially homogeneous system. Prepare, in separated flasks the following three solutions in distilled water: (1) 0.60 mol l^{-1} sodium bromate, (2) 0.24 mol l^{-1} malonic acid and 0.60 mol l^{-1} sulfuric acid, (3) 12×10^{-3} mol l^{-1} ferroin (redox color indicator).

One can easily find commercial ready made ferroin solutions. Beware, it is important to avoid chloride ion impurities. Mix equal volumes of each solution. Solutions thus prepared will oscillate from dark red to blue. The same solution, if spread in a petri dish in a thin layer 1-2 mm deep, will spontaneously generate a vivid blue target pattern on a reddish background, as shown in Figure 4.14a. Spirals can be obtained by breaking a ring. Sharper patterns are obtained when the petri dish is covered to prevent air turbulence over the surface.

References

1. E. S. Hedges and J. E. Myers, *The Problem of Physicochemical Periodicity*, Edward Arnold, London (1928).
2. B. Veil, *Les Phénomènes Périodiques de la Chimie*, Hermann, Actualités Scientifiques (1934).
3. G. Nicolis and J. Portnow, *Chem. Rev.* **73**, 365 (1973).
4. G. Runge, *Der Bildungstrieb der Stoffe* (1855).
5. K. H. Stern, *Bibliography of Liesegang rings*, NBS. Miscellaneous Publication 292, Sept I (1967).
6. S. Kai, S. Müller and J. Ross, *J. Chem. Phys.* **76**, 1392 (1982).
7. J, Keiser, P. A. Rock, and Shu-vai Lin, *J. Am. Chem. Soc.* **101**, 5637 (1979).
8. R. J. Field and M. Burger, eds., *Oscillations and Traveling Waves in Chemical Systems*, Wiley-Interscience, New York (1985).
9. P. Glansdorff and I. Prigogine, *Thermodynamic Theory of Structure, Stability and Fluctuation*, Wiley-Interscience, New York (1971).
10. G. Nicolis and I. Prigogine, *Self-Organization in Nonequilibrium Systems*, Wiley-Interscience, New York (1977).
11. R. Aris, *Introduction to the Analysis of Chemical Reactors*, Prentice-Hall, Englewood Cliffs, N.J. (1965).
12. D. D. Perlemutter, *Stability of Chemical Reactors*, Prentice-Hall, Englewood Cliffs, N.J. (1972).
13. A. Uppal, W. H. Ray, and A. B. Poore, *Chem. Eng. Sci.* **29**, 967 (1974).
14. J. F. Griffiths, in: *Oscillations and Traveling Waves in Chemical Systems* (R. J. Field and M. Burger, eds.), p. 529, Wiley-Interscience, New York (1985).
15. C. Vidal, P. De Kepper, A. Noyau, and A. Pacault, C.R. Acad. Sci. Paris **285C**, 357 (1977).
16. C. Vidal and A. Noyau, *J. Am. Chem. Soc.* **102**, 6666 (1980).
17. W. C. Bray, *J. Am. Chem. Soc.* **43**, 1262 (1921).
18. B. P. Belousov, *Sbornik Ref. Radiat. Meditsine*, 1958, Medgiz, Moscow, 145 (1959).
19. A. M. Zhabotinskii, *Biofizika* **9**, 306 (1964).
20. (a) G. M. Franck, *Oscillating Processes in Biological and Chemical Systems*, Science Publ., Moscow (1967). (b) E. E. Sel'kov, A. M. Zhabotinskii, and S. E. Schnoll, Eds., *Oscillating Processes in Biological and Chemical Systems*, Science Publ., Moscow (1971).

21. L. N. Duysens and J. Amesz, *Biochim. Biophys. Acta* **24**, 19 (1957).
22. A. T. Wilson and M. Calvin, *J. Am. Chem. Soc.* **77**, 5948 (1955).
23. B. Chance, E. K. Pye, A. K. Ghosh, and B. Hess, *Biological and Biochemical Oscillators*, Academic Press, New York (1973).
24. (a) R. J. Field, E. Körös, and R. M. Noyes, *J. Am. Chem. Soc.* **94**, 8649 (1972). (b) R. J. Field and R. M. Noyes, *J. Chem. Phys.* **60**, 1877 (1974).
25. M. Orbàn and E. Körös, *J. Phys. Chem.* **82**, 1672 (1978).
26. T. S. Briggs and W. S. Rauscher, *J. Chem. Educ.* **50**, 496 (1973).
27. J. Chopin-Dumas and P. Richetti, in: *Nonlinear Phenomena in Chemical Dynamics* (C. Vidal and A. Pacault, Eds.), Synergetics Vol. 12, p. 213, Springer-Verlag, Berlin (1981).
28. J. Boissonade and P. De Kepper, *J. Phys. Chem.* **84**, 501 (1980).
29. P. De Kepper, I. R. Epstein, and K. Kustin, *J. Am. Chem. Soc.* **103**, 2133 (1981).
30. M. Orbàn, C. Dateo, P. De Kepper, and I. R. Epstein, *J. Am. Chem. Soc.* **104**, 5911 (1982).
31. I. R. Epstein and M. Orbàn, in: *Oscillations and Traveling Waves in Chemical Systems* (R. J. Field and M. Burger, eds.), Wiley-Interscience, New York (1985).
32. M. Orbàn and I. R. Epstein, *J. Am. Chem. Soc.* **107**, 2302 (1985).
33. Q. Ouyang and P. De Kepper, *J. Phys. Chem.* **91**, 6040 (1987).
34. J. H. Jensen, *J. Am. Chem. Soc.* **105**, 2639 (1983).
35. R. J. Field, in: *Oscillations and Traveling Waves in Chemical Systems* (R. J. Field and M. Burger, eds.), p. 55, Wiley-Interscience, New York (1985).
36. M. Orbàn, P. De Kepper, and I. R. Epstein, *J. Am. Chem. Soc.* **104**, 2657 (1982).
37. M. Orbàn and I. R. Epstein, *J. Phys. Chem.* **87**, 3212 (1983).
38. M. Alamgir, P. De Kepper, M. Orbàn, and I. R. Epstein, *J. Am. Chem. Soc.* **105**, 2641 (1983); O. Citri and I. R. Epstein, *J. Am. Chem. Soc.* **108**, 357 (1986).
39. C. E. Dateo, M. Orbàn, P. De Kepper, and I. R. Epstein, *J. Am. Chem. Soc.* **104**, 504 (1982); I. R. Epstein and K. Kustin, *J. Phys. Chem.* **89**, 2275 (1985).
40. P. De Kepper, I. R. Epstein, K. Kustin, and M. Orbàn, *J. Phys. Chem.* **86**, 170 (1982).
41. M. Orbàn, P. De Kepper, and I. R. Epstein, *J. Phys. Chem.* **86**, 431 (1982).
42. M. Alamgir and I. R. Epstein, *Int. J. Chem. Kinet.* **17**, 429 (1985).
43. S. D. Furrow, in: *Oscillations and Traveling Waves in Chemical Systems* (R. J. Field and M. Burger, eds.), p. 171, Wiley-Interscience, New York (1985).
44. E. Edblom, M. Orbàn, and I. R. Epstein, *J. Am. Chem. Soc.* **108**, 2826 (1986).
45. J. Chopin and M. N. Papel, in: *Non-Equilibrium Dynamics in Chemical Systems* (C. Vidal and A. Pacault, eds.), Synergetics Vol. 27, p. 69, Springer-Verlag, Berlin (1984).
46. M. Burger and R. J. Field, *Nature* **307**, 720 (1984).
47. G. J. Kasperek and T. C. Bruice, *Inorg. Chem.* **10**, 382 (1971).
48. K. R. Westerterp, W. R. M. Van Swaay, and A. A. C. M. Beenackers *Chemical Reactor Design and Operation*, Wiley, New York (1983); J. Villermaux, *Genie de la reaction chimique*, Tech. et Doc., Lavoisier (1982).
49. C. Vidal, J. C. Roux, and A. Rossi, *J. Am. Chem. Soc.* **102**, 1241 (1980).
50. J. C. Roux, R. H. Simoyi, and H. L. Swinney, *Physica* **8D**, 257 (1983).
51. J. C. Roux, P. De Kepper, and J. Boissonade, *J. Phys. Lett.* **97**, 168 (1983); M. Menzinger, M. Boukalouch, P. De Kepper, J. Boissonade, J. C. Roux, and H. Saadaoui, *J. Phys. Chem.* **90**, 313 (1986); E. Kumpisky and I. R. Epstein, *J. Chem. Phys.* **82**, 53 (1985).
52. P. De Kepper and J. Boissonade, in: *Oscillations and Traveling Waves in Chemical Systems* (R. J. Field and M. Burger, eds.), Wiley-Interscience, New York (1985).
53. A. Pacault, P. De Kepper, P. Hanusse, and A. Rossi, *C. R. Acad. Sci.* **281C**, 215 (1975); P. De Kepper and A. Pacault, C. R. Acad. Sci. **286C**, 437 (1978).
54. J. Chopin-Dumas, C. R. Acad. Sci. Paris **287C**, 553 (1978).
55. M. Alamgir and I. R. Epstein, *J. Am. Chem. Soc.* **105**, 2500 (1983).
56. P. De Kepper and J. Boissonade, *J. Chem. Phys.* **75**, 189 (1981).
57. M. Alamgir, M. Orbàn, and I. R. Epstein, *J. Phys. Chem.* **87**, 3725 (1983).
58. J. Maselko, M. Alamgir, and I. R. Epstein, *Physica* **D19**, 153 (1986).

59. J. S. Reckley and K. Showalter, *J. Am. Chem. Soc.* **103**, 7012 (1981).
60. P. De Kepper, Q. Ouyang, and E. Dulos, in: *Non-Equilibrium Dynamics in Chemical Systems* (C. Vidal and A. Pacault, eds.), Synergetics Vol. 27, p. 44, Springer-Verlag, Berlin (1984).
61. M. Orbàn and I. R. Epstein, *J. Am. Chem. Soc.* **104**, 5918 (1982).
62. G. Bazsa and I. R. Epstein, *J. Phys. Chem.* **84**, 3050 (1985).
63. P. De Kepper, J. Boissonade, and R. Khalifeh, to be published.
64. D. Ruelle, *Trans. N. Y. Acad. Sci.* II **35**, 66 (1973).
65. P. Berge, Y. Pomeau, and C. Vidal, *L'ordre dans le chaos*, Hermann, Paris (1984).
66. H. L. Swinney and J. C. Roux, in: *Non-Equilibrium Dynamics in Chemical Systems* (C. Vidal and A. Pacault, eds.), Synergetics Vol. 27, Springer-Verlag, Berlin (1984).
67. F. Takens, *Lecture Notes in Mathematics*, **898** (D. A. Rand and S. L. Young, eds.), p. 366, Springer Verlag, Berlin (1981).
68. M. Orbàn and I. R. Epstein, *J. Phys. Chem.* **86**, 3907 (1982).
69. J. S. Turner, J. C. Roux, W. E. McCormick, and H. L. Swinney, *Phys. Lett.* **85A**, 9 (1981).
70. D. Farmer, E. Ott, and J. Yorke, *Physica* **7D**, 153 (1983).
71. M. J. Feigenbaum, *J. Stat. Phys.* **19**, 25 (1978); P. Collet and J. P. Eckman, *Iterated Maps of the Interval as Dynamical Systems*, Birkhauser, Boston (1980).
72. R. H. Simoyi, A. Wolf, and H. L. Swinney, *Phys. Rev. Lett.* **49**, 245 (1982).
73. Y. Pomeau, J. C. Roux, A. Rossi, S. Bachelart, and C. Vidal, *J. Phys. Lett.* **42,** L 271 (1981); J. C. Roux, *Physica* **7D**, 57 (1983).
74. J. C. Roux, P. Richetti, A. Arneodo, and F. Argoul, *J. Mec. Théorique et Appliquée*, Special edition, p. 77 (1984).
75. J. C. Roux and A. Rossi, in *Non-Equilibrium Dynamics in Chemical Systems* (C. Vidal and A. Pacault, eds.), *Synergetics* Vol. 27, p. 141, Springer-Verlag, Berlin (1984).
76. F. Argoul, A. Arneodo, P. Richetti, and J. C. Roux, *J. Chem. Phys.*, **86**, 3325 (1987); P. Richetti, J. C. Roux, F. Argoul and A. Arneodo, *J. Chem. Phys.*, **86**, 3339 (1987).
77. P. De Kepper, I. R. Epstein, K. Kustin, and M. Orbàn, *J. Phys. Chem.* **86**, 170 (1982).
78. A. T. Winfree, *Sci. Am.* **230**(12), 82 (1974).
79. A. M. Turing, *Phil. Trans. R. Soc. London* **237B**, 37 (1952).
80. P. Hanusse, in: *Far From Equilibrium* (A. Pacault and C. Vidal, eds.), Synergetics Vol. 3, p. 70, Springer-Verlag, Berlin (1979).
81. A. Pacault and C. Vidal, *J. Chim. Phys. (Paris)* **79**, 691 (1982); C. Vidal and A. Pacault, in: *Nonlinear Phenomena in Chemical Dynamics*, Springer-Verlag, Heidelberg (1981).
82. M. L. Smoes, in: *Dynamics of Synergetic Systems*, p. 80, H. Haken, Springer-Verlag (1980).
83. A. T. Winfree, in: *Oscillations and Traveling Waves in Chemical Systems* (R. J. Field and M. Burger, eds.), p. 441, Wiley-Interscience, New York (1985).
84. K. I. Agladze and V. I. Krinsky, *Nature* **296**, 424 (1982).
85. A. Saul and K. Showalter, in: *Oscillations and Traveling Waves in Chemical Systems* (R. J. Field and M. Burger, eds.), p. 419, Wiley-Interscience, New York (1985).
86. G. Bazsa and I. R. Epstein, *J. Phys. Chem.* **89**, 3050 (1985).
87. J. J. Tyson, in: *Oscillations and Traveling Waves in Chemical Systems* (R. J. Field and M. Burger, eds.), p. 93, Wiley-Interscience, New York (1985).
88. N. Kopell and L. N. Howard, *Adv. Appl. Math.* **2**, 417 (1981).
89. D. Walgraef, G. Dewel, and P. Borckmans, *J. Chem. Phys.* **78**, 3043 (1983).
90. A. M. Ahabotinsky and A. N. Zaikin, *J. Theor. Biol.* **40**, 45 (1973).
91. M. Orbàn, *J. Am. Chem. Soc.* **102**, 4311 (1980).
92. K. Showalter, *J. Chem. Phys.* **73**, 3735 (1980).
93. P. Moclel, *Naturwissenschaften* **64**, 224 (1977).
94. D. Avnir and M. Kagan, *Nature* **307**, 717 (1984).
95. M. Gimenez and J. C. Micheau, *Naturwissenschaften* **70**, 90 (1983).
96. J. C. Micheau, M. Gimenez, P. Borckmans, and G. Dewel, *Nature* **305**, 43 (1983).
97. I. Nagypal, G. Bazsa, and I. R. Epstein, *J. Am. Chem. Soc.*, **108**, 3635 (1986).
98. E. Dulos and P. De Kepper, *Biophys. Chem.* **18**, 211 (1983).

99. M. Ruoff, *J. Phys. Chem.* **88**, 2851 (1984).
100. M. Marek, in: *Temporal Order* (L. Rensing and N. I. Jaeger, eds.), Springer Series in Synergetics 29, p. 105 (1985).
101. P. De Kepper and W. Horsthemke, in: *Far From Equilibrium* (A. Pacault and C. Vidal, eds.), Springer Series in Synergetics, No. 3, p. 61, Springer-Verlag, Berlin (1979).
102. K. Nakajima and Y. Sawada, *J. Chem. Phys.* **72**, 2231 (1980).
103. I. Stuchl and M. Marek, *J. Chem. Phys.* 77, 2956 (1982).

5

Synergetics—From Physics to Biology

H. Haken

5.1. The Developing World of Physics

This heading is not to be understood in a political sense but in the following way: Let us consider what one might call the industrialized world of physics. It deals with high-energy or high-power physics. Here we are thinking of the huge machines to create new elementary particles or the powerful stellarators, tokomaks, etc., to generate plasmas at very high temperatures in order to eventually produce fusion.

It is a fascinating world, but in order to conduct experiments enormous sums of money are required. To this world there belongs also the world of astrophysics with its telescopes, its rockets, and satellites. But then there is the world of our dimensions, maybe down to the dimensions of molecules. It is here where the greatest mystery of nature is buried, namely, the mystery of life.

Here, in general, only modest amounts of money are required, but the complexity of the problems represents a fascinating challenge to the human mind. In particular, I believe that physicists with their experience in statistical physics, many-body theory, and so on, can make fundamental contributions to this "developing world."

This chapter deals somewhat with the question: What can physics contribute to the understanding of life? Indeed all living beings are composed of atoms and molecules, which quite clearly obey the laws of physics. Therefore, it has always been quite a legitimate question for a biologist to ask a physicist

---

H. Haken • Institut für Theoretische Physik und Synergetik, Universität Stuttgart, 7000 Stuttgart 80, Federal Republic of Germany.

whether he can explain, at least in principle, the formation of ordered structures and eventually the structures of life.

The discipline of physics which is quite generally concerned with order and disorder is thermodynamics, including irreversible thermodynamics. However, it is interesting to note that this simple question has embarrassed physicists for quite a long time. Indeed, there are experiments in physics belonging to our daily experience that seem to contradict the formation of structures. We know that in closed systems the second law of thermodynamics holds according to which entropy increases; this implies that disorder increases or, in other words, that in such a system structures disappear.

Boltzmann, the founder of statistical mechanics, was entirely aware of this difficulty but his statistical mechanics allowed him to make an interesting suggestion. He said: "Life is a giant fluctuation" which is indeed still possible in statistical mechanics and does not violate fundamental laws of physics. But there are two objections. First, such a giant fluctuation is extremely improbable and, in fact, we must admit that its probability is practically vanishing.

And there is a second point. If life is just a giant fluctuation, why does complexity, at least in general, increase? In fact we find in evolution that we have an ever-increasing complexity and there is no reason to expect such a behavior if life is just a giant fluctuation. Now, if closed systems do not provide us with an explanation of increasing structures, at least in principle, we may ask: what happens in open systems? A typical open system is one into which energy is pumped from an energy source and eventually energy is given away into an energy sink.

A typical example for such a system is provided by the earth, which receives a flux of energy from the sun and which, especially during night, dissipates energy into the very cold universe. Such a system is kept far from thermal equilibrium. But then a number of questions immediately arise. Can one define entropy for an open system? This is actually a question which is at least open, if not to say that it is still under dispute. Second, thermodynamics does not offer us any principle or mechanism as to how order arises.

In statistical physics we may define an entropy function. This allows us immediately to check some statements made by people who based their work mainly on thermodynamics. For instance, they tell us that the entropy decreases when a system far from thermal equilibrium goes from a disordered into an ordered state. However, actually more recently I could show in detail that such a statement is wrong: the entropy may even increase.

Quite evidently new principles are needed to explain why order can be created in systems far from thermal equilibrium. But these new principles must not contradict thermodynamics; rather they must be added to it. So in a way the situation is reminiscent of Euclidian geometry. Euclid tried to derive the existence of parallels from his axioms, but we nowadays know that a new axiom—the parallel axiom—actually was needed to establish this fact. I was very lucky to come across a physical system more than 25 years ago which allowed me to find such principles. The system I am referring to is the laser.

5.2. The Laser Paradigm

We shall briefly remind the reader of such a system. On the one hand, a laser is composed of a cavity, which means that two mirrors are mounted on the end faces of the laser device (see Figure 5.1). Standing axial waves can develop in such an arrangement. On the other hand, the laser-active material is assumed to be composed of atoms distinguished by subscript μ, and we assume that each atom has dipole moment p_μ.

The electric field strength of a standing wave is given by

$$E(x, t) = A(t) \sin kx \tag{5.2.1}$$

The amplitude $A(t)$ is decomposed into a rapidly varying part $\exp(-i\omega t)$ multiplied by a slowly varying amplitude $E(t)$ and, of course, the complex conjugate must be added. The frequency ω occurring in the equation

$$A(t) = E(t)\, e^{-i\omega t} + E^*(t)\, e^{i\omega t} \tag{5.2.2}$$

is just the optical transition frequency of the atom which supports the laser action. Here and in the following we shall actually assume that there is perfect resonance between the field mode with frequency ω and the optical transition frequency of the individual atoms.

In analogy to the decomposition of $A(t)$, we decompose the dipole moment $\hat{p}_\mu(t)$ according to the expression

$$\hat{p}_\mu(t) = p_\mu(t)\, e^{-i\omega t} + p_\mu^*(t)\, e^{i\omega t} \tag{5.2.3}$$

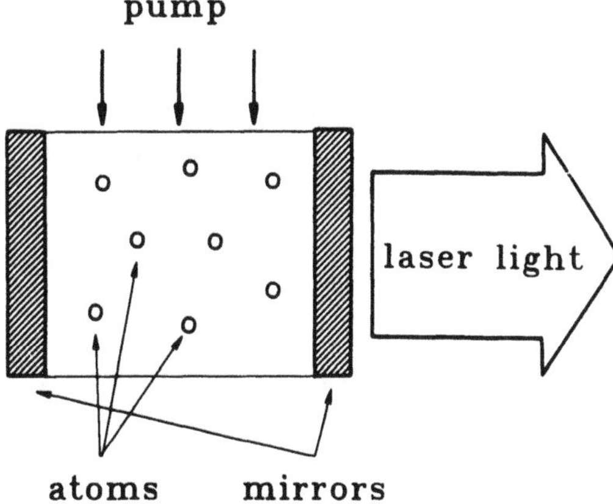

Figure 5.1. Typical setup of a laser.

The derivation of the basic laser equations will not be repeated here; rather, we write down their form in order to be able to draw some important conclusions. The slowly varying field amplitude $E(t)$ changes in time due to various causes that can be expressed in the form

$$\dot{E}(t) = -\kappa E + g \sum_{\mu} p_{\mu} + F(t) \tag{5.2.4}$$

First of all the light field may escape through the mirrors; this leads to a damping $-\kappa E$. On the other hand, the field mode is driven by the oscillating atomic dipole moments which, in the approximation used here, is described by the sum occurring in equation (5.2.4), where g is the coupling constant between E and p_{μ}. Finally, function F represents a fluctuating force. As is known, whenever there is dissipation in a physical system there are also fluctuations. Therefore, for instance, the electrons in the mirrors which cause damping of the field strength cause, at the same time, fluctuations acting on the field strength E. The field reacts on the atomic dipole moments so that we must consider equations for them. According to quantum theory one can show that the dipole moment of the atom μ changes in time due to various causes that can be written as

$$\dot{p}_{\mu} = -\gamma p_{\mu} + gEd_{\mu} + F_{\mu}(t) \tag{5.2.5}$$

Here, owing to the interaction of the atom with its surroundings, including the effects of radiative transitions into all possible modes, the dipole moment will be damped by means of a term $-\gamma p_{\mu}$. On the other hand, the field drives a dipole moment which is described by the second term on the right-hand side of this equation. This introduces a new dynamical variable, namely d_{μ}, and is a consequence of quantum mechanics, which applies in our case to a system of atoms each having two levels only. Now quite clearly energy may flow from the atom into the field if the atom is in its upper state. On the other hand, the atom will absorb energy from the field if it is in its lower state. Therefore there is a specific sign of the energy flow and this sign is determined by the difference of the occupation numbers of the upper and lower levels. This difference is called the inversion, which is given by

$$d_{\mu} = N_{2,\mu} - N_{1,\mu} \tag{5.2.6}$$

As a result of the processes going on in the laser, the inversion also changes as a function of time according to the equation

$$\dot{d}_{\mu} = (d_0 - d_{\mu})\gamma' - g(E^* p_{\mu} + E p_{\mu}^*) + f_{\mu}(t) \tag{5.2.7}$$

Here, the first term describes a relaxation process caused by the pump introduced into the atomic system from the outside and simultaneously by nonradiative transitions or radiative transitions into modes which do not lead to laser action. Thus the first term describes a relaxation of the atomic inversion d_μ toward a prescribed inversion d_0. On the other hand, the atom interacts with the field and exchanges energy with it. This is described by the second term of this equation. The last term then describes still fluctuations connected with the pump processes.

The above set of equations is rather complicated. First, it has very many variables; for instance, in a typical laser there may be 10^{16} degrees of freedom. Furthermore, these equations are nonlinear, as is shown by the term Ed_μ or by the terms occurring in the second set of parentheses in the last equation.

Fortunately, it has been found possible to reduce this complicated system quite considerably. In most lasers, the inequality $\gamma \gg \gamma' \gg \kappa$ holds, which means the damping constant of the dipole moments is much bigger than the damping constant of the inversion, which in turn is much bigger than the damping constant of the field in the cavity.

We now consider equation (5.2.4). Its first term on the right-hand side tells us that the field decays rather slowly; this is described by the damping constant κ. On the other hand, in equation (5.2.5), the field E acts as a driving force. When we are close to a steady state, p_μ will behave similarly to E. Hence, on the left-hand side of equation (5.2.5) we obtain practically a term κp_μ which, of course, is much smaller than γp_μ. But the resulting inequality $\kappa p_\mu \ll \gamma p_\mu$ implies that, in practice, we may set p_μ equal to zero. When doing so, we may immediately solve equation (5.2.5) with respect to p_μ and obtain the result

$$p_\mu = \frac{1}{\gamma} E d_\mu + \frac{1}{\gamma} F_\mu(t) \tag{5.2.8}$$

This result tells us that the dipole moment p_μ is immediately determined by the field strength E. This is, in fact, the slaving principle in its simplest form. Quite generally the slaving principle, at least in the present context, tells us the following: Quickly relaxing variables, namely p_μ, are slaved by slowly relaxing variables, namely E.

Similar statements hold for d_μ, but it is immediately seen that by this slaving principles p_μ and d_μ can be directly expressed in terms of E. In other words, they can be eliminated from equations (5.2.4), (5.2.5), and (5.2.7). Therefore, we can find an equation for E alone in the following form, at least for values of E not too large:

$$\dot{E} = (G - \kappa)E - CEE^2 + F_{\text{tot}}(t) \tag{5.2.9}$$

The constant G is called the gain constant and is proportional to d_0, namely, the inversion described by pumping and relaxation processes. Equation (5.2.9) is not quite trivial because, on the one hand, it is nonlinear and, on the other, it contains a fluctuating force, i.e., it is a nonlinear stochastic equation.

When we resort to mechanics, however, we can easily interpret the meaning of this equation. Let us identify E with a particle coordinate q and add an acceleration term $m\ddot{q}$ to the equation for E. Then this equation assumes the form

$$m\ddot{q} + \dot{q} = -\frac{\partial V}{\partial q} + F_{\text{tot}}(t) \tag{5.2.10}$$

where V is a potential function which can be easily constructed for the driving force on the right-hand side of equation (5.2.10). In order to discuss the behavior of E, it is first assumed that the gain G is smaller than the loss κ. Then the potential V can be represented by Figure 5.2. The effect of the fluctuating force can be represented by individual random pushes exerted on the individual particle. After each push it will relax toward the equilibrium position. When we plot the field strength E vs. time, we find the envelope of the curve shown in Figure 5.3.

In order to find the field strength, we multiply E by $\exp(-i\omega t)$ and then take the real part. This is then indicated in Figure 5.3 by the oscillatory curve. Suppose we now increase the gain constant G so that the potential curve becomes flatter (Figure 5.4). Quite evidently the particle will fall down the potential hill at the lower speed, hence the wave tracks shown in Figure 5.5 now decrease more slowly than those in Figure 5.3.

It is known from optics that the decay constant is responsible for the linewidth $\Delta\nu$. Because the decay constant becomes smaller, the decay actually occurs more slowly, and the linewidth will decrease. So we find a decrease according to the expression

$$\Delta\nu = \frac{\text{const}}{P} \tag{5.2.11}$$

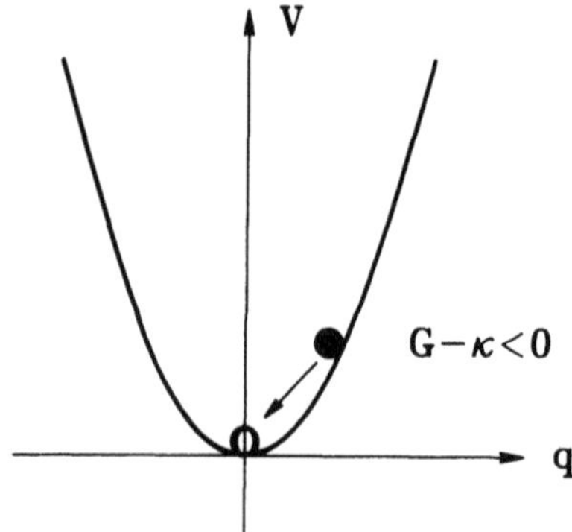

Figure 5.2. Effective potential for $G - \kappa < 0$.

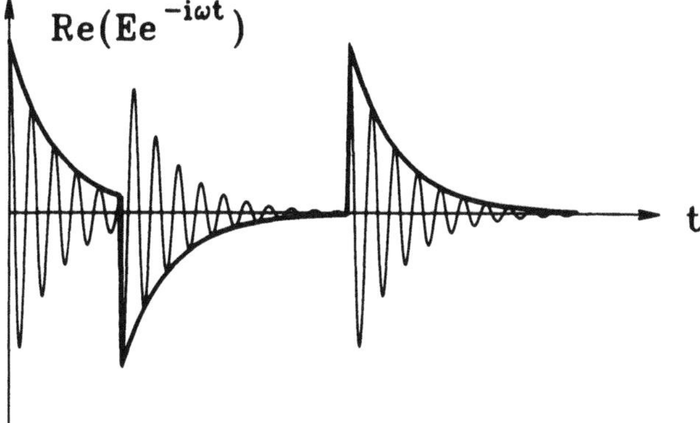

Figure 5.3. The electric field strength of the emitted light consists of individual wave tracks.

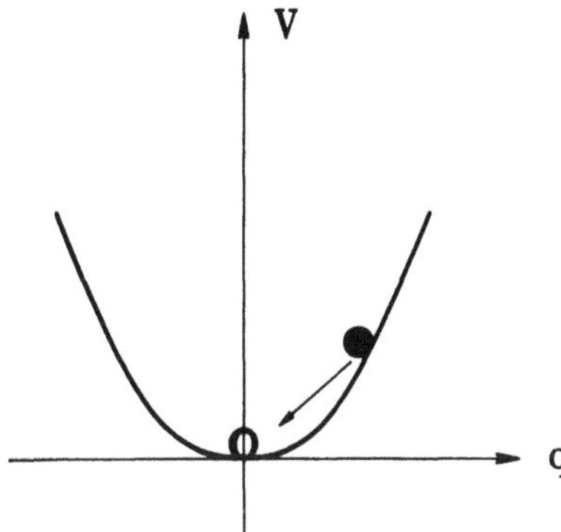

Figure 5.4. With increased gain the potential curve becomes flatter.

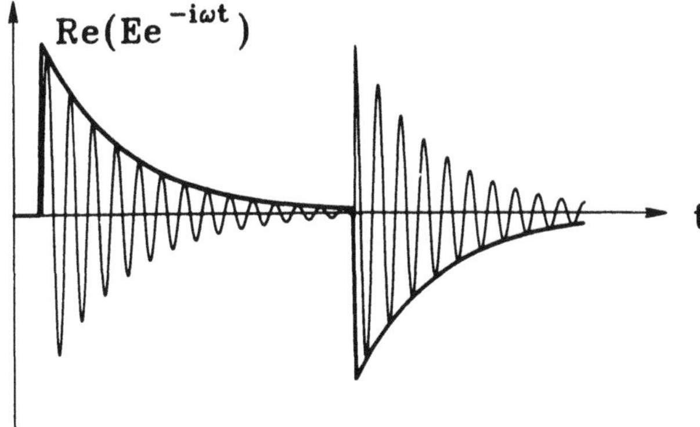

Figure 5.5. The electric field strength for the emitted light consists again of wave tracks which, for increased gain, decay more slowly, however.

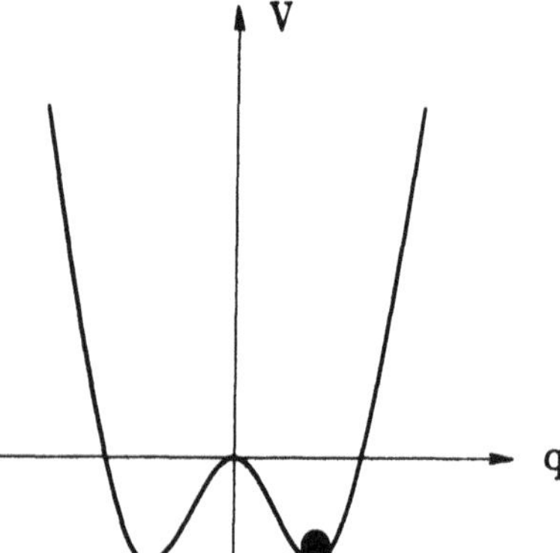

Figure 5.6. Potential curve for $G - \kappa > 0$.

where P is the output power of the laser. In this way we have recovered the famous Townes formula.

We now consider what happens if the gain G becomes bigger than the loss κ. Under this condition the potential function V acquires a form as shown in Figure 5.6. Here apparently a new stable state has arisen that is connected with a stable amplitude of the field E. When we again multiply E by the function $\exp(-i\omega t)$ and take the real part, we obtain the solid line of Figure 5.7.

So we are immediately led to the idea that we have here a field with a stabilized amplitude. That means we expect a very pronounced coherence. Such a field has also a new kind of photon statistics. Owing to the fluctuating forces there will be small amplitude and phase fluctuations. It is rather straightforward to calculate the intensity fluctuations stemming from the amplitude fluctuations. If we divide the intensity fluctuations by the average output intensity and plot this ratio vs. the gain, Figure 5.8 results.

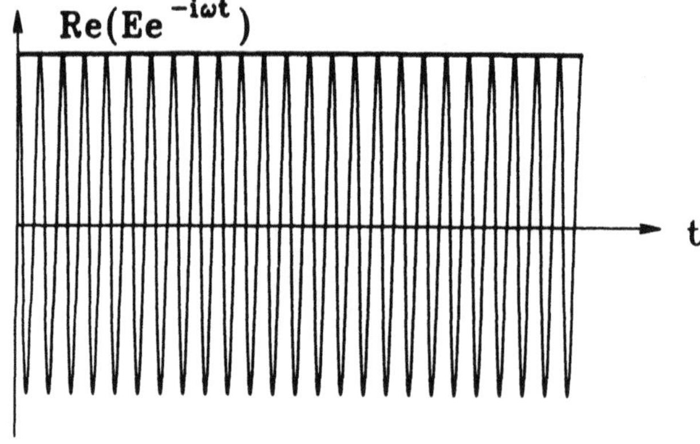

Figure 5.7. The electric field strength has become an amplitude stabilized wave.

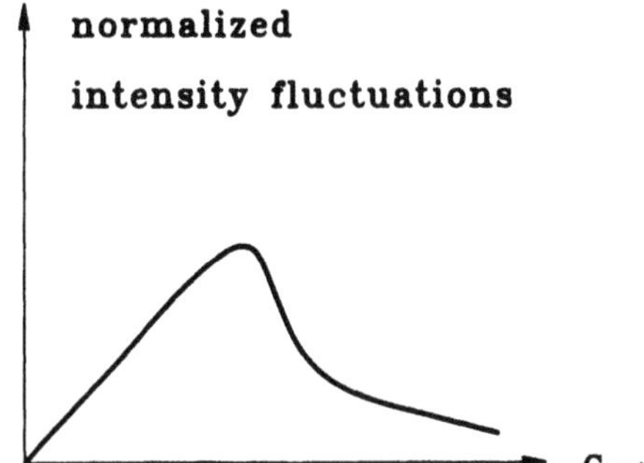

Figure 5.8. Normalized intensity fluctuations of a laser vs. pump power.

In addition, one may calculate the linewidth, which now turns out to be given by the expression

$$\Delta\nu = \frac{\text{const}}{2P} \tag{5.2.12}$$

When I derived the result shown in Figures 5.7 and 5.8, I found myself in opposition to all previously published papers and therefore I asked some of the experts whether such behavior could occur. And they told me: No, this cannot be the case because the laser will approach a potential curve as shown in Figure 5.4. The only fact happening is that this potential curve becomes flatter and flatter, which is connected with a higher and higher output of the laser and the linewidth becoming narrower and narrower. But I was so very convinced of the correctness of my idea that I published my results and fortunately enough two American physicists, Armstrong and Smith, became interested in my results and conducted experiments on semiconductor lasers measuring the intensity fluctuations. Their measurements entirely confirmed my predictions.

From that time on, I have learned that one must not trust experts. Rather one really has to convince oneself whether concepts or theories are right or wrong. Why did I tell you this old story about the laser? The reason is that it will guide us into all sorts of problems of self-organization. But first let me consider the relations to other concepts.

5.3. Relations to Other Concepts

When we plot Figures 5.3 and 5.6 together, we obtain Figure 5.9 where the curve with one minimum refers to a region below threshold and the curve with two minima to the region above threshold, $G > \kappa$. Actually, it should be mentioned that in reality Figure 5.9 is somewhat more complicated, because

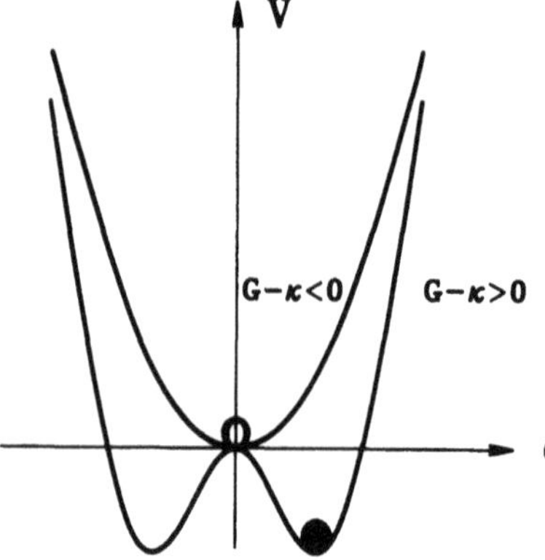

Figure 5.9. The potential curve for $G - \kappa < 0$ and $G - \kappa > 0$.

we have not only two minima but some sort of a circular valley obtained by rotating the whole curve around the V axis. We now consider the more simplified case with two minima only.

Quite evidently the right-hand position is equivalent to the left-hand position of the minima. Clearly, when we change from the curve with $G - \kappa < 0$ to $G - \kappa > 0$ the position at $q = 0$ becomes unstable, but at the same time the system must break the symmetry between the by now possible left- and right-hand positions of the newly generated minima. Such a phenomenon is therefore called "symmetry breaking instability." Furthermore, it has been seen before that when we increase the gain below threshold, the particle relaxes more slowly, a phenomenon called critical slowing down or, equivalently, one says that a soft mode occurs.

Finally, we may observe that with a flatter curve the restoring force becomes smaller and in this way the fluctuations become bigger. These are the so-called critical fluctuations close to threshold. All these phenomena are strongly reminiscent of phenomena of systems close to phase-transition points, but in such a case the phase transitions occur in systems in thermal equilibrium. Here, on the other hand, we have a system, namely the laser, which is driven far from thermal equilibrium. Therefore the transition we found in it can be called a nonequilibrium phase transition and this detailed analogy was demonstrated by Graham and myself in 1968.

Actually, this analogy between phase transitions of systems in thermal equilibrium and far away from equilibrium can be made quite rigorous. For instance, one may derive the distribution function $f(q)$, which has the form

$$f(q) = N \exp[-V(q)/Q] \tag{5.3.1}$$

In this form V corresponds to a free energy, but only in a formal sense. Hence one may establish a formal analogy with the Landau theory of phase transitions, where q or equivalently E plays the role of an order parameter. It must be stressed, however, that there is a basic difference between systems in thermal equilibrium and away from equilibrium. While expression (5.3.1) is exact for the system here far from equilibrium, we know that the Landau theory is not sufficient for systems in thermal equilibrium, where rather the Wilson approach is required.

Another relation may be established when we plot the equilibrium position q_0 vs. the gain parameter G or equivalently vs. $G - \kappa$ as shown in Figure 5.10. Quite clearly, when $G - \kappa < 0$ only one equilibrium position $q_0 = 0$ exists. On the other hand, beyond the critical point $G = \kappa$, the curve splits into two branches, which have the shape of a fork. Therefore this phenomenon is called a bifurcation in mathematics (cf. Chapter 2) and there is a comprehensive literature on bifurcation theory now available. From a physicist's viewpoint, however, one must stress that the concept of a nonequilibrium phase transition including its mathematical treatment is far more general than the concept of bifurcation. For instance, in nonequilibrium phase transitions we consider also the neighborhood of the newly developing stable states and, in particular, we take into account the fluctuations which play a decisive role close to instability points.

Further contact can be made with problems in biology. What is most striking here is the pronounced coordination of the individual parts of a system, such as muscles and neurons in animals. So we consider again the laser under this aspect and first examine the case in which the order parameter E is small. Expression (5.2.8) yields that in such a case the fluctuations will dominate and we will have a situation as depicted in Figure 5.11. Here, the dipole moments point in random directions and we observe microscopic disorder.

Now let us consider the case in which E is comparatively large. In such a case all the dipole moments will point in a specific direction (Figure 5.12). Small fluctuations described by the second term in equation (5.2.8) will be superimposed on these now ordered states. At any rate, we realize that when E increases, the microscopic chaos of Figure 5.11 will be replaced by microscopic order (Figure 5.12). But this microscopic order becomes also manifest on macroscopic scales. Here we have a typical example of self-ordering. The

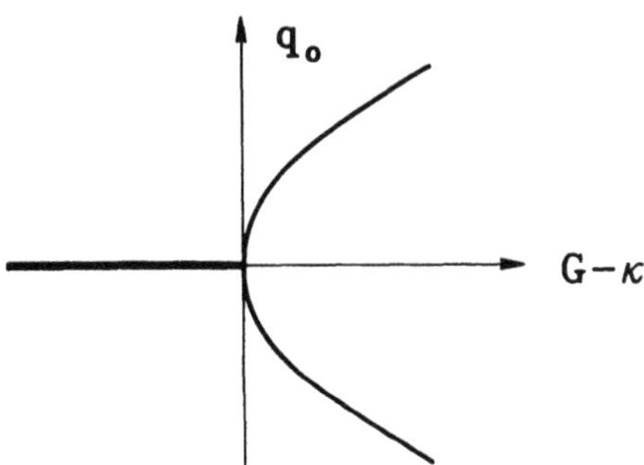

Figure 5.10. Bifurcation diagram.

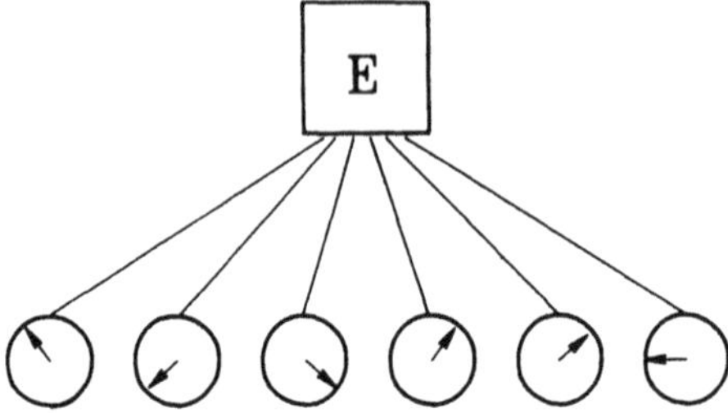

Figure 5.11. If the order parameter E is small, the dipole moments are at random.

order is not imposed on the system from the outside but rather generated by the system itself. Therefore we have here a typical example of self-organization. At the same time we observe an enormous reduction in the degrees of freedom. For instance, a typical laser has 10^{16} atoms and one light mode, but as we have seen here, due to the occurrence of macroscopic order the whole system is governed by the single order parameter E, implying a single degree of freedom.

Finally, contact can be made with information theory. We consider Figure 5.13 where, apparently, the system can stay in the state 0 or 1. In other words our self-organized system possesses two holding states or, in other words, it can store one bit of information. As we shall see elsewhere, we may now establish some kind of information theory based on self-organizing systems.

Before we arrive at the central topic of this chapter, namely the definition of synergetics, we note an additional point again concerned with lasers. It is known that in a laser not only a single mode may exist, but we may have several modes as indicated in Figure 5.14. We now consider another arrangement, namely, that shown in Figure 5.15, where instead of the standing waves of Figure 5.14 we have running waves, and one may construct the device in such a way that the waves are all running in one direction. We know from quantum mechanics that the individual waves are occupied by specific numbers

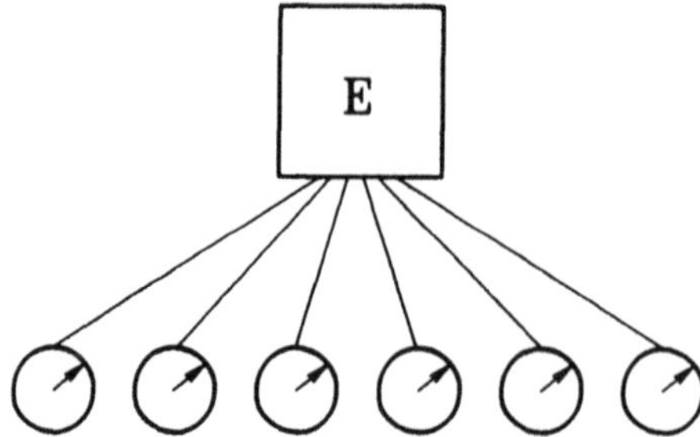

Figure 5.12. If the order parameter E is sufficiently large, the dipole moments become ordered.

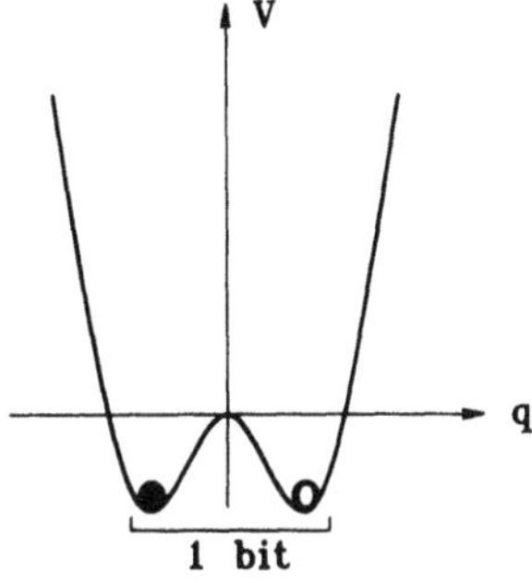

Figure 5.13. A double-well potential can store one bit.

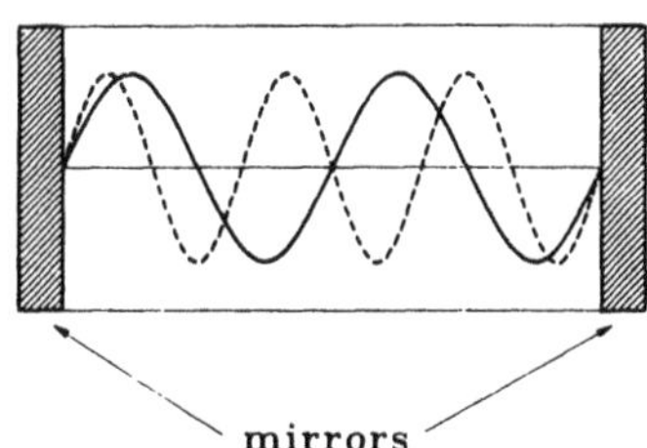

Figure 5.14. Standing waves in between two mirrors.

of photons. So wave with mode number 1 may be occupied by n_1 photons, mode number 2 by n_2 photons, and so on.

Many years ago we studied what happens in a ring laser when we give all the modes initially the same chance, i.e., if we occupy them initially each with the same number of photons. Then what happens is depicted in Figure 5.16. First the individual mode numbers multiply, but their multiplication rate is different depending on the distance of the photon frequencies from the center of the emission line.

Thus, first we find what is called a segregation: one kind of photon multiplies more rapidly than others. However as time elapses, eventually only one kind of photon, namely that which was multiplying fastest, survives and all the others die out. Therefore we observe here a situation strongly reminiscent of Darwinism in the animate world; I drew this analogy nearly 20 years ago.

To come to the decisive step, let me briefly summarize what we have seen so far. First we have seen that there are nonequilibrium phase transitions in lasers which are strongly reminiscent of phase transitions in systems in thermal equilibrium. Phase transitions in thermal equilibrium are a widespread phenomenon. Therefore the question arises whether the nonequilibrium phase

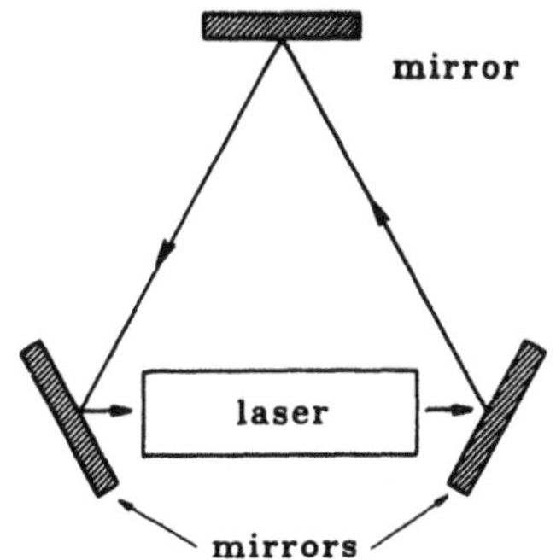

Figure 5.15. Running waves in a ring laser.

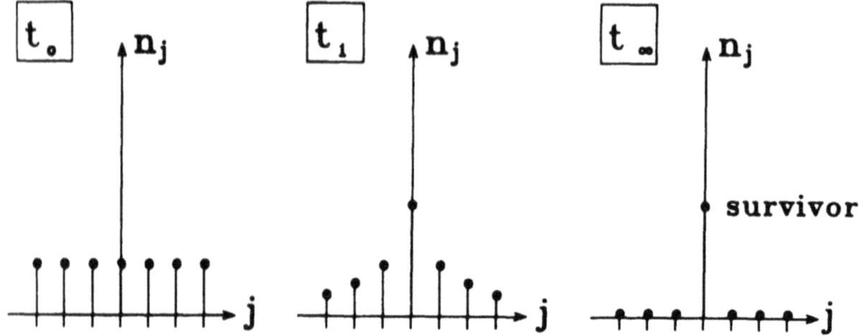

Figure 5.16. Darwinism of laser modes. At time t_0 all modes have the same number of photons. At a later time t_1, a segregation has taken place. At a sufficiently large time t_∞, only one kind of photon has survived.

transitions are not also a widespread phenomenon and the laser is just a specific example!

Second, we have just seen that there is a Darwinism of laser photons. On the other hand, Darwinism of the animate world is a quite general principle of its evolution. Therefore the question arises as to whether Darwinism is not also a quite general principle in the inanimate world. When we put these questions together, we are led to the general question, whether there are general principles in the inanimate and animate world close to phase-transition points, i.e., to points where new structures are formed.

5.4. Synergetics

This brings me directly to the definition of the new interdisciplinary field that I called synergetics. The word is taken from Greek and means: "science of cooperation." I introduced this term in 1969 during a lecture given at the University of Stuttgart. As we all know, most objects of study in science and the humanities can be decomposed into individual parts, or subsystems, or elements. But through the cooperation of these subsystems, a total action on a macroscopic level is produced or macroscopic structures are generated.

The question I then posed is: Are there general principles which govern this generation of macroscopic spatial, temporal, or functional structures, irrespective of the nature of the subsystems?

This question sounds at first sight crazy because the subsystems may be as diverse as atoms, or molecules, or photons, or cells, or organs, or even groups of humans. However, in a way I have been lucky because I could find some general principles, provided we focus our attention on those situations where the system changes its properties qualitatively. But what are now qualitative changes on macroscopic scales?

To this end I have first to explain what I understand by qualitative changes and, second, I must give examples at least of what we mean by macroscopic scales. First I present a counterexample against a qualitative change. To this end we consider Figure 5.17, where two kinds of fish are shown, namely, a porcupine fish and a sunfish. The famous Scottish biologist D'Arcy-Wentworth-Thompson drew one of these fishes on a grid, i.e., a coordinate system, and then deformed that grid like a rubber sheet. Then he could show that one obtains the sunfish by a mere deformation. A mathematician calls such a transformation structurally stable. That means an eye is transformed into another eye, a fin into a fin, and so on, where all the relative positions are maintained. Biology provides us also with numerous examples of transitions which are not structurally stable but give rise to qualitative changes. Figure 5.18 shows the developmental stages of a California newt. Here the fertilized egg starts to develop incensions due to cell division, and so on, until the whole animal develops its extremities. Quite clearly, each state differs from the foregoing one by a qualitative macroscopic change. In synergetics we shall be concerned with such qualitative changes. We shall now explain what we understand by macroscopic scales.

When we look at the sky we occasionally observe cloud streets. Here the molecules of the water vapor organize themselves into streets, which have dimensions of many hundreds of meters if not kilometers. Similar phenomena can be generated in the laboratory. A famous example is provided by the Bénard instability, where a fluid layer is heated from below and develops a specific pattern, such as the form of honeycomb cells, if the temperature difference between the lower and upper surfaces is big enough. Other examples

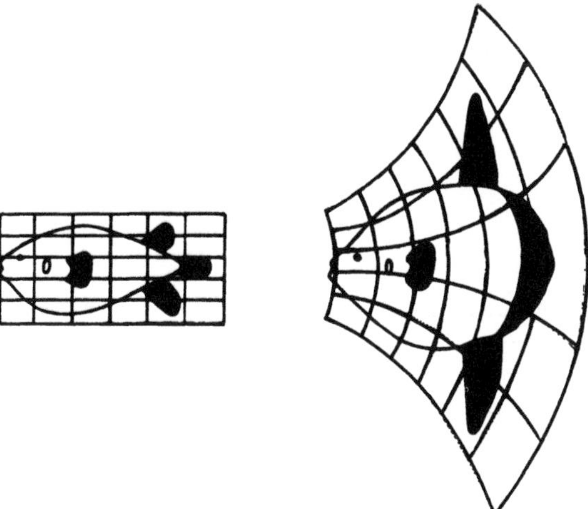

Figure 5.17. The porcupine fish (left) and the sunfish (right) can be transformed one into another by a simple coordinate transformation (after D'Arcy-Wentworth-Thompson).

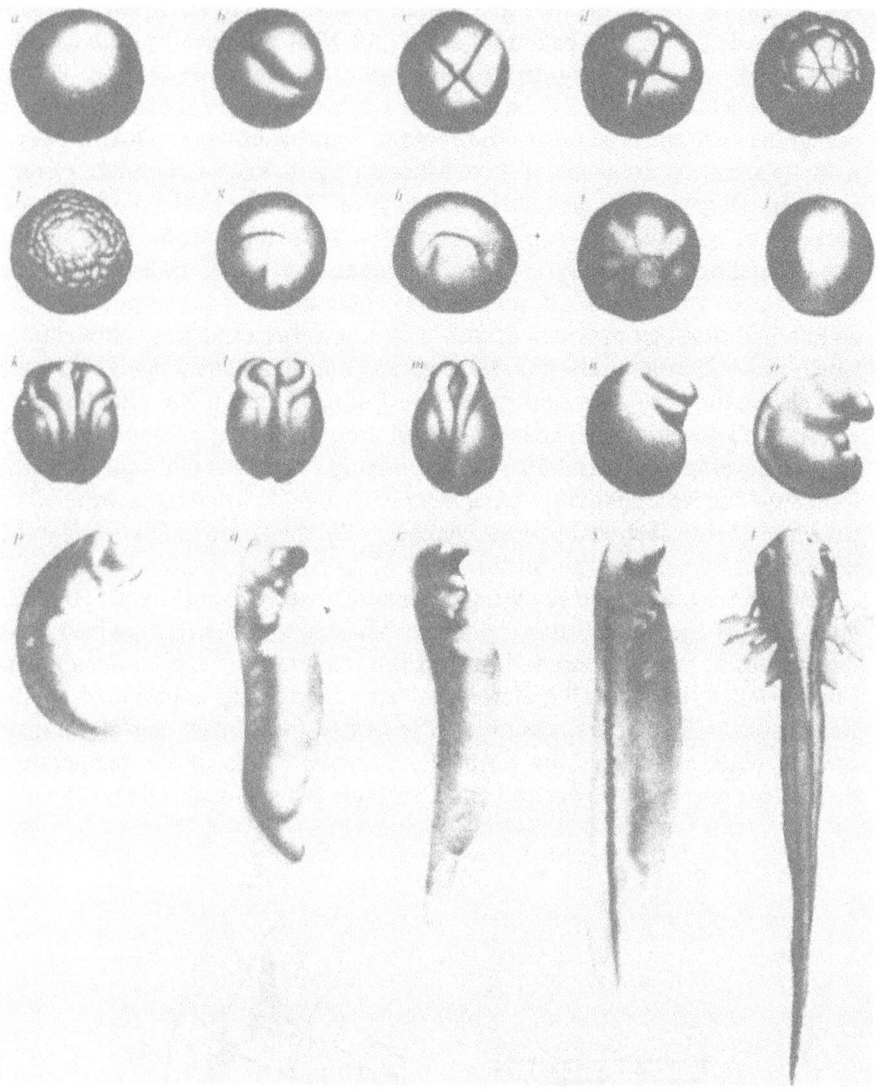

Figure 5.18. Developmental stages of a California newt.

are provided by chemistry, where we may find chemical oscillations, or spiral waves, or concentric waves possessing dimensions of several millimeters to centimeters. Further examples are provided by flames produced by plane burners.

Let us now briefly discuss the general case, though it will be certainly far beyond the scope of this presentation to supply mathematical details. The general case refers to a variety of fields, such as physics, chemistry, biology, and many other fields including specific aspects of economy and sociology. In our quantitative treatment we start with evolution equations for a state vector $\mathbf{q}$. Such a state vector $\mathbf{q}$ may comprise, in the case of the laser, the electric field strength, the atomic polarization, the atomic inversion, or in the case of a liquid it may comprise the density of the liquid, its velocity field, and its temperature field.

Evolution equations are of the general form

$$\dot{\mathbf{q}} = \mathbf{N}(\mathbf{q}, \alpha) + \mathbf{F}(t) \tag{5.4.1}$$

where $\mathbf{N}$ is a nonlinear function of $\mathbf{q}$ which may still depend on control parameters, α, which control the system from the outside, for instance, by temperature, a flux of energy into the system, and so on. Quantity $\mathbf{F}(t)$ represents fluctuations, which are either internal in the system or external. There are still more complicated formulations possible, for instance, where the fluctuations depend on the state vector $\mathbf{q}$ itself.

The idea now is to start from a given control parameter, or a set of control parameters, and assume that for such a set of parameters the solution of a specific solution of equation (5.4.1) is known. The control parameter is now changed and we study whether the system remains stable. For instance, when we heat the liquid only weakly it will remain at rest. When we increase the temperature at the lower surface we find an instability, which eventually gives rise to the formation of rolls or hexagons. Linear stability analysis is a standard procedure, but it allows us to derive the collective modes which may be growing or decaying.

For example, a specific kind of roll may grow in amplitude while other kinds of modes may decay, even if they have been generated by a spontaneous fluctuation or artificially. Those collective modes which have a growing amplitude are called order parameters and allow us to describe the order of a system. In general there are few order parameters. All other modes, namely the decaying modes, can be expressed by means of the slaving principle in terms of the order parameters. This allows us to reduce appreciably the total number of degrees of freedom of the system.

Therefore we may establish order parameter equations possessing a rather simple structure, in particular, because close to instability points the order parameters are small and there are symmetries available. This allows us to cast the order parameter equations into universality classes which explains the fact that, for instance, a single mode laser, a roll pattern in a liquid, and a certain chemical oscillation behave in quite the same way, though they belong to quite different physical or chemical phenomena. The formalism then allows us to calculate the formation of spatial, temporal, or functional structures. When we increase the control parameter more and more, the system may run through an instability hierarchy, and quite often after a few steps we arrive at states which may be called deterministic chaos. The whole formalism has been applied by us to a great number of different systems and the next few figures illustrate a few examples of its application to physics and chemistry. In Section 5.6 we shall discuss one of its applications to biology.

5.5. Some Examples from Physics and Chemistry

Figure 5.19 shows the calculated evolution of an initially random velocity field of a fluid heated from below in a regular hexagonal pattern. Figure 5.20 shows the temperature field of a liquid layer (atmosphere) on a sphere, heated at the sphere and cooled from the outside, and subject to a rotation-symmetric gravitational field. Here a static pattern evolves. Figure 5.21 is the same as Figure 5.20, but here a rotationary temperature wave evolves. Figure 5.22 is

Figure 5.19. Computer calculation of the development of pattern in a rectangular container of fluid (after Bestehorn and Haken).

Figure 5.20. Temperature field on a sphere (see text; after Friedrich and Haken).

the same as Figure 5.20, but here a *chaotic* temperature field evolves. Figure 5.23 finally shows the calculated patterns of flame fronts on plane burners.

5.6. Modeling of Complex Systems by Means of the Order Parameter and Slaving Concepts: An Example from Biology

When we observe the motion of humans or animals, we are struck by their precise movement which is based on an extremely high coordination between the activities of neurons and muscles. One may speculate that this high coordination is produced by specific programs, so-called motor programs, of the neuronal net. More recent experiments and their interpretation by means of synergetics indicate rather different mechanisms, however.

To this end let us consider a rather simple, though still quite surprising, experiment conducted by Kelso. He asked test persons to move their fingers in parallel and then asked them to move their fingers more quickly (Figure 5.24). Surprisingly, it turned out that when the frequency of the finger oscillation was increased, a sudden involuntary change of the mode of motion occurred. Namely, instead of a motion in the form shown in Figure 5.24, a new symmetric motion occurred (Figure 5.25). We shall try to model this abrupt change of

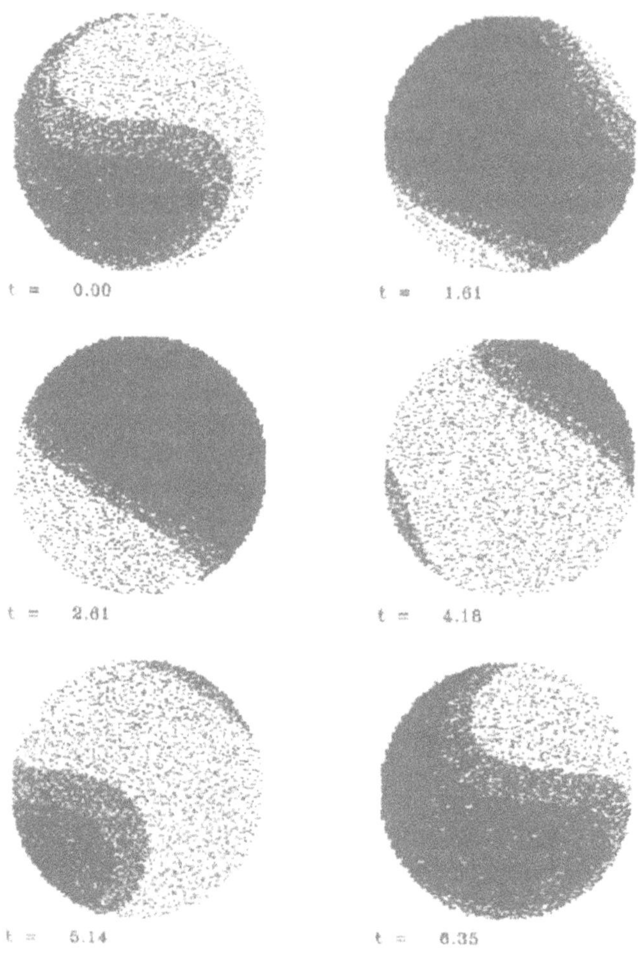

Figure 5.21. Temperature field in the form of a rotating wave (after Friedrich and Haken).

behavior and write the elongations of the finger tips in the form

$$x_1 = r_1 \cos(\omega t + \phi_1) \tag{5.6.1}$$

$$x_2 = r_2 \cos(\omega t + \phi_2) \tag{5.6.2}$$

where ω is the basic frequency of the hand movement, while the amplitudes r_1, r_2 and phases ϕ_1, ϕ_2 are time-dependent quantities, whose time dependence is assumed to be much slower than that defined by the frequency ω.

We define the relative phase by

$$\phi = \phi_2 - \phi_1 \tag{5.6.3}$$

In order to describe the change of phase we adopt our basic ideas introduced

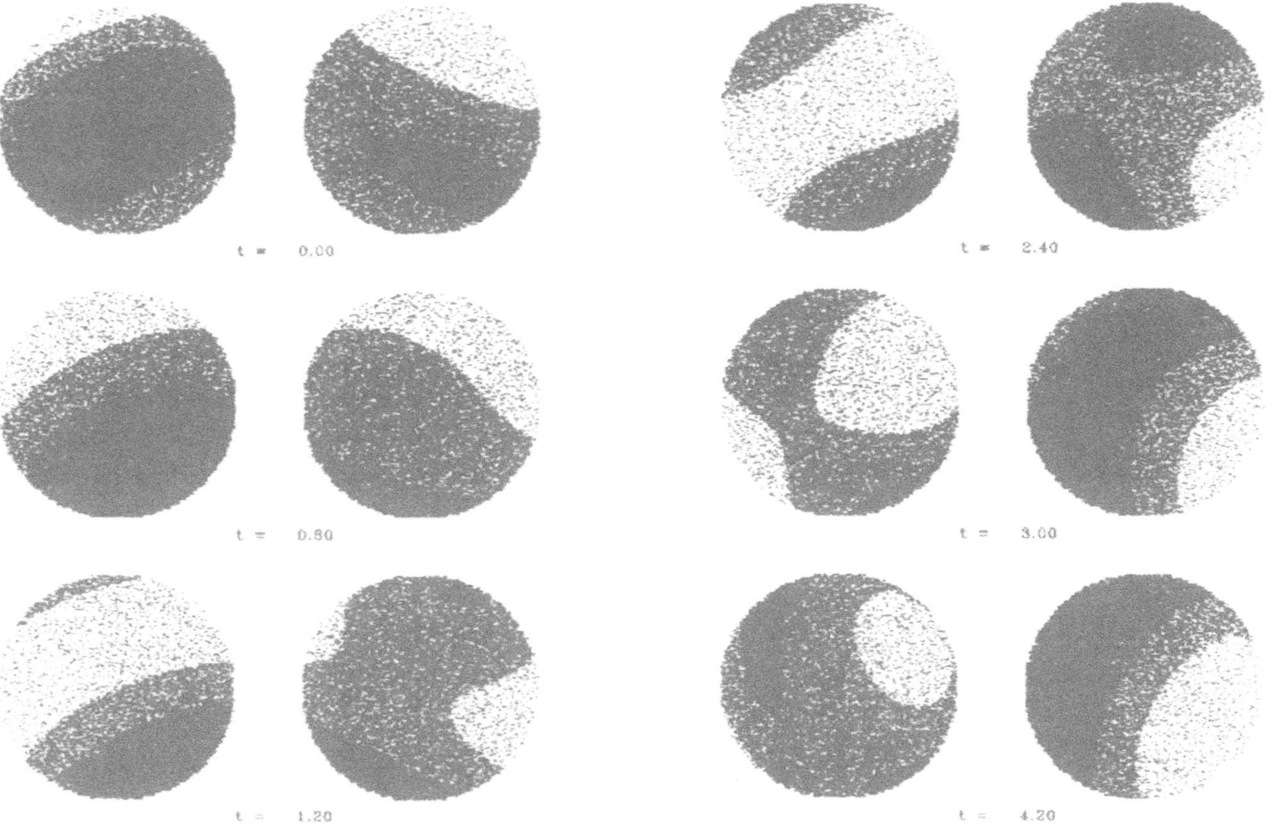

Figure 5.22. Temperature field showing time evolution of chaos (after Friedrich and Haken).

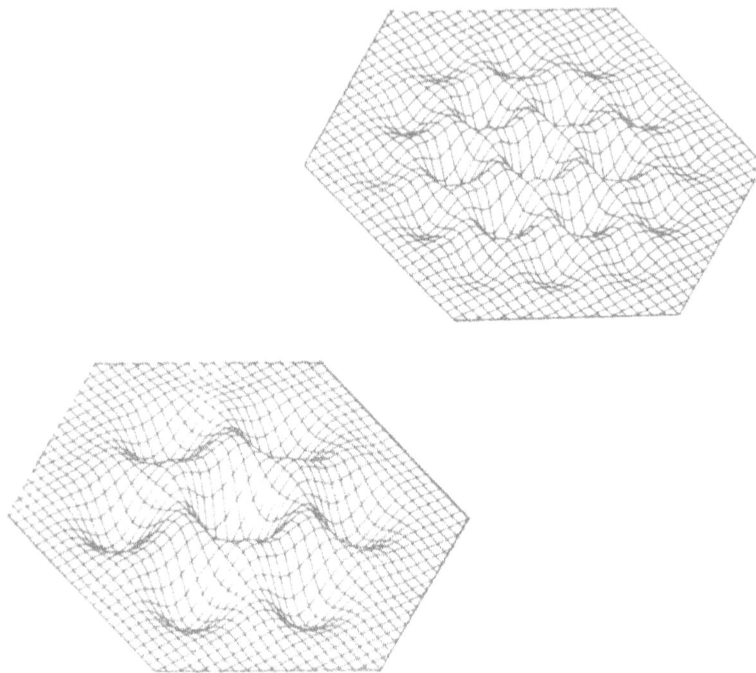

Figure 5.23. Hexagonal wave front of flames (after Schnaufer and Haken).

above. The order parameter equation is expected to have the form

$$\dot{\phi} = -\frac{\partial V}{\partial \phi} \tag{5.6.4}$$

where V is a potential function, in some analogy to that introduced above. In our search for a model we make a few rather obvious assumptions about V. Since ϕ occurs under cosine or sine functions, the properties of a physical

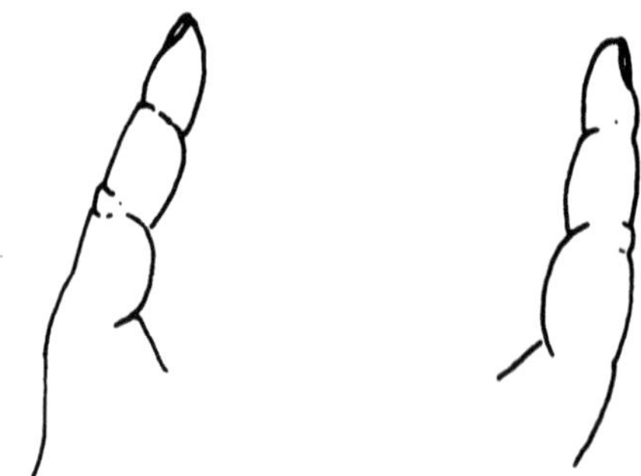

Figure 5.24. Parallel arrangement of fingers.

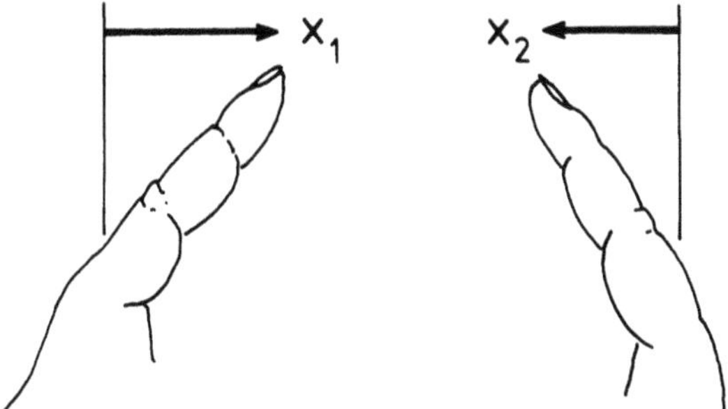

Figure 5.25. Symmetric arrangement of fingers.

system must not change when ϕ is replaced by $\phi + 2\pi$. Consequently, we shall postulate that the potential V is periodic:

$$V(\phi + 2\pi) = V(\phi) \tag{5.6.5}$$

We furthermore introduce the assumption that both hands play a symmetric role. In such a case the behavior of the system must not depend on the way we label the right and left hands. This means that V must remain unchanged when we exchange subscripts 1 and 2 of the two fingers. This in turn means that the potential V is symmetric:

$$V(\phi) = V(-\phi) \tag{5.6.6}$$

We assume that V obeys conditions (5.6.5) and (5.6.6) in the simplest possible form, which explains the above-mentioned experimental results. To this end we express V as a superposition of two cosine functions:

$$V = -a \cos \phi - b \cos 2\phi \tag{5.6.7}$$

It was seen above that the behavior of the system obeying equation (5.6.4) can be easily described by identifying ϕ with the coordinate of a particle moving in an overdamped fashion in the potential V. This potential V is represented in Figure 5.26 for various values of the ratio b/a, where b/a is assumed to decrease with increasing frequency. Quite evidently, at a critical value ω_c the ball makes a transition from the state $\phi = \pi$ to $\phi = 0$, or equivalently we may say that the antisymmetric hand movement makes a transition to the symmetric hand movement.

On the other hand, when we decrease ω starting from high values, the system remains all the time in the $\phi = 0$ state, even if ω drops below ω_c. This "hysteresis" phenomenon, which easily follows from our simple model, is also found in the real experiments.

This model has been refined in our original paper to treat the oscillatory motion of the hands explicitly. However, in the context of this chapter we dwell on the analogy with the phase transition, in particular with respect to the phenomenon of critical fluctuations mentioned above.

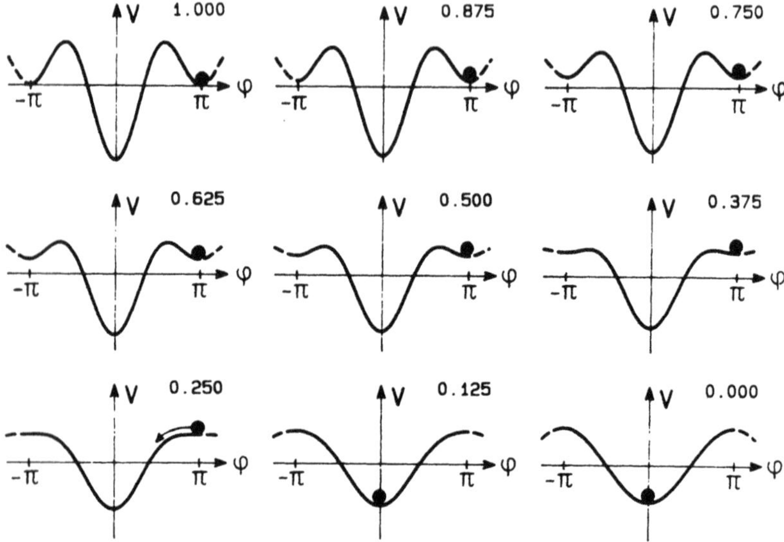

Figure 5.26. The behavior of the phase ϕ can be visualized as the overdamped motion of a particle in a potential which is exhibited in this figure. The shape of the potential changes if the control parameters a or b are changed as indicated. Note that the metastable state of the upper left corner becomes destabilized in the lower right corner.

After looking at our model and bearing in mind the typical critical fluctuations of synergetic systems close to their transition points, we suggested to Kelso that he search for such fluctuations. Figure 5.27 shows his experimental results. In this figure both the average phase and the phase fluctuations, i.e., more precisely the root mean square of the phase fluctuations, are plotted. In the case of transition from the antisymmetric to the symmetric hand motion, big critical fluctuations indeed occur. We have modeled this transition under the impact of fluctuations by means of adding a fluctuating force to equation (5.6.4). In other words, we treated the equation

$$\dot{\phi} = -\frac{\partial V}{\partial \phi} + F(t) \tag{5.6.8}$$

Using the Fokker–Planck equation, we studied both the behavior of the root mean square as well as correlation functions, and results are found in excellent agreement with experiment.

What is most interesting, and important, is the consequence of this treatment. Before we discuss it we briefly mention that by extension of our model, which takes into account the oscillatory motion of the hands, we may reproduce the experimental curve shown in Fig. 5.28.

We shall now discuss the important consequences. When we first assumed that the transition between one kind of hand movement to the other is caused

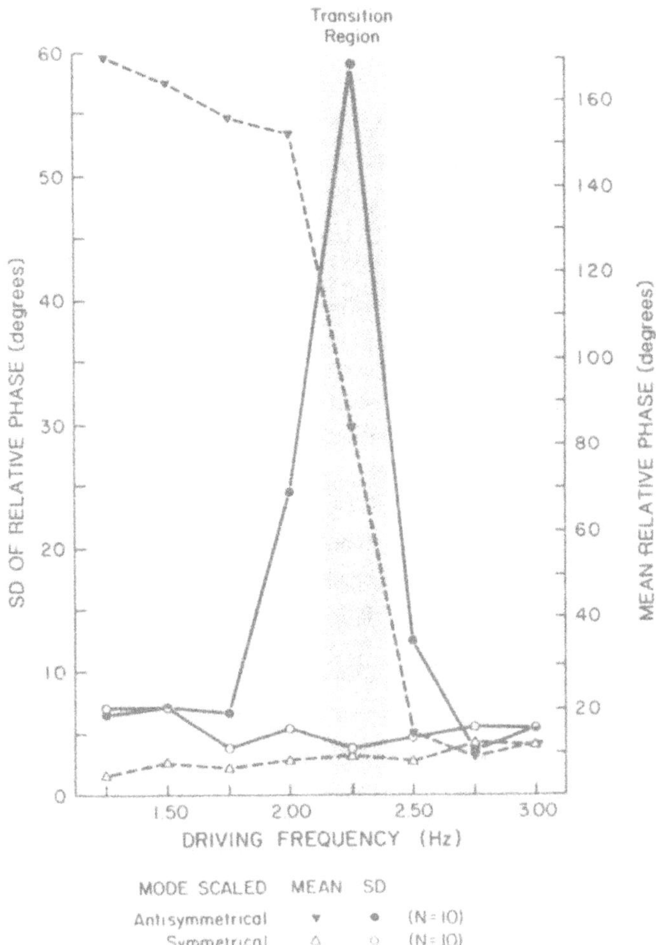

Figure 5.27. Behavior of the mean phase and its variance when the frequency is increased (after Kelso and Scholz).

by the change in the motor program of the neurons, it was very difficult to understand why any fluctuations should occur at all. Indeed, a motor program is a fixed program and no fluctuations should be expected. The way the transition occurs in the hand movement rather indicates that we are dealing here with a typical act of self-organization. This system organizes itself, i.e., the individual neurons and muscles act jointly as if the whole system acts as a total autonomous system.

It is quite clear that this introduces an entirely new paradigm in biology and it can be hoped that similar mechanisms and models apply to more complicated motions, where the next step will be to study the change in the

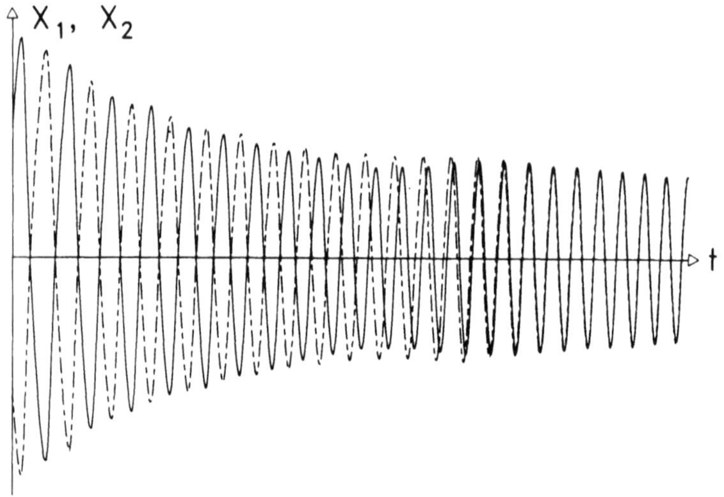

Figure 5.28. Out-of-phase and in-phase motion of the two fingers as a function of time when the frequency is increased at the same time continuously (after Haken, Kelso, and Bunz).

gaits of horses. Other highly coordinated motions may most probably be treated very much the same way, for instance, rhythmic motions like breathing and heartbeat, and their coordination.

Recommended Reading

H. Haken, *Synergetics. An Introduction*, 3rd edn., Springer-Verlag, Heidelberg, New York (1983) (also available in German, Russian, Chinese, Japanese, and Hungarian).

H. Haken, *Advanced Synergetics*, Springer-Verlag, Heidelberg, New York (1983) (also available in Russian and Japanese).

H. Haken, *Synergetics: The Science of Structure*, Van Nostrand Reinhold, New York (1983) (also available in German, Italian, Spanish and Japanese).

M. Bestehorn and H. Haken, *Phys. Lett.* **99A**, 265 (1984); *Z. Phys. B* **57**, 329 (1984).

R. Friedrich and H. Haken, in: *Complex Systems—Operational Approaches* (H. Haken, ed.), Springer-Verlag, Berlin, Heidelberg, New York (1985).

B. Schnaufer and H. Haken, *Z. Phys. B* **59**, 349 (1985).

J. A. S. Kelso and I. P. Scholz, in: *Complex Systems—Operational Approaches* (H. Haken, ed.), Springer-Verlag, Berlin, Heidelberg, New York (1985).

H. Haken, J. A. S. Kelso, and H. Bunz, *Biol. Cybern.* **51**, 347 (1985).

G. Schöner, H. Haken, and J. A. S. Kelso, *Biol. Cybern.* **53**, 247 (1986).

6

Instabilities and Chaos in Lasers: Introduction to Hyperchaos

F. T. Arecchi

6.1. Introduction

In the previous chapter, the laser was utilized as an example of a link between concepts in physics and biology. In the present chapter, the onset of deterministic chaos in lasers is studied by referring to the invariant properties of low-dimensional attractors, in order to isolate the characteristics of chaos from the random fluctuations due to the coupling with a thermal reservoir. For this purpose, attention is focused on single-mode homogeneous line lasers, whose dynamics is ruled by a low number of coupled variables. In the cases examined, experiments and theoretical models will be found to be in close agreement. In particular, when many attractors coexist for the same parameter values (generalized multistability) the presence of random noise induces long-lived transients with $1/f$-like low-frequency spectra. These are due to jumps over different basins of attraction (so-called "hyperchaos").

Quantum optics from its beginning was regarded as the physics of coherent and intrinsically stable radiation sources. Lamb's semiclassical theory[1] showed the role of the electromagnetic field in the cavity in ordering the phases of the induced atomic dipoles, thus giving rise to a macroscopic polarization and making possible a description in terms of very few collective variables. In the case of a single-mode laser and a homogenous gain line this meant just five coupled degrees of freedom, namely, a complex field amplitude E, a complex polarization P, and a population inversion N. A corresponding quantum theory, even for the simplest model laser (the so-called Dicke model, i.e., a discrete collection of modes interacting with a finite number of two-level

F. T. Arecchi • Istituto Nazionale di Ottica and Department of Physics, University of Florence, Florence, Italy.

atoms), does not lead to a closed set of equations; however, the interaction with other degrees of freedom acting as a thermal bath (atomic collisions, thermal radiation), provides truncation of high-order terms in the atom–field interaction.[2-4] The problem may be reduced to five coupled equations (the so-called Maxwell–Bloch equations), but now they are affected by noise sources to account for the coupling with the thermal bath.[5] Being stochastic, or Langevin, equations, the corresponding solution in closed form refers to a suitable weight function or phase space density. Anyway, the average motion matches the semiclassical one, and fluctuations play a negligible role if one excludes the bifurcation points where there are changes of stability in the stationary branches. Omitting the peculiar statistical phenomena which characterize the threshold points and which suggested a formal analogy with thermodynamic phase transitions,[6] the main point of interest is that a single-mode laser provides a highly stable or coherent radiation field.

From the point of view of the associated information, the standard interferometric or spectroscopic measurements of classical optics, relying on average field values or on their first-order correlation functions, are insufficient. In order to characterize the statistical features of quantum optics it was necessary to make extensive use of photon statistics.[7,8]

As discussed in detail in Section 6.2, coherence is equivalent to having a stable fixed point attractor, and this does not depend on details of the nonlinear coupling but on the number of relevant degrees of freedom. Since such a number depends on the time scales on which the output field is observed, coherence becomes a question of time scales. This is the reason why, for some lasers, coherence is a robust quality, persistent even in the presence of strong perturbations, while in other cases coherence is easily destroyed by the manipulations common in the laboratory use of lasers, such as modulation, feedback, or injection from another laser.

Here we review instabilities and chaos in active quantum optics. Section 6.2 is a general presentation of low-dimensional chaos in lasers, including the description of relevant measurements upon which any assessment of chaos has to rely. Sections 6.3–6.6 are respectively devoted to lasers with modulated losses, lasers with injected signals, lasers with feedback, and bidirectional ring lasers. Section 6.7 discusses the "hyperchaos," or $1/f$ noise.

A more comprehensive approach to the problem can be found elsewhere in a monograph on the subject.[9]

6.2. Deterministic Chaos

6.2.1. Historical Aspects

Until quite recently the current viewpoint was that a few-body dynamics was fully predictable, and that only addition of noise sources due to coupling with a thermal reservoir could provide statistical fluctuations. Lack of long-time predictability, or turbulence, was considered as resulting from the interaction of a large number of degrees of freedom, as in a fluid above the critical Reynolds number (Landau–Hopf model of turbulence). It is now known, however, that even in systems with few degrees of freedom nonlinearities may give rise to expanding directions in phase space and this, together with the lack of precision in assigning initial conditions, is sufficient to induce a loss of predictability over long times.

This level of dynamical description was born with the three-body problem in celestial mechanics (Poincaré). Already a three-body dynamic system is very different from the two-body problem, since in general there are asymptotic instabilities. This means a divergence, exponential in time, of two phase-space trajectories stemming from nearby initial points. The uniqueness theorem for solutions of differential systems seems to offer an escape route: be more and more precise in localizing the coordinates of the initial point. However, a fundamental difficulty arises. Only rational numbers can be assigned by a finite number of digits, A "precise" assignment of a real number requires an infinite acquistion time and an infinite memory capacity to store it, and neither of these two infinities is available to the physicist. Hence any initial condition implies a truncation. A whole range of initial conditions, even if small, is usually given and from within it trajectories may arise whose difference becomes sizeable after a given time, if there is an exponential divergence. This means that predictions are in general limited in time and that motions are complex, starting already from the three-body case. In fact we know nowadays from very elementary topological considerations that a three-dimensional phase space corresponding to three coupled degrees of freedom is already sufficient to yield a positive Lyapunov exponent, and accordingly an expanding phase-space direction. This complexity is not due to coupling with a noise source as a thermal reservoir, but to sensitive dependence on initial conditions. It is called deterministic chaos.

The birth of this new dynamics was motivated by practical problems, as fixing the orbit of a satellite or forecasting meteorology,[10] and it was strongly helped by the introduction of powerful computers. The mathematics of multiple bifurcations leading from a simple to a complex behavior is under current investigation. Some regularities, such as the "scenarios" or routes to deterministic chaos, have already been partly explored.[11]

6.2.2. Dynamical Aspects

A dissipative system (i.e., with damping terms) does not conserve the phase-space volume, as already discussed in chapter 1. If we start with initial conditions confined in a hypersphere of radius ε, that is, with an initial phase volume

$$V_0 = \varepsilon^N$$

as time proceeds, the sphere transforms into an ellipsoid with each axis modified by a time-dependent factor. Its volume is given by

$$V_t = \varepsilon^N e^{\Sigma \lambda_i t}$$

where λ_i are Liapunov exponents. Since the volume has to contract, $V_t < V_0$, so

$$\sum_{i=1}^{N} \lambda_i < 0$$

We denote the sequence of λ exponents, starting from the smallest up to the highest, as the Liapunov spectrum. For simplicity, we consider just the signs of nonzero λ_i, retaining the zero for $\lambda_i = 0$. We then describe a sequence of negative, zero, and positive λ_i as, e.g., $(--0+)$. For $N = 1$, we have $(-)$ and a segment $V_0 = \varepsilon$ of initial conditions shrinks to a single point for $t \to \infty$, that is, the attractor is a fixed point. For $N = 2$, the system goes either to a fixed point $(--)$ or to a limit cycle (-0). Chaotic motion $[(-+)$ with $\lambda_+ < |\lambda_-|]$ is forbidden in two dimensions by the Poincaré-Bendixon theorem. For $N = 3$, besides fixed point $(---)$ and limit cycle $(--0)$, we can have motion on a torus with two incommensurate frequencies (-00), but we can also have $(--+)$, namely, a positive λ which gives an expanding direction along which we rapidly get uncertainty.

An example of chaotic motion is offered by the Lorenz model (cf. Chapter 1) of hydrodynamic instabilities[10] that corresponds to the following equations, where the parameter values have been chosen in order to yield one positive Liapunov exponent:

$$\begin{aligned} \dot{x} &= -10x + 10y \\ \dot{y} &= -y + 28x - xz \\ \dot{z} &= -\tfrac{8}{3}z + xy \end{aligned} \tag{6.2.1}$$

The above considerations suggest that one should study low-dimensional chaos with the simplest phase-space topology allowing for the appearance of a positive Liapunov exponent.

Focusing on these situations in quantum optics permits close comparison between experiment and theory. On purpose we do not tackle the vast class of inhomogeneously broadened lasers, where it is extremely difficult to introduce close correspondences between experiment and theory owing to the large number of coupled degrees of freedom.

If we couple Maxwell equations with Schrödinger equations for N atoms confined in a cavity, and expand the field in cavity modes while retaining only the first mode E which goes unstable, this is coupled with the collective variables P and Δ describing the atomic polarization and population inversion as follows:

$$\dot{E} = -kE + gP$$
$$\dot{P} = -\gamma_{\perp}P + gE\Delta \qquad (6.2.2)$$
$$\dot{\Delta} = -\gamma_{\parallel}(\Delta - \Delta_0) - 4gPE$$

For simplicity we consider the cavity frequency at resonance with the atomic resonance, so that we can take E and P as real variables and have three coupled equations. Here k, $\gamma_{\perp}$, and $\gamma_{\parallel}$ are the loss rates for field, polarization, and population respectively, g is a coupling constant, and Δ_0 is the population inversion which would be established by the pump mechanism in the atomic medium in the absence of coupling. While the first equation originates from Maxwell equations, the two others imply the reduction of each atom to a two-level atom resonantly coupled with the field, that is, a description of each atom in an isospin space of spin $\frac{1}{2}$. The last two equations are like Bloch equations which describe the spin precession in the presence of a magnetic field. For this reason equations (6.2.2) are called Maxwell-Bloch equations.

The presence of loss rates means that the three relevant degrees of freedom are in contact with a "sea" of other degrees of freedom. In principle, equations (6.2.2) could be deduced from microscopic equations by statistical reduction techniques.[5]

The similarity of the Maxwell-Bloch equations (6.2.2) with Lorenz equations (6.2.1) would suggest the easy appearance of chaotic instabilities in single-mode, homogeneous-line lasers. However, time-scale considerations rule out the full dynamics for most of the available lasers. Equations (6.2.1) have damping rates within one order of magnitude. On the other hand, in most lasers the three damping rates are very different from one another.

The following classification has been introduced[12]:

Class A (e.g., He-Ne, Ar, Kr, dye): $\gamma_{\perp} \simeq \gamma_{\parallel} \gg k$
The last two equations of (6.2.2) can be solved at equilibrium (adiabatic elimination procedure) and one single nonlinear field equation describes the laser; $N = 1$ means fixed point attractor, hence coherent emission.

Class B (e.g., ruby, Nd, CO_2): $\gamma_{\perp} \gg x \gtrsim \gamma_{\parallel}$
Only polarization is adiabatically eliminated [the second of equations (6.2.2)] and the dynamics is ruled by two rate equations for field and population; $N = 2$ allows also for period oscillations.

Class C (e.g., FIR lasers): $\gamma_{\parallel} \approx \gamma_{\perp} \approx k$
The complete set of equations (6.2.2) must be used, hence Lorenz-like chaos is feasible.

We have carried out a series of experiments on the birth of deterministic chaos in CO_2 lasers (Class B). In order to increase the number of degrees of freedom by at least one, we have tested the following configurations:

1. Introduction of a time-dependent parameter to make the system nonautonomous.[13] To be precise, an electro-optical modulator modulates the cavity losses at a frequency near the proper oscillation frequency Ω provided by a linear stability analysis, which for a CO_2 laser happens to lie in the 50–100 kHz range, easily making it an accurate set of measurements.
2. Injection of a signal from an external laser detuned with respect to the main one, choosing the frequency difference near the above-mentioned Ω. With respect to the external reference the laser field has two quadrature components, which represent two dynamical variables. Hence we reach $N = 3$ and observe chaos.[12]
3. Use of a bidirectional ring, rather than a Fabry–Perot cavity.[14] In the latter case the boundary conditions constrain forward and backward waves, by phase relations on the mirror, to act as a single standing wave. In the former case forward and backward waves have just to fill the total ring length with an integral number of wavelengths, but there are no mutual phase constraints, hence they act as two separate variables. Furthermore, when the field frequency is detuned with respect to the center of the gain line, a complex population grating arises from interference of the two countergoing waves, and as a result the dynamics becomes rather complex, requiring $N > 3$ dimensions.
4. Addition of an overall feedback, besides that provided by the cavity mirrors, by modulating the losses with a signal provided by the output intensity.[15] If the feedback has a time constant comparable with the population decay time, it provides a third equation sufficient to yield chaos.

Note that while methods (1), (2), and (4) require an external device, method (3) provides intrinsic chaos. In any case, since feedback, injection, or modulation are currently used in laser applications, the evidence of chaotic regions introduces some caution as to optimistic trust in the laser coherence.

Of course, the requirement of three coupled nonlinear equations does not necessarily restrict attention to just Lorenz equations. In fact none of the cases (1)–(4) explored corresponds to Lorenz chaos.

6.3. Information Aspects

In this section we discuss what we measure to assess chaos. We plot two of the three (or more) variables on a plane phase-space projection. This way, we build projections of phase-space trajectories on an -x–y oscilloscope.

Simultaneously we can measure the power spectrum. Later we present a sequence of subharmonic bifurcations which eventually leads to an intricate trajectory (strange attractor) and to a continuous power spectrum. But how çan we discriminate between deterministic chaos and noise? After all, noise also would give a continuous spectrum, and the phase-space point would fill ergodically part of the plane, thus covering a two-dimensional set.

In order to discriminate deterministic chaos from order as well as from random noise, we introduce two invariants of the motion, one static the other dynamic.

We partition the phase space into small boxes of linear size ε and give the ith box a probability $p = M$, M equal to the fractional number of times it has been visited by the trajectory. This way, we build a Shannon information $I(\varepsilon)$, and with it an "information dimension" $D_1(\varepsilon)$,[16] which is in general a fractional number, or a "fractal" (see also Chapter 1):

$$I(\varepsilon) = -\sum_i p_i \log p_i \tag{6.3.1}$$

$$D_1(\varepsilon) = -\lim_{\varepsilon \to 0} \frac{I(\varepsilon)}{\log \varepsilon} \tag{6.3.2}$$

To understand the meaning of a fractal, we search for an operational definition of dimension.[17] Three sets are compared: (1) a segment of unit length; (2) the Cantor set, constructed by removing the middle third of the unit segment and repeating the operation on each fragment; (3) the Koch curve, constructed by replacing the middle third by the other two sides of an equilateral triangle and repeating the operation ad infinitum. At every stage of the partition, we cover each set with beads of suitable size to avoid loss of resolution (e.g., diameter $\frac{1}{3}$ at the first partition) and count the number N for each set (at the first partition we need 2 for the Cantor set, 3 for the segment, and 4 for the Koch curve). We define the fractal dimension as the ratio given by (see also Chapter 1)

$$D_0(\varepsilon) = \frac{\log N(\varepsilon)}{\log (1/\varepsilon)} \tag{6.3.3}$$

This definition is independent of the partition. Indeed for the Cantor set and the Koch curve we have $N = 2$, $\varepsilon = \frac{1}{3}$ and $N = 4$, $\varepsilon = \frac{1}{3}$, respectively, at the first partition, yielding

$$D_0(\text{Cantor}) = \frac{\log 2}{\log 3} \simeq 0.63\ldots$$

and

$$D(\text{Koch}) = \frac{\log 4}{\log 3} \simeq 1.218\ldots$$

At the second partition the number of necessary beads goes as N^2 and the diameter of each as ε^2, hence D_0 remains invariant.

Going back to the information dimension $D_1(\varepsilon)$ we see that $\log N$ has been replaced by $I(\varepsilon)$, which is an *average* [for p_i all equal, we recover $I(\varepsilon) = \log N$]. Hence D_1 generalizes D_0 whenever the density of points is not uniform along the trajectory.

As D_0 was independent of the partition stage, similarly D_1 is an invariant, but static (time does not enter). It can be shown that $D_0 \geq D_1$, however, for nonpathological sets the difference is irrelevant. Let us refer for simplicity to an $N = 3$ dimensional phase space. If $D = 0$ (fixed point) or 1 (limit cycle) or 2 (torus) we have an ordered, or coherent, motion. In the other limit of random noise, fluctuations fill ergodically an N-dimensional region of the space, hence $D = 3$. Deterministic chaos has to lie in between, that is,

$$2 < D < 3$$

Hence a fractal dimension is an indicator of chaos. It will be shown later that this indicator is expressed in terms of correlation functions, therefore it requires the same measuring techniques introduced in photon statistics.

These features related to the topology of the attractor have a temporal counterpart in another invariant, which measures how information is dissipated in a motion to maintain knowledge about the system. To build this dynamic invariant, we partition both space and time in boxes of sizes ε and τ that we name $i_1, i_2, \ldots, i_d$ at each of the discrete times $\tau, 2\tau, \ldots, d\tau$, and introduce the joint probability over the d time intervals,

$$p_{i_1 i_2 \ldots i_d} \equiv \{x(t = \tau) \subset i_1; \ldots; x(t = d\tau) \subset i_d\}$$

Correspondingly, we define a joint information

$$I_d(\varepsilon) = -\sum_{\{i_1 \cdots i_d\}} p_{i_1 \cdots i_d} \log p_{i_1 \cdots i_d} \tag{6.3.4}$$

Then, by a limit operation, we define the Kolmogorov entropy as the rate of information loss per unit time:

$$K \equiv \lim_{\tau \to 0} \lim_{\varepsilon \to 0} \lim_{d \to \infty} \frac{1}{d\tau} \sum_{h=1}^{d} (I_{n+1} - I_n) = \lim \frac{1}{d\tau} I_d \tag{6.3.5}$$

Now we have two indicators to gauge the difference among *order*, *random noise* (Brownian motion), and *deterministic chaos*. Referring to K, we easily see that

$K = 0$ for order (no information loss)

$K = \infty$ for random noise (total information loss)

$0 < K < \infty$ for deterministic chaos

The box counting method described above is impractical. It may require 10^6 points for a convergent numerical result. On the other hand, the following method introduced by Grassberger and Procaccia[18] is applicable to only 10^3–10^4 independent data points. We generalize Shannon information defining the order-f information as

$$I_f(\varepsilon) = \frac{1}{1-f} \ln \sum_i p_i^f \tag{6.3.6}$$

For $f \to 1$ we recover the usual definition. Associated with I_f, there is an order-f dimension of the attractor given by

$$D_f = \lim_{\varepsilon \to 0} \frac{I_f(\varepsilon)}{\ln(1/\varepsilon)} \tag{6.3.7}$$

For $f = 0$ and 1 we recover D_0 and D_1. Consider $f = 2$. The sum $\sum p_i^2$ is just the probability that a pair of random points on the attractor fall into the same box, i.e., that two arbitrary points are situated apart a distance less than ε. If this probability is $C(\varepsilon)$, we expect that

$$C(\varepsilon) = \lim_{N \to \infty} \frac{1}{N^2} \sum_{ij} \theta(\varepsilon - |\mathbf{x}_i - \mathbf{x}_j|) \simeq e^{D_2} \tag{6.3.8}$$

where $C(\varepsilon)$ is measured as the number of pairs (i, j) with distance apart satisfying $|\mathbf{x}_i - \mathbf{x}_j| < \varepsilon$. Here θ is the Heaviside step function.

Experimentally, we do not measure at each time the vector $\mathbf{x}(y)$ of phase space, but just one component $x_i(t)$ (for instance, just the light out of a laser). However, in a nonlinear system, any component $x_k(t)$ will influence x_i at a later time (no normal-mode transformation!). Hence, we can build an m-dimensional phase space $\boldsymbol{\xi}(t)$ by just measuring one single component x_i at successive times and considering the m-fold as a single point in m-space:

$$\boldsymbol{\xi}(t) \equiv [x_i(t), x_i(t+\tau), \ldots, x_i(t+m\tau)] \tag{6.3.9}$$

As we evaluate the slope log C vs. log ε from our data, we can stop increasing m when the slope shows saturation. The saturated slope is D_2.

6.4. The Modulated Laser[13,20]

For a single-mode, class B laser tuned at the center line, the phase space becomes two-dimensional. However, introduction of a time-dependent parameter makes the system nonautonomous, adding a third equation and thus making possible the appearance of a positive Liapunov exponent. It is then a practical matter to localize the values of the control parameters (pump, modulation frequency, and amplitude) for which this will occur.

When we apply a time-dependent loss, the laser equations become

$$\begin{aligned} \dot{I} &= 2kI(z-1) \\ \dot{z} &= \gamma_{\parallel}(z_0 - z - zI) \\ \dot{k} &= mk_1\Omega \sin \Omega t \end{aligned} \tag{6.4.1}$$

They are obtained from equations (6.2.2) by eliminating P and rescaling $I = E^2$ and $z = \Delta$. For $m \to 0$, we have small deviations from the equilibrium values

$$\begin{aligned} \bar{I} &= z_0 - 1 \\ \bar{z} &= 1 \end{aligned} \tag{6.4.2}$$

These deviations are linear in m and synchronous with the external frequency Ω. Destabilization of this limit cycle must be dealt with by Floquet theory.[19] It may be shown that even for $m \to 0$ a nonlinear resonance yields a positive Liapunov exponent for Ω around the characteristic frequency[13]

$$\Omega \simeq \sqrt{2k_1\gamma_{\parallel}(z_0 - 1)} \tag{6.4.3}$$

For a CO_2 laboratory laser near threshold $k \simeq 3 \times 10^7\ \mathrm{s}^{-1}$, $\gamma_{\parallel} \simeq 10^4\ \mathrm{s}^{-1}$, and $z_0 - 1 \simeq 0.1$ (10% above threshold), the corresponding frequency $f = \Omega/2\pi$ is in the 50-kHz range, easily accessible.

Thus we have conducted two series of experimental observations, the first[13] devoted to an experimental assessment of chaotic instabilities by phase-space portraits and power spectra, the second[20] to fractal dimensions and Kolmogorov entropy.

The driving frequency f was chosen to vary in the region from $\Omega/2\pi$ to its third harmonic, i.e., from 60 to 190 kHz. We have explored modulation values between 1% and 5%. A complete state diagram would yield the dynamical features for all possible values of the modulation parameters m and Ω. However, the strip $m = 1$–5% does not display m-dependence; therefore we limit ourselves to giving experimental results at $m = 1\%$ for various values of Ω.

The experimental setup consists of a CO_2 laser carefully stabilized against thermal and acoustic disturbances, with the discharge current stabilized better than 1/10. No long-term stabilization was necessary. The electrooptical modulator was a CdTe, antireflex-coated, 6-cm-long crystal, with an absorption less than 0.2%. The laser cavity includes also a $\lambda/4$ plate and a beam expander, both coated to limit the total losses per pass to 20%. The laser output is detected on a fast (2.5 ns rise time) pyroelectric detector whose current, proportional to the photon number $n(t)$, is sent together with its time derivative

$\dot{n}(t)$ to an x–y scope, in order to have the phase-space portrait $(n. \dot{n})$. The detector is also sent to a Rockland spectrum analyzer to measure the power spectra. The limited range (up to 100 kHz) of the spectrum analyzer has limited the frequency range explored in this first run. We show later that interesting bifurcations are also expected in 180-kHz domain.

Figure 6.1 presents experimental data in a narrow region between 62.7 and 64.25 kHz, where various bifurcations occur. This region is limited above and below by wide intervals with stable single-period limit cycles. Figure 6.1a shows the $f/2$ bifurcation at $f = 62.7$ kHz, Figure 6.1b the $f/4$ case for $f = 63.8$ kHz, Figure 6.1c the strange attractor and a broad-band spectrum for $f = 64.0$ kHz, and Figure 6.1d shows the $f/3$ case for $f = 64.2$ kHz.

Furthermore, at $f = 63.85$ kHz a new feature appears, namely, the coexistence of two independent stable attractors, one of period $4(f/4)$ and the other of period $3(f/3)$ (Figure 6.2). This bistable situation has nothing to do with the common optical bistability, where two dc output amplitude values appear for a single dc driving amplitude. We call this coexistence of two attractors "generalized bistability" (see Section 6.7).

Figures 6.3 and 6.4 report the theoretical equivalents of Figures 6.1 and 6.2, respectively, obtained by computer solution of equations (6.4.1) with parameter values in the range of the experiment.

As stated in Section 7, $1/f$-type low-frequency divergences, with power spectra as $f^{-\alpha}$ ($\alpha = 0.6$–1.2), appear when the following conditions are fulfilled: (1) There are at least two basins of attraction; (2) the attractors have become strange and any random noise (always present in a macroscopic system) acts as a bridge, triggering jumps between them. These jumps have the $f^{-\alpha}$ feature. In the region of bistability (see Figure 6.2) we have increased the modulator amplitude m up to the point where the two attractors have become strange. Figure 6.5 shows the sudden increase in the low-frequency spectrum. The divergent part has a power-law behavior $f^{-\alpha}$ with $\alpha \simeq 0.6$.

The above-described first run is still affected by the experimental uncertainties which characterize a phase-space projection or a power spectrum. Does the first one show a self-similar structure beyond the chaotic threshold as theoretically expected for a strange attractor, or does it ergodically just fill a two-dimensional region of the $(n, \dot{n})$ plane, thus being trivial random noise? After all, the latter test (continuous frequency spectrum) is also a common property of random noise and it is not a sufficient characterization of deterministic chaos.

In order to set a more reliable distinction between chaos and random noise, and also to specify the route to chaos (Figure 6.1 is only preliminary evidence of a Feigenbaum, or subharmonic, route), we have digitized the signal by a LeCroy transient recorder with 32,000 samples in memory. On setting the internal clock at 320 ns, we obtained approximately 16 points for each period of the fundamental frequency with light-bit resolution. By synchronizing the sampling time to the external drive period we obtained a projection of the

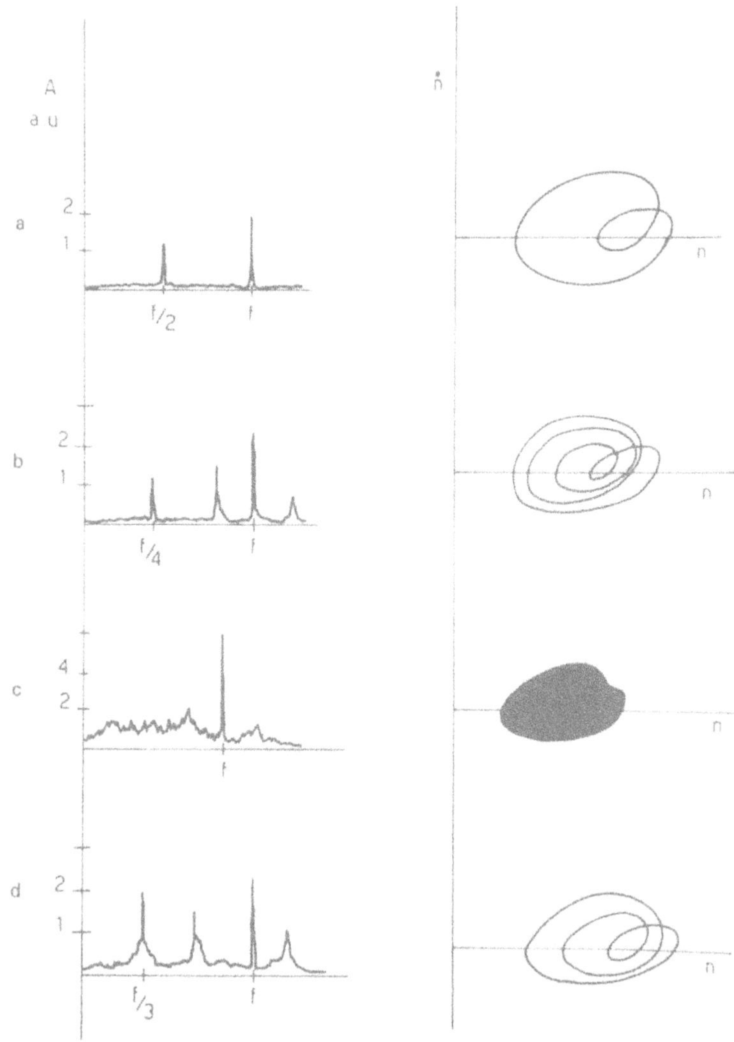

Figure 6.1. Experimental phase-space portraits ($n-\dot{n}$) (right) and corresponding frequency spectra (left) for different modulation frequencies f. (a) $f = 62.75$ kHz. Period-two limit cycle and corresponding $f/2$ subharmonic. (b) $f = 63.80$ kHz. Period-four limit cycle and $f/4$ subharmonic. (c) $f = 64.00$ kHz. The phase-space portrait shows a strange attractor (the oscilloscope spot could not resolve single windings). The power spectrum is quasi-continuous with a small peak at the modulation frequency (see the scale change with respect to the previous figures). (d) $f = 64.13$ kHz. Period-three limit cycle and $f/3$ subharmonic.

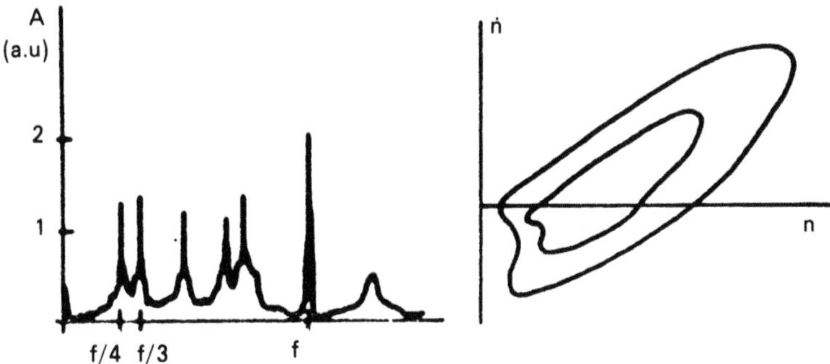

Figure 6.2. $f = 63.85$ kHz. Experimental evidence of generalized multistability (coexistence of two independent attractors). The power spectrum shows that those attractors correspond to $f/3$ and $f/4$ subharmonic bifurcations, respectively; in phase space, the multiple windings merged within the thickness of the phase-portrait contour.

Poincaré section. The projection is onto a one-dimensional space (we measure only the intensity) independent of the other variables. In Figure 6.6 we present the sections and corresponding time series, respectively. The advantage of this signal processing is that we are able to analyze a high number of periods

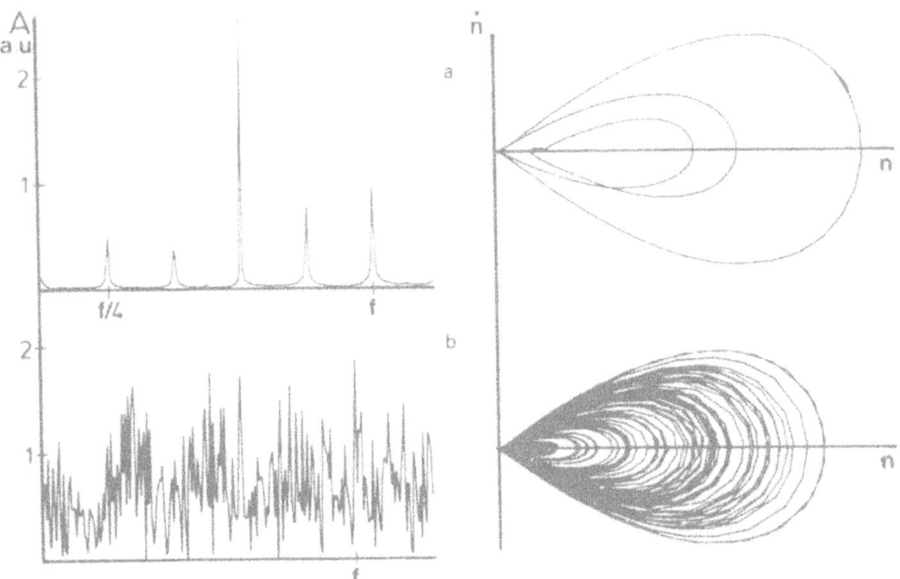

Figure 6.3. Computer plots for the parameter values $\gamma_{\parallel} = 10^3\ \mathrm{s}^{-1}$, $K = 7 \times 10^7\ \mathrm{s}^{-1}$, $m = 2.0 \times 10^{-2}$, $GR = 2.0 \times 10^{11}$. (a) $f = 64.33$ kHz. Subharmonic bifurcation $f/4$, as in the experiment of Figure 6.1b. (b) $f = 78.8$ kHz, $m = 3 \times 10^{-2}$. Strange attractor and broad spectrum corresponding to a chaotic solution.

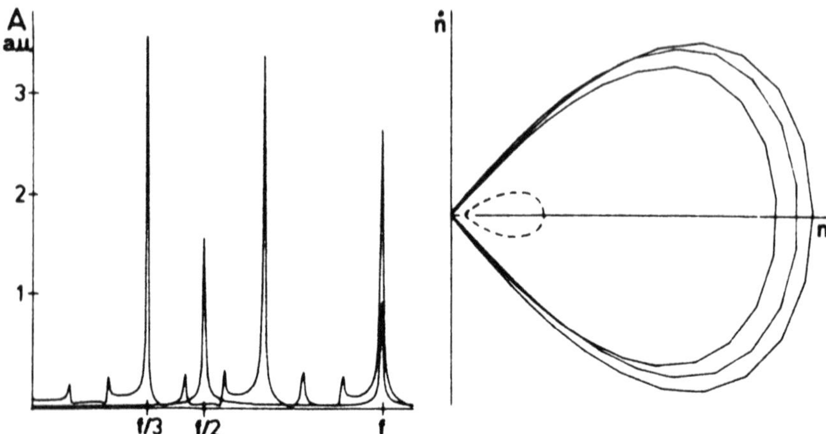

Figure 6.4. Theoretical generalized bistability; $f = 119.0$ kHz, $m = 2.0 \times 10$. The phase-space portrait shows the existence of two independent attractors, corresponding to the subharmonic frequencies $f/2$ (dashed line) and $f/3$ (continuous line); relative spectra are superimposed. It should be noted that one attractor remains inside the other as in the experiment of Figure 6.2. If initial conditions are properly changed, a third attractor is found with a subharmonic frequency $f/10$ (not plotted for the sake of simplicity). Initial conditions: $n_0 = 4 \times 10^8$, $\dot{n} = 0$ (dashed), $n_0 = 2 \times 10^5$, $\dot{n}_0 = -2 \times 10^6$ (continuous).

(32,000 maximum) with a single acquisition. Furthermore, it allows a much larger bandwidth processing of narrow-pulse sequences, which otherwise requires a high sampling rate with the related problems in data storing and processing. In Figure 6.6, on the left-hand side, the bandwidth is 300 kHz, and on the right it is 100 MHz; indeed we can already note in the $f/8$ plot a loss of resolution in the smaller peaks on the left-hand side.

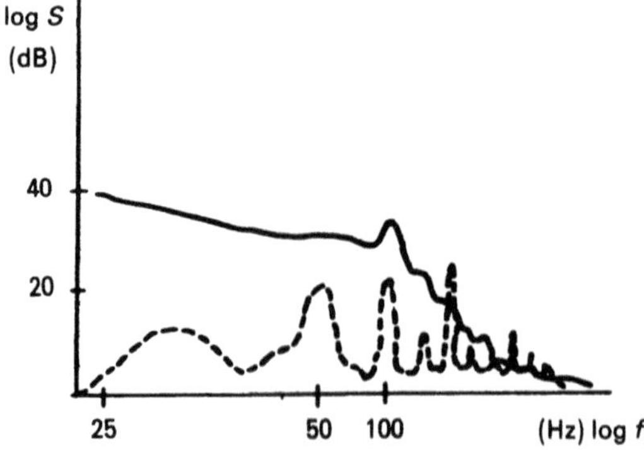

Figure 6.5. Experimental power spectra in the case of two attractors: stable (dashed line) and strange (solid line).

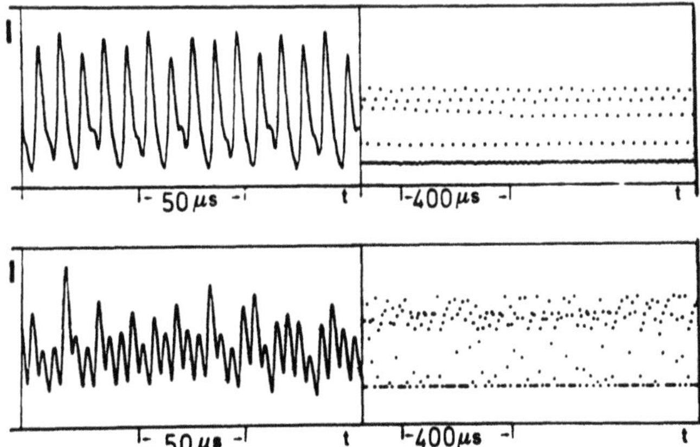

Figure 6.6. Top: Laser intensity vs. time for an $f/8$ subharmonic frequency, and corresponding stroboscopic intensity plot with the time interval between successive points equal to the period of the modulation frequency (191.000 kHz). Bottom: Laser intensity vs. time and stroboscopic plot for chaotic behavior. The period of the modulation is 5.2 μs. We note on the left-hand side the loss of resolution due to the limited acquisition bandwidth. This drawback is absent on the right-hand side owing to the huge increase in bandwidth.

We analyze digitized time sequences of the laser output intensity and reconstruct the attractors with an embedding technique. For the determination of the fractal dimension we follow the Grassberger–Procaccia method.

If we define $N_n(\varepsilon)$ as the number of vectors whose distance apart is less than ε, and if the embedding dimension n is large enough, then $N_n(\varepsilon) \sim \varepsilon^{\nu}$, where ν is the D_2 dimension of the attractor. In Figures 7a–f we plot log $N_n(\varepsilon)$ as a function of log ε for a sequence of bifurcations $f/4$, $f/8$, and chaos. We limit our analysis to the regions where the slope remains constant over a wide region of log ε and where it is independent of n, as it must be from theoretical predictions.

Inspection of Figures 6.7a and b indicates clearly that the slope obtained for the $f/4$ subharmonic saturates at $\nu \sim 1$ in the time series and $\nu \sim 0$ in the Poincaré section. For the $f/8$ subharmonic ν is slightly above 1.5 (Figures 6.7c and d). This result, even though not readily understandable because the time signal still appears periodic, nevertheless agrees with the theoretical prediction for the dimension at the accumulation point (infinite periodicity) of the logistic map ($1.5376 < \nu < 1.53385$). Indeed this dimension has been proven to be universal for those mappings for which the Feigenbaum scaling law holds.[21] We present here a heuristic interpretation based on our data. In our experimental system, the unavoidable noise yields a trajectory wandering over a nonzero range of parameter values, thus "testing" nearby periodic attractors of the subharmonic sequence. For infinite resolution, we would see for the stroboscopic data a staircase of horizontal plateaus each with zero slope, as it appears at higher embedding dimensions in Figures 6.7b and d.

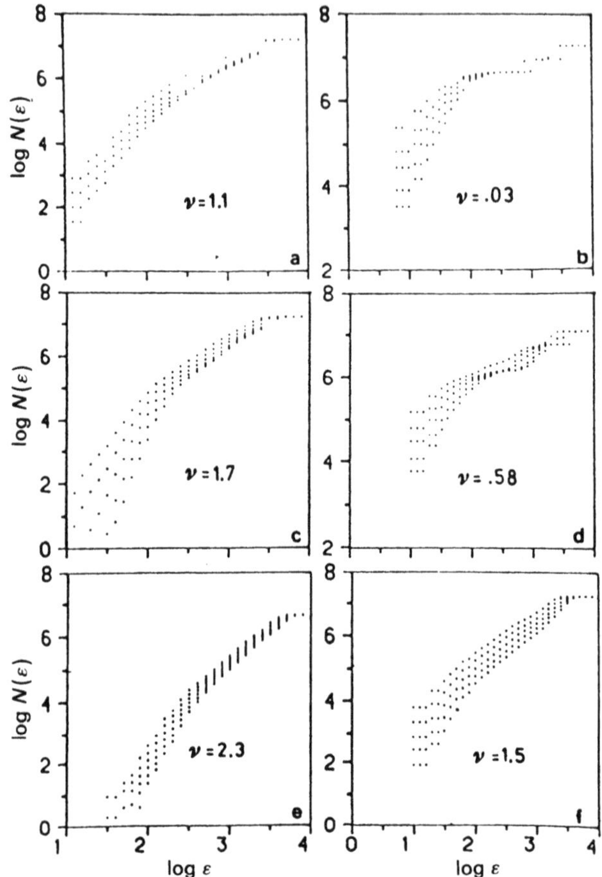

Figure 6.7. Plots of log $N_n(\varepsilon)$ vs. log ε for different values of n calculated from the time series (left-hand panels) and from the stroboscopic sections (right-hand panels) for different subharmonic frequencies: (a), (b) $f/4$; (c), (d) $f/8$; (e), (f) chaotic behavior. All best-fit values of the slope ν are assumed to have an overall estimated error of ±0.1. 6000 points were used. The embedding dimensions for all reported plots run from 5 to 9. Dimensions from 1 to 15 were tested.

However, the finite resolution of the correlation measurements averages over adjacent steps, and thus provides the 0.58 slope, as it appears in Figure 6.7d. This is the first time that the dimension at the accumulation point of a Feigenbaum cascade has been measured in an experimental system.

When the system enters the chaotic region, the fractal dimension suddenly jumps to a higher value ($\nu = 2.4$).

The time behavior of the intensity obtained by numerical integration of equations (6.4.1) was processed in the same manner as the experimental signal. Figure 6.8 shows the results obtained for an $f/8$ solution and a strange attractor. Again, near the accumulation point $n \sim 1.5$. Direct comparison of Figure 6.8 with Figures 6.7c and d shows good agreement between experiment and model.

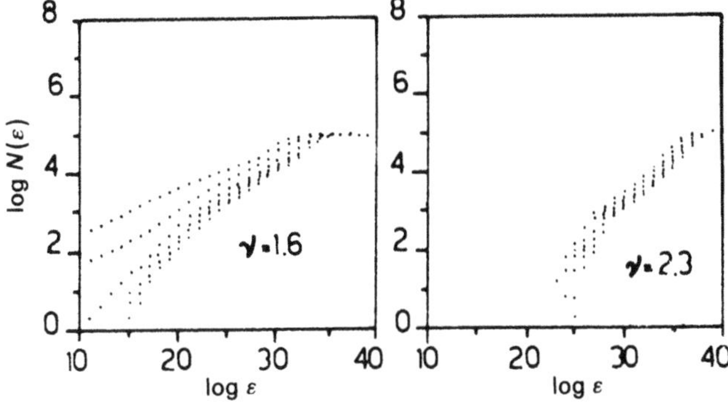

Figure 6.8. Plots of log $N(\varepsilon)$ vs. log ε for different dimensions n obtained from the numerical integration of the model equations for two different cases, $f/8$ subharmonic (left-hand side) and chaos (right-hand side). 6000 points were used.

In Figure 6.9 we report the correlation entropy K_2 vs. the embedding dimension for the $f/4$ and for the chaotic attractor, vs. the embedding dimension, from the data of Figure 6.7. We see that while $K_2 = 0$ for $f/4$, $K_2 \simeq 35$ kHz for the chaotic attractor. As we have a single positive Liapunov exponent and as the embedding time is 5.2 μs, we estimate that the half-loss of information corresponds to 3.8 periods of the modulation frequency.

6.5. The Laser with Injected Signal (LIS)[12,24,25]

Injecting an external signal into a single-mode laser provides an extra degree of freedom. Indeed, in general, the field amplitude x must be decomposed into two dynamical variables, namely, the two quadrature components $x_1 + ix_2 = x$ with respect to the external phase reference. In class C lasers this

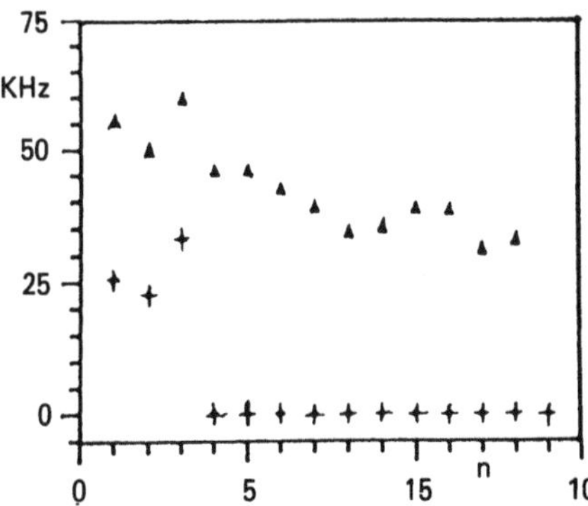

Figure 6.9. Correlation entropy K_2 (in kHz) vs. the embedding dimension for the $f/4$ (crosses) and the chaotic (triangles) attractor.

provides a fourth equation,[22,23] which is more than the necessary requirement for deterministic chaos. A simpler situation is that of a class B LIS, which is governed by the three equations

$$\begin{aligned} \dot{I}/2k &= Iz(1+\delta^2)^{-1} - I + \sqrt{I}x_0 \cos\varphi \\ \dot{\varphi}/k &= -\theta - \delta z(1+\delta^2)^{-1} - x_0 \sin\varphi/\sqrt{I} \\ \dot{z}/\gamma_\parallel &= z_0 - z - zI(1+\delta^2)^{-1} \end{aligned} \tag{6.5.4}$$

where $x_0 = A$ is the external field and $x = \sqrt{I}\,e^{i\varphi}$ the internal one. The frequency relations involving gain line, cavity, external field, and internal laser are given in Figure 6.10.

We have observed two different ways of reaching the locked regime, either by decreasing the oscillation frequency (tangent bifurcation) or by decreasing the amplitude of oscillation (Hopf bifurcation), and two different routes to chaos, either by intermittency or by period doubling.

An extensive linear stability analysis has been reported,[24] together with numerical solutions of equations (6.5.4) for different values of the injected amplitude $A = x_0$ and mutual detuning $\theta - \delta$.

Preliminary experimental data have been obtained by a three-laser setup,[25] the first laser being a ring laser where the dynamics develops, the second the external injecting laser, and the third laser a master oscillator with reference to which the first and second lasers are stabilized. The parameter region explored in this experiment was sufficient to yield oscillatory instabilities, but not enough to reach chaos.

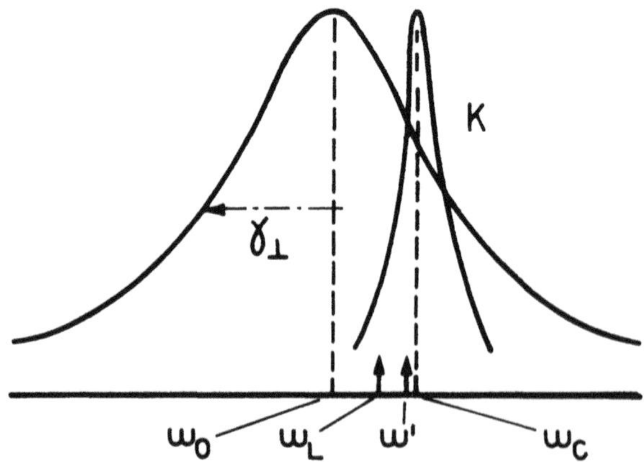

Figure 6.10. Qualitative plot of the frequency relations among atomic resonance (homogeneous width $\gamma_\perp$) centered at ω_0, cavity resonance (width K) centered at ω_c, and injected field at ω_1.

6.6. The Laser with Feedback[15]

In laser applications where high stability is required, an overall negative feedback is currently used, besides that already provided by the electromagnetic cavity, such as by controlling the pump strength with a signal provided by the detected output intensity. However, a fundamental objection to a feedback scheme is that it provides one extra dimension to phase space and hence the modified dynamics can be affected by irregular behavior.

Here we show how feeding the laser output back on an intracavity modulator introduces a third degree of freedom leading to chaotic instability. When the feedback loop is so fast that it practically provides an "instantaneously" adapted loss coefficient, it does not modify the phase-space topology which, in the case of a class B laser, remains two-dimensional. If, however, the time scale of the feedback loop is of the same order as that of the other relevant variables, the system becomes three-dimensional. Such a system is ruled by three first-order equations for the intensity x, population difference z, and modulation voltage V. With suitable normalizations (here x is the intensity, *not* the field) the equations are

$$\begin{aligned} \dot{x} &= -kx(1 + \alpha \sin^2 v - z) \\ \dot{z} &= -\gamma_\parallel(z - A + xz) \\ \dot{v} &= -\beta(v - B + fx) \end{aligned} \tag{6.6.1}$$

where $k(v) = k_0(1 + \alpha \sin^2 v)$ is the loss rate modulated by the voltage v, k_0 is the nonmodulated cavity loss parameter, α the coupling between dector voltage v and modulator, $\gamma_\parallel$ the population decay rate, and β the damping constant of the feedback loop. Furthermore, B is the voltage bias applied to the second input of the modulator amplifier, A is the pump parameter, f is a coupling coefficient between intensity x detected on D and voltage v. We note that x is normalized to the saturation intensity, z and A to the threshold population (without feedback), and v is given in angular units.

The experimental system of Arecchi *et al.*[15] has $k_0 = 1.17 \times 10^7\ (\mathrm{s}^{-1})$, $\gamma_\parallel = 0.98 \times 10^4\ (\mathrm{s}^{-1})$, $\beta = 3.0 \times 10^4\ (\mathrm{s}^{-1})$, and a normalized pump $A = 4.2$.

The stationary solutions $(\bar{x}, \bar{y}, \bar{z})$ of equations (6.6.1) imply the condition

$$B = f\bar{x} + \arcsin\left(\frac{A/\alpha}{1+\bar{x}} - \frac{1}{\alpha}\right)^{1/2}$$

Depending on the feedback coupling f, for different bias values B we can have monostability or bistability (Figure 6.11). In particular, around $f = 0.1$ we expect an ambiguity, since equation (6.6.2) provides a quasi-vertical curve. Indeed, as we show later, this is the region where we observe chaos.

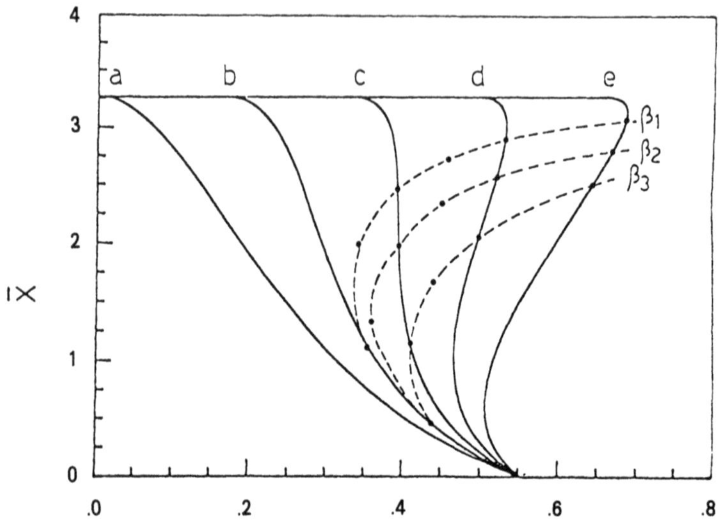

Figure 6.11. Laser with feedback: Plots of normalized stationary intensity x vs. B (the bias voltage B is expressed in angular units) for different values of the feedback coupling constant f. Curves (a) to (e) refer to $f = 0$, 0.052, 0.102, 0.152, and 0.202, respectively. Dashed lines correspond to the loci of the first Hopf bifurcations for three different values of the damping constant (s^{-1}) of the feedback loop: $\beta_1 = 3.5 \times 10^4$, $\beta_2 = 3.0 \times 10^4$, and $\beta_3 = 2.5 \times 10^4$.

By a linear stability analysis around the stationary solution, we evaluate the points where the system starts self-pulsing (Hopf bifurcations). The lines of Hopf bifurcations are drawn in Figure 6.15 (dashed) for three different values of β.

In Figure 6.12 we present the power spectra of the intensity detected in the experiment. Figure 6.12a shows the first Hopf bifurcation, Figure 6.12b the appearance of a subharmonic $f/2$, and Figure 6.12c corresponds to the appearance of chaos. Beyond chaos, there are periodic time windows. In order to get full assurance of the chaotic nature of the time plot of Figure 6.12c, the correlation dimension was measured.

Figure 6.13 shows a fractal exponent $D_2 = 2.6 \pm .1$. While Figure 6.13 comes from the experiment, the same value of D_2 is obtained by solving numerically equations (6.5.1) for $\beta = \beta_2 = 3.0 \times 10^4$ and $B = 0.383$. The theoretical plots, when reported in Figure 6.13, closely follow the experimental ones with uncertainties smaller than the dot sizes.

6.7. The Bidirectional Ring Laser[14]

Finally we consider a longitudinal single-mode CO_2 ring laser in which both directions of propagation are allowed (Figure 6.14). The linewidth being

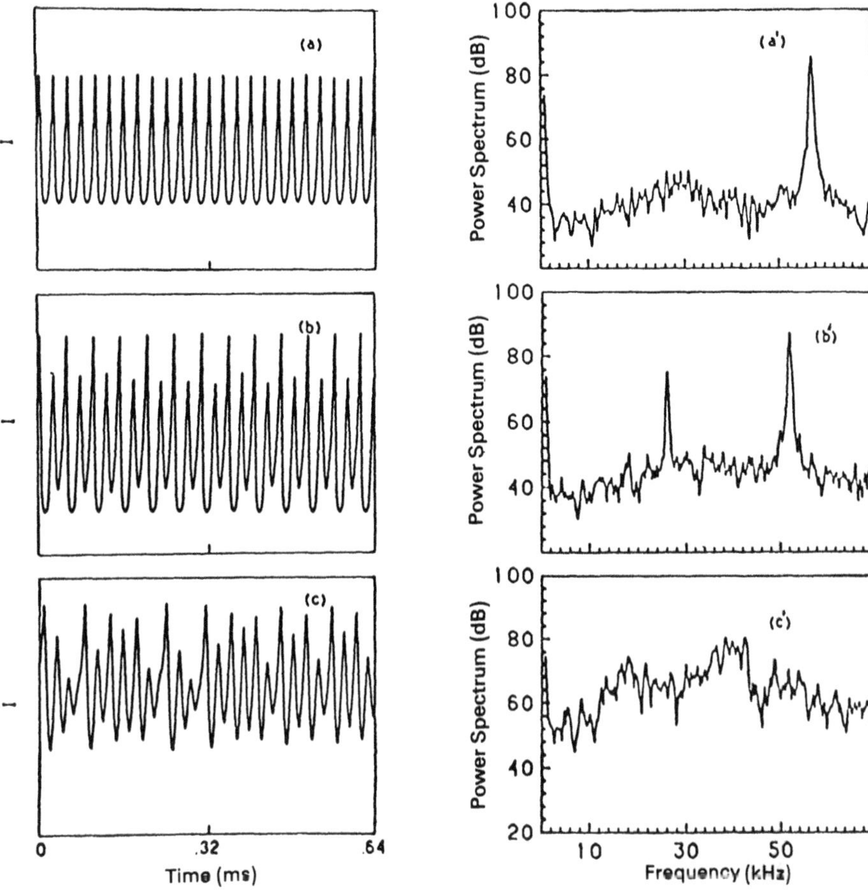

Figure 6.12. Laser with feedback: Digitizer time plots of the experimental laser intensity (left) and the corresponding power spectra (right) for increasing values of the control parameter B. (a) corresponds to the onset of the first Hopf bifurcation, at a frequency $\nu = 57.3$ hHz, $B = 0.364$; (b) shows the appearance of a,subharmonic bifurcation $f/2$ where the fundamental frequency is $\nu = 52.0$ kHz, $B = 0.378$; (c) shows the appearance of chaos, $B = 0.383$.

homogeneously broadened, the two counterpropagating beams cannot work at the same time because they must compete for the same amount of population inversion. Moreover, they are slightly detuned from each other—and with respect to the line center—because, for intrinsic asymmetries, cavity losses are different on the the two propagation directions (K_1 and K_2); this results in a different mode pulling and then different lasing frequency. Indeed, having a gas flow in the laser tube, this already induces a small amount of Doppler shift in the interaction with one or other of the two counterrunning modes. The detuning has been shown experimentally, as well as in the numerical solution, to be essential for breaking the symmetry between the two directions.

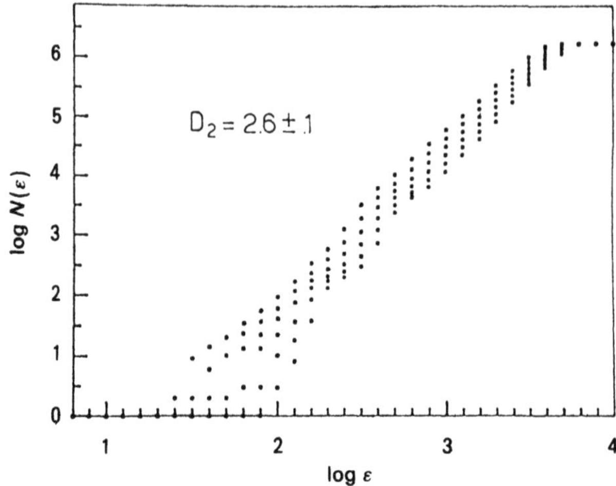

Figure 6.13. Laser with feedback: Plots of log $N_n(\varepsilon)$ vs. log ε for different values of embedding dimension n (n = 10–15). Square dots denote experiment. Theoretical plots coincide with experimental plots within the dot sizes.

A forbidden gap around the center of the molecular line, as well as the interchange of role of forward and backward fields at right and left of the line center, are evidence of such a detuning. If $K_{1,2}$ were the "cold" damping rates, they could not differ for reciprocity (in a passive medium thermodynamics forbids such a symmetry breaking). However, the $K_{1,2}$ in the active medium differ for the above-mentioned gas-flow effect.

Through the grating induced in the population inversion by the interference of the two waves, we have an interchange of energy from one field into the other by backscattering, so that we may consider the system as a LIS (where the injection comes from the counterpropagating mode).

In the experimental parameter space (Figure 6.14) we can distinguish three main different regions showing completely different behavior. In the first we observe a self-pulsing very similar to that of the laser with an injected signal. One mode is also running cw while in the other we observe only spikes, in phase with the main mode, which occur at a repetition rate (ω_s) of the order of $\gamma_\parallel$. In fact, as in the LIS system, the cw working mode injects some energy into the other one and allows population inversion increase up to a level at which a giant pulse takes place (the height may be 500 times greater than the stationary level). During the pulse both modes go above threshold and spike in phase. Superimposed to the decay we see relaxation oscillations typical of CO_2 lasers with a frequency (ω_0) very near to Ω; they are out of phase owing to competition between the two modes.

For higher excitation currents we observe a deterministic switching due to competition between the two fields with low frequency (30 Hz). During

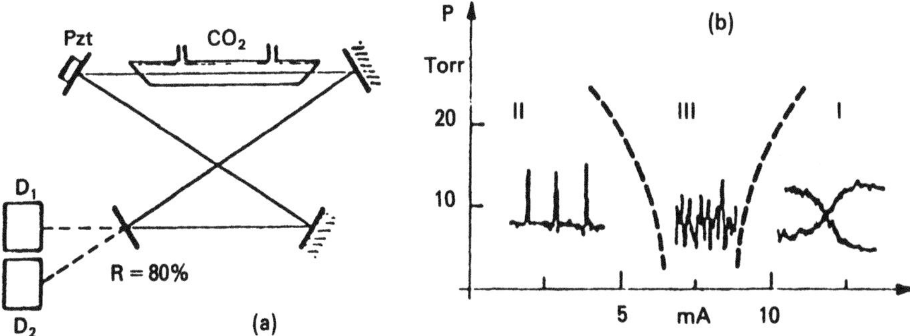

Figure 6.14. Bidirectional ring laser: (a) Experimental set-up: Gain cell with partial pressures: CO_2 1, N 1.5 Torr, He variable. Cavity length 4.2 m; PZT denotes piezoelectric mirror translator; D_1 and D_2 denote detectors for forward and backward intensities. (b) Phase diagram for total pressure (P) and discharge current (i). Regions are: (I) mode alteration, (II) self-Q-switch, (III) irregular pulsation.

interchange jumps we observe again the two frequencies of Figure 6.15, but with the lower one increased due to a larger value of $\gamma_{\parallel}$ (higher current) while the higher one can be varied also by adjusting the cavity length and alignment by moving a mirror mounted on a piezoelectric crystal.

The transition between these two regimes is not abrupt and takes place through a region which shows chaotic behavior. Here both phenomena related to population inversion, spiking (lower currents) and oscillation (higher currents), take place; effective output frequency results also as a combination of the two others $(\omega_s + \omega_0)$. At the same time, if we adjust the cavity mirror position so that we bring $\Omega \simeq (\omega_s + \omega_0)$, we obtain a competition of two different variables (population inversion and field) on the same time scale. The result is a fully developed chaos (Figure 6.16).

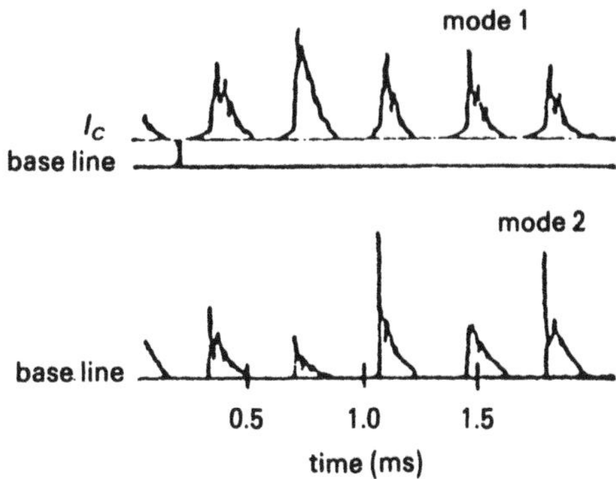

Figure 6.15. Bidirectional ring laser—Region (II): Output intensity vs. time. The upper signal has a cw baseline, while the lower oscillates over a zero level. Oscillations relax to the steady state before a new jump (interchange between cw action) takes place.

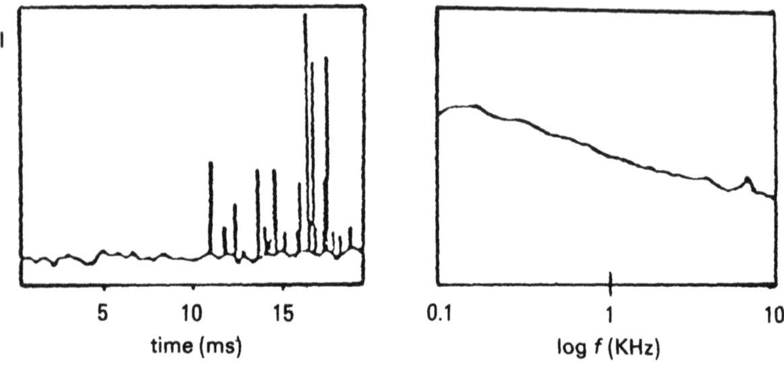

Figure 6.16. Bidirectional ring laser—Region (III): Output intensity vs. time for a wholly chaotic signal (left); log-log power spectrum with low-frequency divergence $f^{-\alpha}$, $\alpha \simeq 0.6$ (right).

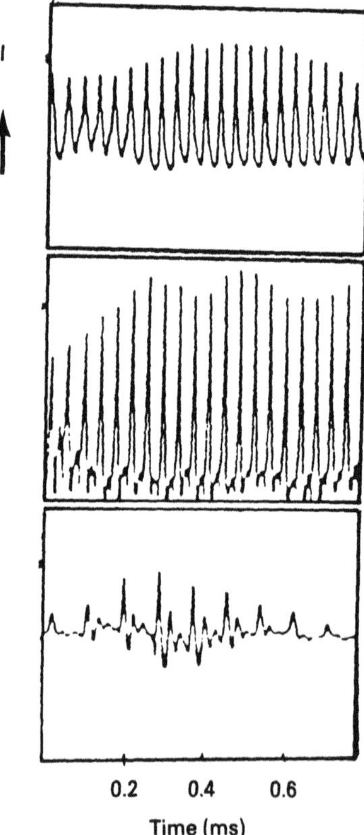

Figure 6.17. Bidirectional ring with extra mirror to reinject the forward mode into the backward one: Output intensity vs. time. Bifurcation sequence in analogy with a laser with injected signal.

If we now inject back one field into the laser with an external mirror (a fifth mirror in the configuration of Figure 6.14), we obtain stabilization of self-spiking, stable laser action instead of switching between the two modes, and chaotic behavior. At the boundary between the spiking and chaotic regions we observe a phenomenology, typical of a laser with an injected signal (Figure 6.17). It means that in this situation we have parameters practically equal to those responsible for such behavior in the LIS case, although the system here is more complicated.

6.8. Noise-Induced Trapping at the Boundary between Two Attractors: Hyperchaos and $1/f$ Spectra[26]

In this section we show how addition of random noise in a nonlinear dynamical system with more than one attractor may lead to $1/f$ spectra, provided the basin boundary is fractal. This shows that combining the features leading to deterministic chaos with a random noise is somewhat equivalent to a double randomness and we call "hyperchaos" such a situation. Indeed random–random walks in ordinary space, as diffusion in disordered systems, have shown a $1/f$ behavior.[27,28] Thus, hyperchaos here introduced is a random–random walk in phase space, where in fact one of the two sources of complex behavior is due to the fractal structure arising from deterministic dynamics.

To evaluate the impact of the following arguments, we premise some historical remarks on $1/f$ spectra in nonlinear dynamics.

Some years ago it was discovered[29] that in a nonlinear dynamical system with more than one attractor, introduction of random noise induces a hopping between different basins of attraction, giving rise to a low-frequency spectral divergence, resembling the $1/f$ noise well known in many areas of physics.[30] Such a discovery was confirmed by a laser experiment implying two coexisting attractors[13] already reported in Section 6.4, and later the effect was observed in other areas, such as Josephson tunnel junctions.[31,32]

The effect was questioned with two objections:

1. A noise-induced jump across a boundary leads to a telegraph signal, hence to a single Lorentzian spectrum.[33a]
2. A computer experiment yieleded a power law only over a limited spectral range.[33b]

The questions were answered[33c] with a statement of the empirical conditions under which the $1/f$ spectra appeared, namely:

1. Coexistence of at least two attractors (so-called "generalized multistability"[13]).

2. Presence of noise.
3. Some "strangeness" in the attractors.

As a matter of fact, this third condition was rather vague. To make it more precise, two theoretical models were explored: a one-dimensional cubic iteration map with noise[34] and a forced Duffing equation with noise.[35] Both these papers disclose interesting features, shedding more light on the above assumption (3). Figure 2 in the paper of Arecchi *et al.*[34] shows that the size of the $1/f$ spectral region increases with the root mean square of the applied noise, that is, with the probability of crossing the basin boundary by a noise-induced jump (Figure 3 of the same paper shows that the Liapunov exponent approaches the crisis value for increasing noise).

The numerical evaluation of Arecchi *et al.*[35] showed that for some control parameters, the boundary between basins of attraction was an intricate set of points through which it was impossible to draw a simple line. In such cases the noise was most effective in yielding low-frequency $1/f$-like spectra.

On the other hand, a fundamental logical approach to the $1/f$ problem was based on the composition of a large number of Lorentzians (or elementary Markov processes with exponential decay) whose weights are lognormally distributed,[36] thus fulfilling the relation

$$\int_{\gamma_1}^{\gamma_2} \frac{\gamma}{\omega^2 + \gamma^2} p(\gamma)\, d\gamma \simeq \text{const} \times \frac{1}{\omega} \tag{6.8.1}$$

provided $p(\gamma) \sim 1/\gamma$, and for the frequency range $\gamma_1 \ll \omega \ll \gamma_2$.

Motivated by the rate-processes considerations of Beasley *et al.*,[33a] who yielded a single Lorentzian for two attractors, we developed a kinetic model[34] based on a single transition rate for each pair of attractors. In the case of M attractors, this yielded $M - 1$ Lorentzians. In order to approximate the integral (6.8.1) by a sum (5% accuracy in fitting a $1/f$ law would require about one pole per decade), a large number $M \gg 2$ of attractors is necessary and hence the integral in equation (6.8.1) would be replaced by the sum over the $M - 1$ Lorentzians corresponding to the eigenvalues of the kinetic model. However, there is no reason to weigh the Lorentzians according to their reciprocal widths, hence no satisfactory reconstruction of a $1/f$ spectrum was possible. In fact, an experiment on a forced and noisy Duffing oscillator with an increasing number of attractors[37] did not offer clear evidence of the expected scaling of the spectral exponent with the number of attractors. On the contrary, we showed[35] that the boundary region between just two attractors was sufficient to yield $1/f$-like spectra, at variance with the many-attractor model. Hence this suggested that the boundary structure was really responsible for a large number of decay constants (possibly lognormally distributed).

In the meantime, the fractal structure of a basin boundary was explored in some examples.[38] This involves the following. As the phase point wanders within one basin of attraction, if we draw a sphere around the point defining its distance from the other basin of attraction, the radii of these spheres are distributed with all scale lengths, according to the self-similar structure of the fractal boundary. If we consider two-dimensional projections of the phase space as in Figures 3-6 of Arecchi *et al.*,[35] the spheres will be circles.

On the basis of the above considerations, we have constructed an elementary cellular automaton, which models the motion of the phase point within a fractal basin boundary in the presence of random noise. We model the boundary region of two basins of attraction A and B as two adjacent one-dimensional lattices of sites. Suppose we start from site i. At each discrete time step, if i belongs to A ($i = i_A$) it moves one step forward on the same lattice $[i_A \to (i_A + 1)]$ and if it belongs to B it goes one step backward $[i_B \to (i_B - 1)]$. In the absence of noise, once the motion has started on one basin, it will remain on it forever. In the presence of noise, at each time step there is a finite probability of a "cross" jump at the same lattice site, from stripe A to B: $i_A \to i_B$.

We call L the maximum size of the boundary region and $l_i \leq L$ any of the possible sizes of the fractal set. At each time step, the probabilities of permanence and jump are respectively

$$\begin{aligned} P_{AA} &= P_{BB} = l_i/L \\ P_{AB} &= P_{BA} = 1 - l_i/L \end{aligned} \tag{6.8.2}$$

To construct a self-similar structure we allow l_{i_k} to scale as $l_{i_k}/L = (1/2)^{V(i_k)}$ where $V(i_k)$ is a natural number sorted randomly for each site i_k ($i = -\infty$ to ∞, $k = $ A, B). A real numerical experiment can be treated by considering finite sequence of N sites (e.g., $N = 10^3$) and truncating the fractality by imposing $0 \leq V(i_k) < F$, where F is a finite integer denoting the maximum partitioning $(1/2)^{F-1}$, that is, the ultimate resolution of the measuring device in appreciating the fractal structure of our set. With all this in mind, for each evolution we extract a double sequence of N integers randomly distributed between 0 and $F - 1$, and denote each site i_k by the corresponding number $V(i_k)$. This means that we have attributed to each site an "area of respect," that is, a specific separation l_{i_k} from the other attractor, with l_{i_k} depending on $V(i_k)$ as shown above. We start, e.g., on the basin A from $i_A = N/2$.

At this step, we account for a suitable noise yielding the permanence and jump probabilities (6.8.2) by generating a random number y distributed uniformly between 0 and 1. If $y \leq (1/2)^{V(i_A)}$, then at the next time the point goes to $i_A + 1$ on attractor A; if $y > (1/2)^{V(i_A)}$, then the point jumps instantaneously to site i_B and at the next time it goes to $i_B - 1$ on attractor B.

By measuring the position coordinate, taking the Fourier transform, and squaring it, we can build the power spectra, that is, the transforms of the position correlation functions.

Figure 6.18 presents two power spectra for $F = 4$ and 14. In fact, we have measured spectra for all integer values of F between 4 and 14, but we just report two samples over slightly more than three frequency decades. The sequence shows that, as the fractality increases, the slope of the log-log plot goes from about 2 (single Lorentzian) to about 1 ($1/f$ spectrum). This appears better in Figure 6.19, where the slope α of the $f^{-\alpha}$ spectral law is plotted vs. the fractality F. The Lorentzian ($\alpha = 2$) of the random telegraph model is easily recovered for $F = 1$, thus showing that noise-induced jumps across a

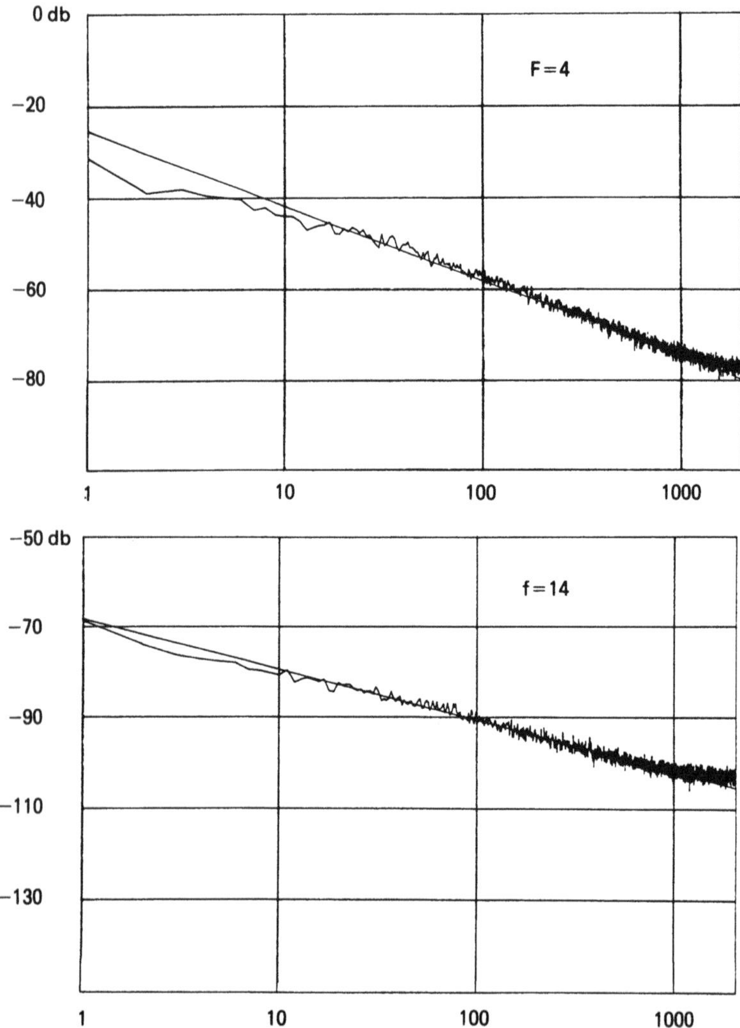

Figure 6.18. Power spectra (vertical) vs. frequency (horizontal) in log-log scale. Wavy lines: measured spectra, straight lines: best fits, whose slopes α are reported in the next figure. The two samples shown refer to $F = 4$ and 14.

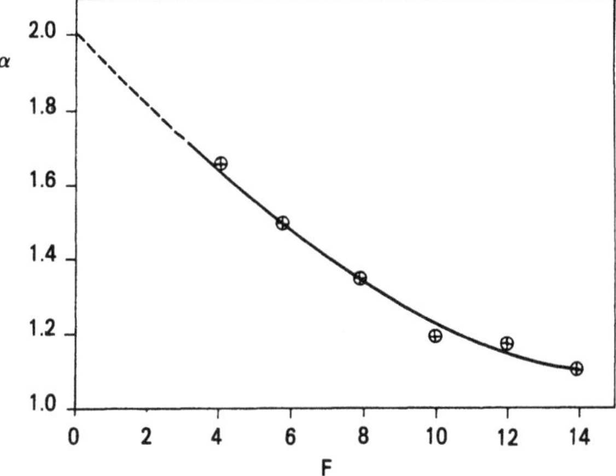

Figure 6.19. Exponents α of the power law $f^{-\alpha}$ vs. fractality F.

regular line boundary fulfil the intuitive expectation of a single decay rate. An analogy with the random-random walk[27,28] is easily drawn. Indeed, our motion is bound with a root-mean-square deviation from about $\sqrt{t}$ to $|\log t|^2$ as the fractality F increases from 4 to 14, according to Sinai.

For comparison we mention other approaches leading to $1/f$ or at least non-Lorentzian low-frequency spectra:

1. Pomeau-Manneville type-3 intermittency corresponds to slowly diverging trajectories with a $1/f$ power spectrum.[39] This behavior is intrinsic to the dynamics, hence it occurs without noise.
2. A deterministic diffusion process may occur beyond "crisis"[40] when two otherwise disjoint attractors merge into a single one. Here again no noise is required, and a comparison of this behavior with noise-induced jumps was given elsewhere.[35]
3. Another comparison of intrinsic vs. noise-induced intermittency was carried on for a damped driven pendulum, which models a Josephson junction.[41] This last paper offers numerical evaluations of spectra, showing a $1/f$ region extending over two decades, but to our knowledge nobody has tried so far to analyze the role of fractality and draw a comparison with Sinai subdiffusive motion.

Among other things, the results of this paper may strongly affect our current understanding of Optical Bistability (OB) phenomena. OB is described in terms of two fixed point attractors which, however, are the result of a collective dynamics implying many degrees of freedom. There are no exhaustive analyses of the structure of the basin boundary, thus possible fractal structures may appear if the dynamics is evaluated in detail. On the other hand, in order to reduce the signal power necessary to drive the OB device from one state to the other, the system is usually set very near to the boundary. Thus, unavoidable random noise might induce low-frequency spectra of the type described above.

6.9. Conclusion: Long Memory in Statistical Physics

We conclude with some speculation on the role of the long-time terms in nonequilibrium statistical mechanics.

It has been shown that, whenever in nonlinear dynamics more than one attractor is present, there are two distinct power spectra: (1) a high-frequency one, corresponding to the decay of correlations within one attractor; (2) a low-frequency one, corresponding to noise-induced jumps. Based upon (1), the usual transport coefficients for macroscopic equations of evolution have been constructed. Effect (2) has been overlooked so far. Here, I wish to consider an example showing the relevance of (2) with respect to multiphoton molecular excitation.

Let us consider a molecule with two isomeric states (*cis* and *trans*) of almost equal energy, separated by an energy barrier, say, of 1 eV (e.g., rhodopsin molecule in the retina of vertebrates).

We known that an IR laser, such as a CO_2 laser ($\lambda = 10\ \mu$m, $h\nu \simeq 0.1$ eV), may give rise to a multiphoton absorption process if it is powerful enough to provide 10 photons within one coherence time of the "*cis*" valley, so that 10 small photons pile up to 1 eV excitation. (We are considering a molecule large enough so that the barrier is already a classical one, and so quantum tunneling is possible.) We know that a vibrational-mode IR active decays by intramolecular relaxations toward the thermal bath of all other modes, in a time of the order of 10^{-12} s = 1 ps. For a multiphoton isomeric transition to arise we should have a laser power of 10 photons/1 ps $\simeq 10^{-7}$ W over a cross section of about $(1\ \text{Å})^2 = 10^{-16}\ \text{cm}^2$, and thus a laser intensity of $10^9\ \text{W/cm}^2$. But this was a Markovian point of view, based on a memory time of 1 ps related to the high-frequency spectral broadening. A double-potential valley dynamics is described by a Duffing equation (see Arecchi *et al.*[29,35]), and the presence of an IR laser illumination as a forcing term yields a motion on an attractor not necessarily confined in one valley, even for very low intensities (see Figure 3 of Arecchi and Lisi[29]).

Such a chaotic motion may pass near the boundry, hence requiring an activation energy much less than the barrier of 1 eV to be introduced into an Arrhenius-type law. For instance, in Figure 2 we have seen a high-frequency spectrum around 100 kHz, and in Figure 5 the corresponding low-frequency jump spectrum at 1 Hz (5 decades below). By the same reasoning, we might expect that an intensity 5 or 6 decasdes lower (that is, 10 photons/μs or just $10^3\ \text{W/cm}^2$) might be sufficient for a multiphoton isomerization process.

If we could use such a large enhancement factor in most activation processes of biochemical relevance, the consequence would be that the times necessary for biochemical evolution on Earth could be correspondingly reduced. This is just a guess, to show how the introduction of the long memory processes here described for the first time may open new routes in the physics of complex systems.

References

1. W. E. Lamb Jr., *Phys. Rev.* **134**, A1429 (1964).
2. H. Haken, *Laser Theory*, in: Encyclopedia of Physics, Vol. XXV/2c (S. Flügge, ed.), Springer-Verlag, Berlin (1970).
3. M. Scully and W. E. Lamb Jr., *Phys. Rev. Lett.* **16**, 853 (1966); *Phys. Rev.* **159**, 208 (1967); *Phys. Rev.* **166**, 246 (1968).
4. J. P. Gordon, *Phys. Rev.* **161**, 367 (1967).
5. H. Haken, *Synergetics*, 3rd edn., Springer-Verlag, Berlin (1983).
6. F. T. Areccji, in: *Order and Fluctuations in Equilibrium and Nonequilibrium Statistical Mechanics* (G. Nicolis *et al.*, eds.), Proc. XVII Solvay Conf. on Physics, Wiley, New York (1981).
7. R. J. Glauber, in: *Quantum Optics and Electronics* (C. De Witt *et al.*, eds.), Gordon and Breach, London (1965).
8. F. T. Arecchi, in: *Quantum Optics* (R. J. Gluber, ed.), Academic Press, New York (1969).
9. F. T. Arecci and R. G. Harrison, eds., *Instabilities and Chaos in Quantum Optics*, Springer-Verlag, Berlin (1986).
10. E. M. Lorenz, *J. Atmos. Sci.* **20**, 130 (1963).
11. J. P. Eckmann, *Rev. Mod. Phys.*, **53**, 643 (1981); J. P. Eckmann and D. Ruelle, *Rev. Mod. Phys.* **57**, 617 (1985).
12. F. T. Arecchi, G. L. Lippi, G. P. Puccioni, and J. R. Tredicce, *Opt. Commun.* **51**, 308 (1984).
13. F. T. Arecchi, R. Meucci, G. P. Puccioni, and J. R. Tredicce, *Phys. Rev. Lett.* **49**, 1217 (1982).
14. G. L. Lippi, J. R. Tredicce, N. B. Abraham, and F. T. Areccji, *Opt. Commun.* **53**, 129 (1985).
15. F. T. Arecchi, W. Gadomski, and R. Meucci, *Phys. Rev. A* **43**, 1617 (1986).
16. J. D. Farmer, *Physics* **4D**, 366 (1982).
17. B. B. Mandelbrot, *The Fractal Geometry of Nature*, Freeman, San Francisco (1982).
18. P. Grassberger and I. Procaccia, *Phys. Rev. Lett.* **50**, 346 (1983).
19. G. Ioos, and D. D. Joseph, *Elementary Stability and Bifurcation Theory*, Springer-Verlag, Berlin (1980).
20. G. P. Puccioni, A. Poggi, W. Gadomski, J. R. Tredicce, and F. T. Arecchi, *Phys. Rev. Lett.* **55**, 339 (1985).
21. P. Grassberger, *J. Stat. Phys.* **26**, 173 (1981).
22. L. Lugiato, L. M. Narducci, D. K. Bandy, and C. A. Pennise, *Opt. Commun.* **46**, 64 (1983).
23. D. K. Bandy, L. M. Narducci, and L. Lugiato, *J. Opt. Soc. Am.* **B2**, 248 (1985).
24. J. R. Tredicce, F. T. Arecchi, G. L. Lippi, and G. P. Puccioni, *J. Opt. Soc. Am.* **B2**, 173 (1985).
25. J. L. Boulnois, P. Cottin, A. Van Lenberghe, F. T. Arecchi, and G. P. Puccioni, *Opt. Commun.* **58**, 124 (1986).
26. F. T. Arecchi, and A. Califano, *Europhys. Lett.* **3** (1986).
27. Ia. G. Sinai, in: *Proc. Berlin Conf. on Math. Problems in Theoretical Physics* (R. S. Schrader *et al.*, eds.), p. 12, Springer-Verlag (1982).
28. E. Marinari, G. Parisi, D. Ruelle, and P. Windey, *Phys. Rev. Lett.* **50**, 1223 (1983).
29. F. T. Arecchi and F. Lisi, *Phys. Rev. Lett.* **49**, 94 (1982).
30. P. Dutta and P. M. Horn, *Rev. Mod. Phys.* **53**, 497 (1981).
31. R. F. Miracky, J. Clarke, and R. H. Koch, *Phys. Rev. Lett.* **50**, 856 (1983).

32. R. L. Kautz, *J. Appl. Phys.* **58**, 424 (1985).
33. (a) M. R. Beasley, D. D' Humieres, and B. A. Huberman, *Phys. Rev. Lett.* **50**, 1328 (1983).
(b) R. Voss, *Phys. Rev. Lett.* **50**, 1329 (1983).
(c) F. T. Arecchi and F. Lisi *Phys. Rev. Lett.* **50**, 1330 (1983).
34. F. T. Arecchi, R. Baddi, and A. Politi, *Phys. Rev. A* **29**, 1006 (1984).
35. (a) F. T. Arecchi, R. Badii, and A. Politi, *Phys. Lett. A* **103**, 3 (1984).
(b) F. T. Arecchi, R. Badii, and A. Politi, *Phys. Rev. A* **32**, 402 (1985).
36. E. W. Montroll and M. F. Shlesinger, *Proc. Natl. Acad. Sci. U.S.A.* **79**, 3380 (1982).
37. F. T. Arecchi and A. Califano, *Phys. Lett. A* **101**, 443 (1984).
38. G. Grebogi, E. Ott, and J. A. Yorke, *Phys. Rev. Lett.* **50**, 935 (1983); S. M. McDonald, C. Grebogi, E. Ott, and J. A. Yorke, *Physica* **17D**, 125 (1985).
39. Y. Pomeau and P. Manneville, *Commun. Math. Phys.* **74**, 189 (1980); I. Procaccia and H. Schuster, *Phys. Rev. A* **28**, 1210 (1983).
40. T. Geisel and S. Thomae, *Phys. Rev. Lett.* **52**, 1936 (1984); T. Geisel, J. Nierwetberg, and A. Zacherl, *Phys. Rev. Lett.* **54**, 616 (1985).
41. E. G. Gwinn and R. M. Westervelt, *Phys. Rev. Lett.* **54**, 1613 (1985).

7

Nonlinear Optics of Bistability and Pulse Propagation

R. Loudon

7.1. Introduction

The field of nonlinear optics embraces a wide variety of phenomena that can only partially be covered even in a lengthy book. Some individual nonlinear optical processes have been studied in such detail that entire monographs and review articles are needed to provide adequate accounts of them.

The material selected for the present review is intended to introduce some of the processes important for studies of optical stability and chaos. Even with this restriction, there still remains a broad range of nonlinear processes, and the material is further limited to phenomena that can occur with only a single quasi-monochromatic light beam. A more general review has been given by Ackerhalt *et al.*,[1] while the book by Shen[2] is recommended as a basic text covering the whole range of nonlinear optical processes.

The main effects of optical nonlinearities on the propagation of a monochromatic light beam are modifications of the refractive index and absorption coefficient by the addition of terms that depend on the optical intensity. The light beam itself changes the properties of the material through which it propagates, to produce several kinds of self-inflicted alterations of its propagation characteristics. The refractive-index and absorption-coefficient changes both provide mechanisms for optical bistability, the topic of Section 7.2. Only the refractive-index change is important for propagation in low-loss optical fibers, whose pulse-propagation capabilities are treated in Section 7.3. Some brief discussion of the laser is also given, since its grosser features can be obtained by changing the intensity-dependent absorption coefficient to an

R. Loudon • Physics Department, Essex University, Colchester CO4 3SQ, England.

intensity-dependent gain. These systems are all capable of producing optical chaos under appropriate operating conditions.[1,3]

7.1.1. Linear Susceptibility

The linear optical properties of an isotropic material are described by its frequency-dependent linear susceptibility $\chi^{(1)}(\omega)$. Thus with a single-frequency field

$$E(t) = E(\omega)\exp(-i\omega t) + E^*(\omega)\exp(i\omega t) \quad (7.1.1)$$

the associated electrical polarization is

$$P(t) = \varepsilon_0\chi^{(1)}(\omega)E(\omega)\exp(-i\omega t) + \varepsilon_0\chi^{(1)}(-\omega)E^*(\omega)\exp(i\omega t) \quad (7.1.2)$$

where ε_0 is the permittivity of free space. The susceptibility is generally a complex quantity with the property

$$\chi^{(1)}(-\omega) = \chi^{(1)}(\omega)^* \quad (7.1.3)$$

The calculations that follow use complex fields, represented by the first terms in expressions like (7.1.1) and (7.1.2). Thus the dielectric displacement field is

$$D(t) = \varepsilon_0 E(t) + P(t) = D(\omega)\exp(-i\omega t) \quad (7.1.4)$$

where

$$D(\omega) = \varepsilon_0[1 + \chi^{(1)}(\omega)]E(\omega) \equiv \varepsilon_0\varepsilon(\omega)E(\omega) \quad (7.1.5)$$

and $\varepsilon(\omega)$ is the dielectric function.

The relation between the wave vector k and the frequency ω of a plane electromagnetic wave is most conveniently expressed in terms of the refractive index $\eta_0(\omega)$ and the extinction coefficient $\kappa_0(\omega)$, defined according to

$$k^2c^2/\omega^2 = \varepsilon(\omega) = 1 + \chi^{(1)}(\omega) \equiv [\eta_0(\omega) + i\kappa_0(\omega)]^2 \quad (7.1.6)$$

The usual plane-wave factor is then

$$\exp(ikz - i\omega t) = \exp[i\omega(\eta_0 z/c - t) - \omega\kappa_0 z/c] \quad (7.1.7)$$

where η_0 and κ_0 are evaluated at frequency ω. The cycle-averaged intensity of a monochromatic light wave with field $E(\omega)$ is given by

$$I = 2\varepsilon_0 c\eta_0(\omega)|E(\omega)|^2 \quad (7.1.8)$$

and according to (7.1.7) this has a spatial dependence

$$I(z) = I(0)\exp(-\alpha_0 z) \quad (7.1.9)$$

where

$$\alpha_0(\omega) = 2\omega\kappa_0(\omega)/c \tag{7.1.10}$$

is the absorption coefficient.

The susceptibility can be calculated for the various kinds of materials, at least in principle. Thus for the simplest example of a gas of N two-level atoms in a volume V

$$\chi^{(1)}(\omega) = \frac{N|X_{12}|^2}{3\varepsilon_0 \hbar V}\left(\frac{1}{\omega_1 - \omega - i\gamma} + \frac{1}{\omega_1 + \omega + i\gamma}\right) \tag{7.1.11}$$

where X_{12} is the transition dipole-moment matrix element, ω_1 is the transition frequency, and 2γ (equal to the Einstein A-coefficient) is the spontaneous decay rate of the upper level.

7.1.2. Nonlinear Susceptibility

The linear relation (7.1.2) between field and polarization is an approximation valid for sufficiently small fields. It is generalized for larger fields to an expression with the symbolic form

$$P = \varepsilon_0[\chi^{(1)}E + \chi^{(2)}E^2 + \chi^{(3)}E^3 + \cdots] \tag{7.1.12}$$

where $\chi^{(2)}$, $\chi^{(3)}$, etc. are components of the nonlinear susceptibility. These have forms analogous to equation (7.1.11) but they have increasing numbers of dipole matrix elements and frequency denominators, and they include increasing numbers of terms.[2] They depend on the frequencies of the fields that they multiply, for example, $\chi^{(3)}$ is a function of three frequencies, and the resulting polarization has contributions at all their sums and differences (22 distinct frequencies in all![4]). The susceptibility components also depend upon the orientations of the fields that they multiply, giving several different contributions of the same order for anisotropic crystals. Since P and E both have the character of polar vectors, it is clear from expression (7.1.12) that $\chi^{(2)}$ and the higher even-order susceptibility components must vanish for materials invariant under spatial inversion.

We consider here only those nonlinear processes in which a single-frequency field (7.1.1) modifies the optical properties of a material at its own frequency. It is obvious that only combinations of odd numbers of contributions $\pm\omega$ can produce a total of $\pm\omega$, and only the odd-order terms in relation (7.1.12) survive to give

$$P(\omega) = \varepsilon_0[\chi^{(1)}E(\omega) + \chi^{(3)}|E(\omega)|^2E(\omega) + \chi^{(5)}|E(\omega)|^4E(\omega) + \cdots] \tag{7.1.13}$$

The material is assumed to be optically isotropic, for example, a gas or a glass, and the field polarization is not explicitly included in the notation. The higher-order susceptibility components are generally complex, similar to $\chi^{(1)}(\omega)$, and it is convenient to define nonlinear refractive index and extinction coefficients η and κ by analogy with relation (7.1.6),

$$1 + \chi^{(1)} + \chi^{(3)}|E(\omega)|^2 + \chi^{(5)}|E(\omega)|^4 + \cdots = (\eta + i\kappa)^2 \qquad (7.1.14)$$

where the various optical functions are all evaluated at frequency ω. The refractive index and the absorption coefficient have the forms

$$\eta = \eta_0 + \eta_2|E(\omega)|^2 + \eta_4|E(\omega)|^4 + \cdots \qquad (7.1.15)$$

and

$$\alpha = 2\omega\kappa/c = \alpha_0 + \alpha_2|E(\omega)|^2 + \alpha_4|E(\omega)|^4 + \cdots \qquad (7.1.16)$$

This last expression for the absorption coefficient is essentially an expansion in powers of the beam intensity. The saturable two-level absorbing atoms treated in Section 7.2.1 produce an absorption coefficient of this general form, although the atomic model is sufficiently simple that α can be written as a closed expression [see expression (7.2.12)] and it is not necessary to use the series expansion explicitly.

As for the nonlinear refractive index, we shall be interested in Section 7.2.5 and Section 7.3 in transparent media, where ω is far removed from any transition frequencies and the susceptibility components are accordingly real. Thus with attention restricted to the first nonlinear term,

$$\eta = \eta_0 + \eta_2|E(\omega)|^2 \qquad (7.1.17)$$

and comparison with equation (7.1.14) gives

$$2\eta_0\eta_2 = \chi^{(3)} \qquad (7.1.18)$$

The intensity-dependent term in the refractive index gives rise to the optical Kerr effect,[2] a birefringence induced by the linearly polarized optical field. However, the single linearly-polarized light beam assumed here maintains its initial polarization. The coefficient η_2 can be either positive or negative and it has magnitudes of order

$$\eta_2 \approx 10^{-21}\text{–}10^{-23} \qquad (\mathrm{mV}^{-1})^2 \qquad (7.1.19)$$

for optical glasses.

The distribution of intensity in a laser light beam typically has a Gaussian cross section. According to equation (7.1.17), if η_2 is positive, the beam experiences a higher refractive index at its center than at its edges, and the velocity at the center is correspondingly smaller. The result is a self-focusing effect in which the beam converges inward. The converse effect of self-defocusing occurs when η_2 is negative.[2] These focusing effects will be avoided in the following analysis by the assumption of isotropic transverse intensity distributions.

7.2. Optical Bistability and Laser Action

With very minor changes in the theoretical description (but with more drastic changes in the corresponding experimental realizations!), the same physical model can be operated either as a bistable optical switch or as a laser light source. We treat here the simplest system that has the necessary properties, based on the nonlinearities associated with saturable optical absorption. The optical Kerr effect is ignored until Section 7.2.5.

7.2.1. Saturable Absorption

Consider a collection of N absorbing atoms contained in a volume V. The atoms are assumed to have a pair of nondegenerate energy levels separated by energy $\hbar\omega$, where N_1 of the atoms are in the lower level (the ground state) and N_2 are in the upper excited level such that

$$N_1 + N_2 = N \tag{7.2.1}$$

The equations that describe the interaction of light with the atomic transition take different forms depending on the spectral distribution of the light.[4] The optical Bloch equations are appropriate for monochromatic light, while rate equations apply for light whose frequency components spread uniformly across the linewidth of the transition. The latter regime is assumed here, since it can be treated without any detailed discussion of the quantum mechanics of the coupling between light and atom. The rate equations are based on the simple phenomenological theory due to Einstein.

Figure 7.1 shows the atomic energy levels with representations of the three kinds of transition that can occur. These are described as follows:

1. In the absence of any light, an excited atom can decay to its ground state with spontaneous emission of a photon at a rate equal to A, the Einstein A-coefficient. The emitted photon has energy $\hbar\omega$ and its propagation direction has a random spatial orientation.

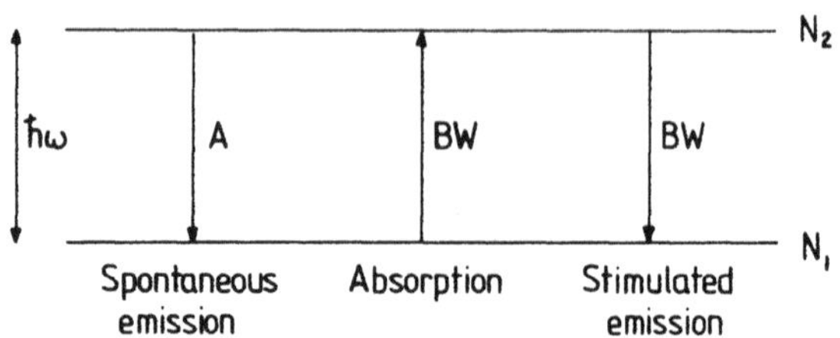

Figure 7.1. The three basic kinds of optical transition.

2. In the presence of light of energy density W, an atom in its ground state is excited at a rate BW, where B is the Einstein B-coefficient. A photon $\hbar\omega$ is removed from the light.
3. The presence of light also enhances the decay rate of an excited atom by an amount BW corresponding to stimulated emission. The photon emitted in the stimulated process has its propagation direction parallel to that of the incident light.

The quantity W is the energy of the light in unit volume and unit angular frequency range at the atomic transition frequency ω. The light is in the form of a parallel beam for the applications of interest here.

The above transition rates can be used to set up rate equations for the atomic level populations,

$$dN_1/dt = -dN_2/dt = N_2A - N_1BW + N_2BW \tag{7.2.2}$$

In the steady state

$$N_2(A + BW) = N_1BW \tag{7.2.3}$$

and with the use of equation (7.2.1), the level populations can be written as

$$N_1 = \tfrac{1}{2}N(W + 2W_S)/(W + W_S) \qquad \text{and} \qquad N_2 = \tfrac{1}{2}NW/(W + W_S) \tag{7.2.4}$$

where the quantity defined by

$$W_S = A/2B \tag{7.2.5}$$

is known as the saturation energy density of the atomic transition. The level populations have a linear variation with energy density for $W \ll W_S$, but the dependence becomes nonlinear for $W \approx W_S$, and both populations tend to $\frac{1}{2}N$ for $W \gg W_S$, as illustrated in Figure 7.2. This nonlinear behavior is called saturation of the atomic transition.

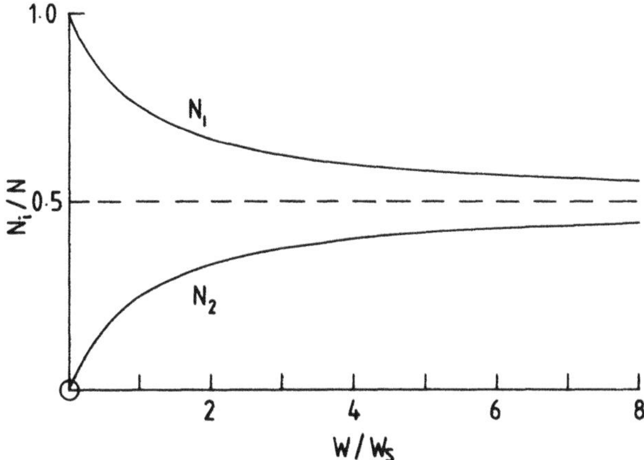

Figure 7.2. Steady-state atomic-level populations.

For illumination of the atoms by parallel light, each absorption event removes energy $\hbar\omega$ from the beam and each stimulated emission event puts energy $\hbar\omega$ back into the beam. The net rate of change of beam energy in volume V is

$$V dW/dt = -(N_1 - N_2)BW\hbar\omega \tag{7.2.6}$$

Experiments on light beams usually study the variation of beam intensity I (given by Poynting's vector) with propagation distance z, rather than the variation of energy density W with the time t. The two variations are related by a continuity equation of the usual form,

$$dW/dt = dI/dz \tag{7.2.7}$$

The two magnitudes are related by

$$W = I/c \tag{7.2.8}$$

where the refractive index η is assumed sufficiently close to unity that the free-space velocity of light can be used. These substitutions convert equation (7.2.6) to the form

$$dI/dz = -(N_1 - N_2)BI\hbar\omega/Vc \tag{7.2.9}$$

The solution of this equation is complicated by the dependences of N_1 and N_2 on the intensity I. However, its application to optical bistability requires only the solution to first order in the propagation distance z, given by

$$I(z) = (1 - \alpha z)I \tag{7.2.10}$$

where I is the intensity at $z = 0$ and the absorption coefficient α is

$$\alpha = (N_1 - N_2)B\hbar\omega/Vc \tag{7.2.11}$$

Expressions (7.2.4) and (7.2.8) enable this to be written as

$$\alpha = \alpha_0/[1 + (I/I_S)] \tag{7.2.12}$$

where

$$\alpha_0 = NB\hbar\omega/Vc \tag{7.2.13}$$

and

$$I_S = cW_S = cA/2B \tag{7.2.14}$$

The absorption coefficient is equal to α_0 for low incident intensity, but it falls to $\frac{1}{2}\alpha_0$ at the saturation intensity I_S, and tends to zero for still higher intensities. The absorber is said to be saturable; it becomes progressively more transparent with increasing intensity, a phenomenon known as bleaching. The absorption coefficient (7.2.12) can of course be expanded in a power series in I, or equivalently in $|E|^2$, and the coefficients are related to increasing orders of the nonlinear susceptibility, as in relations (7.1.14) and (7.1.16).

7.2.2. Fabry–Perot Cavity

Consider a one-dimensional optical cavity of length L confined between two identical mirrors whose intensity reflection and transmission coefficients are R and T. The mirrors are assumed lossless, so that

$$R + T = 1 \tag{7.2.15}$$

The cavity is initially taken to be empty, with a light beam of field E_0 and intensity I_0 incident from the left. It is necessary to work with both fields and intensities, the two being related by

$$I = 2\varepsilon_0 c|E|^2 \tag{7.2.16}$$

Figure 7.3 shows the notation for the various fields of the waves that impinge on the left-hand mirror and for the wave that emerges from the

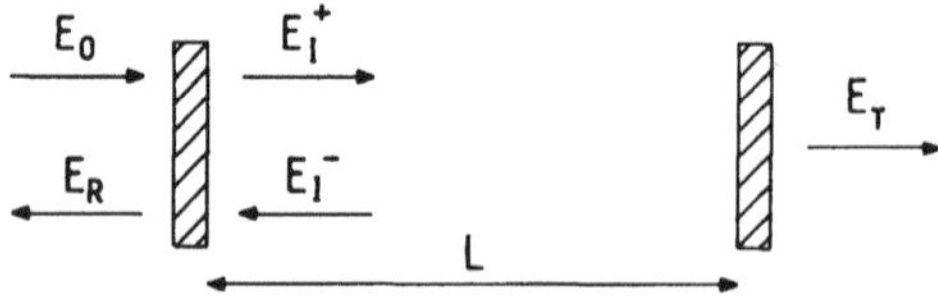

Figure 7.3. Notation for field amplitudes of the light waves associated with a Fabry–Perot cavity.

right-hand mirror. The field reflection and transmission coefficients at each mirror have magnitudes $R^{1/2}$ and $T^{1/2}$. These coefficients are generally complex, since phase changes occur on reflection and/or transmission at a mirror. However, the phase changes are here ignored because they add nothing of significance to the present discussion. The boundary conditions at the left-hand mirror then give

$$E_{\mathrm{I}}^{+} = T^{1/2}E_0 + R^{1/2}E_{\mathrm{I}}^{-} \qquad \text{and} \qquad E_{\mathrm{R}} = R^{1/2}E_0 + T^{1/2}E_{\mathrm{I}}^{-} \quad (7.2.17)$$

Propagation of the light waves across the cavity produces a change of phase kL, where k is the optical wave vector. Matching of fields at the right-hand mirror therefore gives

$$E_{\mathrm{T}} = T^{1/2}E_{\mathrm{I}}^{+}\exp(ikL) \qquad \text{and} \qquad E_{\mathrm{I}}^{-} = R^{1/2}E_{\mathrm{I}}^{+}\exp(2ikL) \quad (7.2.18)$$

The internal and transmitted fields obtained by solution of these equations are

$$\begin{aligned} E_{\mathrm{I}}^{+} &= T^{1/2}E_0/[1 - R\exp(2ikL)] \\ E_{\mathrm{I}}^{-} &= R^{1/2}T^{1/2}E_0\exp(2ikL)/[1 - R\exp(2ikL)] \\ E_{\mathrm{T}} &= TE_0\exp(ikL)/[1 - R\exp(2ikL)] \end{aligned} \quad (7.2.19)$$

Standard Fabry-Perot theory is mainly concerned with the relation between output and input intensity. This is obtained from expressions (7.2.19), with the help of condition (7.2.15), in the form

$$I_{\mathrm{T}} = \frac{T^2 I_0}{T^2 + 4R\sin^2(kL)} \quad (7.2.20)$$

The maximum transmitted intensity is

$$I_{\mathrm{T}} = I_0 \qquad \text{for } kL = n\pi, \qquad \text{with } n = 0, 1, 2, \cdots \quad (7.2.21)$$

and the condition for half the maximum transmission is

$$I_{\mathrm{T}} = \tfrac{1}{2}I_0 \qquad \text{for } \sin(kL) = T/2R^{1/2} \quad (7.2.22)$$

Figure 7.4 shows the first two transmission maxima for mirrors of fairly modest 90% reflectivity. Practical cavities often have mirrors of higher reflectivity than this and the transmission maxima are then very narrow. The empty cavity conserves energy, and the incident intensity that is not transmitted all goes into the reflected beam. The internal intensities obtained from expressions (7.2.19) become very large at the condition for maximum transmission (7.2.21), with values

$$I_{\mathrm{I}}^{+} = I_0/T \qquad \text{and} \qquad I_{\mathrm{I}}^{-} = RI_0/T \quad (7.2.23)$$

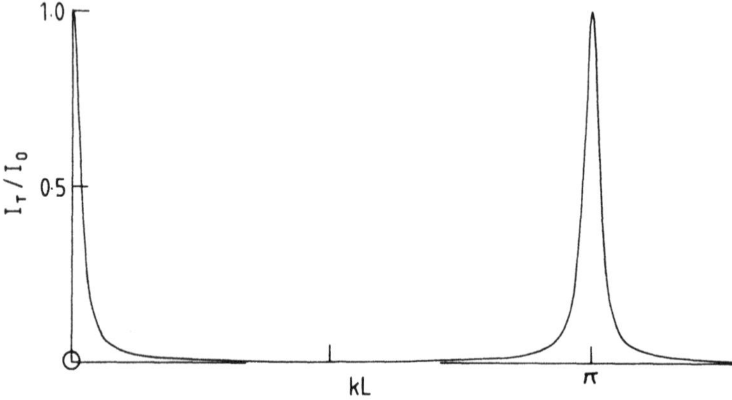

Figure 7.4. First two transmission maxima of a Fabry–Perot cavity for $R = 0.9$ and $T = 0.1$.

Now suppose that the cavity is filled with a saturable absorber of the kind treated in Section 7.2.1, whose refractive index η is still assumed to be close to unity. The absorption for a single transit of the cavity is assumed to be small,

$$\alpha L \ll 1 \tag{7.2.24}$$

so that a linear approximation as in equation (7.2.10) is adequate, and the form (7.2.12) of the absorption coefficient can be used. The phase factor that appears in the electric field for a single transit is modified as

$$\exp(ikL) \to (1 - \tfrac{1}{2}\alpha L)\exp(ikL) \tag{7.2.25}$$

and this substitution converts the field expressions (7.2.19) to the absorbing case. For the treatment of absorptive bistability that follows in Section 7.2.3, it is sufficient to consider the special case of a cavity length L that produces maximum transmission, as in equation (7.2.21). The modified fields (7.2.19) are then

$$\begin{aligned} E_I^+ &= T^{1/2}E_0/(T + R\alpha L) \\ E_I^- &= R^{1/2}T^{1/2}(1 - \alpha L)E_0/(T + R\alpha L) \\ E_T &= \pm T(1 - \tfrac{1}{2}\alpha L)E_0/(T + R\alpha L) \end{aligned} \tag{7.2.26}$$

The transmitted intensity is obtained as before from the square of the field. The contribution αL in the numerator of E_T can be neglected in view of condition (7.2.24). However, it must be retained in the denominator since T is also much smaller than unity. Thus

$$I_T = T^2 I_0/(T + R\alpha L)^2 \tag{7.2.27}$$

and the 100% transmission of the empty cavity is much reduced for absorption coefficients so large as to satisfy

$$1 \gg R\alpha L \gg T \tag{7.2.28}$$

The corresponding expressions for the internal intensities are

$$I_{\rm I}^{+} = TI_0/(T + R\alpha L)^2 \qquad \text{and} \qquad I_{\rm I}^{-} = RTI_0/(T + R\alpha L)^2 \tag{7.2.29}$$

and the total internal intensity, irrespective of propagation direction, is

$$I_{\rm I} = T(1 + R)I_0/(T + R\alpha L)^2 = (1 + R)I_{\rm T}/T \tag{7.2.30}$$

The internal intensity is always much larger than the transmitted intensity for $T \ll 1$.

7.2.3. Absorptive Optical Bistability

The relation (7.2.27) between cavity input and output intensities can be put in a more explicit form by substitution of the expression (7.2.12) for the saturable absorption coefficient. The absorption and stimulated emission processes occur with equal probabilities for the two counterpropagating components of the internal light beams, and the total internal intensity (7.2.30) determines the saturation of the atomic transition. Thus taking $R \approx 1$, equation (2.27) can be rearranged in the form

$$I_0 = I_{\rm T}\left[1 + \frac{\alpha_0 L/T}{1 + (2I_{\rm T}/TI_{\rm S})}\right]^2 \tag{7.2.31}$$

It is convenient to study the relation between input and output fields, rather than intensities, the two being related by equation (7.2.16). The square root of equation (7.2.31) then gives

$$|E_0| = |E_T|\left[1 + \frac{\alpha_0 L/T}{1 + (2|E_{\rm T}|^2/T|E_{\rm S}|^2)}\right] \tag{7.2.32}$$

where $|E_{\rm S}|$ is the field magnitude associated with the saturation intensity $I_{\rm S}$.

Figure 7.5 shows the relation between the dimensionless scaled input and output fields for three values of the parameter $\alpha_0 L/T$. The equality of the fields in the absence of absorption is represented by the dashed line. The equation (7.2.32) that determines $|E_{\rm T}|$ for a given value of $|E_0|$ is essentially a cubic, and it can have either one or three real roots. The condition for three real roots, easily found by differentiation of equation (7.2.32), is

$$\alpha_0 L/T > 8 \tag{7.2.33}$$

The critical curve, for $\alpha_0 L/T = 8$, has a vertical tangent at

$$|E_0|/T^{1/2}|E_S| = (27/2)^{1/2}, \qquad |E_T|/T^{1/2}|E_S| = (3/2)^{1/2} \tag{7.2.34}$$

marking the onset of the s-shaped curves that occur for larger values of the parameter combination. It can be shown that the part of an s-shaped curve with negative slope corresponds to an unstable state of the cavity system. Thus in the range defined by condition (7.2.33) the system shows optical bistability, with two stable output fields $|E_T|$ for a range of values of the input field $|E_0|$.

Figure 7.6 shows more details of the bistable behavior. As the incident intensity is increased from zero, the absorbing atoms in the cavity remove much of the energy from the light beam and only a small fraction is transmitted. With further increase in input intensity to the lower bend of the s-curve, the output field has to jump discontinuously to the upper branch of the curve, which corresponds to a saturated atomic state. The cavity is largely bleached and the output intensity is only a little smaller than the input intensity. If the input intensity is now reduced, the system remains in its transmitting condition to the upper bend in the s-curve, when a discontinuous downward transition to the lower branch occurs, and the system again absorbs most of the input light. Thus when condition (7.2.33) is satisfied, there is a hysteresis loop in the variation of $|E_T|$ with $|E_0|$. This behavior is similar to that in a first-order phase transition and there are indeed close analogies between the bistable

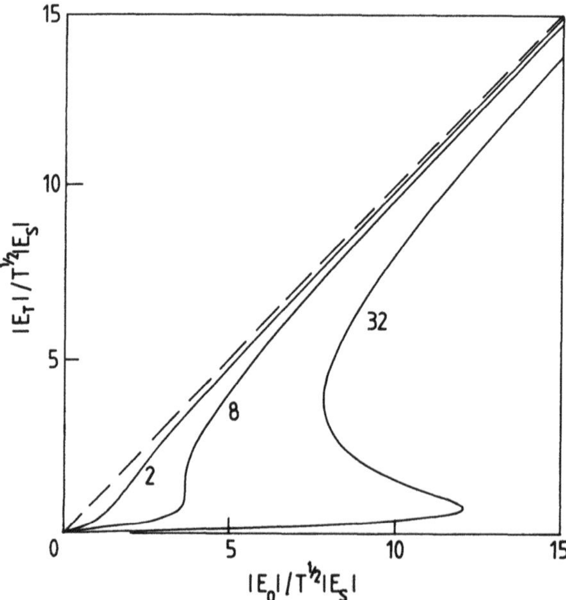

Figure 7.5. Relations between cavity input and output fields for the values of $\alpha_0 L/T$ shown beside the curves.

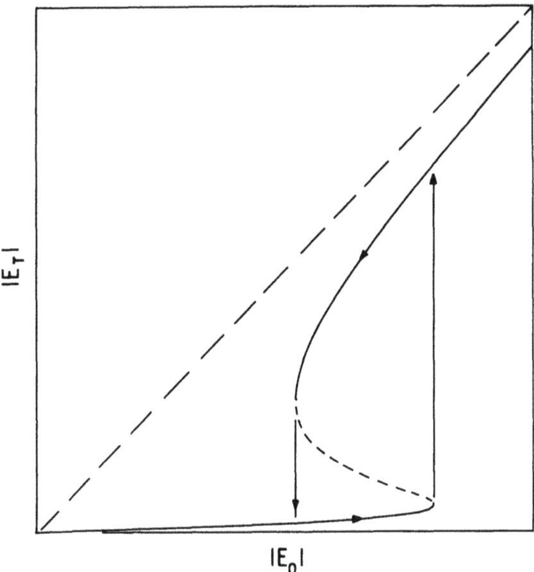

Figure 7.6. Hysteresis behavior of the bistable cavity showing the discontinuous transitions. The dashed part of the s-curve corresponds to unstable states of the cavity.

cavity and thermodynamic systems that undergo phase transitions. For example, the curve in Figure 7.5 for $\alpha_0 L/T = 8$ is analogous to the critical isotherm in the van der Waals theory of the gas–liquid system.

Within the hysteresis region the bistable cavity can act as an optical switch. With an input "holding" intensity whose field $|E_0|$ lies in the center of the bistable region, the cavity can be switched from its absorbing to its transmitting state by a momentary increase in intensity to take $|E_0|$ beyond the lower bend in the s-curve. A momentary decrease in intensity similarly returns the cavity from its transmitting to its absorbing state. There are important potential applications of the bistable switch to systems for processing optical signals.[5]

We return to the topic of optical bistability in Section 7.2.5, where the limitations of the present simple calculation are discussed. In particular, the stability effects of the intensity dependence of the refractive index, here ignored, are described and illustrated.

7.2.4. The Laser

The theory of the previous section also provides a description of laser action if the absorbing atoms in the optical cavity are replaced by emitting atoms. The absorption coefficient (7.2.11) becomes negative, corresponding to a gain of intensity by the light beam, for interaction with a pair of levels that show population inversion, with $N_2 > N_1$. The absorption coefficient α is then replaced by a gain coefficient

$$g = (N_2 - N_1)B\hbar\omega/Vc \tag{7.2.35}$$

The theory of Section 7.2.1 (see particularly Figure 7.2) shows that population inversion cannot be achieved by interaction of light with only two levels of an atom. The active levels of the atoms in a laser are normally part of a more extensive energy-level scheme and the inversion is usually achieved by pumping the atoms into excited states by some external source of energy, perhaps by an electrical discharge or optical pulse. The three-level laser is easy to treat[(4)] and it gives a dependence of gain on intensity of the form

$$g = g_0/[1 + (I/I_S)] \tag{7.2.36}$$

analogous to the absorption coefficient (7.2.12). The gain is equal to g_0 for $I \ll I_S$, but falls off owing to saturation effects when I is comparable to or greater than the saturation intensity I_S. This quantity is given by an expression analogous to (7.2.14), except that A is replaced by a function of the various Einstein A-coefficients for the three levels. The quantity g_0 is proportional to the external pumping rate which can be varied to examine the laser characteristics.

The basic equations for the laser intensity and field are given by equations (7.2.31) and (7.2.32) but with α_0 replaced by $-g_0$. The field equation is therefore

$$|E_0| = |E_T|\left[1 - \frac{g_0L/T}{1 + (2|E_T|^2/T|E_S|^2)}\right] \tag{7.2.37}$$

The output field is now larger than the input field and the cavity system acts as an optical amplifier. We consider first the case in which there is no input light beam. With $|E_0|$ set equal to zero, there are two possible solutions of equation (7.2.37),

$$|E_T| = 0 \tag{7.2.38}$$

and

$$\frac{|E_T|}{T^{1/2}|E_S|} = \frac{1}{2^{1/2}}\left(\frac{g_0L}{T} - 1\right)^{1/2} \tag{7.2.39}$$

The second of these solutions is physically significant only when $g_0L \geq T$; it corresponds to laser emission sustained by the pumping energy. The laser threshold condition is

$$g_0L = T \tag{7.2.40}$$

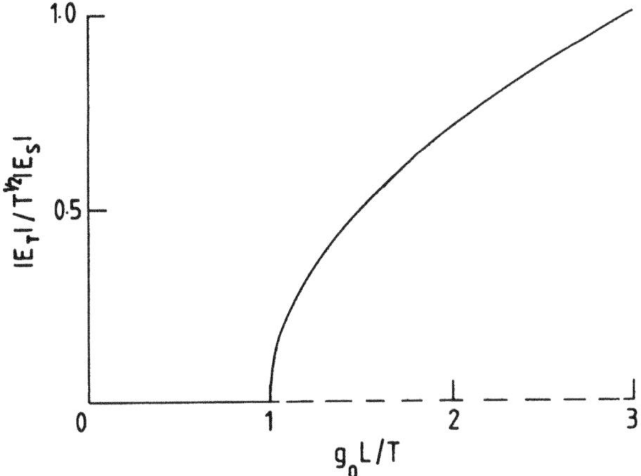

Figure 7.7. Laser output field as a function of the pumping rate.

It can be shown that the first solution above is stable for pumping rates that give smaller gains than condition (7.2.40), while the second solution is stable for gains that exceed the threshold value. Laser action occurs in the latter case because the gain g_0L in optical intensity on a single traverse of the cavity exceeds the loss T at each encounter with a mirror. The two solutions for the output field are illustrated in Figure 7.7, with the unstable solution shown dashed. The model considered here is strictly a two-ended laser, but only trivial changes occur if one of the mirrors is made perfectly reflecting to represent the usual single-ended laser arrangement.

The behavior of the laser at threshold is similar to that of a second-order thermodynamic phase transition at the critical temperature. Thus the vertical axis in Figure 7.7 corresponds to the order parameter, the horizontal axis to the temperature, and threshold to the critical temperature. In the laser system it is the phase angles of the contributions of the different atoms to the output light that order to produce a large total field $|E_T|$. Below threshold the different atoms emit with random phases, and their contributions interfere destructively to give zero average total field.

Now consider the behavior of the laser when there is a nonzero input field $|E_0|$. For small values of $|E_0|$, it is appropriate to evaluate the differential amplification defined by

$$\Xi = (d|E_T|/d|E_0|)_{|E_0|=0} \tag{7.2.41}$$

The inverse derivative is easily found from equation (7.2.37), and substitution respectively of the values of $|E_T|$ from solutions (7.2.38) and (7.2.39) then gives

$$\Xi = \frac{1}{1-(g_0L/T)} \qquad \text{for } g_0L < T \tag{7.2.42}$$

and

$$\Xi = \frac{g_0L/T}{2[(g_0L/T)-1]} \qquad \text{for } g_0L > T \tag{7.2.43}$$

The gain dependence of the amplification is plotted in Figure 7.8. There is large differential amplification in the region of the laser threshold, but its value falls below unity for gains that exceed twice the laser threshold. The curves shown in Figure 7.8 closely resemble those for the temperature dependence of the susceptibility of a ferromagnet.

The laser theory outlined here is extremely simplified in that, for example, it is based on rate equations for the optical intensity and atomic populations rather than the Maxwell-Bloch equations for the optical field and atomic density matrix, and it ignores the effects of the position-dependent field distribution in a Fabry-Perot cavity. More sophisticated theories[1,3] show that the single-mode laser can become unstable at a second threshold, with pumping rates much larger than that at the first threshold given by condition (7.2.40). The instability leads to an irregular pulsating output corresponding to deterministic optical chaos. Such chaos is difficult to achieve experimentally with the single-mode laser whose atoms show only homogeneous spontaneous-emission broadening, but a rich variety of instability phenomena leading to optical chaos occurs in practical systems with inhomogeneous Doppler broadening and multimode excitation.[1,6,7]

7.2.5. Dispersive Optical Bistability

Several comprehensive reviews of optical bistability phenomena are available,[8-11] and the discussion here is limited to a few remarks intended to place the derivations of the present section in perspective.

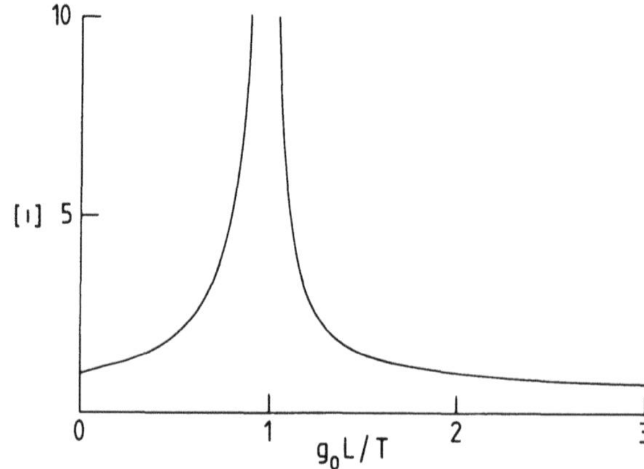

Figure 7.8. Differential optical amplification as a function of laser gain.

The absorptive optical bistability mechanism described in Section 7.2.3 was originally suggested by Szöke *et al.*,[12] and measurements that clearly show hysteresis cycles similar to Figure 7.6 have been made, for example, by Gozzini *et al.*[13] and Rosenberger *et al.*[14] The relations (7.2.31) and (7.2.32) between input and output intensities and fields need some generalizations to allow, for example, for the nonuniform optical fields experienced by the absorbing atoms in the cavity, but the bistability mechanism responsible for the observed intensity variations in these experiments remains the saturable atomic absorption.

However, strong optical fields affect not only the absorption coefficient, as in expression (7.1.16), but also the refractive index, as in expression (7.1.15), and the latter effect can also produce an optical bistability associated with atoms in a cavity. Indeed, this dispersive mechanism was responsible for the bistability seen in the first experimental observation,[15] and in many subsequent studies. While absorptive bistability occurs in conditions where the input frequency ω is in resonance with the atomic transition and with the Fabry-Perot cavity, purely dispersive bistability occurs in nonresonant conditions where absorption can be ignored and kL is not an integer multiple of π.

With only the first of the nonlinear terms retained, the refractive index (7.1.15) of the atomic medium is

$$\eta = \eta_0 + \eta_2 |E_I|^2 \tag{7.2.44}$$

where the total squared field $|E_I|^2$ in the cavity from relations (7.2.19) is

$$|E_I|^2 = |E_I^+|^2 + |E_I^-|^2 = (1 + R)|E_T|^2/T \tag{7.2.45}$$

The relation between input and output intensities for dispersive bistability is given by the empty cavity expression (7.2.20) after modification of the phase factors to include the nonlinear refractive index. Thus we must set

$$kL = \eta\omega L/c = k_0 L + (|E_T|^2/|E_D|^2) \tag{7.2.46}$$

where

$$k_0 = \eta_0 \omega / c \tag{7.2.47}$$

and E_D is a characteristic field associated with the dispersive nonlinearity, defined so that

$$\eta_2 |E_D|^2 = T\eta_0/(1 + R)k_0 L \tag{7.2.48}$$

The relation between input and output field magnitudes obtained from the square root of expression (7.2.20) can now be written in the form

$$|E_0| = |E_T| \left[1 + \frac{4R}{T^2} \sin^2 \left(k_0 L + \frac{|E_T|^2}{|E_D|^2} \right) \right]^{1/2} \tag{7.2.49}$$

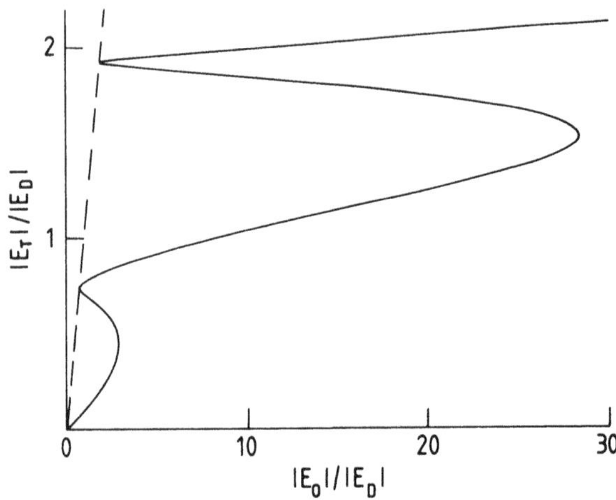

Figure 7.9. Relation between cavity input and output fields for $R = 0.9$, $T = 0.1$, and $k_0L = n\pi + 2.6$.

Figure 7.9 shows the form of relation between input and output fields for dispersive bistability. Indeed the phenomenon is now more accurately described as optical multistability, because there may be several stable output field magnitudes for a given input field. The curve has an unlimited number of oscillations and it touches the dashed line, representing perfect transmission with $|E_T| = |E_0|$, for all values of $|E_T|$ such that the sine in equation (7.2.49) vanishes. The characteristic curves for absorptive and dispersive bistability therefore differ in their detailed forms, although both display the hysteresis effects needed for applications to optical switching. A more general theory is required for systems in which the saturable absorption and optical Kerr effect have comparable importance in determining the cavity transmission.

Much more detailed theories of stability in optical cavities are available in the literature,[8-11] including treatments of the possibilities of chaotic behavior from such systems.[1,3,11]

7.3. Optical Propagation in Nonlinear Fibers

Nonlinear optical processes in fibers are interesting for several reasons. Optical fibers are of course widely used for communication systems, and nonlinear processes are beginning to affect the signal propagation at power densities currently considered for practical use. From the standpoint of basic nonlinear optics, fibers enable quite weak nonlinearities to be studied since they confine the optical power to a narrow cross section that can be maintained over enormous propagation lengths, measured in kilometers or tens of kilometers rather than the millimeters or at most meters of the conventional laboratory experiment. Single-mode fibers, to which this discussion is limited, have only one permitted transverse field variation, and they offer an optical system with the one-dimensional character often assumed in calculations but not always justified in practice.

We treat here the effects of the nonlinear refractive index, or optical Kerr effect, on the propagation of pulses through fibers. All optical materials have the property that monochromatic waves of different frequencies travel with slightly different group velocities. This group velocity dispersion spreads apart the different frequency components that make up the pulse, and serious pulse broadening can occur over long propagation distances. In favorable circumstances, the broadening can be compensated by the effects of the nonlinear refractive index to produce stable soliton pulses that propagate without change of shape. These are clearly of potential practical importance, and they provide laboratory systems in which the basic features of soliton behavior can be studied experimentally. Pulse propagation in an optical fiber constructed in the form of a ring cavity also shows phenomena characteristic of deterministic chaos when the fiber is excited by injection of a series of pulses.

7.3.1. Nonlinear Pulse Propagation

The propagation of electromagnetic waves in optical fibers is of course described by Maxwell's equations. With field variations perpendicular to the fiber axis ignored, the one-dimensional wave equation is

$$\frac{\partial^2 E}{\partial z^2} - \frac{1}{c^2}\frac{\partial^2 E}{\partial t^2} = \mu_0 \frac{\partial^2 P}{\partial t^2} \tag{7.3.1}$$

where μ_0 is the magnetic permeability of free space. It is convenient to divide the polarization, given by equation (7.1.13) into its linear and nonlinear parts,

$$P = P^{\mathrm{L}} + P^{\mathrm{NL}} \tag{7.3.2}$$

The linear contribution can be taken across to the left-hand side of equation (7.3.1) where it combines with the second term to form the linear displacement, defined as in equation (7.1.4),

$$\frac{\partial^2 E}{\partial z^2} - \mu_0 \frac{\partial^2 D}{\partial t^2} = \mu_0 \frac{\partial^2 \mathrm{P}^{\mathrm{NL}}}{\partial t^2} \tag{7.3.3}$$

Consider a field excitation of the form shown in Figure 7.10, where a monochromatic wave is modulated by a relatively slowly-varying envelope,

$$E(z, t) = \exp(ik_0 z - i\omega_0 t)\mathscr{E}(z, t) \tag{7.3.4}$$

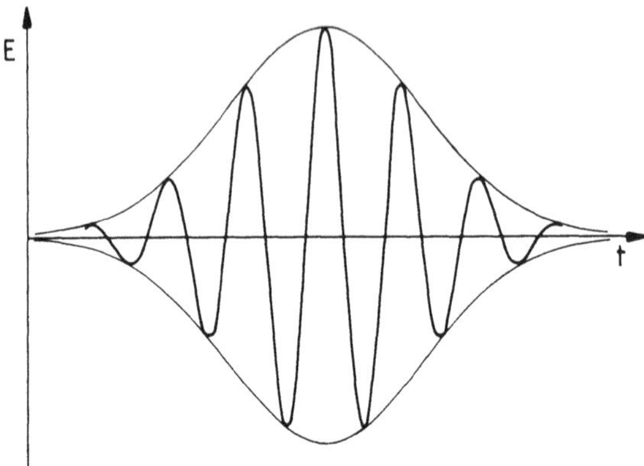

Figure 7.10. Monochromatic input pulse with slowly-varying Gaussian envelope. Wave period exaggerated for ease of drawing.

Optical fibers have absorption coefficients as low as $0.05\ \text{km}^{-1}$ and the wave vector k_0 can be taken to be real,

$$k_0 = \eta_0\omega_0/c \tag{7.3.5}$$

where η_0 is the linear refractive index. The envelope has a Fourier integral

$$\mathscr{E}(z, t) = \int \mathscr{E}(z, \nu) \exp(-i\nu t)\, d\nu \tag{7.3.6}$$

where the slow variation implies a Fourier transform that is significant only for $|\nu| \ll \omega_0$. The displacement is given by a generalization of equation (7.1.5) to the case of a pulse excitation,

$$D(z, t) = \varepsilon_0 \exp(ik_0 z - i\omega_0 t) \int \varepsilon(\omega_0 + \nu)\mathscr{E}(z, \nu) \exp(-i\nu t)\, d\nu \tag{7.3.7}$$

where ε is the linear dielectric function as before. As in relation (7.1.6), it determines the wave vectors $k_0 + \kappa$ associated with the frequency components $\omega_0 + \nu$ that make up the pulse according to

$$(k_0 + \kappa)^2 c^2 = (\omega_0 + \nu)^2 \varepsilon(\omega_0 + \nu) \tag{7.3.8}$$

Substitution of the assumed field (7.3.4) in the first term of the wave equation (7.3.3) gives

$$\frac{\partial^2 E}{\partial z^2} = \exp(ik_0 z - i\omega_0 t)\left(-k_0^2\mathscr{E} + 2ik_0\frac{\partial \mathscr{E}}{\partial z} + \frac{\partial^2 \mathscr{E}}{\partial z^2}\right) \tag{7.3.9}$$

and substitution of the displacement (7.3.7) into the second term gives

$$-\mu_0 \frac{\partial^2 D}{\partial t^2} = \mu_0 \varepsilon_0 \exp(ik_0 z - i\omega_0 t) \int (\omega_0 + \nu)^2 \varepsilon(\omega_0 + \nu) \mathscr{E}(z, \nu) \exp(-i\nu t)\, d\nu \tag{7.3.10}$$

The integrand can be written in terms of the wave vector with the use of relation (7.3.8), and the wave vector can be expanded around k_0 according to the Taylor series

$$k_0 + \kappa = k_0 + \nu k_0' + \tfrac{1}{2}\nu^2 k_0'' + \cdots \tag{7.3.11}$$

where

$$k_0' \equiv dk/d\omega|_{\omega_0} \equiv 1/v_G \tag{7.3.12}$$

is the reciprocal of the group velocity and

$$k_0'' \equiv d^2k/d\omega^2|_{\omega_0} = d(1/v_G)/d\omega \tag{7.3.13}$$

provides a measure of the group velocity dispersion. Thus with the use of relations (7.3.8) and (7.3.11), equation (7.3.10) becomes

$$-\mu_0 \frac{\partial^2 D}{\partial t^2} = \exp(ik_0 z - i\omega_0 t) \int (k_0 + \nu k_0' + \tfrac{1}{2}\nu^2 k_0'')^2 \mathscr{E}(z, \nu) \exp(-i\nu t)\, d\nu \tag{7.3.14}$$

where the series expansion can be truncated because ν is small. If the squared term in the integrand is also truncated at order ν^2, the above expression can be converted to

$$-\mu_0 \frac{\partial^2 D}{\partial t^2} = \exp(ik_0 z - i\omega_0 t)\left[k_0^2 \mathscr{E} + 2ik_0 k_0' \frac{\partial \mathscr{E}}{\partial t} - (k_0'^2 + k_0 k_0'') \frac{\partial^2 \mathscr{E}}{\partial t^2}\right] \tag{7.3.15}$$

where equation (7.3.6) has been used.

Expressions (7.3.9) and (7.3.15) provide the forms of the two terms on the left-hand side of the wave equation (7.3.3) when the field excitation has the form (7.3.4) of a monochromatic wave modulated by a slowly-varying envelope. Substitution into the wave equation gives

$$2ik_0 \frac{\partial \mathscr{E}}{\partial z} + \frac{\partial^2 \mathscr{E}}{\partial z^2} + 2ik_0 k_0' \frac{\partial \mathscr{E}}{\partial t} - (k_0'^2 + k_0 k_0'') \frac{\partial^2 \mathscr{E}}{\partial t^2} = \mu_0 \exp(-ik_0 z + i\omega_0 t) \frac{\partial^2 P^{NL}}{\partial t^2} \tag{7.3.16}$$

For a slowly-varying envelope, the first derivatives of E are much larger than the second derivatives. If the second derivatives and the nonlinear polarization are ignored for the moment, equation (7.3.16) reduces to

$$\frac{\partial \mathscr{E}}{\partial z} + \frac{1}{v_G}\frac{\partial \mathscr{E}}{\partial t} = 0 \tag{7.3.17}$$

whose solution produces a pulse envelope that travels without distortion at the group velocity v_G. In the remaining terms of equation (7.3.16) it is a good approximation to put

$$\frac{\partial^2 \mathscr{E}}{\partial z^2} = \frac{1}{v_G^2}\frac{\partial^2 \mathscr{E}}{\partial t^2} \quad \text{and} \quad \frac{\partial^2 P^{NL}}{\partial t^2} = -\omega_0^2 P^{NL} \tag{7.3.18}$$

and the complete equation then reduces to

$$i\frac{\partial \mathscr{E}}{\partial z} + \frac{i}{v_G}\frac{\partial \mathscr{E}}{\partial t} - \frac{k_0''}{2}\frac{\partial^2 \mathscr{E}}{\partial t^2} = -\frac{\mu_0\omega_0^2}{2k_0}\exp(-ik_0 z + i\omega_0 t)P^{NL} \tag{7.3.19}$$

This is the equation that describes optical pulse propagation in the slowly-varying envelope approximation. The distortionless propagation produced by its first two terms is modified by the two following terms that allow for the group velocity dispersion and optical nonlinearity of the medium.

7.3.2. Group Velocity Dispersion

Before considering the solution of the complete equation (7.3.19), it is instructive to evaluate separately the effects of the two additional terms that modify the simple propagation equation (7.3.17). Let us therefore neglect the nonlinear polarization in equation (7.3.19) but retain the term that involves the group velocity dispersion.

The left-hand side of equation (7.3.19) is put into a more convenient form by transformation to a time variable τ measured with respect to the time taken to cover the distance z at the velocity v_G,

$$\tau = t - (z/v_G) \tag{7.3.20}$$

The propagation equation is thus reduced to

$$i\frac{\partial \mathscr{E}}{\partial z} = \frac{k_0''}{2}\frac{\partial^2 \mathscr{E}}{\partial \tau^2} \tag{7.3.21}$$

This is identical in form to the time-dependent Schrödinger equation for a particle in free space, except that the time and position variables are interchanged, and we can make use of the well-known solutions of the quantum-mechanical problem.

Consider a pulse whose shape at the fiber input at $z = 0$ is given by the Gaussian function

$$\mathscr{E}(0, \tau) = \mathscr{E}(0, 0) \exp(-\tau^2/2\tau_0^2) \tag{7.3.22}$$

where τ_0 is a measure of the pulse time duration. Adaptation of the standard theory of wave packet behavior (see, e.g., Davydov[16]) then produces the general position-dependent solution of equation (7.3.21) in the form

$$\mathscr{E}(z, \tau) = \frac{\tau_0 \mathscr{E}(0, 0)}{(\tau_0^2 - ik_0''z)^{1/2}} \exp\left[-\frac{\tau^2}{2(\tau_0^2 - ik_0''z)}\right] \tag{7.3.23}$$

The time duration of the pulse at coordinate z, given by the real part of the exponent, is

$$[\tau_0^2 + (k_0''z/\tau_0)^2]^{1/2} \tag{7.3.24}$$

The pulse shape thus remains Gaussian, but its duration is increased by the group velocity dispersion.

7.3.3. Self-Phase Modulation

We now neglect the group velocity dispersion in the propagation equation (7.3.19) but retain the nonlinear polarization. The form of nonlinear polarization associated with the optical Kerr effect is obtained from relations (7.1.13) and (7.1.18) as

$$P^{\mathrm{NL}} = 2\varepsilon_0 \eta_0 \eta_2 |E(z, t)|^2 E(z, t) \tag{7.3.25}$$

Use of expressions (7.3.4) and (7.3.5) and the time transformation (7.3.20) converts the propagation equation for $\mathscr{E}(z, \tau)$ to the form

$$i\partial\mathscr{E}/\partial z = -(\omega_0 \eta_2/c)|\mathscr{E}|^2 \mathscr{E} \tag{7.3.26}$$

It is easy to verify that this equation has the solution

$$\mathscr{E}(z, \tau) = \mathscr{E}(0, \tau) \exp[i\omega_0 \eta_2 |\mathscr{E}(0, \tau)|^2 z/c] \tag{7.3.27}$$

and the complete modulated field obtained from equation (7.3.4) is

$$E(z, t) = \mathscr{E}(0, \tau) \exp[ik_0 z - i\omega_0 t + i\omega_0 \eta_2 |\mathscr{E}(0, \tau)|^2 (z/c)] \tag{7.3.28}$$

This result shows that the input pulse shape, given for example by the Gaussian function (7.3.22), is not changed by the nonlinearity. The optical Kerr effect does however affect the phase of the wave such that the more intense parts of the pulse travel more slowly for positive η_2. The effect of this self-phase modulation is to change the apparent frequency of the wave at a given coordinate z to

$$\omega_0\left[1-\eta_2\frac{\partial}{\partial\tau}|\mathscr{E}(0,\tau)|^2\frac{z}{c}\right] \tag{7.3.29}$$

The frequency is thus reduced in the front rising part of the pulse and is increased in its rear falling part, as illustrated schematically in Figure 7.11. The resulting frequency chirp corresponds to a spectral broadening of the pulse. The spectrum can have a complicated semiperiodic structure caused by the interference of the pairs of contributions in the chirp that have the same frequencies.[2]

7.3.4. Soliton Propagation

We now return to the complete propagation equation (7.3.19) and consider the combined effects of group velocity dispersion and the optical Kerr effect.

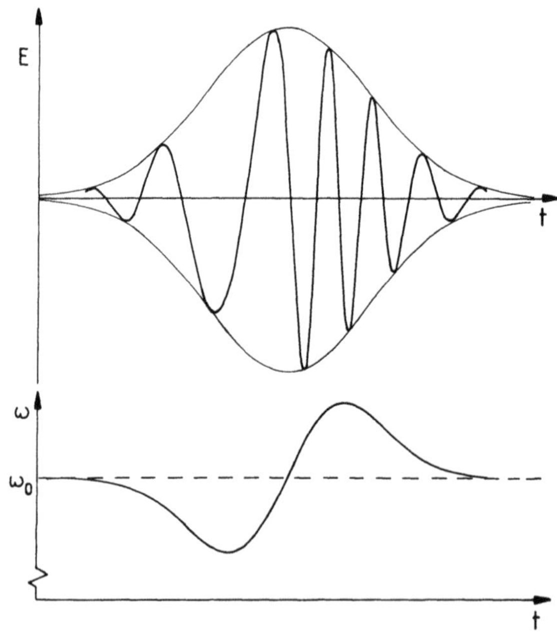

Figure 7.11. Gaussian pulse subjected to self-phase modulation, showing the frequency chirp, greatly exaggerated.

The appropriate generalization of equations (7.3.21) and (7.3.26) is

$$i\frac{\partial \mathscr{E}}{\partial z} = \frac{k_0''}{2}\frac{\partial^2 \mathscr{E}}{\partial \tau^2} - \frac{\omega_0 \eta_2}{c}|\mathscr{E}|^2\mathscr{E} \tag{7.3.30}$$

Suppose η_2 is positive but that the group velocity dispersion is negative, that is,

$$k_0'' = d(1/v_G)/d\omega < 0 \tag{7.3.31}$$

or alternatively

$$dv_G/d\omega > 0 \qquad \text{and} \qquad dv_G/d\lambda < 0 \tag{7.3.32}$$

where λ is the optical wavelength. The higher-frequency components of a pulse then travel faster than the lower-frequency components. The optical Kerr effect therefore tends to counteract the group velocity dispersion, since it puts the more-rapidly-traveling higher frequencies at the rear of the kind of pulse shown in Figure 7.11 and the more-slowly-traveling lower frequencies at its front. The combined result can be a net increase or a net decrease in the initial pulse width.

For certain forms of input pulse, the optical Kerr effect produces an exact compensation of the broadening effects of the group velocity dispersion acting alone. In order to examine these solutions it is convenient to convert equation (7.3.30) to dimensionless form in terms of a redefined τ and space and field variables given by

$$\tau = \frac{t - (z/v_G)}{t_0}, \qquad \aleph = \frac{z}{z_0} \qquad \text{and} \qquad u = \frac{\mathscr{E}}{\mathscr{E}_0} \tag{7.3.33}$$

where t_0, z_0, and $\mathscr{E}_0$ are scaling constants chosen to satisfy

$$k_0'' z_0/t_0^2 = -1 \qquad \text{and} \qquad \omega_0 \eta_2 \mathscr{E}_0^2 z_0/c = 1 \tag{7.3.34}$$

The propagation equation now takes the final form

$$i\frac{\partial u}{\partial \aleph} = -\frac{1}{2}\frac{\partial^2 u}{\partial \tau^2} - |u|^2 u \tag{7.3.35}$$

known as the nonlinear Schrödinger equation.

A number of solutions of this equation are known,[17,18] and we consider those that result from input pulse shapes of the form

$$u(0, \tau) = N \operatorname{sech} \tau \tag{7.3.36}$$

The pulses propagate as solitons for integer N, that is, they retain their shape even after colliding with one another. The shape never changes for the $N = 1$ fundamental soliton, while the initial shape is restored periodically, with period $\frac{1}{2}\pi$ in the dimensionless distance $\aleph$, for $N \geq 2$. The explicit form of the first-order solution is

$$u(\aleph, \tau) = \exp\left(\tfrac{1}{2}i\aleph\right) \operatorname{sech} \tau \tag{7.3.37}$$

and it is easily verified that this satisfies the nonlinear Schrödinger equation (7.3.35). It represents the most perfect state of balance between the effects of group velocity dispersion and the nonlinear refractive index. The importance of this solution for pulse propagation in optical fibers was first pointed out by Hasegawa and Tappert.[19] The higher-order solitons with $N \geq 2$ show initial narrowing from the input shape (7.3.36), with subsequent broadening to restore this shape at $\aleph = \frac{1}{2}\pi$. The intermediate pulse shapes also show some structure, whose complexity increases with increasing values of N.[20-22] The soliton behavior expected for suitable input pulse shapes has been verified by experiments on optical fibers.[20]

All the calculations of the present section have been based on the assumption of a real linear dielectric function, embodied in the use of a real wave vector k_0 defined in relation (7.3.5). The attenuation in optical fibers is indeed very small, but it becomes significant in the long lengths of fiber used in communications systems. Retention of the linear absorption α_0 modifies the nonlinear Schrödinger equation to

$$i\frac{\partial u}{\partial \aleph} = -i\Gamma u - \frac{1}{2}\frac{\partial u^2}{\partial \tau^2} - |u|^2 u \tag{7.3.38}$$

where

$$\Gamma = \tfrac{1}{2}z_0\alpha_0 \tag{7.3.39}$$

The equation no longer has soliton solutions and pulses are irretrievably broadened after propagation down sufficient lengths of fiber. Remnants of the soliton behavior remain however, particularly for low loss, and input pulse shapes of the forms (7.3.36) retain some advantages in terms of comparatively modest broadenings.[21]

7.3.5. Optical Chaos in Ring Cavities

Both the Fabry-Perot cavity considered in Section 7.2 and the optical ring cavity form bases for systems that show chaotic behavior when the optical propagation is supported by some kind of nonlinear material.[1,3] We do not attempt to review here the many aspects of these phenomena, but confine the

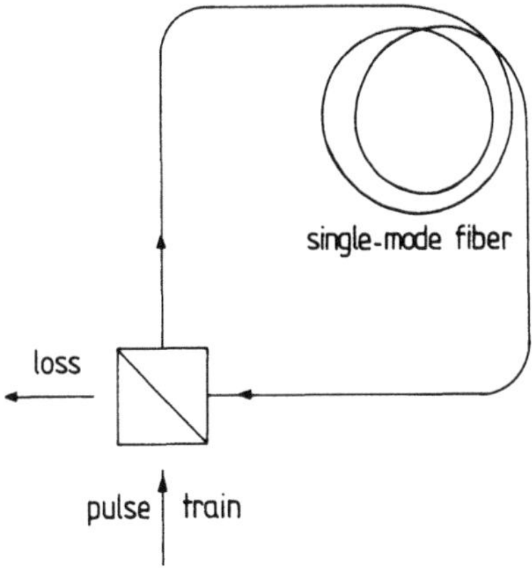

Figure 7.12. Fiber ring cavity.

discussion to a conceptually simple experiment[23] that relies on the ideas of the previous subsection.

Figure 7.12 shows a ring cavity constructed from a length of single-mode optical fiber. The cavity is excited by a series of input pulses of the form $A \operatorname{sech} \tau$, similar to equation (7.3.36) but with an amplitude A that is varied continuously to study the dependence of the optical state established in the cavity. The dimensionless length of the fiber is $\aleph = \frac{1}{4}\pi$ and the input pulse train has a period equal to the pulse travel time round the cavity. The separation between pulses is long compared to their width, so that they do not overlap. The pulse propagation in the fiber is described by the damped nonlinear Schrödinger equation (7.3.38). The $(n + 1)$th pulse leaving the beam splitter in Figure 7.12 is made up of a new input contribution $A \operatorname{sech} \tau$ and a reflected fraction of the nth pulse as modified by its propagation once round the cavity. Energy is also lost from the cavity in the fraction of this pulse transmitted through the beam splitter.

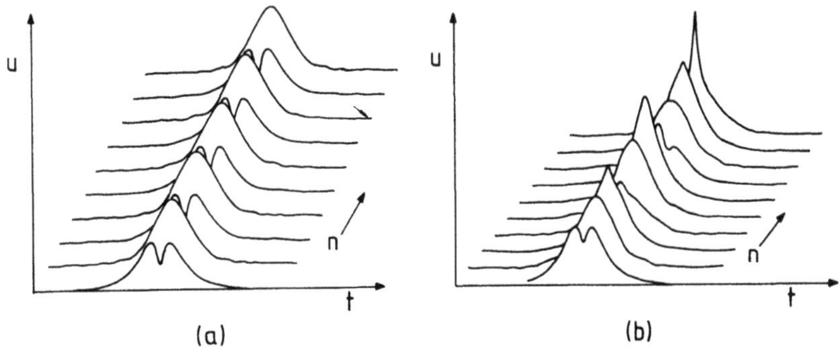

Figure 7.13. Series of pulses in a ring cavity for (a) $A = 1.75$ and (b) $A = 2.25$ (after Blow and Doran[23]).

The nature of the pulse train established in the cavity depends sensitively on the size of the input amplitude A. A constant pulse shape occurs for small values of A, but when A is increased beyond 1.526, the pulse shape becomes periodic in n with period two. Figure 7.13a shows the results of a computer experiment[23] for $A = 1.75$, with quite distinct alternating pulse shapes. This period-doubling bifurcation (see Chapters 1 and 2 also) is followed by further bifurcations to higher-order periodicities as A is further increased. For $A > 1.97$ the series of pulses has an apparently-random chaotic behavior, as in the results for $A = 2.25$ shown in Figure 7.13b. This simple system shows many of the features that are generally characteristic of chaotic behavior.[3]

References

1. J. R. Ackerhalt, P. W. Milonni, and M. L. Shih, *Phys. Rep.* **128**, 205 (1985).
2. Y. R. Shen, *The Principles of Nonlinear Optics*, John Wiley and Sons, New York (1984).
3. F. T. Arecchi, Chapter 6, this volume p. 193.
4. R. Loudon, *The Quantum Theory of Light*, Clarendon Press, Oxford (1983).
5. B. S. Wherrett and S. D. Smith (eds.), *Optical Bistability, Dynamical Nonlinearity and Photonic Logic*, Royal Society, London (1985).
6. N. B. Abraham, L. A. Lugiato, and L. M. Narducci, *J. Opt. Soc. Am. B* **2**, 7 (1985).
7. R. G. Harrison and D. J. Biswas, *Prog. Quantum Electron.* **10**, 147 (1985).
8. E. Abraham and S. D. Smith, *Rep. Prog. Phys.* **45**, 815 (1982).
9. R. Bonifacio and L. A. Lugiato, in: *Dissipative Systems in Quantum Optics* (R. Bonifacio, ed.) pp. 61-92, Springer-Verlag, Berlin (1982).
10. S. L. McCall and H. M. Gibbs, in: *Dissipative Systems in Quantum Optics* (R. Bonifacio, ed.) pp. 93-109, Springer-Verlag, Berlin (1982).
11. L. A. Lugiato, *Contemp. Phys.* **24**, 333 (1983).
12. A. Szöke, V. Daneu, J. Goldhar, and N. A. Kurnit, *Appl. Phys. Lett.* **15**, 376 (1969).
13. A. Gozzini, F. Maccarrone, and I. Longo, *Nuovo Cimento* **1D**, 489 (1982).
14. A. T. Rosenberger, L. A. Orozco, and H. J. Kimble, *Phys. Rev. A* **28**, 2569 (1983).
15. H. M. Gibbs, S. L. McCall, and T. N. C. Venkatesan, *Phys. Rev. Lett.* **36**, 1135 (1976).
16. A. S. Davydov, *Quantum Mechanics*, Pergamon Press, Oxford (1976).
17. V. E. Zakharov and A. B. Shabat, *Sov. Phys. JETP* **34**, 62 (1972).
18. J. Satsuma and N. Yajima, *Prog. Theor. Phys., Suppl.* **55**, 284 (1974).
19. A. Hasegawa and F. Tappert, *Appl. Phys. Lett.* **23**, 142 (1973).
20. L. F. Mollenauer, R. H. Stolen, and J. P. Gordon, *Phys. Rev. Lett.* **45**, 1095 (1980).
21. N. J. Doran and K. J. Blow, *IEEE J. Quantum Electron.* **19**, 1883 (1983).
22. L. F. Mollenauer, *Phil. Trans. R. Soc. London, Ser. A* **315**, 437 (1985).
23. K. J. Blow and N. J. Doran, *Phys. Rev. Lett.* **52**, 526 (1984).

8

Electron and Phonon Instabilities

P. N. Butcher

8.1. Introduction

Electron and phonon instabilities of many types occur in solids. Displacive phase transitions in which an atom changes its position in the unit cell of a crystal come about because a phonon mode goes soft and makes the original atomic arrangement unstable.[1] The Peierl's transition leading to dimerization in one-dimensional atomic chains has a similar origin.[2] Electrons in n-type semiconductors, together with the ionized donors, constitute a plasma and show aspects of a large number of familiar plasma instabilities but with very significant modifications because of the strong electron scattering.[3] Semiconductors are particularly interesting because both electrons (or holes) and phonons in them can be driven unstable by applying high electric fields. The ultimate form of the instability is then of particular interest and its determination presents a formidable problem. To avoid superficial coverage of many different mechanisms we choose to discuss only two which are reasonably well understood at both small and large signal levels: the Gunn effect[4-7] and acoustoelectric domain propagation.[7,8]

The Gunn effect requires a particular type of conduction band structure which is common in III–V semiconductors and gives rise to an N-shaped velocity-field characteristic. Acoustoelectric domains, on the other hand, arise most commonly when the electron drift velocity exceeds the sound velocity in a piezoelectric semiconductor with a linear velocity-field characteristic. In the Gunn effect the uniform distribution of electrons becomes unstable, i.e., it is an electron instability, and in acoustoelectric domains the uniform distribution of phonons becomes unstable, i.e., we are concerned with a phonon instability.

P. N. Butcher • Department of Physics, University of Warwick, Coventry CV4 7AL, England.

Small deviations from a uniform steady state may be analyzed into traveling-wave components of the form $\exp[i(kx - \omega t)]$, where x is the coordinate in the direction of propagation and t is time. We may suppose that when $t = 0$ the disturbance is localized in a finite region of space. Then only real wave numbers arise in its Fourier decomposition. The frequency $\omega(k)$ which is associated with each value of k may be either real or complex. It is determined by a dispersion relation $d(\omega, k) = 0$, which reflects the physics of the system. In zero field, when the system is in thermal equilibrium, there are usually dissipative processes present which give ω a negative imaginary part so that the initial disturbance decays to zero when $t \to \infty$. In conservative systems, which are the exception rather than the rule at the macroscopic level, we may find real values of ω. Lattice vibrations in the harmonic approximation are an obvious example. We shall be concerned here with cases in which an electric field is applied to a semiconductor that is large enough to drive the system so far from equilibrium that the imaginary part of ω becomes positive for some values of k. Then the initial disturbance grows as $t \to \infty$.

In Section 8.2 we discuss the small signal theory of the electron instability produced by a negative differential mobility. This is the mechanism of the Gunn effect and in Section 8.3 we outline the theory of the propagating high field domains which arise at high signal levels. The physics of the velocity-field characteristic is discussed in Section 8.4. In Section 8.5 we set out the basic equations required for a discussion of the acoustoelectric instability. Section 8.6 is devoted to the small signal theory of this instability. In Section 8.7 we discuss the solution of the space charge equations for a nonlinear traveling wave in a piezoelectric semiconductor. Finally, in Section 8.8 we consider the large-signal behavior of the acoustic gain and the local dc electric field on the assumption that the electrons are tightly bunched in potential minima.

8.2. The Effect of a Negative Differential Mobility

Consider an electric field F applied in the negative x direction. It will make the electrons in an n-type semiconductor drift in the positive x direction with a mean drift velocity $v(F)$. When $F \to 0$, $v(F)$ is linear in F and we write $v = \mu F$ where μ is the electron mobility. As F increases $v(F)$ becomes nonlinear and we define $\mu_d = dv(F)/dF$ to be the differential mobility at the bias field F. We show immediately that a homogeneous electron density becomes unstable in a uniform static field F if $\mu_d < 0$.

To simplify the algebra we suppose that all the variables in the problem depend on just one spatial coordinate x. Then the electron conservation equation is

$$\frac{\partial n}{\partial t} + \frac{\partial j}{\partial x} = 0 \tag{8.2.1}$$

In equation (1) n denotes the electron density and j is the electron current density in the x direction:

$$j = nv(F) - D_n \, \partial n/\partial x \tag{8.2.2}$$

The second term in equation (8.2.2) is an elementary representation of the electron diffusion current density in which we take the diffusivity D_n to be constant. The last equation which we need is Poisson's equation

$$\frac{\partial F}{\partial x} = \frac{e}{\varepsilon}(n - n_0) \tag{8.2.3}$$

where n_0 is the equilibrium value of n which we assume to be uniform and ε is the electric permittivity of the semiconductor.

To linearize these equations we write

$$F = F_0 + F_1 \tag{8.2.4a}$$

$$n = n_0 + n_1 \tag{8.2.4b}$$

where F_0 is a static uniform bias field. Then we have from equations (8.2.1) and (8.2.2)

$$\frac{\partial n_1}{\partial t} + \frac{\partial}{\partial x}\left(n_0 v_0 + n_0 v_1 + n_1 v_0 + n_1 v_1 - D_n \frac{\partial n_1}{\partial x}\right) = 0 \tag{8.2.5}$$

where

$$v_0 = v(F_0) \tag{8.2.6a}$$

and

$$v_1 = \mu_d(F_0) F_1 \tag{8.2.6b}$$

Hence, by neglecting the second-order term $n_1 v_1$ in equation (8.2.5) and using equation (8.2.3), we obtain

$$\frac{\partial n_1}{\partial t} + \omega_{cd} n_1 + v_0 \frac{\partial n_1}{\partial x} - D_n \frac{\partial^2 n_1}{\partial x^2} = 0 \tag{8.2.7}$$

where

$$\omega_{cd} = e n_0 \mu_d / \varepsilon \tag{8.2.8}$$

is the "differential" dielectric relaxation frequency at the bias field F_0. When a traveling wave of the form $\exp[i(kx - \omega t)]$ is assumed for n_1 in equation (8.2.7), we immediately obtain the dispersion relation

$$\omega = kv_0 - i(\omega_{cd} + D_n k^2) \tag{8.2.9}$$

Equation (8.2.9) has a very simple interpretation. The term kv_0 on the right-hand side shows that the disturbance propagates at the electron drift velocity v_0. The term $-i(\omega_{cd} + D_n k^2)$ shows that, as it propagates, the Fourier component of the disturbance with wave number k decays in time with a decay constant $\omega_{cd} + D_n k^2$. When $\mu_d > 0$ so is ω_{cd} and we have a positive decay constant for all k. On the other hand, when $\mu_d < 0$ so that $\omega_{cd} < 0$ we still have a positive decay constant for large k *but* we have instability when $k < (|\omega_{cd}|/D_n)^{1/2}$. Thus long-wavelength disturbances become unstable when $\mu_d < 0$ while short-wavelength disturbances are stabilized by diffusion.

8.3. The Large Signal Form of the Gunn Effect Instability

The ultimate form of the instability which arises when $\mu_d < 0$ depends on the particular form of $v(F)$. Since $\mu_d > 0$ for small fields and we require $\mu_d < 0$ to achieve an instability, the velocity-field characteristic must have a peak in it. We show a typical characteristic in Figure 8.1. The peak drift velocity v_T occurs at the "threshold field" F_T and then falls through a region of negative differential mobility to a saturated value $v_s \sim 0.5 v_T$ at high fields. For GaAs: $v_T \sim 10^7$ cm s^{-1} and $F_T \sim 3$ kV cm^{-1}.[3] The physical mechanisms which underlie this characteristic are discussed in the next section.

To establish an instability we subject the semiconductor to a uniform bias field $F_0 > F_T$ so that the uniform electron density n_0 becomes unstable. The large signal form of the instability is known to consist of a uniformly propagating high field domain of the type shown in Figure 8.2a.[4-7] In the figure, F_R denotes the uniform field outside the domain and F_D is the peak domain field. The domain propagates with the electron drift velocity $v_R = v(F_R)$ in the

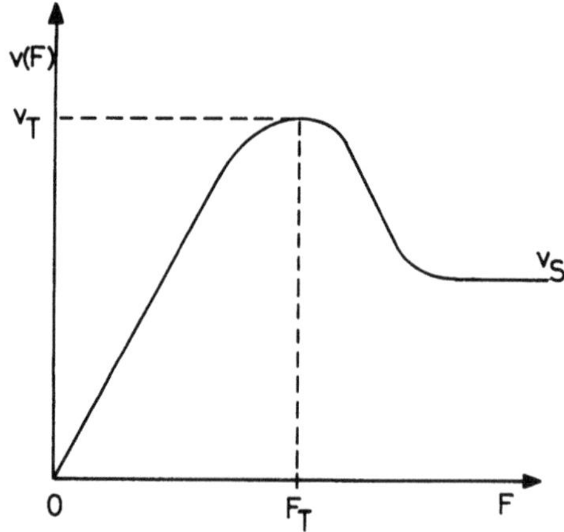

Figure 8.1. Typical velocity-field characteristic for a semiconductor exhibiting the Gunn effect.

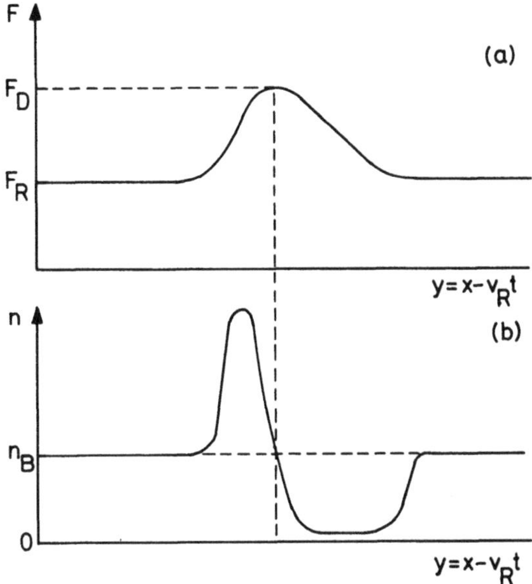

Figure 8.2. Variation of (a) F and (b) n with $x - v_R t$ in a uniformly propagating Gunn domain.

outside field. The corresponding electron density shown in Figure 8.2b may be derived immediately from Poisson's equation (8.2.3).

To determine the shape of these curves we return to the fundamental equations (8.2.1), (8.2.2), and (8.2.3) and suppose that both F and n depend on $y = x - v_R t^3$. Then we have

$$-v_R \frac{dn}{dy} + \frac{d}{dy}\left[nv(F) - D_n \frac{dn}{dy}\right] = 0 \tag{8.3.1}$$

and

$$\frac{dF}{dy} = \frac{e}{\varepsilon}(n - n_0) \tag{8.3.2}$$

We may integrate equation (8.3.1) immediately and obtain the constant of integration by noting that outside the domain $n = n_0$ and $v(F) = v_R$. Hence we have

$$D_n \frac{dn}{dy} = n[v(F) - v_R] \tag{8.3.3}$$

To proceed it is convenient to eliminate y from equations (8.3.2) and (8.3.3) by division to obtain

$$\frac{dn}{dF} = \frac{\varepsilon}{eD_n} \cdot \frac{n[v(F) - v_R]}{n - n_0} \tag{8.3.4}$$

The variables n and F may now be separated in equation (8.3.4) and the constant of integration is determined by the requirement that $n = n_0$ when $F = F_R$ as shown in Figure 8.2. Thus we find that

$$\frac{n}{n_0} - \ln\frac{n}{n_0} - 1 = \frac{\varepsilon}{eD_n n_0} \int_{F_R}^{F} [v(F') - v_R]\, dF' \tag{8.3.5}$$

We see from Figure 8.1 that, provided v_R lies between v_S and v_T, we have $F_R < F_T$ and the integral in equation (8.3.5) increases as F rises above F_R. It approaches zero again when $F \to F_D$ such that

$$\int_{F_R}^{F_D} [v(F') - v_R]\, dF' = 0 \tag{8.3.6}$$

The function on the left-hand side of equation (8.3.5) vanishes when $n = n_0$ and increases indefinitely as n increases or decreases from n_0. It is therefore clear that equation (8.3.5) provides two values of n for every F between F_R and F_D. They are the values appropriate to the accumulation and depletion layers shown in Figure 8.2b. These values of n come together again and both equal n_0 when $F = F_D$ as defined by equation (8.3.6). We see from Figure 8.2a that F_D is the peak domain field and we also see from Figure 8.1 that it is determined by the equality of the two areas which lie between the velocity-field characteristic and a horizontal line drawn between F_R and F_D at the velocity v_R.[3]

To complete the calculation of the curves in Figure 8.2 it is necessary to substitute for n as a function of F in equation (8.3.2) and integrate over y. This calculation must be done numerically when D_n is finite, but we can obtain the result analytically in the limiting case $D_n \to 0$. Then, we see from equation (8.3.5) that $n \to 0$ in the depletion layer, which is therefore fully depleted over some distance d. Moreover, $n \to \infty$ in the accumulation layer which must take the form of a Dirac δ-function so as to contain a finite number of electrons per unit area in the yz plane. Since $F = F_R$ on either side of the domain, we see by integrating equation (8.3.2) from one side to the other that the total number of electrons per unit area of the yz plane in the accumulation layer is just $n_0 d$. Hence we conclude that, when $D_n \to 0$, the rounded high-field domain shape shown in Figure 8.2a becomes a right-angle triangle. Quantity F increases vertically from F_R to F_D through the accumulation layer and drops back to F_R again linearly through the fully depleted layer, whose width d is given immediately by equation (8.3.2) as $d = (F_D - F_R)\varepsilon/en_0$.

8.4. Physical Mechanisms Which Determine the Shape of the Velocity-Field Characteristic

A full calculation of the velocity-field characteristic involves a numerical solution of Boltzmann's transport equation at high fields for a material with an appropriate conduction band structure.[5,6] Fortunately, we can arrive at an understanding of the basic mechanisms involved by using elementary physical arguments. The first requirement is a conduction band structure like that shown in Figure 8.3, which is a schematic plot for GaAs. The central valley has a light effective mass m_1, the satellite valleys have a heavy effective mass m_2, and the energy separation Δ between the central and satellite minima is much greater than $k_B T$. For GaAs $m_1 \sim 0.07m_0$, $m_2 \sim 0.3m_0$, and $\Delta \sim 0.35$ eV. The second requirement is a dominant scattering mechanism in the central valley which is unable to maintain a stable electron distribution there at room temperature when F exceeds a critical value in the order of F_T. In GaAs, this mechanism is scattering off the longitudinal optic phonons. In weak fields, almost all the electrons are in the central valley and they exhibit a relatively high mobility owing to their light effective mass. However, when $F > F_T$ the electrons in the central valley rapidly acquire sufficient energy for most of them to transfer to the satellite valleys, so that intervalley scattering is present

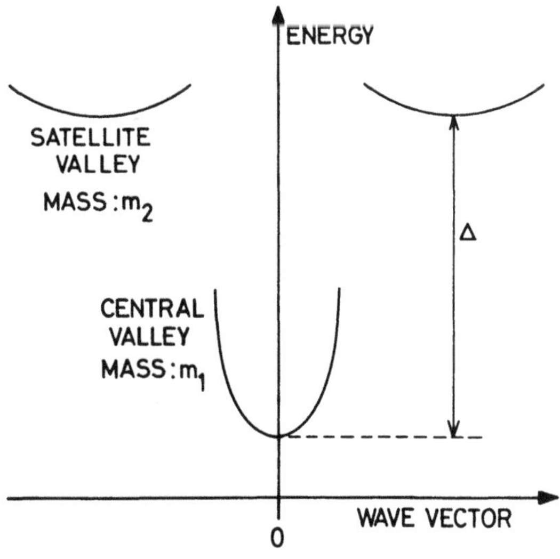

Figure 8.3. Schematic plot of the conduction band structure of a material exhibiting the Gunn effect, $m_2 \gg m_1$ and $\Delta \gg k_B T$.

and becomes the mechanism which stabilizes the overall electron distribution. The heavy effective mass in the satellite valleys ensures that they have a high density of states, so that most electrons reside in them at high fields.

Further insight into this "electron transfer" mechanism of negative differential mobility can be obtained by taking moments of the Boltzmann equation for an individual valley. Thus we obtain equations of momentum and energy conservation,[3] which we write here in their simplest form:

$$\frac{dv}{dt} = \frac{eF}{m^*} + \left(\frac{dv}{dt}\right)_c \tag{8.4.1a}$$

$$\frac{dE}{dt} = eFv + \left(\frac{dE}{dt}\right)_c \tag{8.4.1b}$$

Here, v and E denote the mean electron velocity and energy, respectively; m^* is the effective mass in what we have assumed is a parabolic spherical valley with a minimum energy of zero. The final terms in both equations describe the effects of collisions. A simple but useful theory can be constructed by writing[9]

$$\left(\frac{dv}{dt}\right)_c = -\frac{v}{\tau_M} \tag{8.4.2a}$$

$$\left(\frac{dE}{dt}\right)_c = -\frac{E - E_0}{\tau_E} \tag{8.4.2b}$$

where τ_M and τ_E are momentum and energy relaxation times and E_0 is the equilibrium value of E.

The solution of equations (8.4.1) and (8.4.2) under steady-state conditions is

$$v = \frac{e\tau_M}{m^*} F \tag{8.4.3a}$$

$$E = E_0 + \frac{e^2}{m^*} \tau_M \tau_E F^2 \tag{8.4.3b}$$

The high-field behavior is therefore controlled by the dependence of τ_M and τ_E on E. For polar mode scattering $\tau_M = aE^{1/2}$ and $\tau_E = bE^{3/2}$, where a and b are constants.[5] Consequently equation (8.4.3b) is a quadratic equation for E with the appropriate root

$$E = \frac{m^*}{2e^2abF^2}\left[1 - \left(1 - \frac{4E_0e^2abF^2}{m^*}\right)^{1/2}\right] \tag{8.4.4a}$$

which is real only when

$$F < F_T = \left(\frac{m^*}{4E_0 e^2 ab}\right)^{1/2} \tag{8.4.4b}$$

The physical interpretation of this result for GaAs is that the electron distribution function in the central valley becomes unstable when $F > F_T$. The electrons are then forced to increase their energy and transfer to the satellite valleys, where they have a relatively heavy effective mass. Consequently $v(F)$ falls off when $F > F_T$. Moreover, because there are several heavy-mass satellites, most of the electrons are in these valleys when $F \gg F_T$ and transitions between them replaces polar mode scattering as the dominant scattering mechanism. For intersatellite-valley scattering $\tau_M \propto (E - \Delta)^{-1/2}$ and $\tau_E \propto (E - \Delta)^{1/2}$.[9] Consequently $\tau_M \tau_E$ is independent of E in equation (8.4.3b), so that $E - E_0$ is proportional to F^2. It then follows from equation (8.4.3a) that v saturates when $F \to \infty$, as indicated in Figure 8.1.

8.5. The Basic Equations of the Acoustoelectric Instability

Elemental semiconductors like Ge and Si are centrosymmetric, which means that they have no piezoelectric properties. On the other hand, binary semiconductors like GaAs, CdS, and ZnO lack a center of inversion symmetry and are piezoelectric. That is to say, a polarization vector (of rank 1) appears in them when the material is subjected to a strain (of rank 2).[10] The proportionality factor is the piezoelectric tensor which has rank 3 (and consequently vanishes when inversion is a symmetry operation of the crystal). The constitutive relations between the stress T, the displacement D (in the negative x direction), the strain S, and the electric field F (in the negative x direction) are

$$T = cS + e_p F \tag{8.5.1a}$$

$$D = -e_p S + \varepsilon F \tag{8.5.1b}$$

The coefficients in these equations are the elastic stiffness c, the electric permittivity ε, and the piezoelectric constant e_p. The appearance of e_p in both equations is a consequence of time-reversal symmetry.[10] Strictly speaking, all the quantities in equation (8.5.1) are tensors and an elaborate subscript notation is required to handle them properly. Fortunately, we can retain the essential physics of the problem while greatly simplifying the algebra by treating them all as scalars.[11]

Newton's equation of motion for acoustic waves in the material is

$$\rho \frac{\partial^2 S}{\partial t^2} = \frac{\partial^2 T}{\partial x^2} \tag{8.5.2}$$

where ρ is the mass density. To put equation (8.5.2) in its most convenient form we introduce the normalized strain

$$S' = e_p S/\varepsilon \tag{8.5.3a}$$

to obtain, with the aid of equation (8.5.1a),

$$\frac{\partial^2 S'}{\partial t^2} - v_s^2 \frac{\partial^2 S'}{\partial x^2} = K^2 v_s^2 \frac{\partial^2 F}{\partial x^2} \tag{8.5.3b}$$

where $v_s = (c/\rho)^{1/2}$ is the velocity of sound when $F = 0$ and $K^2 = e_p^2/\pi\rho v_s^2$ is the "electromechanical coupling constant" which is nondimensional.[11]

Equation (8.5.3) is the equation of motion for the atoms in the material. To describe the behavior of the electrons when the material is an n-type semiconductor we again have the electron number conservation equation (8.2.1), which is repeated here for convenience:

$$\frac{\partial n}{\partial t} + \frac{\partial j}{\partial x} = 0 \tag{8.5.4}$$

In equation (8.5.4), j denotes the electron current density in the x direction and it is assumed to be given by the ohmic form of equation (8.2.2):

$$j = n\mu F - D_n \frac{\partial n}{\partial x} \tag{8.5.5}$$

The last equation which we need is Poisson's equation (8.2.3) generalized to allow for the piezoelectric polarization in D as specified by equation (8.5.1b). Thus we have, in view of equation (8.5.3a),

$$\frac{\partial}{\partial x}(F - S') = \frac{e}{\varepsilon}(n - n_0) \tag{8.5.6}$$

where n_0 is the equilibrium value of n which we assume to be a constant.

8.6. The Small Signal Regime in a Piezoelectric Semiconductor[11]

The basic equations (8.5.3)–(8.5.6) admit the homogeneous steady-state solution

$$S' = 0, \qquad n = n_0, \qquad F = F_0 \tag{8.6.1}$$

We study the small signal regime by considering a small perturbation about this steady state:

$$S' = S'_1, \qquad n = n_0 + n_1, \qquad F = F_0 + F_1 \tag{8.6.2}$$

where S'_1, n_1, and F_1 are all small and have a common traveling-wave form $\exp[i(kx - \omega t)]$. Then the wave equation (8.5.3b) may be written in the form

$$\omega = kv_s + \frac{K^2 v_s^2 k^2}{\omega + kv_s}\left(\frac{F_1}{S'_1}\right) \tag{8.6.3}$$

This form is convenient for our present discussion in which we assume *weak* electromechanical coupling. Then $K^2 \to 0$ and we work to first order in K^2 by iterating equation (8.6.3).

When $K^2 = 0$ we have $\omega = kv_s$, which is the dispersion relation for a sound wave traveling in the positive x direction. To first order in K^2 we may replace k by ω/v_s on the right-hand side of equation (8.4.3). Then we have the corrected dispersion relation

$$\omega = kv_s + \frac{\omega}{2} K^2\left(\frac{F_1}{S'_1}\right) \tag{8.6.4}$$

To evaluate F_1/S'_1 we simply eliminate n_1 between the linearized forms of equations (8.5.4), (8.5.5), and (8.5.6). Thus we have

$$-i\omega n_1 + ik[n_0\mu F_1 + n_1\mu F_0 - D_n ikn_1] = 0 \tag{8.6.5a}$$

and

$$ik(F_1 - S'_1) = \frac{e}{\varepsilon} n_1 \tag{8.6.5b}$$

so that, after a little algebra, we find that

$$\frac{F_1}{S'_1} = \frac{i(kv_0 - \omega) + D_n k^2}{i(kv_0 - \omega) + D_n k^2 + \omega_c} \tag{8.6.6}$$

where $v_0 = \mu F_0$ is the electron drift velocity in the bias field F_0 and $\omega_c = e\mu n_0/\varepsilon$ is the dielectric relaxation time. To express the final result in conventional notation we put $k = \omega/v_s$ in equation (8.6.6) and write D_n in the form v_s^2/ω_D, where ω_D is a characteristic frequency that provides an inverse measure of the size of the diffusion constant D_n. Then, upon substituting equation (8.6.6) into equation (8.6.4) we find the perturbed dispersion relation:

$$\omega = kv_s + \frac{\omega K^2}{2} \frac{i\gamma - \omega/\omega_D}{i\gamma - (\omega/\omega_D + \omega_c/\omega)} \tag{8.6.7a}$$

$$= kv_s + \Delta\omega_R + i\Delta\omega_I \tag{8.6.7b}$$

where

$$\gamma = 1 - v_0/v_s \tag{8.6.8}$$

and the real and imaginary parts of the perturbation of ω are

$$\Delta\omega_R = \frac{\omega K^2}{2} \cdot \frac{\omega/\omega_D(\omega/\omega_D + \omega_c/\omega) + \gamma^2}{(\omega/\omega_D + \omega_c/\omega)^2 + \gamma^2} \tag{8.6.9}$$

and

$$\Delta\omega_I = -\frac{\omega_c K^2 \gamma}{2}\left[\left(\frac{\omega}{\omega_D} + \frac{\omega_c}{\omega}\right)^2 + \gamma^2\right]^{-1} \tag{8.6.10}$$

On the right-hand side of these equations ω is to be set equal to kv_s when the real wave number k is specified.

The interpretation of the results (8.6.9) and (8.6.10) is very straightforward. Quantity $\Delta\omega_R$ contributes to a small change in the phase velocity of the acoustic wave which does not concern us here. The more interesting quantity is $\Delta\omega_I$, which is negative when $\gamma > 0$ and positive when $\gamma < 0$. In the former case the electron drift velocity is less than the velocity of sound and the wave is stable. In the latter case the electron drift velocity is greater than the velocity of sound and the wave is unstable. We have here the acoustic analogue of the Cerenkov effect for electromagnetic radiation.[8] It occurs at fields on the order of 300 V cm^{-1}, which are an order of magnitude down on the fields required to initiate the Gunn effect (3000 V cm^{-1} in GaAs). For fixed positive γ, $\Delta\omega_I$ reaches a maximum when $\omega = (\omega_c\omega_D)^{1/2}$. A trivial manipulation of this relation shows that it implies that the acoustic wavelength $2\pi v_s/\omega$ is 2π times the Debye screening length for the electrons $L_D = (\varepsilon k_B T/n_0 e^2)^{1/2}$, where we have employed the Einstein relation $\mu = eD_n/k_B T$. At the peak gain frequency, the value of γ which produces an overall maximum gain is $-2(\omega_c/\omega_D)^{1/2}$, strongly dependent on the value of the electron mobility. Amplification and oscillation resulting from this instability mechanism have been observed in many piezoelectric semiconductors at frequencies running all the way from a few megahertz to a few gigahertz.[7,8,12-14]

8.7. Solution of the Space Charge Equations for a Nonlinear Traveling Wave in a Piezoelectric Semiconductor

The large signal regime has a dynamic character and is difficult to analyze in detail. A spatially narrow domain of acoustic flux builds up and propagates through the specimen.[7,8,12] The flux domain is accompanied by a domain of high electric field which moves with it. Reviews of experimental data are given elsewhere.[12,13] Up conversion, parametric generation of subharmonics, and continuous shifts of peak gain frequency all occur in the propagating domain.

We can gain some insight into the behavior of high flux domains by seeking a nonlinear traveling wave solution of the fundamental equations (8.5.3)-(8.5.6) which is the approach followed by Butcher and Ogg in a series of three papers. In the first paper the longitudinal component D of the displacement vector is regarded as the basic variable describing the electron system and the other variables are expressed in terms of it. In the second paper the theory is reworked with D replaced by the electron density n as the basic variable and some elementary results are obtained for a symmetrical pulse with exponential tails. In the final paper it is recognized that the electrons form Gaussian bunches in a large-amplitude acoustic wave and numerical results are evaluated for this case. We concentrate in this section on the solution of the space charge equations for a nonlinear traveling wave when n is taken as the basic variable. The behavior of the large signal acoustic gain when the electrons form Gaussian bunches is discussed in Section 8.

We see from equation (8.5.6) that

$$n = n_0 + \frac{\varepsilon}{e}\frac{\partial}{\partial x}(F - S') \tag{8.7.1}$$

When this expression is substituted into the electron number conservation equation (8.5.4) together with the constitutive relation (8.5.5) for j we readily find that the total current density

$$I = ne\mu F + \varepsilon\frac{\partial}{\partial t}(F - S') - eD_n\frac{\partial n}{\partial x} \tag{8.7.2}$$

is independent of x. For a traveling wave it must also be independent of t and therefore plays the role of a numerical parameter in the theory. We express all the other field variables in the nonlinear traveling-wave form

$$g(x, t; \theta) = \sum_{m=-\infty}^{\infty} g_m(x, t) \exp(im\theta) \tag{8.7.3}$$

where the phase angle is given by

$$\theta = \omega\left(\frac{x}{v_s} - t\right) \tag{8.7.4}$$

and the Fourier coefficients $g_m(x, t)$, with m equal to an integer, are slowly-varying functions of x and t compared to $\exp(im\theta)$. By taking the Fourier transform of equation (8.7.3) with respect to θ we see that

$$g_m(x, t) = \frac{1}{2\pi}\int_{-\pi}^{\pi} g(x, t; \theta)\exp(-im\theta)\, d\theta$$

$$= \langle g(x, t; \theta)\exp(-im\theta)\rangle \tag{8.7.5}$$

In this equation and in the following analysis we use angle brackets to denote a phase average.

In lowest order we may treat the Fourier coefficients of the field variables as constants in equations (8.7.1) and (8.7.2). Then

$$\frac{\partial}{\partial x} = \frac{\omega}{v_s}\frac{\partial}{\partial \theta} \tag{8.7.6a}$$

$$\frac{\partial}{\partial t} = -\omega\frac{\partial}{\partial \theta} \tag{8.7.6b}$$

so that $\partial(F - S')/\partial\theta$ may be eliminated immediately to yield an expression for F in terms of n and I:

$$F = \frac{1}{ne\mu}\left[I + ev_s(n - n_0) - \frac{eD_n}{v_s}\frac{\partial n}{\partial \theta}\right]$$

$$= F_s\left[1 - \left(1 - \frac{I}{I_s}\right)\frac{n_0}{n} + \frac{\omega}{\omega_D}\frac{\partial}{\partial\theta}\ln\left(\frac{n}{n_0}\right)\right] \tag{8.7.7}$$

In the second form, which is the most convenient, we have written $F_s = v_s/\mu$ for the "synchronous field, $I_s = en_0v_s$ for the synchronous current density, and $\omega_D = v_s^2/D_n$ as in Section 8.6.

The total current appears in equation (8.7.7) as a parameter. It is more usual to use the local dc electric field $F_0 = \langle F\rangle$. To do so we have only to take the phase average of equation (8.7.7) to obtain

$$F_0 = F_s\left[1 - \left(1 - \frac{I}{I_s}\right)\left\langle\frac{n_0}{n}\right\rangle\right] \tag{8.7.8}$$

When this equation is solved for I we may write the result in a physically transparent form:

$$I = \sigma F_0(1 - f) + I_s f \tag{8.7.9}$$

where $\sigma = en_0\mu$ is conductivity and

$$f = 1 - \left\langle \frac{n_0}{n} \right\rangle^{-1} \tag{8.7.10}$$

to which we refer as the "trapping factor." The interpretation of f and equation (8.7.9) is discussed in Section 8.8. Finally, by eliminating I between equations (8.7.7) and (8.7.9) and using equation (8.7.10), we obtain the desired final expression for F in terms of n and F_0:

$$F = F_0 + F_{ac} \tag{8.7.11}$$

where the ac electric field is given by

$$F_{ac} = F_s\left[\gamma\left(1 - \frac{n^{-1}}{\langle n^{-1}\rangle}\right) + \frac{\omega}{\omega_D}\frac{\partial}{\partial\theta}\ln\frac{n}{n_0}\right] \tag{8.7.12}$$

in which γ is given by equation (8.6.8), i.e., $\gamma = 1 - F_0/F_s$.

In the next section we discuss the acoustic gain in the nonlinear regime. For this purpose we require the Fourier coefficients of the ac electric field and in writing them down it is convenient to introduce the following two frequencies:

$$\omega'_{cm} = -\omega_c \frac{n_m}{n_0}\frac{\langle n^{-1}\rangle}{\langle n^{-1}\exp(-im\theta)\rangle} \tag{8.7.13a}$$

$$\omega'_{Dm} = -\omega_D \frac{\langle n^{-1}\exp(-im\theta)\rangle}{\langle n^{-1}\rangle\langle \ln(n/n_0)\exp(-im\theta)\rangle} \tag{8.7.13b}$$

where ω_c is the dielectric relaxation frequency and n_m, with $m \neq 0$, is the mth Fourier coefficient of n. The significance of ω'_{cm} and ω'_{Dm} is discussed in Section 8.8. With this notation a trivial integration by parts shows that

$$\begin{aligned} F_m &= \langle F_{ac}\exp(-im\theta)\rangle \\ &= F_s\frac{\omega_c}{\omega'_{cm}}\left(\gamma + \frac{im\omega}{\omega'_{Dm}}\right)\frac{n_m}{n_0} \end{aligned} \tag{8.7.14}$$

We need two more relations before we can discuss the acoustic gain: expressions for the normalized strain $S' = e_p S/\varepsilon$ and its Fourier coefficients in terms of n. The dc strain must vanish if the lattice displacement is to have a periodic traveling-wave form. With this boundary condition we find, by integrating equation (8.7.1) with $\partial/\partial x$ given by equation (8.7.6a), that

$$S' = F_{ac} - F_s\frac{\omega_c}{\omega}(N - \langle N\rangle) \tag{8.7.15}$$

where

$$N(x, t; \theta) = \int_0^\theta \left(\frac{n(x_1 t_1 \theta')}{n_0} - 1 \right) d\theta' \qquad (8.7.16)$$

is the integrated, fractional electron density which, we note, is a periodic function of θ since $\langle n \rangle = n_0$. The Fourier coefficients of S' follow immediately from these equations on integrating by parts. When $m \neq 0$ they are given by

$$S'_m = F_m + \frac{iF_s\omega_c}{m\omega} \cdot \frac{n_m}{n_0} \qquad (8.7.17)$$

where F_m is given in equation (8.7.14).

8.8. Nonlinear Acoustic Gain and Acoustoelectric Current

In discussing the small signal regime in Section 8.6 we simplified the algebra by assuming that the electromechanical coupling constant K^2 is small. The analysis culminated in the dispersion relation (8.6.7) in which the real and imaginary parts, $\Delta\omega_R$ and $\Delta\omega_I$ of the perturbation of ω away from kv_s, where k is real, are given by equations (8.6.9) and (8.6.10). We have set up the results derived from the space charge equations in Section 8.7 in a way which makes it very easy to generalize the small signal expressions to the nonlinear regime.

The additional equation required is derived from the wave equation (8.5.3b) by substituting the Fourier expansions of S' and F, retaining only first derivatives of S'_m on the left-hand side (because S'_m is slowly varying) and dropping all derivatives of E_m on the right-hand side (because E_m is slowly varying *and* K^2 is small). The result is

$$\frac{\partial S'_m}{\partial t} + v_s \frac{\partial S'_m}{\partial x} = -i\Delta\omega_m S'_m \qquad (8.8.1)$$

where $\Delta\omega_m = \Delta\omega_{Rm} + i\Delta\omega_{Im}$ is just the perturbation term in the small signal equation (8.6.4) with ω, F_1, and S'_1 replaced by $m\omega$, F_m, and S'_m because we are now concerned with the mth harmonic. Moreover, upon substituting for F_m and S'_m from equations (8.7.14) and (8.7.17), we find that n_m/n_0 cancels out to leave an expression for $\Delta\omega_m$ which is just the perturbation term in the small signal equation (8.6.7a) with ω, ω_c, and ω_D replaced by $m\omega$, ω'_{cm}, and ω'_{Dm}, respectively. The reason for the complicated definitions (8.7.13) adopted for the latter two frequencies is now clear: the nonlinear gain equation at frequency $m\omega$ is identical to the small signal gain equation at frequency ω with ω, ω_c, and ω_D replaced by $m\omega$, ω'_{cm}, and ω'_{Dm}, respectively. This simplification makes it easy to interpret the results obtained in the nonlinear regime.

We begin with equation (8.7.9), which is reproduced here for convenience:

$$I = \sigma F_0(1 - f) + I_s f \tag{8.8.2}$$

where the "trapping factor" f is defined by equation (8.7.10): $f = 1 - \langle n_0/n \rangle^{-1}$. We may readily verify that $0 \leq f \leq 1$.[16] We see from equation (8.8.2) that I behaves as though a fraction f of the electrons are forced to move at the velocity of sound while the rest exhibit the normal dc response to the local dc electric field. In the small signal regime, $n \to n_0$ and $f \to 0$, so that equation (8.8.2) reduces to Ohm's law: $I = \sigma F_0$. On the other hand, in the severely nonlinear regime we expect very tight electron bunching in the potential energy minima of the traveling wave with $n \to 0$ at the potential energy maxima. Consequently, $\langle n^{-1} \rangle \to \infty$ and $f \to 1$ so that I saturates at $I_s = en_0 v_s$.

We may rewrite equation (8.8.2) in another form which is instructive. Direct algebraic manipulation shows that[16]

$$I = \sigma F_0 + I_{ae} \tag{8.8.3}$$

where the "acoustoelectric current density" is given by

$$I_{ae} = -\frac{2\mu \Delta\omega_I}{v_s^2}\Phi \tag{8.8.4}$$

In this equation

$$\Phi = 2\rho v_s^3 \langle S^2 \rangle$$
$$= \sum_m \Phi_m \tag{8.8.5a}$$

is the total acoustic flux, with Φ_m denoting the contribution from the mth harmonic and

$$2\Delta\omega_I = 2\Phi^{-1} \sum \Delta\omega_{Im} \Phi_m / \Phi \tag{8.8.5b}$$

is the growth constant of Φ.

We note that equation (8.8.5a) involves S and not S', and that the 2 appears in these equations because we wish to introduce the power gain $2\Delta\omega_I$ so as to allow an immediate comparison of equations (8.8.3) and (8.8.4) with the Weinreich relation,[18] which may be derived by elementary arguments. Thus the rate of increase in phonon density N_{ph} is $2\Delta\omega_I N_{ph}$ and $\Phi = N_{ph} v_s$. It follows that the rate of increase in phonon momentum density is $2\Delta\omega_I(\Phi/v_s)(\hbar k/\hbar\omega)$. The transfer of x-directed momentum to the phonons means that the electrons experience an acoustoelectric field F_{ae} in the $-x$ direction which is given by

$$-n_0 e F_{ae} = \Phi 2\Delta\omega_I / v_s^2 \tag{8.8.6}$$

and which enhances the ohmic current density by σF_{ae}. This is the relation given by equation (8.8.4). The detailed analysis developed here[16] shows the sense in which this relation holds good in the nonlinear regime.

Equations (8.8.3) and (8.8.4) provide a rough picture of the behavior of propagating acoustoelectric domains. The build-up of flux in the domain implies that $\Delta\omega_I > 0$. Hence, from equation (8.8.2), $I_{ae} < 0$ and, since I in equation (8.8.3) is constant, F_0 must increase where the flux increases to counteract the transfer of electron momentum to the phonons. The high flux domain is therefore accompanied by a high field domain. Outside the domain F_0 approaches the synchronous value F_s and the acoustoelectric gain is insufficient to overcome the acoustic loss mechanisms which are inevitably present but which we do not discuss here.[13]

The behavior of the flux-dependent frequencies ω'_{cm} and ω'_{Dm} is easier to determine than might appear from the complicated expressions (8.7.13).[16] In the small signal regime $\omega'_{cm} \to \omega_c$ and $\omega'_{Dm} \to \omega_D$. At intermediate and large signal levels ω'_{cm} and ω'_{Dm} are generally complex. However, they are real when n is an even function of θ and we confine our attention to this case. To be specific we suppose that

$$\frac{n}{n_0} = c \exp(-b\theta^2) \tag{8.8.7}$$

where the "bunching parameter" b determines the width of the bunch centered on $\theta = 0$ and c is determined by the normalization condition $\langle n/n_0 \rangle = 1$. Our main concern is with tight bunches for which $b \gg 1$, but we use equation (8.8.7) to extrapolate down to the small signal limit. For tight bunches Butcher and Ogg[17] show that

$$b = e|\phi''|/2k_B T \tag{8.8.8}$$

where ϕ is the ac potential and ϕ'' is its second derivative with respect to θ calculated at the peak of the bunch.

The trapping factor f is easily calculated and is shown as a function of b in figure 8.4. We see that trapping is essentially complete when $b > 1$. When ϕ is sinusoidal this inequality reduces to the intrinsically appealing condition that the extent of the potential energy oscillations should be greater than $4k_B T$.

Plots of ω'_{cm} and ω'_{Dm} against b are given in Figures 8.5 and 8.6. The behavior is complicated for intermediate values of b, but for $b \gg 1$ we see by inspection that $\omega'_{cm} \to \omega_c(-1)^{m-1}$ and $\omega'_{Dm} \to \omega_D m^2/2b$.[17] We see from the nonlinear extrapolation of equation (8.6.10) for $\Delta\omega_I$ for the mth harmonic that positive gain occurs when $\gamma > 0$ only when $\omega'_{cm} > 0$. Thus odd harmonics continue to build up in the nonlinear regime while even harmonics decay. We saw in our discussion of small signal behavior that peak gain occurs at the frequency $(\omega_c\omega_D)^{1/2}$. It therefore follows that in the nonlinear regime the peak

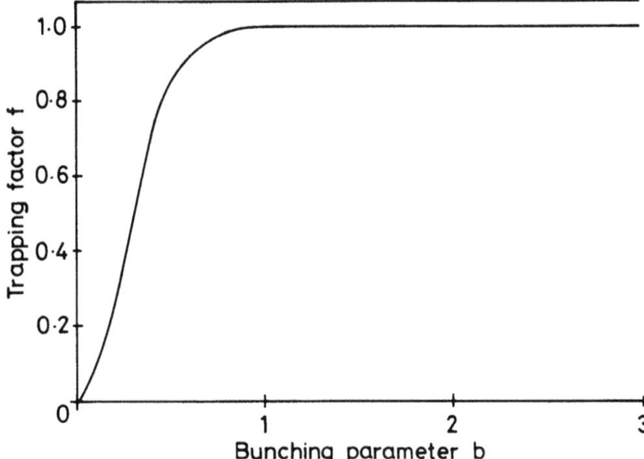

Figure 8.4. Plot of the trapping factor f against the bunching parameter b for Gaussian pulses (after Butcher and Ogg[17]).

gain frequency of the mth harmonic is at $(\omega'_{cm}\omega'_{Dm})^{1/2}$, which falls off as $m/b^{1/2}$ with increasing b. Butcher and Ogg[10,11] calculate the peak gain frequency ω_{pk} for the total flux gain given by equation (8.8.5). The result is plotted as a function of b in Figure 8.7. As expected from the behavior of the harmonics, ω_{pk} falls off as b increases. The physical reason is easy to ascertain. In the small signal regime $\omega_{pk} = (\omega_c\omega_D)^{1/2}$ which, as we saw in Section 8.6, implies that the Debye length is equal to the acoustic wavelength over 2π. In

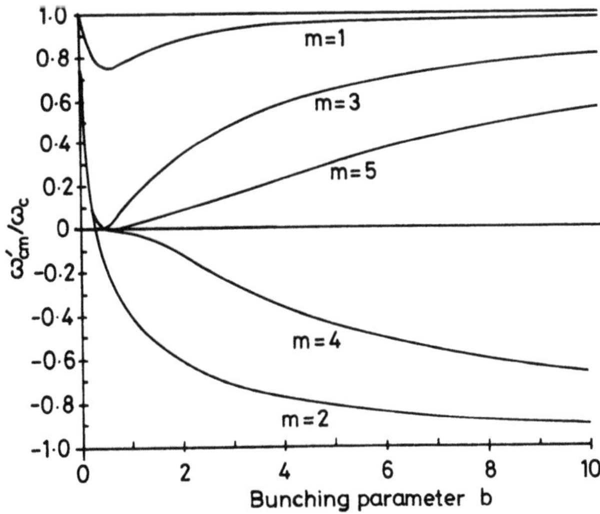

Figure 8.5. Plot of ω'_{cm} against b for various harmonics m (after Butcher and Ogg[17]).

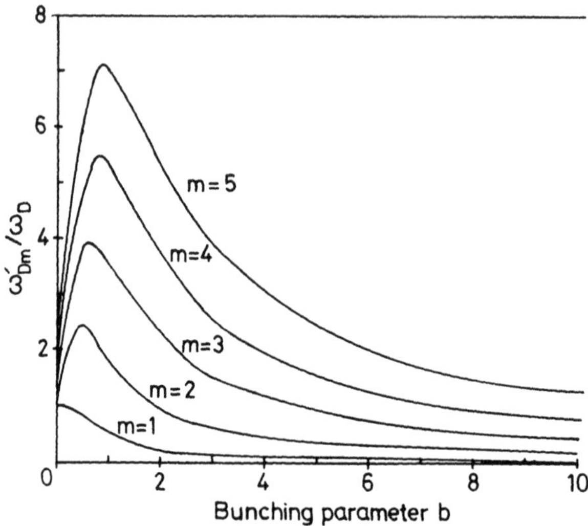

Figure 8.6. Plot of ω'_{Dm} against b for various harmonics m (after Butcher and Ogg[17]).

the large signal regime, the bunch width replaces the acoustic wavelength in this condition for maximum gain. Since the bunch width is proportional to $b^{-1/2}$, so is ω_{pk}. Figure 8.8 shows the behavior of the total flux at the peak gain frequency as a function of b for various values of the parameter $a = -\gamma(\omega_D/\omega_c)^{1/2}$. We conclude that, as the flux increases and b decreases, the peak gain frequency moves down continuously. We would therefore expect a continuous down-shifting of the frequencies involved in a uniformly propaga-

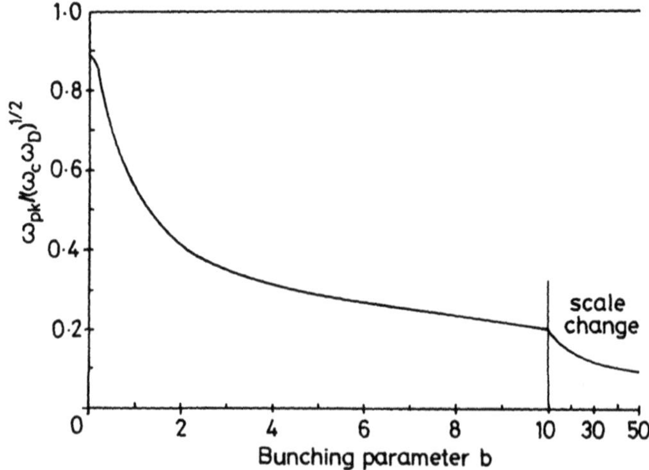

Figure 8.7. Plot of the ratio of the frequency of peak total flux gain to $(\omega_c\omega_D)^{1/2}$ against b (after Butcher and Ogg[17]).

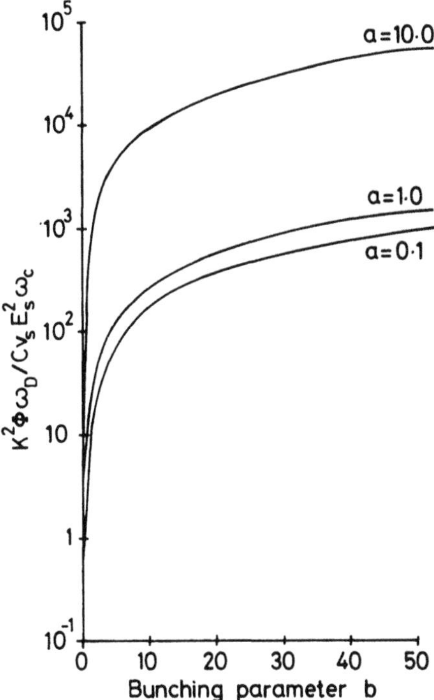

Figure 8.8. Plot of the total flux Φ in reduced units at the peak gain frequency against b for various values of $a = -\gamma(\omega_D/\omega_c)^{1/2}$ (after Butcher and Ogg[17]).

ting domain. There is some experimental evidence for this,[12,13] but the data are difficult to interpret owing to the mixing and parametric amplification effects also present and the wide bandwidth and phase randomness of the phonons excited in the domain. A study of these complications would take us too far afield.

8.9. Conclusion

In this article we have given an introduction to the theory of Gunn domains and acoustoelectric domains. We have chosen to take a macroscopic viewpoint because of its conceptual simplicity. It is easy to find experimental situations in which a microscopic view of the physics becomes essential. In the Gunn effect, for example, our tacit assumption that the electrons can follow the field variations without lagging is seriously at fault in microwave oscillators. The study of hot electron phenomena, including the Gunn effect, on short time scales may be approached by Monte Carlo simulation of electron trajectories in momentum space.[19] The calculation of the static velocity-field characteristic has often been conducted in the same way.[5] There has recently been increased interest in the calculation of noise fluctuations about nonequilibrium steady states near the threshold field for the Gunn effect.[20]

Our analysis of the acoustoelectric instability assumes that the acoustic wavelength is much greater than the mean free path. The same approach has been used to investigate the behavior of acoustoelectric oscillators.[21] A detailed study for hexagonal crystals using macroscopic ideas has been given recently by Westera.[22] At high frequencies the macroscopic approach breaks down. When the acoustic wavelength is in the order of the mean free path, it is necessary to describe the behavior of the electrons by a Boltzmann equation[23] and when the wavelength becomes shorter still a full quantum-mechanical treatment is necessary even in the linear regime.[24] The quantum physics is very interesting in its own right, but it makes the development of a nonlinear theory for many frequencies very difficult. Even at the macroscopic level the theory of the acoustoelectric instability involving many waves with random phases remains controversial[12,25,26] and is in urgent need of development. A multiple time and space treatment of propagating solitary waves has been given by Pawlik and Rowlands.[27]

References

1. W. Cochran, *The Dynamics of Atoms in Crystals*, Edward Arnold, London (1973).
2. A. J. Heeger, in: *Highly Conduction One-Dimensional Solids* (J. T. Devreese, R. P. Evrard, and V. E. van Doren), pp. 69–145, Plenum Press, New York (1979).
3. J. Pozhela, *Plasma and Current Instabilities in Semiconductors*, Pergamon Press, Oxford (1981).
4. M. P. Shaw, H. L. Grubin, and P. N. R. Solomon, in: *Gunn–Hilsum Effect Electronics* (L. Marton and C. Marton, eds.), Vol. 51, pp. 310–433, Academic Press, New York (1980).
5. W. Fawcett, in: *Non-Ohmic Transport in Semiconductors, Electronics in Crystalline Solids* (A. Salam, ed.), pp. 531–618, IAEA, Vienna (1973).
6. P. N. Butcher, *Rep. Prog. Phys.* **30**, 97 (1967).
7. V. L. Bonch-Bruevich, I. P. Zvyagin, and A. G. Mironov, *Domain Electrical Instabilities in Semiconductors*, Plenum Press, New York (1975).
8. H. Kuzmany, *Phys. Status Solidi A* **25**, 9 (1974).
9. P. N. Butcher, in: *The Theory of Electron Transport in Crystalline Semiconductors, Crystalline Semiconducting Materials and Devices* (P. N. Butcher, N. March, and M. Tosi, eds.), pp. 131–194, Plenum Press, New York (1986).
10. J. F. Nye, *Physical Properties of Crystals and Their Representation by Tensors and Matrices*, Clarendon Press, Oxford (1957).
11. D. L. White, *J. Appl. Phys.* **33**, 2547 (1962).
12. E. M. Conwell and A. K. Ganguly, *Phys. Rev. B* **4**, 2535 (1971).
13. M. Eizenberger and B. Fisher, *J. Appl. Phys.* **49**, 5260 (1978).
14. H. M. Janus, *J. Phys. D* **3**, 1993 (1970).
15. P. N. Butcher and N. R. Ogg, *Br. J. Appl. Phys., Ser. 2* **1**, 1271 (1968).
16. P. N. Butcher and N. R. Ogg, *Br. J. Appl. Phys., Ser. 2* **2**, 333 (1969).
17. P. N. Butcher and N. R. Ogg, *J. Phys. C* **3**, 706 (1970).
18. G. Weinreich, *Phys. Rev.* **104**, 321 (1956).
19. H. L. Grubin, D. K. Ferry, G. J. Iafrate, and H. L. Barker, in: *VLSEI Electronics* (N. G. Einsprach, ed.), Vol. 3, pp. 197–300, Academic Press, New York (1982).

20. A. Guilera and J. M. Rubi, *Physica A* (1985), to appear.
21. P. N. Butcher and H. Janus, *J. Phys. C* **5**, 567 (1972).
22. W. Westera, *Physica B* **113**, 149 (1982).
23. H. N. Spector, *Phys. Rev.* **165**, 562 (1968).
24. E. Mosekilde, *Phys. Rev. B* **9**, 682 (1974).
25. P. N. Butcher and N. R. Ogg, *Phys. Lett.* **30A**, 66 (1969).
26. P. N. Butcher and H. Slechta, *J. Phys. C* **4**, 870 (1971).
27. M. Pawlik and G. Rowlands, *J. Phys. C* **8**, 1189 (1975).

9

Fractals in Physics: Introductory Concepts

L. Pietronero

9.1. Introduction

Having treated lasers, nonlinear optics, and electron and phonon instabilities at some length in previous chapters, we now return in more detail to the concept of fractals introduced in Chapter 1.

"Fractal geometry is one of those concepts which at first sight invites disbelief but on a second thought becomes so natural that one wonders why it has only recently been developed. . . ." .These words by Berry[1] from his review of Mandelbrot's book[2] provide an apt summary of the present activity on fractals: great interest but also a degree of skepticism. Both are well justified as we shall see.

The aim of this chapter is to provide an introduction to this field from the standpoint of physics. Fractals continue to be a very popular subject in physics, but it is not easy to find introductory papers to this field. In fact there are already several conference proceedings that give an idea of the state-of-the-art,[3] but the material in these books is often fragmented and sometimes confusing. The present chapter will attempt to provide some orientation in this material. So there will not be much technical discussion, since this can be found in the original papers, but we will try to focus on the reasons for the success of this concept, its limitations, and the main open problems.

Fractal geometry describes systems in which increasing detail is revealed by increasing magnification, and the newly revealed structure looks similar to what one can see at lower magnification. Concepts of this type were defined

L. Pietronero • Solid State Physics Laboratory, University of Gröningen, 9718 EP Gröningen, The Netherlands. Present Address: Dipartimento di Fisica, Università di Roma, "La Sapienza," 00185, Rome, Italy.

as mathematical entities long ago by Poincaré, Hausdorff, and others. However, it was Mandelbrot who first realized that familiar shapes in nature ranging from coastlines, trees, and turbulence to lightnings and galaxy clusters actually possess properties of self-similarity and can be described by means of this new type of geometry, which he named fractal from the Latin "fractus" or fragmented. This concept was clearly lacking for the description of complex structures in nature (actually most natural structures are complex) and this is one of the reasons for its present success. In this respect it is interesting to compare this field with that of particle physics, in which the situation is just the opposite. In particle physics theoretical results are often available and enormous experimental efforts are involved in confirming or rejecting them. In the case of fractals the "experimental data" have always been present but, given the absence of an appropriate concept, they have not been considered part of the scientific problematic. Mandelbrot has introduced such a concept providing us with a beautiful playground of new problems. It should be made clear, however, that the introduction of a concept does not imply the formulation of a theory for it. Mandelbrot in fact has not produced a theory to explain how these structures actually arise from physical laws. A deepening of the interrelations between fractal geometry and physical phenomena is what may be called the theory of fractals and forms the objective of present activity in the field.

With respect to the impact of this new concept in physics there are two main streams of activity. The first tries to answer the question: where do fractals come from? That is, to try to understand how it has come about that many shapes in nature manifest these remarkable properties. This will be by and large the approach discussed in this chapter. The other point of view is to assume as a statement of fact that fractal structures exist and to study their physical properties.(4) This consists in assuming a simple fractal model as a starting point and studying, for example, vibrations, electronic states, and diffusion on this structure. For some problems, of course, these two approaches can have points in common.

It should be noted that the concept of self-similarity is certainly not new in physics. It has been well known since the detailed study of critical properties of phase transitions (see chapter 10), and has been instrumental in the formulation of the renormalization group approach (Chapter 10, Section 10.11) that has essentially solved this problem.(5) In this case, however, self-similarity was considered a peculiarity of the competition between order and disorder at a particular temperature for equilibrium phase transitions. We can now see that this property is much more common and appears in many equilibrium and nonequilibrium phenomena apparently unrelated to the problem of phase transitions. Maybe a relation can be found in some critical competition between order and disorder, but the theory developed for phase transitions, or variations of it, is not able to provide an answer to some new problems that have arisen more recently. It is, however, plausible to think that the popularity of fractals in physics is related to the previous success of scaling concepts in phase transitions. A peculiarity of this new field is in the role played by the computer. Until quite recently, in standard problems the computer was mostly used to refine simple analytical results that could already provide the basic concepts, or to extend them to more complex cases. The computer was almost never an essential tool for a conceptual discovery. In fractals (and also in chaos) the situation is quite different for two reasons: the first is that, since the theory of fractal structures is at the moment in a very preliminary stage, computer simulations can be considered as a sort of mathematical experiment that represents an essential guide for the development of theoretical ideas. The second reason is that, even if a theory were available, it could only produce a subset of information, such as the fractal dimension of a given structure, or its universality classes. Since we are dealing with the understanding of complex shapes, in addition to these "numbers," one would also like to actually see the full structure corresponding to a given model. For this reason the computer is going to remain an essential element in this field.

This new situation also has its pitfalls, because a number of papers may give the impression of computer games rather than scientific contributions. In this respect, however, we do not share the pessimistic opinion of Kadanoff,[6] who has limited his analysis to this negative feature. It is certainly useful to warn the reader about these problems but, in our opinion, attention should also be given to the successful results that have made this new field so popular.

9.2. Self-Similarity and Nonanalyticity

Most standard mathematical methods in physics are based on analytic functions such that, at every point, one can define a single tangent line as well as higher derivatives. In geometry also one tends to use idealized smooth objects such as lines, planes, and spheres. With these methods physics has approached the description of nature from the standpoint of smooth structures and deviations from these shapes were considered as imperfections. On the other hand, it is easy to note that trees, clouds, lightnings, and mountains, for example, do not possess these properties of smoothness and one may conclude that irregularities are the norm rather than the exception. Fractal geometry deals with the study of these irregularities as essential elements.

We will consider as an example the common shape of a lightning pattern as shown in Figure 9.1. The traditional methods cannot describe such a structure. It is clear in fact that it is not possible to define tangents or derivatives. We can also see that a small detail of the pattern, if magnified, looks rather similar to a larger portion of the picture. This is a manifestation of the property of self-similarity. It is then clear that self-similarity cannot be compatible with analyticity. This would require in fact that, on a very small scale, a curve should become so smooth that it can be approximated by a single tangent line. Self-similarity implies just the opposite. At smaller and smaller scales new structure appears with the same complexity as the entire structure. Fractal

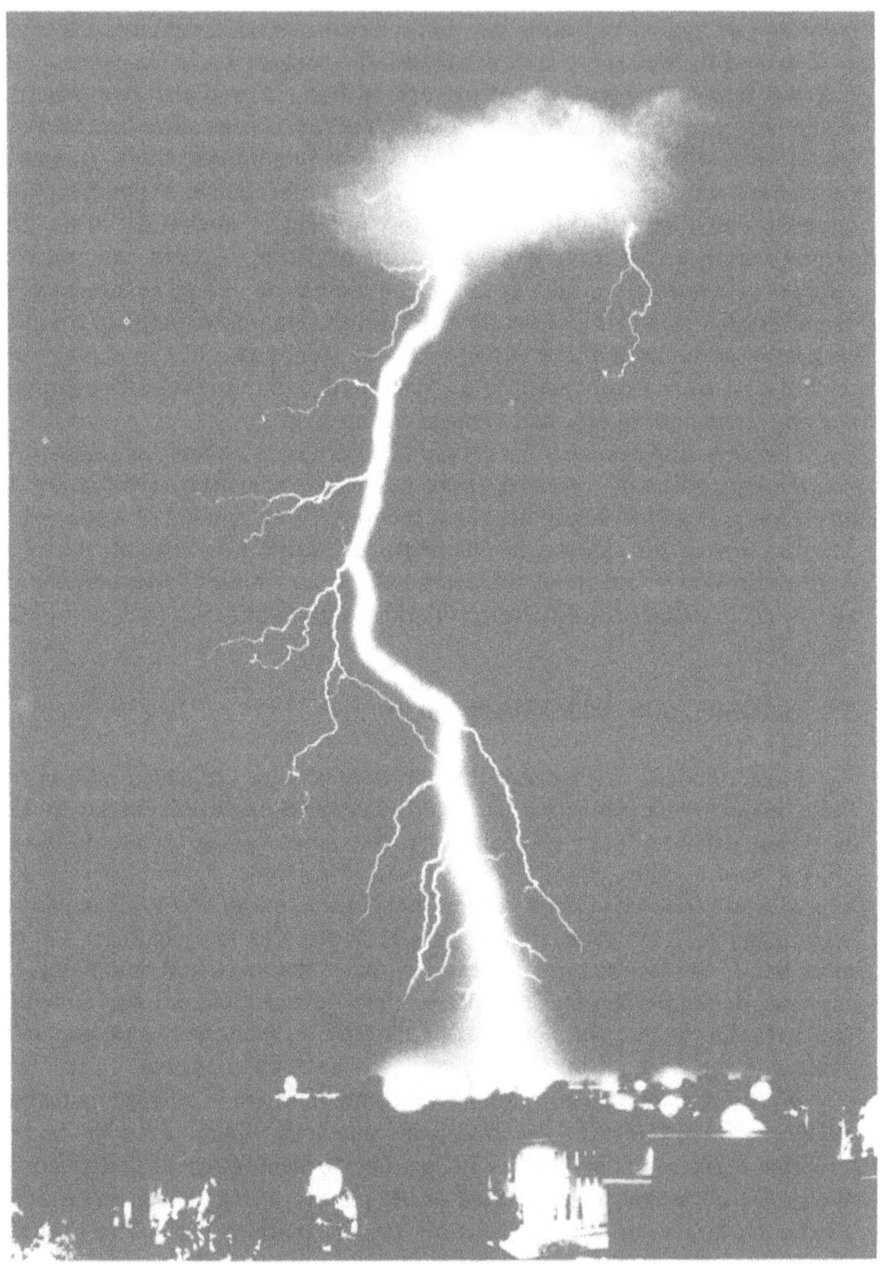

Figure 9.1. Self-similarity implies that a small detail of the pattern is similar in shape to a larger portion of the structure. With a bit of optimism such a characteristic can be detected in this picture.

geometry provides the appropriate tools for the analysis of structures like that of Figure 9.1.

It is useful to consider now a simple, idealized example of fractal structure, like the Sierpinski gasket shown in Figure 9.2. This figure can be constructed by dividing a triangle into four subtriangles of equal shape, then leaving the central one empty and dividing the remaining three in a similar fashion, and so on. It is easy to see that this structure possesses properties of scale invariance and self-similarity. Strictly speaking, self-similarity would require repeating the subdivision an infinite number of times. In Figure 9.2 we have stopped after a few iterations and the small dark triangles represent the lowest cutoff elements of our structure. If we take our unit length to be defined by the size of these smallest triangles, we can consider how many unit triangles are contained in a structure of size L starting from the upper vertex. This number $N(L)$ corresponds to a generalized volume for which one easily constructs the table shown in Figure 9.2.

One way to define the dimension is through the power that relates volume to length. Generalizing this concept we can define a fractal dimension D as the power that relates $N(L)$ to L:

$$N(L) \cong L^D \tag{9.2.1}$$

From the table of Figure 9.2 we then obtain

$$D = \frac{\ln N(L)}{\ln L} = \frac{\ln 3}{\ln 2} = 1.58496\ldots \tag{9.2.2}$$

Volume and length are then related by a noninteger power for an object like the Sierpinski gasket. It should be noted that this is not the original definition of the fractal dimension,[2] but it is often in this sense that it is used in physics.

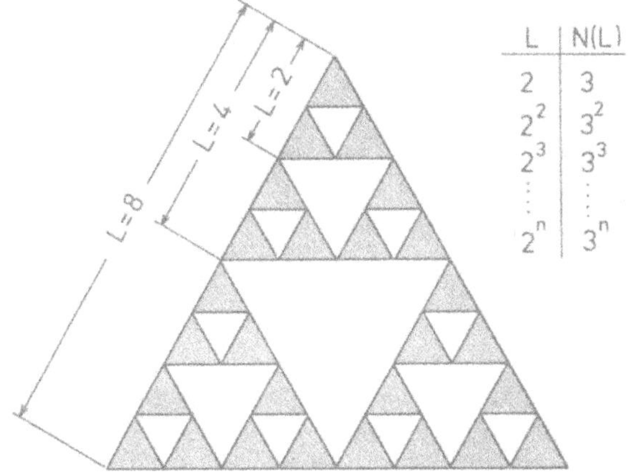

Figure 9.2. The Sierpinski gasket. A simple geometric construction of a fractal structure.

After this simple example we may ask: what has the Sierpinski gasket to do with lightnings? It is indeed an important problem to find out which features of the lightning's propagation are responsible for the supposed-to-be fractal structure of its pattern. These are the basic questions of the physics of fractals that cannot be answered by simply constructing idealized fractal structures like the Sierpinski gasket.

9.3. Power Laws and Self-Similarity

We have seen in the previous section why self-similarity implies nonanalyticity. We show here what type of nonanalytic functions are associated with this property. We shall see that in general one obtains a power law with a noninteger exponent, but complications may also occur.

We start by considering the correlation function for the occupation number $n(\mathbf{r})$

$$\Gamma(r) = \langle(\mathbf{r}_0)n(\mathbf{r}_0 + \mathbf{r})\rangle_0 \tag{9.3.1}$$

This function defines how many occupied points can be found at a distance r (scalar, for simplicity) from an occupied point at $\mathbf{r}_0$. The average $\langle\cdot\cdot\cdot\rangle_0$ refers to all possible choices of $\mathbf{r}_0$. For random fractals one may also average over different realizations of the whole pattern. By integrating equation (9.3.1) we obtain a relation between average volume and length scale of the type discussed in the previous section. For a homogeneous fractal of dimension D this is given by

$$N(R) = \int_0^R d\mathbf{r}\,\langle n(\mathbf{r}_0)n(\mathbf{r}_0 + \mathbf{r})\rangle_0 \cong R^D \tag{9.3.2}$$

We consider now the mathematical restrictions imposed on the function $\Gamma(R)$ [or $N(R)$] by the property of self-similarity. This implies that the physical properties of the system should have the same structure under dilatation or contraction of the length scale. This means that the scale transformation

$$R \to R' = bR \tag{9.3.3}$$

should give rise to the same type of correlation function [and therefore $N(R)$]. This leads to the functional relation

$$N(R') = N(bR) = A(b) \cdot N(R) \tag{9.3.4}$$

where A is a prefactor that does not depend on R but may depend on b. This means that a rescaling of R gives rise to exactly the same function if N is also rescaled by the constant A. If we require equation (9.3.4) to be satisfied for any value of b, the allowed functions are simple power laws

$$N(R) = R^D, \qquad A(b) = b^D \tag{9.3.5}$$

where the exponent D can take any value. There are cases, however, like the Sierpinski gasket of Figure 9.2, where the condition of self-similarity is satisfied only for particular values of b. In the case of Figure 9.2, we have $b = 2^n$ with n an integer. We consider therefore the discrete set of scale transformations defined by

$$b(n) = b_0^n \qquad (n = 1, 2, 3, \ldots.) \tag{9.3.6}$$

Equation (9.3.4) only holds now at these discrete values of b,

$$N(R') = N(b_0^n \cdot R) = A(b_0^n) \cdot N(R) \tag{9.3.7}$$

If we try to solve equation (9.37) with functions of type

$$N(R) = f(R) \cdot R^D \tag{9.3.8}$$

we obtain

$$N(b_0^n \cdot R) = f(b_0^n \cdot R) \cdot b_0^{nD} \cdot R^D = A(b_0^n) \cdot f(R) \cdot R^D \tag{9.3.9}$$

This equation defines the value of A as

$$A(b_0^n) = b_0^{nD} \tag{9.3.10}$$

and gives the following condition for the function f:

$$f(b_0^n \cdot R) = f(R) \tag{9.3.11}$$

We now introduce the new variable

$$y = \ln R \tag{9.3.12}$$

We have

$$y' = \ln R' = n \cdot \ln b_0 + y \tag{9.3.13}$$

and therefore

$$f(n \cdot \ln b_0 + y) = f(y) \tag{9.3.14}$$

This relation is satisfied by all functions that are periodic with respect to the variable $y = \ln R$ with period $\ln b_0$. This means that in the case of discrete scaling we should expect, in general, a power law modulated with a periodic function of ln R. A nice example of this behavior can be found elsewhere.[7,8]

We show now that this periodic modulation of the power law can be related to an imaginary component $i\delta$ of the fractal dimension D. This leads us to the possibility of *complex fractal dimensions.* Consider the generalization

$$N(R) = \mathrm{Re}\,(R^{D+i\delta}) = R^D \cdot \cos(\delta \cdot \ln R) \tag{9.3.15}$$

It is clear that δ corresponds to the periodicity with respect to the variable ln R and it is related to the smallest rescaling factor b_0 by

$$\delta = \frac{2\pi}{\ln b_0} \tag{9.3.16}$$

We note that the cosine in equation (9.3.15) is just one particular function that satisfies the condition of periodicity with respect to $\ln b_0$. In general, one can also have all the higher harmonics, and a more general relation[9] (the most general relation is given by Doucot *et al.*[7]) for discrete scaling is therefore

$$N(R) = \mathrm{Re}\left[\sum_{m=0,1,2,\ldots} A(m) \cdot R^{(D+im\delta)}\right] \tag{9.3.17}$$

where δ is the wave vector corresponding to the shortest wavelength allowed by the periodicity with respect to ln R. It is easy to check that in the case of continuous scaling $b_0 \to \infty$ and $\delta \to 0$, leaving only the real fractal dimension D. For the Sierpinski gasket of Figure 9.2 we have $b_0 = 2$ and

$$\delta = \frac{2\pi}{\ln 2} = 9.06472\ldots \tag{9.3.18}$$

The Fourier components $A(m)$ determine finally the detailed shape of the periodic modulation.

Before concluding this section in which we have seen how self-similarity implies power laws, we can consider the inverse problem, namely, whether a power law implies self-similarity. This question will be examined with a simple example. We consider a number density $\rho(r)$ that behaves as the Coulomb potential in three dimensions,

$$\rho(r) = 1/r \tag{9.3.19}$$

One could argue that the generalized volume is given by

$$N(R) \cong \int_0^R \rho(r) \cdot r^2 \cdot dr \simeq R^2 \tag{9.3.20}$$

and therefore interpret $D = 2$ as the fractal dimension of this distribution. This would be wrong, because equation (9.3.20) holds only for one specific origin ($r = 0$) while for a fractal it should hold for *any origin.* For any other origin one obtains

$$N(R) \cong \int_0^R \rho(\mathbf{r}_0 + \mathbf{r})\, d^d\mathbf{r} \tag{9.3.21}$$

where the exponent $d = 3$ is the dimension of the embedding euclidean space. For $|\mathbf{r}_0| > R$ we can approximate

$$\rho(\mathbf{r}_0 + \mathbf{r}) \cong \rho(\mathbf{r}_0) = \rho_0 \tag{9.3.22}$$

and therefore

$$N(R) \cong \int_0^R \rho_0 r^{(d-1)}\, dr \cong \rho_0 R^d \tag{9.3.23}$$

This gives the standard dimension of the embedding space that shows we are dealing with a smooth (except for the point $\mathbf{r} = 0$), nonfractal distribution. For a homogeneous fractal one should obtain the same nontrivial exponent $D \neq d$ no matter when one starts the integration, and therefore also for average quantities like the correlation function of equation (9.3.1). Since we have mentioned the possibility of fractal properties of distributions (instead of simple sets), one should remark that a distribution that is really self-similar [therefore unlike equation (9.3.19)] in general cannot be described by a single fractal dimension, but the more general concept of multifractals is needed. We will return to this point in Section 9.5.

9.4. Why Fractal Dimensions and Not Just Critical Exponents

Most of the concepts examined until now do not really need the geometric interpretation of the power-law exponent in terms of fractals. Hence one may wonder: what is the advantage of talking about fractal dimensions instead of just critical exponents as is the case for phase transitions? Indeed in some cases this is just a matter of taste; however, in several problems, especially those dealing with irreversible growth, the geometric interpretation can provide a much better intuitive picture of the phenomenon and in this sense it is useful.

We consider as an example the law of codimension additivity. Two planes A and B in a space of euclidean dimension $d = 3$ cross in a set of points whose dimension d_c is given by

$$d_c = d_A + d_B - d = 1 \tag{9.4.1}$$

that is, they have a common line of dimension one. This relation is now generalized to fractal structures. We wish to find the properties of the set of points in which two ideal polymers (A and B) cross in a three-dimensional space. An ideal polymer can be represented by the trajectory of a simple random walk. The linear size of the coil R is then related to the number of monomers $N(R)$ by

$$R \cong N^{1/2} \tag{9.4.2}$$

We can interpret this relation by saying that an ideal polymer has fractal dimension $D = 2$ because

$$N(R) \cong R^2 \tag{9.4.3}$$

The dimension D_c of the set of crossing points is then directly obtained by the generalization of expression (9.4.1) to fractal dimensions,

$$D_c = D_A + D_B - d = 1 \tag{9.4.4}$$

The dimension of the set of crossing points therefore has dimension one; it grows linearly with the size of the volume considered. This result can also be derived from the properties of random walks without using equation (9.4.4), but it would imply a rather lengthy calculation. Equation (9.4.4) can, of course, be used for less trivial problems, such as determining the dimension of the set of points in which a self-avoiding polymer in three dimensions crosses a plane. Using the Flory value for the fractal dimension of the self-avoiding polymer $D = \frac{5}{3}$ [9] we obtain

$$D_c = \tfrac{5}{3} + 2 - 3 = \tfrac{2}{3} \tag{9.4.5}$$

for the set of crossing points. This is a less trivial result to obtain with other methods.

Equation (9.4.4) also suggests the possibility of generalizing the concept of dimension to the case of *negative fractal dimensions.* Let us consider, for example, the case of the crossing points of two lines in three dimensions. We obtain $D_c = -1$. We can now give an interpretation to this curious result in terms of the probability to have a crossing point as a function of the linear size of the volume considered. Take a volume $V \sim L^3$ of linear size L that contains two lines. Each line occupies a number of points of order L, so the probability that a given point of one line meets one of the points of the other line is $L/V \sim L^{-2}$. To find the probability $p(L)$ that the two lines meet at any one point, we have to multiply this result by the total number of points L of the first line. This gives

$$p(L) \sim L^{-1} \tag{9.4.6}$$

and in this sense we can interpret the previous results as a negative fractal dimension. It is noteworthy that fractal dimensions contain, as a special case, also integer dimensions. The important feature when talking about fractals is, however, the requirement of self-similarity that is satisfied by a random walk (polymer) as well as by planes and lines. It is not satisfied, however, by structures with a well-defined length scale, such as a sphere of fixed radius R_0 or a stick with fixed length L_0.

9.5. Fractals and Multifractals

We introduce here the fractal dimension from the more canonical point of view of covering a set with unit volumes and regarding the volume as a function of the size of the measuring unit. This approach implies a coarse graining of the structure that not always coincides with the more physical point of view that we have discussed in Section 9.3. We consider first an euclidean structure of volume V covered with subunits of linear size l. The total volume V is given by the product of the volume of each subunit $v(l)$ multiplied the number of units $N(l)$,

$$V = N(l) \cdot v(l) \tag{9.5.1}$$

We note that N is employed now with a different meaning than that used in the previous sections. If the size of the subunits is changed, we have

$$v(l) \cong l^d \tag{9.5.2}$$

and

$$N(l) \cong l^{-d} \tag{9.5.3}$$

The same covering can be performed for a fractal structure and equation (9.5.3) is generalized to

$$N(l) \cong l^{-D} \tag{9.5.4}$$

which implies

$$V(l) \cong l^{(d-D)} \tag{9.5.5}$$

when D can now be a fractal dimension. The Sierpinski gasket of Fig. 9.2 is again taken as an example. Suppose the figure has a linear size 1. We need only one triangle of size l to cover it completely. This gives $N(l = 1) = 1$. Now take triangles of size $l = \frac{1}{2}$, of which we need $N(l = \frac{1}{2}) = 3$. In general, therefore, for $l = (\frac{1}{2})^n$ we have $N[(\frac{1}{2})^n] = 3^n$, and equation (9.5.4) yields

$$D = \frac{\ln N(l)}{\ln l} = -\frac{\ln 3^n}{\ln (\frac{1}{2})^n} = \frac{\ln 3}{\ln 2} \tag{9.5.6}$$

We have therefore obtained the same result as equation (9.2.2) derived with the volume-size scaling argument.

Let us consider another elementary example, the Koch curve of Figure 9.3. The construction consists in dividing a segment into three equal parts and replacing the central part by a cusp whose sides have the same length as the subsegment we eliminate. The operation is then repeated for each of the new segments, and so on. Suppose the length scale of the whole figure is one and consider a coverage of this structure by circles. If we use circles of diameter one third, we need four of them and, in general, for circles of diameter $l = (\frac{1}{3}^n$ we have $N[(\frac{1}{3})^n] = 4^n$. This corresponds to a fractal dimension

$$D = \frac{\ln 4}{\ln 3} = 1.2628\ldots \quad (9.5.7)$$

This structure can now be analyzed from the point of view of the relation between volume and length scale. The volume corresponds in this case to the total length of the curve as a function of its linear size. We consider as a lower cutoff length l_0 the smallest linear segment shown in the figure. It can be seen that, starting from the point $x = 0$, the structure corresponding to a linear length $l = 3l_0$ has total length $N(3l_0) = 4l_0$. On generalizing this relation we obtain $N(3^n l_0) = 4^n l_0$, corresponding to $D = \ln 4/\ln 3$ as before. Also, in this case therefore the two approaches give the same result.

We shall now examine a more complex case in which different methods will give different results. Consider the Brownian profile generated by the end point of a random walk (in one dimension) as a function of time, that we interpret here as the space variable x. This generates a profile $h(x)$ as shown schematically in Figure 9.4a. Suppose we wish to cover this structure with elements of linear size b. Each of these elements corresponds to an advancement along the x axis of an amount l. Let us note that for smaller and smaller b we encounter more and more structure in the profile, because the step of the random walk along the x axis is supposed to be infinitesimal. Owing to the properties of fluctuations of a random walk we have

$$h(l) \cong l^{1/2} \quad (9.5.8)$$

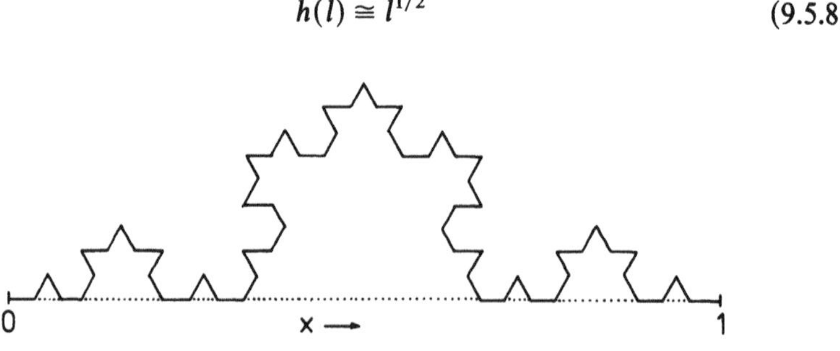

Figure 9.3. Another example of simple geometric construction of a fractal structure: the Koch curve.

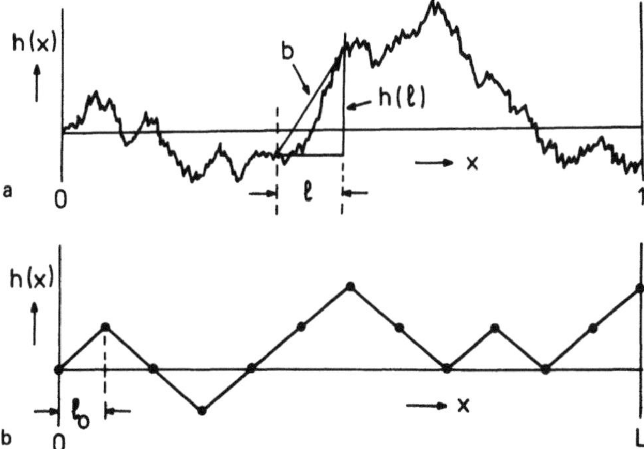

Figure 9.4. Brownian profile obtained by plotting the end point of a one-dimensional random walk (h) as a function of time that we interpret as the space variable x. Fig. 4a describes schematically the case in which the steps along the x axis are finite. In Fig. 4b instead a finite cut-off along the x is considered.

and in the limit of small l

$$b(l) \cong \{l^2 + [h(l)]^2\}^{1/2} \cong (l^2 + l)^{1/2} \cong l^{1/2} \tag{9.5.9}$$

Therefore $l \sim b^2$ and the number of elements of size b needed to cover the structure is

$$N(b) \cong l(b)^{-1} \cong b^{-2} \tag{9.5.10}$$

which in view of equation (9.5.4) implies $D = 2$.

We can now look at the same process from the standpoint of the scaling of the total length of the profile with respect to its linear size. As seen in Figure 9.4b, we indicate by l_0 the elementary step along the x axis. It is clear that for each of these steps the increase in profile step is of the order of l_0. For a distance L along the x axis we have $n = L/l_0$ steps and the total length of the profile (denoted by V in the sense of a generalized volume) is

$$V(L) \cong n \cdot l_0 = L \tag{9.5.11}$$

implying therefore that $D = 1$, contrary to equation (9.5.10).

There is actually a third possibility of fractal analysis of this profile, one to which the literature usually refers. Let us again examine Figure 9.4a and consider a coarse graining of this figure by a step l along the x axis. Our profile consists now only of segments of type b. We note that this is different from the previous coverage approach, in which we intended to cover the real profile with elements of linear size b that are therefore not just straight lines. The total length of this coarse-grained profile can be easily computed. We have l^{-1} segments of length $b(l) \cong l^{1/2}$, therefore

$$V(l) \cong l^{-1/2} \tag{9.5.12}$$

and, in view of equation (9.5.5), we obtain

$$D = \tfrac{3}{2} \tag{9.5.13}$$

the value usually found in the literature.[2]

It can indeed be disorienting to see that for this case three different methods of analysis give rise to three different values of the fractal dimension. This is due to the fact that the Brownian profile is not a homogeneous fractal but a self-affine one.[10,11] This is an expression to indicate that the system has anisotropic scaling, contrary to the cases discussed before. In fact $h(x) \cong x^{1/2}$, so that the height of the profile does not scale as its length. This gives rise to the ambiguity in the definition of its fractal dimension.

There are actually even more complex cases defined as multifractals that need for their characterization a continuous spectrum of fractal dimensions. In particular, this generalization is necessary if one considers self-similar distributions instead of self-similar sets. In a set, a point is either part of the structure or not, as in all cases we have discussed. A distribution instead associates with each point a continuous value. This situation occurs often in physical problems where one deals with distributions of probability, charge, potential, etc. We do not discuss here the case of multifractals. A pedagogical introduction to this topic can be found elsewhere,[12] while original papers include Benzi *et al.*[13] and references cited therein. Finally, we remark that multifractals appear to develop quite naturally in random multiplicative process that often arise in physical problems involving disorder.[14]

9.6. Laplacian Fractals

Until now we have discussed fractal structures from a purely geometric point of view and their relation with natural complex shapes lies only in the fact that these idealized structures resemble in some cases real patterns. It is not clear however which are the physical properties that give rise to these structures. In this section we mention briefly a *well-defined physical model* that gives rise to *random fractal structures.* It is based on a suitable combination of Laplace equation and probability field and it appears to contain the necessary elements to understand the development of fractal structures in a variety of different phenomena, like dielectric breakdown,[15] dendritic growth (in the sense of diffusion-limited aggregation),[16] viscous fingering in fluids,[17] and fracture of materials.[18]

We describe now the growth rules from the viewpoint of dielectric breakdown for which Figure 9.5 presents a schematic lattice description of the process. The black pattern represents the plasma lines at a given time, while the dashed bonds connecting a black to a white point represent the possible candidates for the propagation of the pattern. Considering that the pattern discharge comprises plasma channels with high conductivity, we assume the whole black pattern to be equipotential with the same potential as the central electrode $\phi = 0$. The local field around this pattern can be computed by solving the Laplace equation

$$\nabla^2\phi = 0 \tag{9.6.1}$$

on the whole lattice with boundary conditions $\phi = 0$ on all the points of the black pattern and $\phi = 1$ on a far away circle that represents the other electrode. This problem can be solved by discretizing the Laplace equation on the lattice and using numerical iterative methods. This provides the potential ϕ at all points of the lattice. We then compute the local field around the pattern in the form

$$E_j = |\nabla\phi| \tag{9.6.2}$$

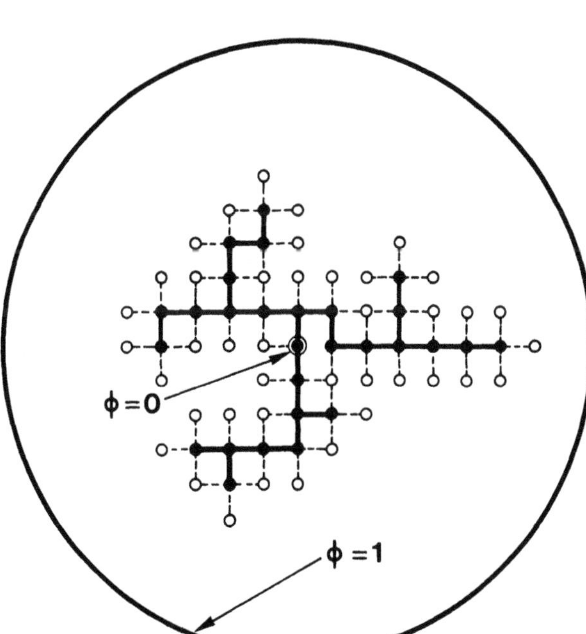

Figure 9.5. Schematic illustration of the stochastic model for dielectric breakdown. The central point represents one of the electrodes while the other electrode is modeled as a circle at large enough distance. The discharge pattern is indicated by the black dots connected with thick lines and it is considered equipotential ($\phi = 0$). The dashed bonds indicate all the possible growth processes. The probability for each of these processes is proportional to the local electric field.

where j labels all the dashed bonds connecting a black to a white point. Given the local field we have to assign a stepping rule for the growth of the pattern. In view of the randomness intrinsic to this process, the relation between growth and local field should be *probabilistic* rather than *deterministic.* The simplest assumption is then to define a normalized growth probability per dashed bond given by

$$p_j = \frac{(E_j)^\eta}{\sum_j (E_j)^\eta} \tag{9.6.3}$$

This model is able to generate random fractals whose fractal dimension D explicitly depends on the exponent η. In Figure 9.6 we show a few examples. Therefore, a suitable combination of *Laplace equation* and a *stochastic field* is able to generate stochastic fractal structures. This is one of the clearest cases in which the physical origin of fractal structure has been, at least in part, understood. Concerning the possibility of applying this model also to problems that appear totally unrelated to dielectric breakdown, this is due to the fact that Laplace equation occurs also in diffusion problems (diffusion-limited aggregation),[16] in the description of the fluid pressure (viscous fingers),[17] and in the elastic equation (fracture).[18]

The open theoretical questions for this model consist in the definition of its universality classes and in the possibility of analytical calculations of its fractal dimensions. Despite its apparent simplicity, the process is actually quite complex from a mathematical point of view. It consists in fact of a stochastic process with infinite memory and long-range interaction. The entire previous structure represents the boundary condition for the definition of the growth probabilities via the solution of the Laplace equation. At the moment there are no mathematical tools available to handle this class of problems that represent one of the most challenging fields in modern theoretical physics. More information about these models can be found elsewhere.[14-18]

9.7. A Note about Experiments

It has been seen how self-similarity implies power laws with non integer exponents. Of course this does not mean that the exponent cannot be an integer, but this occurrence would just be a coincidence as, for example, Plate 143 of Mandelbrot's book[2] in which the structure has fractal dimension $D = 2$ and is embedded in a euclidean space of dimension $d = 2$. One may be led to think therefore that the experimental observation of a power law with noninteger exponent should imply some property of self-similarity of the system. With respect to this kind of reasoning there are some possible complications to take into account. First, in real systems self-similarity develops only

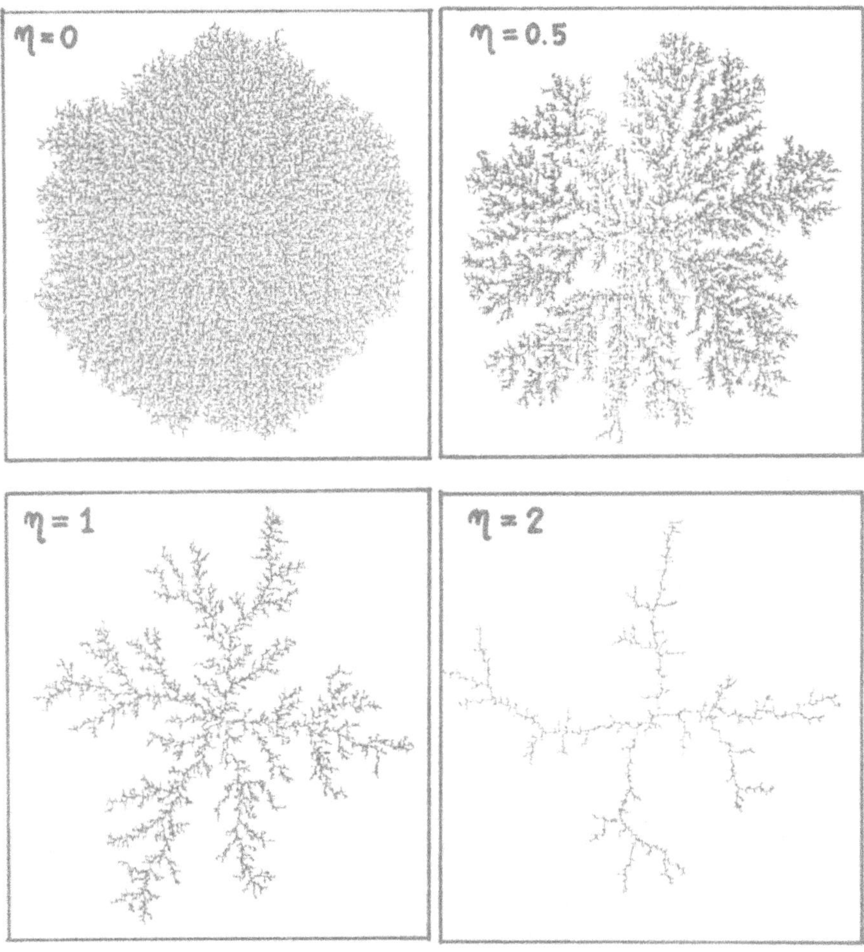

Figure 9.6. Examples of computer-generated discharge patterns for different values of η. The fractal dimensions are $D(\eta = 0) = 2$; $D(\eta = 0.5) \cong 1.92$; $D(\eta = 1) \cong 1.70$, and $D(\eta = 2) \cong 1.43$. More details on this model are given elsewhere.[15]

in a certain range of length scales and there are always lower and upper cutoff lengths. If for these or other reasons one cannot probe a range larger than one decade, it is difficult to draw any reasonable conclusion. Often, in fact, a crossover behavior may be well fitted by a noninteger power law for a limited range. Another fact to take into account is that a non-self-similar system may also give rise to noninteger but rational powers. For this reason it is important to compare the behavior of the quantity measured in a system that may be self-similar with the behavior of the same quantity in normal systems.

Acknowledgments

It is a pleasure to thank A. Coniglio, C. Evertz, B. Mandelbrot, G. Paladin, G. Parisi, L. Peliti, A. P. Siebesma, R. Voss, and A. Vulpiani for many discussions on these topics.

References

1. M. V. Berry, *New Scientist* (27 Jan. 1983).
2. B. B. Mandelbrot, *The Fractal Geometry of Nature*, Freeman, New York (1982).
3. L. Pietronero and E. Tosatti (eds.), *Fractals in Physics*, North-Holland, Amsterdam, New York (1986). H. E. Stanley and N. Ostrowsky (eds.), *On Growth and Form*, Nijhoff Publ., Dordrecht (1986). F. Family and D. P. Landau (eds.), *Kinetics of Aggregation and Gelation*, North-Holland, Amsterdam, New York (1984). *J. Stat. Phys.* **36**, Nos 5/6 (1984).
4. A. Aharony, *Europhys. News* **17**, 41 (April 1986).
5. D. J. Amit, *Field Theory, the Renormalization Group and Critical Phenomena*, McGraw-Hill, New York (1978).
6. L. P. Kadanoff, *Phys. Today* **6** (Feb. 1986).
7. B. Doucot, W. Wang, J. Chaussy, B. Pannetier, R. Rammal, A. Vareille, and D. Henry, *Phys. Rev. Lett.* **57**, 1235 (1986).
8. J. Bernasconi and W. R. Schneider, in: *Fractals in Physics* (L. Pietronero and E. Tosatti, eds.), p. 409, North-Holland, Amsterdam, New York (1986).
9. P. G. de Gennes, *Scaling Concepts in Polymer Physics*, Cornell University Press (1979).
10. B. B. Mandelbrot, in: *Fractals in Physics* (L. Pietronero and E. Tosatti, eds.), p. 3, North-Holland, Amsterdam, New York (1986).
11. R. F. Voss, in: *Fundamental Algorithms for Computer Graphics* (R. A. Earnshaw, ed.), NATO ASI Series, Vol. F17, p. 805, Springer-Verlag, Heidelberg (1985).
12. L. Pietronero, C. Evertsz, and C. P. Siebesma, in: *Stochastic Processes in Physics and Engineering* (S. Albeverio, Ph. Blanchard, L. Streit, and M. Hazewinkel, eds.), D. Reidel, Dordrecht, Boston (1988).
13. R. Benzi, G. Paladin, G. Parisi, and A. Vulpiani, *J. Phys. A* **17**, 3521 (1984). T. C. Halsey, M. H. Jensen, L. P. Kadanoff, I. Procaccia, and B. I. Shraiman, *Phys. Rev. A* **33**, 1141 (1986).
14. L. Pietronero and A. P. Siebesma, *Phys. Rev. Lett.* **57**, 1098 (1986).
15. L. Niemeyer, L. Pietronero, and H. J. Wiesmann, *Phys. Rev. Lett.* **52**, 1033 (1984). L. Pietronero and H. J. Wiesmann, *J. Stat. Phys.* **36**, 909 (1984).
16. T. A. Witten and L. M. Sander, *Phys. Rev. Lett.* **47**, 1400 (1981); *Phys. Rev. B* **27**, 5686 (1983). P. Meakin, *Phys. Rev. A* **27**, 1495 (1983).
17. L. Paterson, *Phys. Rev. Lett.* **52**, 1621 (1984). J. Nittmann, G. Daccord, and H. E. Stanley, *Nature* **314**, 141 (1985). J. Nittmann and H. E. Stanley, *Nature* **321**, 663 (1986).
18. E. Louis, F. Guinea, and F. Flores, in: *Fractals in Physics* (L. Pietronero and E. Tosatti, eds.), p. 177, North-Holland, Amsterdam, New York (1986).

10

Phase Transitions

R. B. Stinchcombe

10.1 Introduction: Emphasis and Layout

This chapter is designed to lead up to, and discuss, with a minimum of detail, modern viewpoints in the theory of phase transitions.

In little more than a decade the subject has shown a remarkable evolution, particularly in the area of critical phenomena, in the range of viewpoints employed, and in the range of transitions considered. An example is the recent emphasis on configurations, and on geometric concepts and transitions (such as percolation). This evolution has resulted in approaches and detailed techniques (for instance, scaling), which are applicable in a much wider context than the thermal transitions for which they were originally developed. Such approaches can be useful in discussing, for example, transitions in dynamical systems; or, on the other hand, the techniques, especially the recursive ones such as the renormalization group transformation, can themselves introduce interesting new features (e.g., those of nonlinear iterative maps). Therefore, emphasis is given here to modern concepts such as geometry, scaling, and recursion. However, the older approaches and techniques (such as mean field theory) remain very valuable when used within their limitations, and we still need to cover these and other basic topics and to refer to basic illustrative systems to give a coherent account of the subject. So we briefly describe basic concepts—order, mean field and Landau approaches, fluctuations, criticality, homogeneity, and universality—before the scaling and other modern techniques can be developed. These are therefore the topics discussed, in roughly that order.

R. B. Stinchcombe • Department of Theoretical Physics, University of Oxford, Oxford OX1 3NP, England.

If further detail is desired, it can be found in the following articles by the author:

1. "Phase transitions and dimensionality," in: *Polymers, Liquid Crystals, and Low-Dimensional Solids* (N. H. March and M. P. Tosi, eds.), Plenum Press, New York (1985).
2. "Introduction to scaling concepts," in: *Scaling Phenomena in Disordered Systems* (R. Pynn and A. Skjeltorp, eds.), Plenum Press, New York (1985).
3. "Introduction to renormalization group methods," in: *Magnetic Phase Transitions* (R. J. Elliott and M. Ausloos, eds.), Springer-Verlag, Berlin (1983).
4. "Dilution-induced criticality," in: *Scaling Phenomena in Disordered Systems* (R. Pynn and A. Skjeltorp, eds), Plenum Press, New York (1985)

Parts of these four references give greater detail of the topics treated in Sections 10.2–10.5, 10.7–10.10, 10.6, and 10.11 and 10.12, respectively, of the present development. A more extensive view of phase transitions is provided by the second edition of the book: H. E. Stanley, *Introduction to Phase Transitions and Critical Phenomena*, Oxford University Press, to be published.

10.2. Basic Considerations

Phase transitions usually take place in systems of many constituents (or many states) because of statistics, or interactions, or both. The "cocktail party effect," in which for a large number N of participants the sound increases rapidly through a threshold, has the interaction and cooperative features; and in the sense that each person speaks more loudly to be heard above the average background provided by the voices of all the other people a "mean field" (the background noise) plays a part. The large N aspect is the so-called "thermodynamic limit."

Two other basic concepts in phase transitions, "order" and "broken symmetry," are illustrated by the "dinner table effect," where each of a group of diners, sitting equally spaced around a circular table each midway between two napkins and not knowing which is his, simply copies his neighbor ("interaction") all (hopefully) taking the left, or all the right, napkin; these are the two ordered "broken-symmetry" states of the system. The chain of diners is a low-dimensional system in which order is easily destroyed (see Section 10.2.1): it only needs one to behave differently from his neighbor to break the pattern. The chain of diners is actually equivalent to an Ising spin system (in one dimension) which we now discuss.

10.2.1. Ising Spin Systems: Ordering

The spin-$\frac{1}{2}$ Ising system[1] consists of spins σ_i, each with the two possible values +1, −1 ("up or down"), at sites i of a (normally) regular lattice, interacting via a pair interaction such that the full interaction energy (Hamiltonian) is

$$H = -\sum J_{ij}\sigma_i\sigma_j \tag{10.2.1}$$

For the rest of this section we suppose $J_{ij} = J$ if i, j are nearest neighbors and J_{ij} is zero otherwise. Then, for the special case of one-dimension ($d = 1$) we have a chain of two-state spins which can represent the left–right states of the diners in the previous discussion and with J representing the influence between neighbors. With J positive, each spin pair has the lowest energy when the pair is parallel. Thus, for the chain, the preferred (lowest energy, i.e., ground) state has all spins up *or* all spins down (compare the diners), so we obtain ordering at temperature $T = 0$ by virtue of the interaction. The following argument[2] shows that for $d = 1$ this order does not persist at $T \neq 0$: a single break ("domain wall") in the spin configuration, dividing the $N + 1$ (say) spins into all spins up to the left of the break and all spins down to the right, is greater in energy than the ground state by $\Delta U = 2J$; and, since the break can be in any of N positions, it is associated with entropy $\Delta S = K \ln N$. So the free energy F of the single domain-wall states is different from that of the ground state by

$$\Delta F = 2J - KT \ln N \tag{10.2.2}$$

Thus for a macroscopic system (N arbitrarily large, i.e., "thermodynamic limit") at any $T \neq 0$ the system prefers to decrease F by introducing domain walls, which removes the order of the ground state.

For the Ising system in two dimensions, the corresponding argument[3] compares the ground-state free energy with that of a state in which all the spins to the right (say) of some domain wall are reversed from their orientation in the ground state. If the wall has length L, that results in

$$\Delta S \propto KL, \qquad \Delta U \propto JL \tag{10.2.3}$$

So, in this case, order persists up to $T \sim J/K$. The model thus provides a mechanism for spontaneous ordering in, e.g., magnets up to a finite transition temperature, and is indeed an excellent description of classes of real magnets.[4–8] Our discussion illustrates the way in which fluctuations (here the domain wall) may destroy order, and that they do so less easily in higher dimensions.

In the above example, the order can be characterized by the magnetization M (excess of spin up over spin down) which, in common with all order parameters, is nonzero in the ordered state and zero above the transition.

10.2.2. Ising Spin Systems: Phase Diagrams

As the above discussion suggests, Ising systems of dimensionality $d \geq 2$ order at a finite temperature T_c, the critical temperature. The phase boundary in the field (h) and temperature plane is a line at $h = 0$ extending from $T = 0$ to $T = T_c$, across which the magnetization has a discontinuity corresponding to the difference of the spontaneous magnetizations in the two broken symmetry "states." The transition is "continuous," in the sense that the order parameter (spontaneous magnetization) goes smoothly to zero as T approaches T_c from below. The susceptibility is

$$\chi \propto \partial M/\partial h \propto \partial^2 F/\partial h^2 \tag{10.2.4}$$

This diverges at T_c, which reflects a divergence of spin fluctuations (see Section 10.2.3) and implies that the transition is of second order (divergence in a second derivative of the free energy).

The liquid–gas transition has many analogies with this magnetic one,[9] showing a diverging compressibility due to a divergence of density fluctuations. However, we continue to discuss the simpler magnetic model.

10.2.3. Correlation Functions, Correlation Length

The (pair) spin correlation function of the Ising model is defined by[10]

$$\Gamma_{ij} \equiv \langle \sigma_i \sigma_j \rangle - \langle \sigma_i \rangle \langle \sigma_j \rangle \tag{10.2.5}$$

where the angle brackets denote thermal averages. The right-hand side of relation (10.2.5) clearly vanishes if σ_i and σ_j are uncorrelated. In zero field above the transition, where the magnetization $\langle \sum \sigma_i \rangle$ vanishes, the average of the total energy (10.2.1), hence also the specific heat, is related to Γ_{ij} for the case where i, j are nearest neighbors, i.e., to a short-range spin correlation. On the other hand, the "sum rule"

$$\chi = \beta \sum_{ij} \Gamma_{ij} \qquad (\beta \equiv 1/KT) \tag{10.2.6}$$

shows that the susceptibility involves long-range spin correlations. The sum rule (10.2.6) can be deduced[11] by taking the derivative of the magnetization $\langle \sum_i \sigma_i \rangle$ with respect to the field h, using the fact that h occurs in the Hamiltonian through a term $-h\Sigma\sigma_j$, so

$$\chi = (\partial/\partial h)\left\langle \sum_i \sigma_i \right\rangle = \beta\left\{\left\langle\left(\sum_i \sigma_i\right)^2\right\rangle - \left\langle \sum_i \sigma_i \right\rangle^2\right\} \tag{10.2.7}$$

For large separation r_{ij} between the spins σ_i, σ_j, the pair correlation function (10.2.5) falls off typically like[10]

$$\Gamma_{ij} \sim \ldots \exp(-r_{ij}/\xi) \tag{10.2.8}$$

where ξ is the characteristic "correlation" length. If $\xi \to \infty$, the long-range spin correlations are not suppressed, and the sum in relation (10.2.6) diverges, leading to a divergent susceptibility. A diverging correlation length, i.e., the occurrence of spin fluctuations of arbitrarily long range, can thus explain the "critical behavior" manifested by the divergence of χ at a continuous transition; also there will then normally be associated singularities in other properties, such as the specific heat, related to short-range correlations, though these need not diverge. The nature of these singularities is specified by "critical exponents" defined in the next section.

This view, of the diverging correlation length driving the critical effects, is the by now standard view of criticality at continuous phase transitions; there is much experimental evidence to demonstrate the divergence of ξ at such transitions (e.g., Als-Nielsen *et al.*[12] and Birgeneau *et al.*[13]) and it provides the basis of the highly successful modern theories of critical phenomena (see Sections 10.6–10.12).

10.2.4. Definition of Critical Exponents

We define here, for the case of a magnet, the exponents[10] characterizing the singularities in the specific heat C, magnetization M, susceptibility χ, and correlation function $\Gamma(r)$. In terms of the field h and reduced temperature variable $t \equiv |T - T_c|$:

1. For $h = 0$ and t small

$$C \propto t^{-\alpha} \tag{10.2.9}$$

$$M \propto t^{\beta} \qquad (T < T_c) \tag{10.2.10}$$

$$\chi \propto t^{-\gamma} \tag{10.2.11}$$

$$\xi \propto t^{-\nu} \tag{10.2.12}$$

where

$$\Gamma(r) \propto r^{2-\eta-d} e^{-r/\xi}, \qquad r \to \infty \tag{10.2.13}$$

2. For $T = T_c$, h small

$$M \propto h^{1/\delta} \tag{10.2.14}$$

Analogous definitions hold for the liquid–gas transition (where, e.g., compressibility plays the role of χ above) and for percolation[14–16] [where a quantity $P(p)$ replaces M and p replaces T, etc.; see Section 10.2.7]. Some other transitions, e.g., thermal transitions at zero temperature, or the Kosterlitz–Thouless transition[17,18] require different variables and/or different characterizations.

10.2.5. Resumé of Basic Points

We now list the main points which have occurred so far:

1. A phase transition is a cooperative effect caused by interactions or statistics.
2. The transition marks the onset of spontaneous ordering, characterized by an order parameter which is nonzero in the ordered phase.
3. As temperature increases, the order disappears when entropy overcomes the energy gained by ordering.
4. Transitions, and cooperative effects in general, are less easy in low dimensions.
5. Fluctuations are large at (continuous) transitions.
6. Singular ("critical") behavior occurs at continuous transitions, and is characterized by critical exponents and driven by long-range correlations (divergent correlation length ξ).

The next section gives an example of a transition caused by statistics alone (Bose–Einstein condensation) and Section 10.2.7 gives an example of a nonthermal (geometric) transition, percolation, for which the terminology of (3) above is inappropriate.

10.2.6. Bose–Einstein Condensation

As is well known, a free Bose gas (no interactions) can undergo the phase transition known as Bose–Einstein condensation, in which a macroscopic number of particles condense into the lowest single particle energy state (of zero wave vector k).[19,20] The chemical potential μ, which occurs in the occupation number n_k, is determined by the total number N of particles through

$$N = \sum n_k \to (2s+1)(2\pi)^{-3} V \int d^3k \{\exp[\beta(\hbar^2 k^2/2m - \mu)] - 1\}^{-1} \tag{10.2.15}$$

The replacement of the sum by an integral, normally appropriate to a large system because of the negligible spacing of k-values, can fail at small k if $\mu \to 0$. Because n_k is positive for all k, including $k = 0$, μ is negative, and so the integral in expression (10.2.15) is less than or equal to $V(2mKT/\hbar^2)^{3/2}\gamma$, where γ is a dimensionless constant, and this upper limit is approached as

$\mu \to 0$ from below. Thus, as T is reduced to T_c where

$$(N/V) = \gamma(2mKT_c/\hbar^2)^{3/2} \tag{10.2.16}$$

μ must tend to zero and the ground-state occupation $N_0 = [\exp(-\beta\mu) - 1]^{-1}$ becomes macroscopic (proportional to N), actually being the difference between N and the maximum value (i.e., the $\mu = 0$ value) of the integral.

This transition is caused by statistics alone. It is characterized by order parameter N_0/N. It occurs for dimensionality d greater than 2. For $d \leq 2$, the integral in expression (10.2.15) diverges as $\mu \to 0$ and N_0/N remains zero in the thermodynamic limit (no transition). The λ-transition in He^4 is similar to Bose-Einstein condensation, except that the interactions here modify the critical behavior,[21] and the superfluid properties require more sophisticated considerations.[19]

10.2.7. Bond Percolation[14–16,22,23]

Suppose that bonds are removed from a square lattice until a concentration p remains. For p greater than a threshold value p_c the system is made up of finite clusters of bonds and a large cluster which, since it spans the lattice even in the infinite system limit, is called the infinite cluster (Figure 10.1a). As p is reduced below p_c, the infinite cluster disappears and only finite clusters remain (Figure 10.1b). p_c is a critical concentration at which the "percolation transition" occurs, in which the order parameter is the probability $P(p)$ of an arbitrarily chosen bond being part of the infinite cluster. The transition is continuous since $P(p)$ goes continuously to zero as $p \to p_c$ from above. Site percolation is very similar, except that dilution is by removal of a site and all the incident bonds, and p is then the site concentration. In the percolation

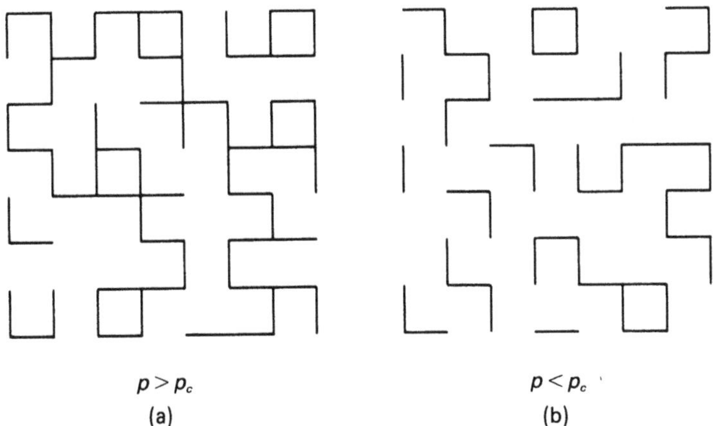

Figure 10.1. Bond percolation on the square lattice: (a) infinite and finite clusters ($p > p_c$), (b) only finite clusters ($p < p_c$).

transition p plays a role analogous to that of T in thermal transitions. These geometric transitions are important in the description of dilute magnets[24] and other dilute systems[25,26] and are very useful because of their "simplicity" (despite which we still await an exact solution for $d > 1$) and for illustration of geometric concepts which have very wide application (see Section 10.3). Bond percolation is actually a special case ($q = 1$) of q-state Potts models,[27,28] while the Ising model is the case $q = 2$.

10.3. Geometric Viewpoints

Geometrical viewpoints are useful in phase transitions in connection with the nature of ordered states, e.g., the geometry of spin configurations in an Ising system. Such visualizations can indicate the wide variety of possible phases or order parameters to allow for (see, for example, Section 10.3.1), or can clarify the concept of scale-invariance (self-similarity) which is crucial for critical phenomena at continuous phase transitions (Sections 10.3.2 and 10.7). They provide more information, albeit qualitatively, than is contained in a pair correlation function, for example, indicating in some examples where crossover occurs (Section 10.8) the role of additional lengths competing with the correlation length.

For such discussions, and indeed for most of this chapter, we shall use the Ising system and percolation as illustrative examples because they are relatively easy to visualize, while still processing a rich variety of possibilities.

10.3.1. Ground State and Other Configurations in Ising Systems

Consider the spin-$\frac{1}{2}$ Ising model, with Hamiltonian as given in equation (10.2.1), first for the case of nearest-neighbor couplings on a square lattice.

1. For the case where the nearest-neighbor exchange constant J is positive, the ground-state spin configuration (Figure 10.2a) is all spins up (or all spins down: such reverse spin and other obvious symmetry-related possibilities are not mentioned hereafter). This is the ferromagnetic ground state.
2. For J negative the system takes up the usual two-sublattice antiferromagnetic ground-state configuration shown in Figure 10.2b in which each spin has the opposite orientation to all its neighbors.
3. In the case of a triangular lattice, the antiferromagnetic case J negative is more difficult, since the lattice cannot be divided into two sublattices such that a spin on one sublattice has all its neighbors on the other. This results in a highly degenerate ground state[2] and the concept of frustration,[29,30] crucial in spin-glass models,[26,31] illustrated by the triangle of nearest-neighbor spins in Figure 10.2c, in which it is impossible to have all pairs antiparallel. We consider next a spin-$\frac{1}{2}$

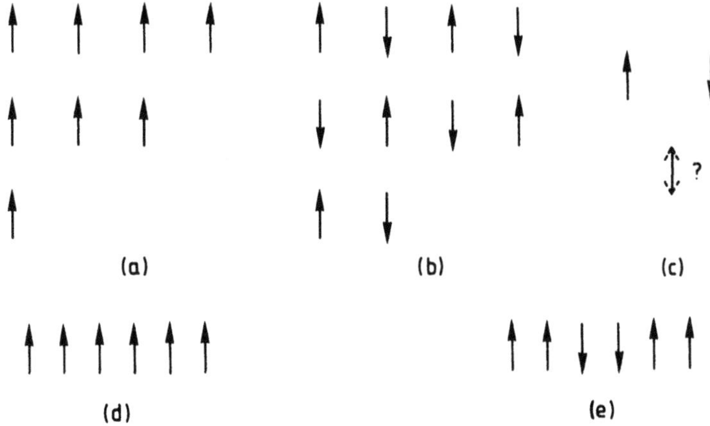

Figure 10.2. Ground-state configurations of spin-$\frac{1}{2}$ Ising systems: (a) ferromagnetic, (b) antiferromagnetic, (c) frustrated, (d) chain with $J > 2|J'|$, (e) chain with $J < 2|J'|$ (with notation of Section 3.1).

chain with positive nearest-neighbor interaction J and negative next-nearest-neighbor interaction J' (a one-dimensional ANNNI model.)[32]

4. In this model, for the case $J > 2|J'|$ the ground state is the ferromagnetic configuration shown in Figure 10.2.d, which clearly has the least energy.
5. For the case $J < 2|J'|$ the model has the ground state shown in Figure 10.2e, which has a repeat length of 4 lattice spacings. This state makes the best of the interaction J' while still gaining some exchange energy from the interaction J (every next-nearest-neighbor pair is antiparallel while every other nearest-neighbor pair is parallel). In the limit $T \to 0$ with $(J - 2|J'|)/T$ held fixed at value γ, say, the limiting state depends on the value of γ, and can have a very long repeat length. Such considerations are used in current theories of polytypism.[33]

A somewhat similar situation can occur in the nearest-neighbor Ising model near its surface: using the label n, increasing from the surface, to label successive layers, the average magnetization M_n in the nth layer satisfies, in a mean field theory of the type discussed in Section 10.4.1, a nonlinear second-order discrete recursion equation equivalent to an area-preserving map.[34] This can give rise to extremely rich possibilities for the layering configurations of the system.

10.3.2. Configurations at a Phase Transition; Scale Invariance

Figure 10.3 shows the spin configuration at the transition in a two-dimensional Ising model. The white and black areas are the spin-up and

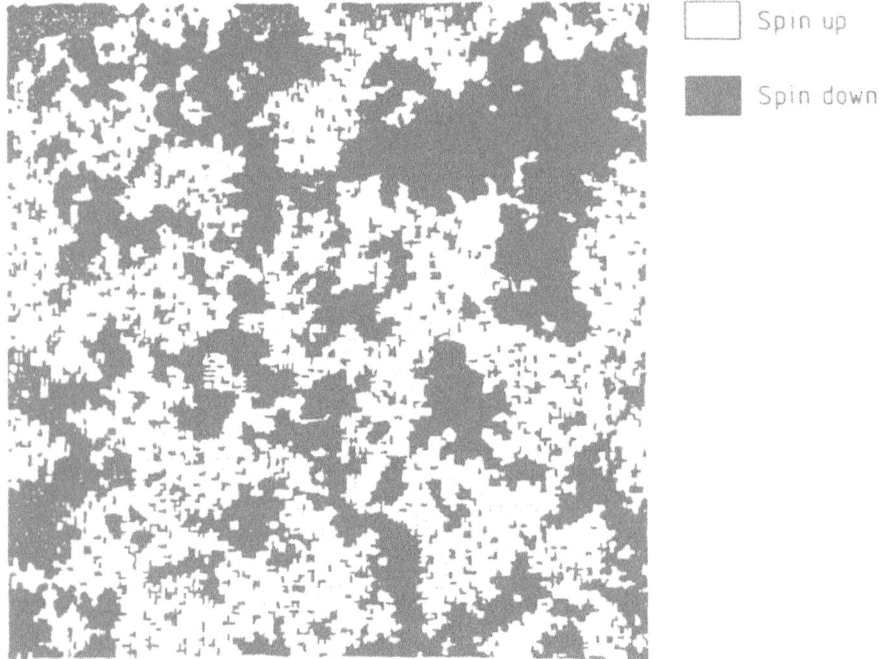

Figure 10.3. Pattern of spin configurations in spin-$\frac{1}{2}$ Ising model at phase transition.

spin-down regions respectively. The difference of the two areas divided by the total area is the specific magnetization which, since we are at the transition, is only nonzero because of the finite size of this system. Its dependence on system size will be exploited later, and is related to a fractal[35] or anomalous dimension[36] (Section 10.7). Slightly easier to interpret is the pattern shown in Figure 10.4, which is the infinite cluster backbone,[25] in bond percolation, at p_c (i.e., the infinite cluster with dangling ends discarded). The scale of the holes in the figure is roughly the percolation correlation length ξ. The appearance of holes in the figure up to the system size is a manifestation of the fact that here, at the transition, the correlation length is divergent. The lack of a finite value of ξ to set the scale of the pattern means that the system is scale-invariant at the transition, where ξ diverges. This scale invariance, or self-similarity, can be exhibited by magnifying Figure 10.4 and noting that, within the scaling "window" limited by the system size and the lattice spacing, the hole pattern is statistically indistinguishable from that in the unmagnified figure. Another geometric aspect of Figure 10.4 is its ramified nature, which is due to the fact that the infinite cluster is just appearing. The ramified nature and the appearance of arbitrarily large holes are related, and they are related further to the fact that in the figure the infinite cluster does not appear to be space-filling. This is in fact so: the scale-invariance of the incipient infinite

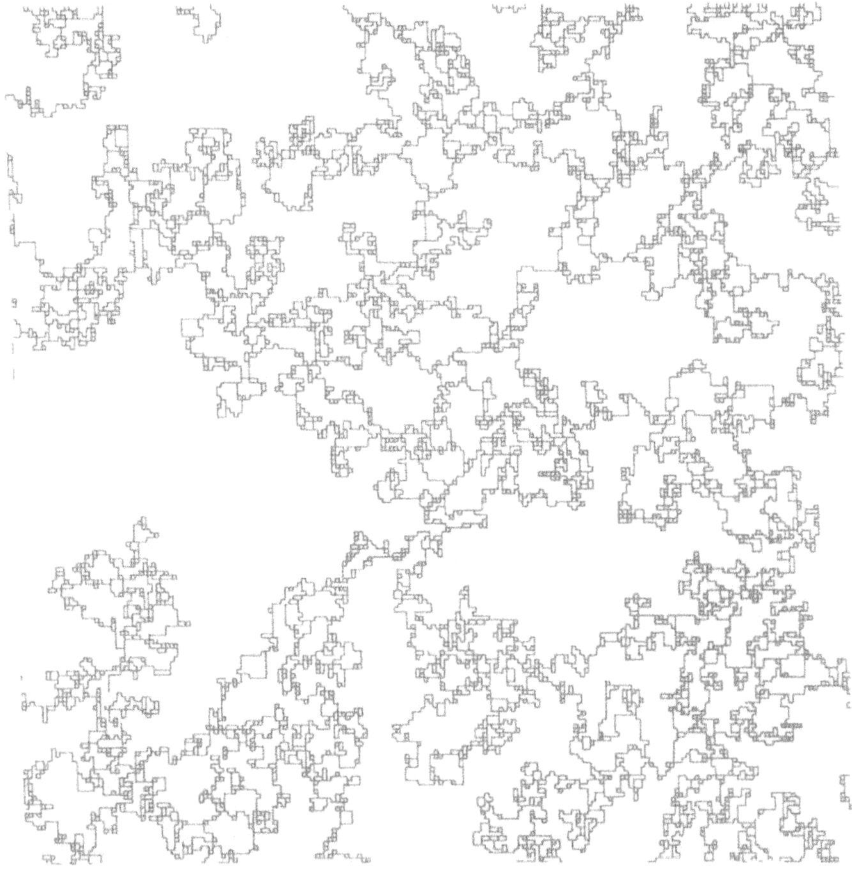

Figure 10.4. Bond percolation infinite cluster backbone at p_c (after Kirkpatrick[25]).

cluster at p_c allows it to have a dimensionality intermediate between 1 and 2. This anomalous or fractal[35] dimension (see also Chapter 9) is associated[36] with critical exponents in the way described in Section 10.7.

10.4. Investigation of Phase Transitions; Mean Field and Landau Theory, Fluctuations

We list below the main approaches to phase transition behavior.

1. *Experiment.* Bulk measurements of susceptibility, compressibility, magnetization, specific heat, etc. and determinations of phase diagrams, critical parameters (T_c, etc.). Scattering (particularly neutron scattering) measurements of correlation functions.

2. *Exact theories* of, for example, the Ising model[1] in $d = 1$, and $d = 2$ ($h = 0$), by transfer matrix[37] and other[38] approaches; or exact relationships (mappings) between models[28]; or exact determinations of particular quantities, e.g., T_c, p_c by duality for the square lattice. By way of illustration, the duality argument for p_c in the bond diluted lattice[39] is as follows: from the bond configuration of the original lattice at a specific p construct the corresponding configuration of the dual lattice by placing its vertices at the center of each square cell of the original lattice, and joining adjacent vertices by a (dual) bond if (and only if) the bond of the original lattice which the dual bond would cross is absent. The dual lattice thus has concentration $p_D = 1 - p$. The criticality at p_c in the original lattice thus occurs at $1 - p_c$ in the dual lattice, which is again a bond-diluted square lattice. Thus $p_c = 1 - p_c$ and so $p_c = \frac{1}{2}$.
3. *Approximate theories.* These include mean field[40] and Landau[19] theories and their generalizations (Sections 10.4.1–10.4.5), scaling[41] and renormalization group approaches[42–50,36] (Sections 10.9–10.11), series methods,[51] etc.
4. *Computational approaches,* such as simulation via Monte Carlo techniques,[52] and numerically exact calculations on systems of large but finite size N, extrapolated to $N \to \infty$ via finite-size scaling techniques,[53] etc.

Many of the techniques [such as neutron scattering in (1), scaling in (3) (see Section 10.11), Monte Carlo in (4)] can yield information about dynamic properties as well as statics, but much more is known about the latter.

Though we shall quote results from all types of approach we have space only to develop a few (mostly approximate) theories, selected because of their general applicability and the insight and emphasis they give to central concepts. We begin with mean field and Landau theories, which work very well in many circumstances. After having seen that their main limitation is a neglect of fluctuation effects, resulting in an incorrect description of critical effects in low dimensions, we move on, for the rest of the chapter, to critical phenomena and the class of approximate theories (scaling and renormalization) able to treat those phenomena.

10.4.1. Mean Field Theory[40]

Suppose the system is described by Hamiltonian $H\{\phi_i\}$, involving microscopic variables ϕ_i [e.g., the spins σ_i, at lattice sites i, of the Ising model described by equation (10.2.1)]. In the simplest type of mean field theory, each ϕ_i is replaced in the Hamiltonian by a common self-consistently determined mean value to obtain an effective internal energy $H\{\phi\}$. The free energy is then obtained by adding the appropriate entropy term. For instance, in the case of the spin-$\frac{1}{2}$ Ising model, the mean value ϕ is the average magnetic moment, and it implies that, of the N spins, on average $(1+\phi)N/2$ are up and $(1-\phi)N/2$ are down, and the resulting entropy is then

$$K \ln \{N!/\{[(1+\phi)N/2]![(1-\phi)N/2]!\}\}$$

$$\sim -K[(1+\phi)\ln(1+\phi)+(1-\phi)\ln(1-\phi)]$$

(using Stirling's approximation). Minimization of the resulting free energy with respect to ϕ then yields the self-consistent ϕ:

$$\phi = \tanh J\phi/KT \tag{10.4.1}$$

The procedure is equivalent to the usual replacement of the Hamiltonian by that for noninteracting spins in the effective field $\partial H\{\phi\}/\partial\phi = 2J\phi$. Alternatively, the same result can be arrived at by using the Bogoliubov inequality with a specific class of variational Hamiltonians (in this case a one-particle Hamiltonian). This approach suggests obvious generalizations (such as including pair effects in the effective Hamiltonian). Even at the level of single-particle effective Hamiltonians, generalizations are necessary to allow for different (equal and opposite) quantities ϕ on the two sublattices of an antiferromagnet, or for effects of nonuniformity, etc.

The main consequences of mean field theory are:

1. Phase transition for all dimensionality d.
2. The same critical behavior (characterized by common "mean field" exponents) for all systems. These exponents are the same as those yielded by Landau theory, and are given in the next section.

10.4.2. Landau Theory of Phase Transitions; Uniform Case; Bulk Properties, Exponents

Landau theory[19] is similar to mean-field theory but less specific to a particular system, and set up for the neighborhood of the transition. It relies on the fact that, for a continuous transition near T_c, the order parameter ϕ is small, so that an expansion in powers of ϕ is suggested; and that the equilibrium value of ϕ minimizes the free energy F.

For the case of a uniform magnetic system, provided fluctuations in ϕ are small and provided a power-series expansion of the free energy exists near the transition ($T \sim T_c$ small, i.e., ϕ small), it has to take the form

$$F = a_0(T) + \tfrac{1}{2}a_2(T)\phi^2 + \tfrac{1}{4}a_4(T)\phi^4 + \cdots - h\phi \tag{10.4.2}$$

because of the invariance under reversal of all spins when the (small) field h is zero. Many other systems "have" the same form of free-energy expansion, however for Potts models with $q \neq 2$, ϕ^3 and other higher odd powers can occur.[54] If expansion (10.4.2) is minimized with respect to ϕ,

$$0 = \partial F/\partial\phi = a_2 + a_4\phi^3 + \cdots - h \tag{10.4.3}$$

Thus, *for* $h = 0$, ϕ vanishes or, if $a_2/a_4 < 0$,

$$\phi = \pm(-a_2/a_4)^{1/2} \tag{10.4.4}$$

If a_4 is positive, the zero and nonzero solutions for ϕ are stable (i.e., minimize F) for a_2 positive and negative, respectively. Thus ϕ becomes nonzero, i.e., spontaneous order appears, where a_2 changes sign. In the region $T \approx T_c$ for which the theory is constructed, it is adequate to take $a = c(T - T_c)$ with c positive and $a_4\sim$ positive constant. These forms in expansion (10.4.2) imply the zero-field free energy shown in Figure 10.5, which shows how the equilibrium ϕ becomes nonzero as T decreases below T_c.

For $h \neq 0$, $T > T_c$, $\phi \sim h/a_2 = \chi h$. Thus

$$\chi = 1/[c(T - T_c)] \tag{10.4.5}$$

Also, at $T = T_c$, where $a_2 = 0$, $a_4\phi^3 \sim h$ so

$$\phi \propto h^{1/3} \tag{10.4.6}$$

Comparison of the results (10.4.4)-(10.4.6) with the critical exponent definitions (10.2.10), (10.2.11), and (10.2.14), respectively, shows that the Landau exponents are

$$\beta = \tfrac{1}{2}, \qquad \gamma = 1, \qquad \delta = 3 \tag{10.4.7}$$

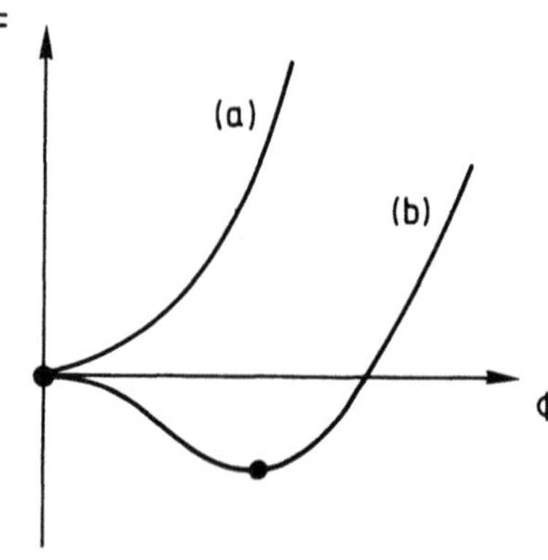

Figure 10.5. Zero-field free energy F vs. order parameter ϕ, in Landau Theory: (a) $T > T_c$, (b) $T < T_c$.

The remaining bulk exponent, defined in relation (10.2.9), is easily seen to be $\alpha = 0$ by the use of expressions (10.4.2) and (10.4.4). These exponents are the same as given by mean-field theory in the usual form described in Section 10.4.1. The Landau values of the correlation exponents ν and η are derived in the next section.

10.4.3. Landau Theory; Space Varying Case; Correlations

Suppose now that the small field h has a slow spatial variation. That produces spatial variations in ϕ which, since they are also slow, can be approximated by the additional term $-\frac{1}{2}(\nabla\phi)^2$ in the free energy (using reduced units for the spatial dependences). With the aid of Fourier transforms the free-energy expansion then becomes

$$F = a_0 + \sum_q \left(\tfrac{1}{2}q^2\phi_q^2 + \tfrac{1}{2}a_2\phi_q^2 + \sum_{q'} \tfrac{1}{4}a_4\phi_q^2\phi_{q'}^2 + \cdots - h_q\phi_q \right) \qquad (10.4.8)$$

where h_q, ϕ_q are the Fourier components of field and order parameters. Minimizing gives

$$0 = \partial F/\partial\phi_q = (q^2 + a_2)\phi_q + a_4\phi_q \sum_{q'} \phi_{q'}^2 + \cdots - h_q \qquad (10.4.9)$$

Thus, for $T > T_c$

$$\phi_q \sim h_q/(a_2 + q^2) \equiv \chi_q h_q, \qquad \chi_q = [c(T - T_c) + q^2]^{-1} \qquad (10.4.10)$$

The "generalized susceptibility" χ_q is the Fourier transform of the pair correlation function (10.2.5), (10.2.13) [as can easily be seen by generalizing the result (10.2.7) to the case of a field with a single Fourier component h_q]. Relation (10.4.10) is the Ornstein–Zernicke "mean field" form for the correlation function[40] and comparison with function (10.2.13) yields the Landau ("mean field") correlation exponents

$$\eta = 0, \qquad \nu = \tfrac{1}{2} \qquad (10.4.11)$$

10.4.4. Fluctuation Effects; Upper Critical Dimensionality

Consider the omissions from Landau theory as described above. The generalization of relation (10.4.10), allowing for the first higher-order term in equation (10.4.9), is

$$\phi_q = h_q \Big/ \left[a_2 + q^2 + a_4 \sum_{q'} \phi_{q'}^2 \right] \equiv \chi_q h_q \qquad (10.4.12)$$

The q-dependent sum rule obtained by generalizing equation (10.2.7), as referred to above, gives $\phi_q^2 = \chi_q$ which, using expression (10.4.12), can be written in the form

$$\phi_q^2 = \chi_q = 1/[(1/\chi) + q^2] \tag{10.4.13}$$

where the q-independent part of the denominator has been written in terms of $\chi = \chi_{q=0}$:

$$1/\chi = a_2 + a_4 \sum_{q'} \phi_{q'}^2 = a_2 + a_4 \sum_q 1/(1/\chi + q^2) \tag{10.4.14}$$

[the last step follows from relation (10.4.13)]. The mean-field results of the last section were obtained neglecting the second term on the right-hand side of equation (10.4.14), which seems to implicitly assume that $\sum_q [(1/\chi) + q^2]^{-1}$ is much less than the left-hand side, $1/\chi$. Actually, in the region of interest (near the transition) where $\chi \to \infty$, the neglected sum gives rise to a constant contribution, which merely shifts the T_c in a_2. The "dangerous" contribution which can arise from the neglected sum is the contribution, from the lower limit of the integral that may be used to replace the sum, which can still retain a dependence on temperature through χ even as $\chi \to \infty$. This contribution, for a d-dimensional system (i.e., a d-dimensional q-space integral), is essentially

$$[q^{d-2}]_{q^2 \sim 1/\chi} \propto (1/\chi)^{(d-2)/2} \tag{10.4.15}$$

This remains small compared to the left-hand side $1/\chi$ of equation (10.4.14) as $\chi \to \infty$ provided

$$d > 4 \equiv d_{\text{upper}} \tag{10.4.16}$$

Thus mean-field values of the exponents are correct, provided the space dimensionality d exceeds a so-called "upper critical dimensionality" d_{upper} equal to 4 for most systems (magnets, liquid gas, etc.). For Potts models with $q \neq 2$ the appearance[54] of a ϕ^3 term in the Landau expansion (10.4.2) makes $d_{\text{upper}} = 6$.

10.4.5. Critique of Mean-Field Theory

The mean-field theories are very valuable, since they show simply how phase transitions can occur (both continuous and first-order ones) and can allow for various types of order parameter. However, in order to use them with confidence, one has to know their limitations. These appear (see end of Section 10.4.1) in the following situations:

1. At low dimensionality ($d < d_{\text{lower}}$), where mean-field theory always gives a transition while the system actually does not order unless $d \geq d_{\text{lower}}$, where, for example, the "lower critical dimensionality" of the Ising model is $d_{\text{lower}} = 2$, from the argument of Section 10.2.1;

2. At intermediate dimensionality, $d_{\text{lower}} \leq d \leq d_{\text{upper}}$, where, though mean-field theory is correct in giving the transition, it gives it with the wrong ("mean-field," or Landau) exponents (Sections 10.4.2–10.4.4).

For $d > d_{\text{upper}}$ mean-field theory correctly gives the transition and the correct exponents.

There are, of course, more sophisticated mean-field approaches than those given in Sections 10.4.1–10.4.3. The "spherical model"[40] is a generalized type of mean-field Gaussian model which includes some of the fluctuation effects occurring in Section 10.4.4 and hence gives different exponents[10] than the Landau ones, and is without the deficiency of always giving order even in low dimensions. The same is true of the "random phase approximation," which is arrived at by a procedure like that in Section 10.4.1 except that the replacement of the variable ϕ_i by its average is used to linearize equations of motion rather than the Hamiltonian.[55,56]

10.5. Review of Systems

It is convenient now to list, for a variety of systems, the features referred to so far, and a few additional ones required for later reference. Table 1 gives, for representative systems, the order of the transition, the order parameter and the number n of its components, the lower and upper critical dimensionalities, and a few special comments.

The Ising, *XY*, and Heisenberg models, differing in the number n of components of the order parameter, are the fundamental spin models. Many other systems, such as the binary alloy, liquid-gas, and cooperative Jahn–Teller systems, have pseudospin representations[57,58] in which they are like a (possibly generalized) one-component (Ising) magnet, and have the same critical exponents as the Ising model. The superfluid[59,60] and superconducting transitions[61] are associated with two-component order parameters (the real and imaginary parts of the complex wave function of the $k = 0$ condensate or the pair condensate, respectively) and have the same critical exponents[10] as the *XY* model, which also has $n = 2$. These $n = 2$ systems exhibit a special transition, the Kosterlitz–Thouless transition,[17,18] in $d = 2$.

The q-state Potts models[27] with Hamiltonian $-\sum_{ij} J_{ij}\delta_{\mu i,\mu j})$, where μ_i are the q states of the microscopic variable at lattice site i, include the Ising and percolation (geometric) models as the special cases $q = 2$, $q = 1$.[28] For higher q, the transition can become first order[27] for high enough d.

The value of d_{upper} (4 or 6) depends on whether the system has the spin-reversal or equivalent symmetry, as suggested by the discussion of Sections 10.4.2 and 10.4.4. The lower critical dimensionality is less for $n = 1$, which corresponds to a discrete symmetry, because softer fluctuations are available[62] to destroy the ordered state[57] when the symmetry is continuous ($n \geq 2$).

Table 10.1. Principal Features of Phase Transitions in Representative Systems[a]

System	Order of transition	Order parameter	Number n of components of order parameter	d_{lower}	d_{upper}	Special comments
Ising magnet	2nd	$\sum_i\langle\sigma_i\rangle$	1	2	4	Fundamental spin model
XY magnet	2nd	$\sum_i\langle\sigma_i\rangle$	2	3	4	Fundamental spin model
Heisenberg magnet	2nd	$\sum_i\langle\sigma_i\rangle$	3	3	4	Fundamental spin model
Binary alloy	2nd	$\sum_i\langle\sigma_i\rangle$, ion sublattice	1	2	4	Pseudospin representation (spin up/down) of two-component alloy or lattice gas: same exponents as Ising (Binary alloy and Liquid–gas)
Liquid–gas	2nd	$\sum_i\langle\sigma_i\rangle$	1	2	4	
Melting	1st ($d \neq 2$)					Poorly understood, except in $d = 2$, $n = 2$ (Kosterlitz–Thouless transition)
Superfluidity	2nd	$\langle a_0\rangle = \sqrt{N_0}e^{i\phi}$	2	3	4	Same exponents as XY; Kosterlitz–Thouless transition for $d = 2$ (Superfluidity and Superconductivity)
Superconductivity	2nd	$\langle a_{-k\downarrow}a_{k\uparrow}\rangle$	2	3	4	
Cooperative Jahn–Teller	2nd		1	2	4	Pseudospin model representation
Percolation	2nd	$P(p)$	1	2	6	Geometric
q-state Potts model	2nd (1st for high q, d)		1	2	4 ($q = 2$); 6 ($q \neq 2$)	Includes Ising ($q = 2$), percolation ($q = 1$)

[a] Further comments are given in Section 10.5.

It will turn out that the dimensionalities of space (d) and of the order parameter (n) are the most important features in the discussion of critical behavior which now follows.

10.6. Critical Behavior: Exponents, Universality, Homogeneity

The critical behavior arising at a continuous phase transition has already been introduced (Sections 10.2.2 and 10.2.3), as well as its characterization by critical exponents (Section 10.2.4). It has also been stated, in the previous section, that some systems share the same critical exponents. We now develop this theme.

Universality is the lack of dependence of critical exponents (and other critical properties) on most details. The evidence for this is our knowledge of exponents coming from:

1. Exact solutions (e.g., the Onsager-Yang-Wu solution[37,63,64] of the $d = 2$ Ising model).
2. Series expansions[51] (e.g., in $1/T$ for $1/\chi$).
3. Experiment (e.g., in liquid-gas and magnetic systems).
4. Scaling and renormalization group analyses.[41-50,36,65-71]
5. Simulation.[52]

Exact solutions and series methods [(1) and (2)] indicated[10] that exponents:

a. Depend on lattice dimension d, but are otherwise independent of lattice type;
b. Depend on the number n of components of the order parameter (see Section 10.5) but are independent of spin magnitude, etc;
c. Are independent of interaction strength and range (if finite), etc.;
d. Satisfy certain "exponent relationships," which were predicted on the basis of the homogeneity hypothesis[72,73] introduced in Section 10.6.1.

Experiment (3) indicated that certain classes of fluids "behave like" (share the same exponents as) certain classes of magnets, etc. (see Section 10.5).

All these conclusions have been further reinforced by the results of (5) (simulation).

Scaling and renormalization approaches (4) have given analytic descriptions explaining why n and d are important in determining exponents and why other details are irrelevant (because they scale away under the governing scaling transformation) and that the homogeneity hypothesis is a consequence of length scaling. At the same time, the scaling methods have shown, as also in some cases indicated by other approaches, that "universality" applies also to amplitude ratios, scaling functions (see Section 10.8), distributions,[74] and so on.

It was also recognized early on that dynamic critical properties show universal behavior, but that the static universality classes may be split for dynamic critical properties[75] (i.e., that d and n do not alone determine the dynamic critical exponents, etc.; in particular, conservation laws play a distinguishing role). It is also known that universality can break down in certain circumstances, both in statics (e.g., Baxter models[76]) or dynamics (e.g., dilute Ising dynamics[77]; see Section 10.12).

10.6.1. Homogeneity Hypothesis, Exponent Relationships, and "Dynamic Scaling"

The homogeneity hypothesis was the assumption that the critical part of the correlation function (and hence the free energy) is a homogeneous function of the basic variables $t \equiv T - T_c$, h, r, or wave vector:[72,73]

$$\langle \sigma_0 \sigma_r \rangle - \langle \sigma_0 \rangle \langle \sigma_r \rangle \equiv \Gamma(r, h, t) = b^{2y} \Gamma(r/b, b^{y_1} h, b^{y_2} t) \qquad (10.6.1)$$

for any b, when r is large and h, t small.

This statement implies the following exponent relationships (see Section 10.6.2):

$$\alpha + 2\beta + \gamma = 2 \qquad (10.6.2)$$

$$\gamma = \beta(\delta - 1) \qquad (10.6.3)$$

$$\gamma = (2 - \eta)\nu \qquad (10.6.4)$$

$$\alpha = 2 - d\nu \qquad (\text{"hyperscaling"}) \qquad (10.6.5)$$

These are satisfied by the exact exponents for the two-dimensional Ising model; and, except for the hyperscaling relation (10.6.5), by the mean-field exponents [which are independent of d and only satisfy expression (10.6.5) at $d = 4$]. They are normally verified within the errors by series, experiment, and simulation, and can be confirmed by renormalization group analyses.

The four exponent relationships (10.6.2)-(10.6.5) imply that, of the six exponents α, β, γ, δ, ν, η, only two [a basic magnetic one, and a basic thermal one; e.g., y_1, y_2 in equation (10.6.1)] are independent. An interesting new development, for which we do not have space to expand on or exploit here, is that, because of conformal invariance, the two basic exponents are themselves related, so only one exponent determines all the rest.[78]

A form of homogeneity hypothesis, called "dynamic scaling," has also been given for critical dynamics,[79] in which the frequency (or time) and wave-vector dependent correlation functions are assumed to be homogeneous functions of h, t, frequency (or time), and wave vector k. A consequence (which can again break down in exceptional circumstances, see Section (10.12) is that the characteristic time or frequency of the critical dynamic process is a function of h, t, k through the combination $k\xi$, where ξ is the correlation length.

10.6.2. Derivation of Exponent Relations from Homogeneity Hypothesis

The static scaling relations (10.6.2)-(10.6.5) can be derived from equation (10.6.1) by the following types of reasoning.[50]

1. Taking first the case $h = 0$, put $b = r$ into equation (10.6.1) to obtain

$$\Gamma(r, 0, t) = r^{2y}\Gamma(1, 0, r^{y_2}t) \tag{10.6.6}$$

Comparison of the right-hand side with relation (10.2.13), with ξ given by relation (10.2.12), gives

$$2y = 2 - \eta - d, \qquad 1/y_2 = \nu \tag{10.6.7}$$

2. Using the sum rule (10.2.6) and the hypothesis (10.6.1),

$$\chi(h, t) = \beta \int d^d r\Gamma(r, h, t) = \beta \int d^d r b^{2y}\Gamma(r/b, b^{y_1}h, b^{y_2}t)$$

$$= b^{2y+d}\chi(b^{y_1}h, b^{y_2}t) \tag{10.6.8}$$

where the last step involves the change of variable $r \to r/b$ in the integral. Putting $h = 0$, $b^{y_2} = 1/t$ in equation (10.6.8) then gives

$$\chi(0, t) = t^{-(2y+d)/y_2}\chi(0, 1) \tag{10.6.9}$$

Comparison with relation (10.2.11) then yields

$$\gamma = (2y + d)/y_2 = (2 - \eta)\nu \tag{10.6.10}$$

where the last step is the replacement of $2y + d$ and y_2 using expressions (10.6.7). This completes the derivation of the exponent relation (10.6.4). The other relations are derived in a similar manner via $M = \int \chi dh$, $F = \int M dh$. These allow the derivation in turn from equation (10.6.8) of homogeneous scaling forms for M, F, of a type similar to equation (10.6.8) but with the b^{2y+d} prefactor replaced in turn by b^{2y+d-y_1}, $b^{2y+d-2y_1}$. The replacement $h = 0$, $b^{y_2} = 1/t$ in M, F then yields β, α in terms of y, y_1, y_2 and the replacement

$t = 0$, $b^{y_1} = 1/h$ in M yields δ. The elimination of y, y_1, y_2 then completes the derivation of the relations (10.6.2)–(10.6.4). The hyperscaling relation (10.6.5) is on a somewhat different footing from the others, and depends on an additional argument regarding "dimensionality" which can be given in several equivalent forms: one is to identify y in equation (10.6.1) with the length-scaling[41] or fractal dimension[35] of the spin variable σ (see Section 10.7.3), and $-y_1$ with that of the field h, and then use the fact that h and σ are conjugate variables, which makes $y - y_1 = 0$, from which hyperscaling then follows; alternatively (and more simply) one can identify the exponent in the prefactor $b^{2y+d-2y_1}$ in the free energy with its dimensionality, namely d, again making $y = y_1$, and yielding hyperscaling in the form (10.6.5). This latter argument is equivalent to a more intuitive discussion given in the next section.

10.7. Self-Similarity, Controlling Length, Anomalous and Fractal Dimensions

The related phenomena of diverging correlation length ξ and scale invariance were introduced in Sections 10.2.3 and 10.3.2. Scale invariance, or self-similarity, occurs when the characteristic length $\xi \to \infty$, for then the scale it sets (to correlation patterns) goes out of the problem and the configuration becomes invariant under a change of scale. It was indicated (Section 10.2.3) how the divergence of ξ can cause the susceptibility to diverge. In this sense ξ is a controlling length for criticality: its divergence causes the critical effects. This point is further illustrated in the next subsection, which also gives an intuitive interpretation of homogeneity and hyperscaling. Later subsections take the interpretation further and discuss the anomalous (or fractal) dimensions that can arise because of the scale invariance.

10.7.1. "Controlling Length" Interpretation of Homogeneity, Hyperscaling, and Criticality

We consider how the divergence of correlation length at the critical point causes the free energy F to become singular. The quantity $F/(KT)$ is extensive, therefore proportional to the system volume V. It is, however, also dimensionless and therefore involves V divided by some length to the power d ($d =$ dimensionality). That length could be the lattice constant a; however, that will not give rise to any singular behavior. An alternative is the correlation length ξ which, because of its singular dependence, (10.2.12), on temperature, will give rise to a singular part, F_s, of F:

$$F_s/(KT) \propto V/\xi^d \propto t^{d\nu} \tag{10.7.1}$$

The last step uses relation (10.2.12). Two derivatives with respect to temperature then produce the most singular part of the specific heat and hence a value for the specific heat exponent α defined by relation (10.2.9):

$$\alpha = 2 - d\nu \tag{10.7.2}$$

This is a nonrigorous derivation of the hyperscaling relation (compare Section 10.6.2).

The "derivation" again shows the role of ξ as a controlling length for criticality. It also shows that a dimensionality appears (as the power to which ξ is raised in the singular part of F). This dimensionality is the length scaling dimension (Section 10.7.2). The dimension for F is "obviously" the space dimension d; actually, it need not have been as simple as that, since we were discussing the singular part F_s of F which could have involved an anomalous dimension of the type referred to in Section 10.3.2, which we now discuss in detail for magnetization and the percolation order parameter.

10.7.2. Anomalous and Fractal[35] Dimensions, and Exponents[36]

Consider again Figure 10.3, the two-dimensional Ising spin pattern at T_c. The total magnetization, nonzero because of the finite size, is the net spin projection, i.e., the difference Δ of white and black areas. Though each of the white and black areas is space filling, their difference is not. It has a dimension less than 2. Such dimensions are defined as follows[35]: draw a square of side length L and work out $\Delta(L)$ for a succession of such squares. Then, if

$$\Delta(L) \propto \mathrm{L}^{d_f} \tag{10.7.3}$$

d_f is the length scaling or "fractal" dimension of Δ. Such dimensions are only anomalous (noninteger) in situations of scale-invariance: if ξ is finite, e.g., below T_c where a spontaneous magnetization has developed making $\Delta(L) \propto L^d$, the behavior (10.7.3) "crosses over" for $L > \xi$ to a simple length scaling (with dimension d in the example just given).

Though crossover will be discussed in detail in Section 10.8, we now use this type of argument to relate the fractal dimension d_f of the magnetization to critical exponents.[36] Suppose T is just below T_c so that ξ is finite but large. Then in a large enough system, i.e., for $L \gg \xi$, the reduced magnetization is

$$M \equiv \Delta(L)/L^d \propto t^\beta \propto \xi^{-\beta/\nu} \tag{10.7.4}$$

using relations (10.2.10) and (10.2.12). This result [$\Delta(L) \propto L^d \xi^{-\beta/\nu}$ for $L \gg \xi$] has to go over to relation (10.7.3) for $L \ll \xi$. A common crossover form, of the type discussed in Section 10.8, linking the two asymptotic regimes $L \gg \xi$, $L \ll \xi$ through a function of the dimensionless ratio L/ξ, is

$$\Delta(L) = L^{d_f} f(L/\xi) \tag{10.7.5}$$

which gives the two asymptotic results if $f(\infty)$ = constant and $f(x) \sim x^{d-d_f}$ for x small, provided

$$d_f = d - \beta/\nu \tag{10.7.6}$$

Expression (10.7.6) relates the magnetization fractal dimension to critical exponents. This anomalous dimension first occurred in Kadanoff's block scaling description,[41] and in field theoretic renormalization procedures. In Kadanoff's scaling discussion, the homogeneity hypothesis (10.6.1) is justified with the interpretation of the quantity b as a length scale factor, and the same interpretation allows the following alternative derivation of the result (10.7.6) using homogeneity. (Compare the discussion at the end of Section 10.6.) At the critical point, equation (10.6.1) becomes

$$\Gamma(r, 0, 0,) = b^{2y}\Gamma(r/b, 0, 0) = \langle\sigma_0\sigma_r\rangle - \langle\sigma_0\rangle\langle\sigma_r\rangle \tag{10.7.7}$$

This shows that the fractal dimension associated with σ (through $\sigma \propto b^{d_\sigma}$) is

$$d_\sigma = y = -\beta/\nu \tag{10.7.8}$$

where the last step uses expression (10.6.7) and other scaling relations quoted and derived in Section 10.6. Allowing for the volume factor and converting σ to a total magnetization (so $d_f = d + d_\sigma$), relation (10.7.8) is equivalent to (10.7.6).

10.7.3. Percolation Order Parameter and Fractal Dimension of Percolation Infinite Cluster

The anomalous dimension associated with the percolation order parameter $P(p)$ can be discussed similarly. This gives the fractal dimension of the percolation infinite cluster.[25,36]

We refer to Figure 10.4 (or rather a figure of similar type but for the full infinite cluster) and follow the discussion in Section 10.7.2 down to relation (10.7.6), but using in place of $\Delta(L)$ the quantity $N(L)$ defined as the number of bonds on the infinite cluster within a square of side length L. The new fractal dimension is now defined by $N(L) \propto L^{d_f}$ when the percolation correlation length ξ is infinite ($p = p_c$). Then, replacing relation (10.7.4) for $L \gg \xi$ ($p > p_c$) by

$$P(p) \equiv N(L)/L^d \propto (p - p_c)^\beta \propto \xi^{-\beta/\nu} \tag{10.7.9}$$

a crossover argument like that given before to derive equation (10.7.6) gives

$$d_f = d - \beta/\nu \tag{10.7.10}$$

where β and ν are now percolation exponents. The definition of d_f makes it the actual fractal dimension of the incipient percolation infinite cluster. This object is a real (random) fractal, being statistically self-similar (at p_c).

10.8. Competition of Lengths; Crossover

The phenomenon of crossover[80] occurs when two or more large lengths compete to control critical behavior. Examples of this were seen already in Sections 10.7.2 and 10.7.3, where the controlling correlation length of the infinite system was competing with system size L. In those examples a function of the ratio of the two competing lengths occurred, multiplied by a power-law prefactor [see expression 10.7.5)]. Such "scaling" functions govern the crossover between the two asymptotic behaviors related to the dominance of either length and indeed relate them [which was how relation (10.7.6) was derived]. The competition of correlation length ξ and system size L is a common crossover which is exploited in finite-size scaling[53] to infer from the L dependence at the transition the critical exponents of the infinite system.

Another example already referred to involving the crossover caused by two competing lengths is the "dynamic scaling" description[79] referred to at the end of Section 10.6.1, where the two lengths are ξ and the wavelength. Examples of the above types of crossover are given in Section 10.8.2. A possibly more fundamental type of crossover, involving two correlation lengths both intrinsic to the system, is described in Section 10.8.1.

Crossover is normally universal in the sense that the scaling functions and associated crossover exponents (see Section 10.8.1) have the same universality (lack of dependence on most details) discussed in Section 10.6. Also the reduction, through the scaling function, of the dependence on, say, two variables to a dependence on one (the argument of the scaling function) is important in allowing reduced "universal plot" descriptions of crossover phenomena.

10.8.1. Crossover from Competition of Two Intrinsic Correlation Lengths

In the case of a Heisenberg model with a very weak uniaxial anisotropy δ, the critical behavior appears to be that of the isotropic Heisenberg model ($n = 3$) until very close to the critical point where it crosses over[80] to Ising behavior ($n = 1$). Renormalization group theory explains the reason for this crossover (the "relevance" of anisotropy); its phenomenological description requires however only the "competition of lengths" viewpoint, using the correlation length $\xi(t)$ of the isotropic Heisenberg model and a second length ξ' associated with anisotropy. The resulting overall correlation length, for example, can then be written as

$$\xi(t, \delta) = \xi f(\xi'/\xi) \qquad (10.8.1)$$

If converted to the basic thermal and anisotropy variables (Section 10.2.4) this becomes of the form

$$\xi(t, \delta) = t^{-\nu}\Phi(\delta^{\phi}/t) \tag{10.8.2}$$

where ν is the isotropic Heisenberg exponent, and ϕ is a new "crossover" exponent. A consequence is that the transition is no longer at $t = 0$ but is where $\xi(t, \delta)$ diverges because of a pole in Φ, i.e., where

$$t \propto \delta^{\phi}. \tag{10.8.3}$$

Other intrinsic crossovers, caused by two "relevant" variables, occur in dilute magnets[24] at the percolation threshold [where they are characterized by a thermal and a percolation correlation length, or alternatively a low-temperature thermal variable and $(p - p_c)$], and because of applied fields of various types.

10.8.2. Crossover and Scaling Forms Involving Extrinsic Lengths

Examples of crossover involving the competition of ξ and an extrinsic length are:

1. *ξ versus system size.* If a quantity Q (such as C, M, or χ) has a singularity of the form t^x in the infinite system, in a finite system of size L its behavior can be written [using relation (10.2.12) in a crossover formulation] in any of the forms

$$Q = t^x f(Lt^{\nu}) = \xi^{-x/\nu} f(L/\xi) = L^{-x/\nu} g(L/\xi) \tag{10.8.4}$$

This was used earlier in equation (10.7.5). Another case is the number n_s of finite clusters of s sites in site percolation, in the form[16]

$$n_s = s^{-\tau}\Phi(s/\xi^{d_1}) \tag{10.8.5}$$

where d_1 will here be the space dimensionality unless large clusters have a fractal character, in which case it becomes their fractal dimension.

2. *ξ versus separation.* Relation (10.2.13) is an example of a quantity (the correlation function) written in a scaling form. It should be noted that homogeneity results in a form (10.6.6), which is equivalent apart from the more general nature of its scaling function.

3. *ξ versus wavelength.* One example is given by the characteristic frequencies of modes of wave vector k in critical dynamics in the "dynamic scaling description:"[79]

$$\omega = k^z f(k\xi) \tag{10.8.6}$$

where z is here the dynamic exponent. Equation (10.8.6) describes the crossover between normal and anomalous dynamics as $k\xi$ varies from small to large values (e.g., in a diluted lattice as $p \to p_c$). The normal dynamics (ω proportional to a simple power of k) can be obtained from a continuum viewpoint, but that breaks down if $k\xi \gg 1$, where $\omega \propto k^z$.

A different sort of crossover is:

4. *Controlling length (other than ξ) versus size.* For example, an L step interacting walk with (controlling) persistence length L_0 (up to which the walk is essentially straight) has end-to-end distance R of form

$$R \propto L^{1/2} F(L/L_0) \tag{10.8.7}$$

crossing over between rigid rod ($R \propto L$) and random walk behavior ($R \propto L^{1/2}$) as L/L_0 increases. Quantity L_0 is entirely analogous to a correlation length in the role it plays here (indeed it is exactly the correlation length when the walk is reformulated as an equivalent spin model).

The emphasis, in Sections 10.7 and 10.8, on the role of lengths in critical behavior leads naturally to length scaling, which is now developed.

10.9. Length Scaling: Renormalization Group Transformation by Decimation or Blocking

Length scaling[41-43] is carried out by changing a reference length while maintaining a critical property on a gross scale. The usual procedure for lattice-based phase transitions is to change the lattice constant, $a \to ba$, maintaining ξ, the correlation length; b is the "dilatation factor."[41-50,36,65-71,81]

This process will be illustrated below with the simplest of all examples: percolation.[49,36,81] For that case relation (10.2.12) is replaced by

$$\xi/a = C|p - p_c|^{-\nu} \tag{10.9.1}$$

This shows that if ξ is held constant and a varied, the basic parameter p must in general change: $p \to p'$. Conversely, if this scaling of parameters is known for a specific dilatation, we can derive equation (10.9.1) and, in particular, the critical concentration p_c and exponent ν.

We now consider the effect of scaling the system recursively.

1. If $\xi = \infty$, the left-hand side of equation (10.9.1) is infinite and remains so under scaling, so p starts at p_c and remains there. There is thus no change in parameters in this special case, as would be expected from its scale invariance.

2. If ξ is large, we start with a difficult situation (criticality) in which the system involves all scales of lengths from a to ξ. Successive scalings $a \to ba \to b^2 a \to \cdots \to b^m a$, $p \to p' \to \cdots \to p^{(m)}$ can eventually result in a scaled lattice

constant $b^m a$ of order ξ in which case the situation is simple again, not having a large range of length scales. What has happened is that p scales away from p_c (toward the points $p = 0$ or $p = 1$) unless it starts exactly at p_c when, according to (1), it remains there. In other words, p_c is an unstable fixed point of the scaling transformation. Points $p = 0$ and $p = 1$ are also (attractive non-critical) fixed points of the scaling transformation, again requiring no change of parameter, because there $\xi = 0$.

Having outlined the basic approach and its interpretation we now show how this is implemented in the "decimation" approach.[47-49]

10.9.1. Decimation: Original and Scaled Systems Illustrated for Bond Percolation

Figure 6a, b shows typical configurations of a bond diluted square lattice for $p > p_c$ and $p < p_c$, respectively. Consider the application of a decimation process in which these original configurations are taken to scaled equivalents, Figure 6A, B, by removal of half the sites (those on a sublattice of the original

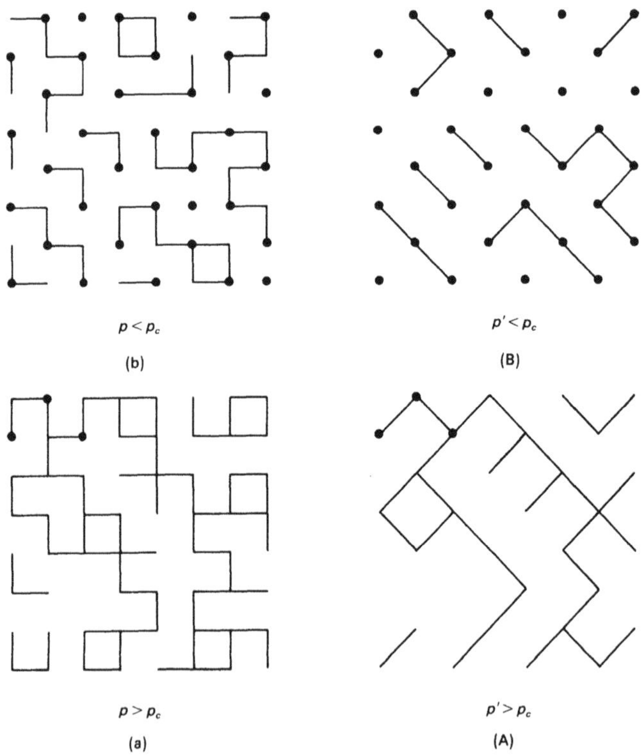

Figure 10.6. Original and decimated configurations (a, A) of a bond diluted lattice above the percolation threshold, and below (b, B).

system.[49,36,81] The remaining sites are joined by a bond ("renormalized bond present") whenever a path of bonds in the original square cell joins its retained diagonal sites. This approximate procedure results in the configurations in Figure 6A, B of the scaled system which closely model, on a gross scale, the configurations of the original system, in particular having (within the approximation) the same absolute correlation length, and lattice spacing larger by a factor $b = \sqrt{2}$. They correspond to scaled concentrations p' greater and less than p_c respectively (Figure 6A having an infinite cluster and Figure 6B only finite clusters). The resulting relationship between p' and p is derived and exploited in the next subsection.

10.9.2. Decimation: "Renormalization Group Transformation" and Extraction of Critical Condition, Exponents[81,36]

Figure 10.7 illustrates in more detail the procedure just introduced. In the original lattice, on the left of the figure, p is the bond concentration, i.e., the probability that nearest-neighbor sites are linked. Its transformed value, p', is thus the probability of linkage of nearest-neighbor sites, such as AB, in the scaled lattice (on the right of the figure), and (by our procedure for approximately preserving absolute correlations) is the same as the probability that A and B are connected via paths on the associated square cell $A1B2$ of the original lattice. Combining the probabilities of these linkage paths (from A to B via 1, or 2, or both 1 and 2) gives[49,36,81]

$$p' = 2p^2 - p^4 \equiv R(p) \tag{10.9.2}$$

This is the renormalization group (RG) transformation of the parameter p under dilatation by $b = \sqrt{2}$. It is exploited as follows:

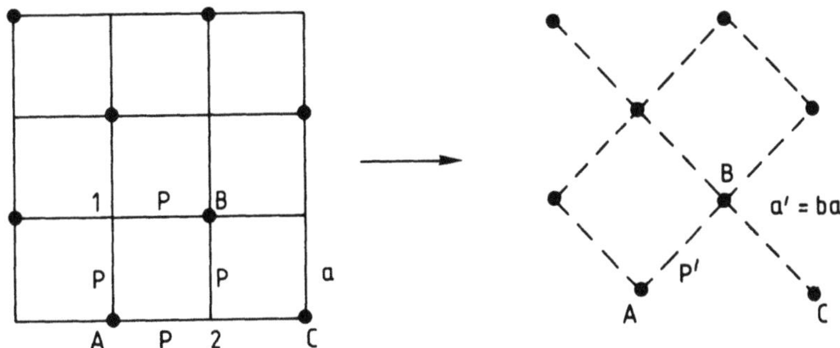

Figure 10.7. Decimation transformation of a square lattice (full lines) to a scaled one (dashed lines) larger by a factor $b = \sqrt{2}$ by removing sites 1, 2, etc.

1. *Scale-invariant situations; fixed points.* When the system is scale-invariant no change of parameters occurs, so p is a fixed point p^* of the transformation, such that

$$p^* = R(p^*) \tag{10.9.3}$$

The fixed points are $p^* = 0, 0.618, 1$. Of these, 0 and 1 are "attractive" fixed points in the sense described at the beginning of Section 10.9 and are hence associated with $\xi = 0$. The unstable (critical) fixed point associated with $\xi = \infty$ (transition) is $p_c = 0.618$.

2. *Near the transition: critical behavior.* Here equation (10.9.1) applies together with a similar equation for the scaled lattice in which p is replaced by $p' = R(p)$, ξ by ξ' $(=\xi)$, and a by $a' = ba$. The ratio of these two equations is therefore

$$b = \lambda^{\nu}, \qquad \lambda \equiv \left.\frac{dR}{dp}\right|_{p_c} \tag{10.9.4}$$

where we have used the fact that p (and therefore p') is close to p_c in order to replace $(p' - p_c)/(p - p_c)$ by the derivative, λ, which is the so-called "eigenvalue" of the (linearized) renormalization group transformation. Since $R(p)$ has been determined for a given b, we can easily calculate λ and hence, using relation (10.9.4), obtain the critical exponent ν.

Thus the eigenvalue gives the exponents and [as explained in (1)] the unstable fixed point gives the critical condition.

The use of the transformation is not limited to the neighborhood of the transition. It can also be used away from the transition, though here long-range effects are not completely dominant and so it loses some accuracy by its smoothing of short-range effects. The procedure, which is illustrated in detail for a special case in Section 10.10.2 and also in Section 10.11.2, is to use the transformation $p' = R(p)$ (without linearization about the fixed point) to obtain a functional equation for the quantity of interest, such as $\xi(p)$.

10.9.3. Decimation: Results and Further Discussion

The decimation transformation (10.9.2) for percolation on the square lattice yields [using (1) and (2) of the previous subsection]:

1. $p_c = 0.618$, while the exact result is $p_c = \frac{1}{2}$ by duality[39] (Section 10.4).
2. $\nu = 0.818$, while the exact result is, according to a conjecture of den Nijs, $\nu = \frac{4}{3}$.

The inaccuracies in these two quantities arise from the neglect of:

(a) Further neighbor connections which appear under decimation of the original lattice (e.g., a renormalized next-nearest-neighbor bond AC of the scaled lattice of Figure 10.7 arising from the bonds $A2$, $2C$ when the site 2 is decimated in the original lattice).

(b) Correlations which arise under decimation (e.g., between the bonds AB, BC of the renormalized lattice in Figure 10.7, because, for instance, bond $B2$ of the original lattice influences the presence of both).

(c) Any ingredient, in decimation, which can give rise to the proper fractal dimension d_f of Section 10.7.3. This omission becomes obvious when decimation is applied to spin models,[47,48] since the retained spins are then spins of the original lattice and therefore with the same magnitude, so the procedure effectively associates a zero value with d_σ [equation (10.7.8)]. The most severe consequences of this are on magnetic exponents: it does not greatly alter the thermal exponent ν. The "blocking" procedure[41,45,46,65,71,82] given in the next section allows for a rescaling of spin magnitude (or its equivalent in the case of percolation) and hence does not suffer from this deficiency.

The two omissions (a) and (b) [though not (c)] can be rectified by carrying out decimation with larger "clusters" than the square cell of Figure 10.7, allowing for the appearance and subsequent scaling of all parameters which can be identified on the cluster. This procedure gives, with quite small clusters [e.g., a $2a \times 2a$ (nine-site) square cluster on the square lattice, or a seven-site hexagonal cluster on the triangular lattice), results for ν accurate to about one percent.[49]

10.9.4. Blocking: Original and Scaled Systems, Renormalization Group Transformation (Example: Site Percolation[82,26,36])

Figure 10.8a shows a particular configuration of occupied (full circle) and unoccupied (open circle) sites for site percolation on the triangular lattice. The bonds drawn, joining only nearest-neighbor occupied sites, are to give an impression of the related connectivity. "Blocking" on this lattice[82,26,36] consists of grouping triangles of three sites (shown shaded in Figure 10.8a) into a single renormalized site. Such a renormalized site is taken to be occupied if a majority (two or three) of the sites of the associated block of the original system are occupied. That procedure results in Figure 10.8b, which is a configuration of a scaled triangular lattice (scale factor $b = \sqrt{3}$) equivalent to the original lattice in its connectivity properties on a gross scale (i.e., for separations of $a\sqrt{3}$ or more). The resulting occupation probability p' of a site of the renormalized lattice is

$$p' = p^3 + 3p^2(1 - p) \qquad (10.9.5)$$

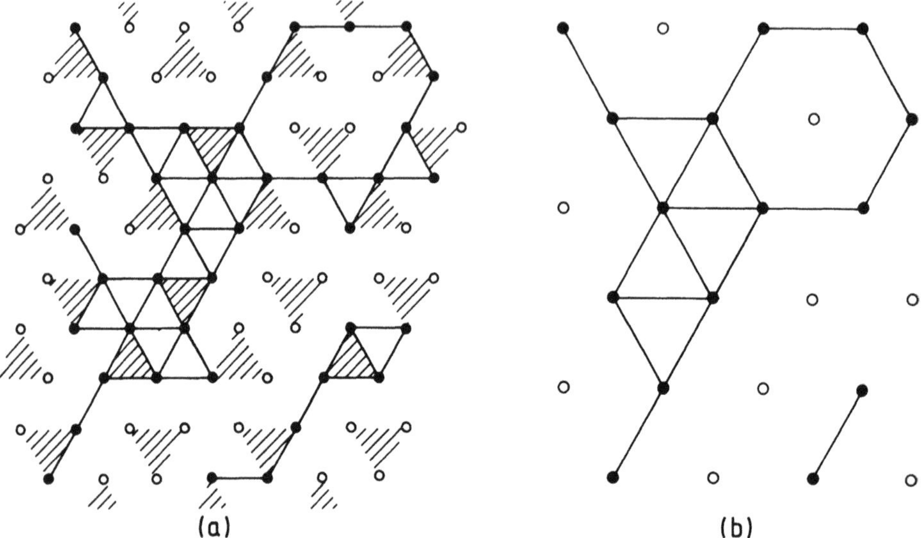

Figure 10.8. (a) Original and (b) scaled configurations in a blocking transformation for site percolation.

(the probability of occupation of two or more sites of the three-site block of the original lattice).

With use of the fixed point and eigenvalue procedures, given in Section 9.2, the scaling transformation (10.9.5) results in the following fortuitously good values for critical condition and exponent:

$$p_c = \tfrac{1}{2}\,(\text{exact}) \qquad \text{and} \qquad \nu = \ln b/\ln \lambda = \ln \sqrt{3}/\ln \tfrac{3}{2} = 1.35 \tag{10.9.6}$$

Blocking is carried out in an analogous way for the Ising model,[45,46,71] where, by the "majority rule," a block of spins is replaced by a renormalized spin which is up or down depending on whether the net spin projection of the block is positive or negative, and the probabilities are now Boltzmann probabilities. This corresponds to an explicit implementation of Kadanoff's original blocking idea[41] and, like it, allows for the scaling of spin magnitude omitted in decimation. The blocking method can become very accurate when the appearance under scaling of more parameters ("extension of parameter space") is allowed for by the use of larger blocks. This procedure is usually carried out using Monte Carlo sampling of configurations ("Monte Carlo renormalization").[83]

We have not treated field-theoretic renormalization group techniques. These are the most controlled of all and cannot be adequately described in the brief space available here. Instead we refer the reader to the original and review literature.[42–44,50,66–70]

10.10. Further Topics in Length Scaling: Flow, Universality; Transformation as Iterative Map; Fractals

The way in which the number of parameters increases under scaling was discussed in Sections 10.9.3 and 10.9.4. This proliferation of parameters leads to the use of enlarged parameter spaces which help in the understanding of universality, as we now discuss.

10.10.1. Extension of Parameter Spaces, Flow, Universality

For illustration, consider again the decimation description of bond percolation (Sections 10.9.2 and 10.9.3), but now allowing only for the appearance under scaling of a next-nearest-neighbor bond [as discussed in (a), Section 10.9.3], whose probability (concentration) we denote by q. The renormalization group transformation is then a transformation of two variables p, q and its effect is to move the point (p, q) in the two-dimensional parameter space to a new position (p', q'). Under further applications of the transformation the point moves along a trajectory or "flow line." Such flow lines are illustrated schematically in Figure 10.9, where the arrows give the direction of flow. The fixed points (no flow) are indicated by "stars," and whether they are attractive (points A or C) or unstable (B) is obvious from the direction of the arrows in their vicinity. The line DBE contains points which flow into the unstable fixed point B and therefore is the "critical line," whose intersection (D) with the "physical line" $q = 0$ corresponding to the physical system of interest (a

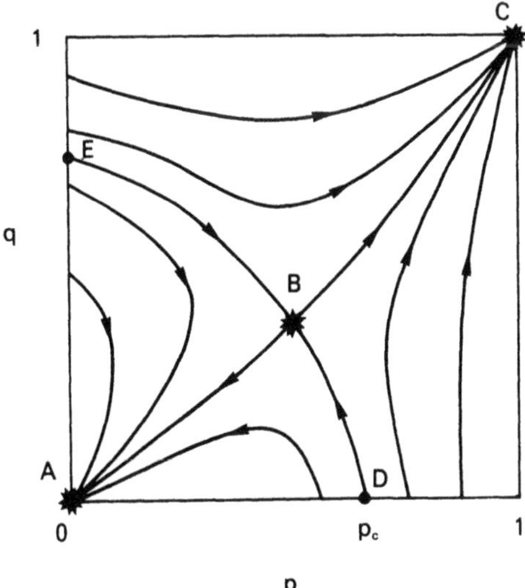

Figure 10.9. Renormalization group flow in a two-parameter space, where A and C are attractive and B unstable fixed points; the critical "surface" is EBD.

lattice with only nearest-neighbor bonds) gives the critical concentration p_c for that system. (Actually, in the present example and point E should also give this critical concentration, since a square lattice with only next-nearest-neighbor bonds is equivalent to two unconnected square lattices with only nearest-neighbor bonds.) The critical behavior is determined by the flow in the vicinity of the unstable fixed point B, in particular by the rate of outward flow of points near B along the line BC or BA. This rate is characterized by the larger of the two eigenvalues [cf. λ in relation (10.9.4)] of the 2×2 matrix resulting from linearizing the two-variable transformation about B. This eigenvalue is greater than 1. The other, less than 1, characterizes the inward flow at B. This inward flow describes the scaling away of "irrelevant" parameters which do not play a role in critical behavior.

This simple two-variable illustration is indicative of what happens in general, except that a full description normally needs very many parameters, of which all except one or two are irrelevant. The scaling away of the irrelevant variables is the explanation of universality.[44]

In the case of a simple thermal transition like the magnet, the relevant variables are, essentially, h and $t \equiv T - T_c$. The linearized renormalization group transformation then results in

$$t' \sim \lambda_T t, \qquad h' \sim \lambda_h h \tag{10.10.1}$$

(where λ_T, λ_h are both greater than 1) from which it follows [with relation (10.9.4)] that

$$\lambda_T = b^{1/\nu} \, (=b^{y_2}), \qquad \lambda_h = b^{\beta\delta/\nu} \, (=b^{y_1}) \tag{10.10.2}$$

which provides the length-scaling interpretation of homogeneity and allows the extraction of all static exponents via the scaling relations (Section 10.6.2). In the thermal case it is also possible to identify the free energy as a normalization parameter in the Boltzmann distribution and to follow its scaling.[84] This allows a check on hyperscaling, and the construction of the full free energy, not just its critical behavior.

10.10.2. One-Parameter Transformation; Simple, Cyclic, and Chaotic Scalings; Discrete Scale Invariance, Fractals

We return to the simpler descriptions involving the renormalization group transformation of just one parameter, such as relation (10.9.2) or (10.9.5). These transformations, in common with nearly all RG transformations of static variables, involve monotonic transformation functions $R(p)$ [or $R(T)$ in a thermal case] which result in a monotonic increase or decrease of p (from p_c) under scaling. From the beginning of Section 10.9 this can be seen to be related to the usually expected monotonic dependence of $\xi(p)$ on p for $p > p_c$ or

$p < p_c$. However, scaling procedures for dynamics (Section 10.11) often result in nonmonotonic transformation functions which can give rise to cyclic or chaotic behavior of the parameter under scaling. In the examples in Section 10.11 this is due to the scaling parameter (frequency) not being a monotonic function of a basic length.

We now illustrate the use of the transformation $R(p)$ in a nonlinearized form (see end of Section 10.9.2), as is required to discuss behavior away from the transition. This leads to the concept of discrete scale-invariance, which actually plays an important role in the behavior of fractals.

As an example, consider the extraction[36] of the full dependence

$$\xi / a = f(p) \tag{10.10.3}$$

of ξ on p for bond percolation in one dimension, starting from the exact $b = 2$ scaling transformation for that case, namely,

$$p' = R(p) \equiv p^2 \tag{10.10.4}$$

(from grouping two bonds, each with probability p, into a single renormalized bond of probability p'). Using relation (10.10.3) and a similar expression for the scaled lattice [where $\xi' = \xi$, $a' = 2a$, and p' is given by equation (10.10.4)] we obtain the following functional relation for $f(p)$:

$$f(p^2) = \tfrac{1}{2} f(p) \tag{10.10.5}$$

The general solution of this equation is

$$f(p) = \frac{A(p)}{\ln p} \tag{10.10.6}$$

where $A(p)$ is any periodic function of $\ln \ln p$ with period $\ln 2$. Now we could have carried out the decimation with any other integer value of the dilatation factor b, such as $b = 3$, which would have resulted in the same result (10.10.6), but with A having period $\ln b$. Since, e.g., $\ln 2$ and $\ln 3$ are incommensurate, A has to be a constant, which yields the exact result $\xi \propto 1/\ln p$. This argument is related to the group structure of the RG transformation, which normally requires

$$R_{bb'} = R_b R_{b'} \tag{10.10.7}$$

(where R_b denotes the transformation function for dilatation b), and is the reason for the power-law b-dependence of eigenvalues [see, e.g., relation (10.10.2)].

An immediate consequence of the above discussion is that if our choice of possible scale factors b had been more limited, we would not have been able to exclude periodic dependences of functions [e.g., of $A(p)$ on $\ln \ln p$].

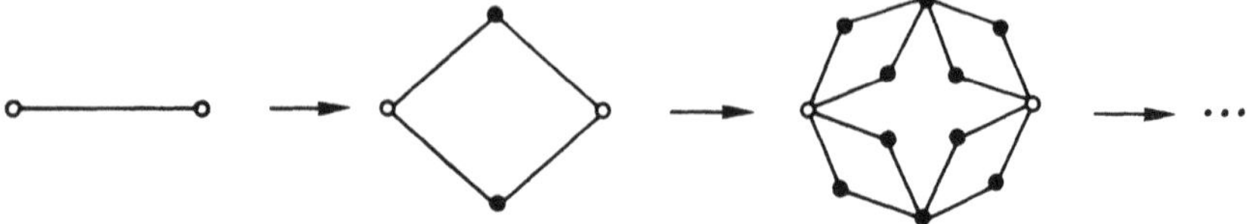

Figure 10.10. Recursive generation of a Berker lattice fractal.

This is exactly what occurs in many fractals, which only have a limited "discrete" scale invariance.

Soon after the introduction of the decimation procedure presented in Section 10.9.2, it was realized that this is exact on the structure generated by the recursion process illustrated in Figure 10.10. This object, the Berker lattice fractal,[85] is generated by forever replacing each bond by a diamond structure as shown in the first step. The decimation process of Section (10.9.2), which removes two sites from a square cluster to replace it by a single diagonal bond (or equivalent procedures in the Migdal–Kadanoff scaling method[86,87]) is the reverse of the fractal construction. Exact results for the fractal are therefore provided by the specific decimation procedure discussed ("square into diagonal") and no other decimations than that (or iterates of it) are applicable. This shows the special nature of the transformation though not of the scale factor, which is ambiguous for the Berker lattice. Another fractal showing discrete scale invariance is the Sierpinsky gasket,[35] generated recursively as illustrated in Figure 10.11. This construction process is the reverse of a decimation (removal of open-circle sites) taking a into $a' = ba$ with $b = 2$. This discrete scale invariance (limited to $b = 2$ or, more generally, 2^n with n any integer) allows periodic dependences in properties of the fractal and these do actually occur (see Section 10.11.4).

10.11. Length Scaling for Critical Dynamics of Chains and Fractals[77,81,88,89]

This section deals with the treatment of dynamical properties by generalizations of the "decimation" length-scaling methods introduced in Section 10.9

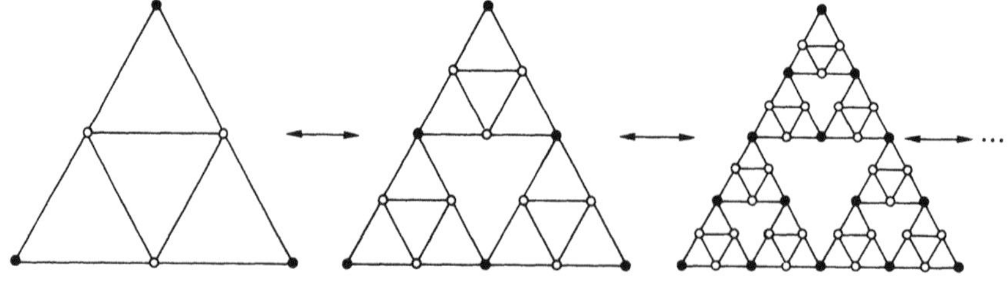

Figure 10.11. Construction or decimation of a Sierpinsky gasket fractal.

for static properties. In this section we consider only the simple linear dynamics which describes, for example, lattice vibrations, spin waves, and diffusion behavior. Even that dynamics can become "anomalous" or "critical" on regular fractals[88,90-92] (Sections 10.11.3 and 11.4) or statistically self-similar structures like the incipient percolation infinite cluster in, for example, dilute magnets at the percolation threshold[24,25,92-94] (Section 10.12). Scaling techniques are necessary to treat such situations.[77,81,88,89-92,94-96]

We begin, however, with a system, the pure chain, to which scaling techniques are applied as an illustration of the procedure, rather than of necessity, since it can be trivially treated by elementary k-space methods.

10.11.1. Length Scaling for Dynamics: Decimation for Simple Chain[77,81,89,95]

For a pure chain, the equations of motion of ferromagnetic spin waves, lattice vibrations, diffusion, etc., are all of the form

$$(2 - \Omega)u_n = u_{n+1} + u_{n-1} \tag{10.11.1}$$

Here u_n is the dynamical variable (transverse spin component, atomic displacement, etc.) at site n and Ω is a reduced "frequency" variable proportional to ω, ω^2, $i\omega$ for ferromagnetic spin waves, phonons, and diffusion, respectively, where ω is the actual frequency.

The decimation of the equations of motion proceeds as follows: use the corresponding equations for $u_{n\pm1}$ (in terms of u_n, $u_{n\pm2}$) to eliminate $u_{n\pm1}$ from equation (10.11.1), hence arriving at

$$[(2 - \Omega)^2 - 2]u_n = u_{n+2} + u_{n-2} \tag{10.11.2}$$

This is an equation of precisely similar form to equation (10.11.1), but relating u_n to $u_{n\pm2}$ rather than $u_{n\pm1}$ (i.e., corresponding to the decimation of every other site), and with a left-hand side which can be regarded as a scaled version of the left-hand side of equation (10.11.1) in which Ω has been replaced by Ω', where

$$\Omega' = 4\Omega - \Omega^2 \tag{10.11.3}$$

The decimation of every other site corresponds to the lattice constant scaling $a \to a'$, where

$$a' = ba \qquad \text{and} \qquad b = 2 \tag{10.11.4}$$

10.11.2. Extraction of Dynamic Critical Exponent, Dispersion Relation, Density of States, and Dynamic Response from Dynamic Length Scaling Transformation

The transformation of frequency (10.11.3) under the dilatation (10.11.4) may be exploited in much the same way as the static renormalization group transformation (Section 10.9.2). Near the fixed point $\Omega^* = 0$ of equation (10.11.3), the transformation can be linearized to the form

$$\Omega' = \lambda\Omega \tag{10.11.5}$$

where $\lambda = 4$. This, together with relations (10.11.4), implies

$$\Omega \propto a^z, \qquad z = \ln\lambda/\ln b \tag{10.11.6}$$

[cf. equations (10.9.1) and (10.9.4)]. In the case under consideration, the dynamic critical exponent z defined by relation (10.11.6) is $z = 2$. This corresponds to the usual (low-frequency) quadratic "dispersion" of spin waves and diffusion, and the linear relationship ω and k for phonons (where $\Omega \propto \omega^2$).

The general relationship between Ω and a,

$$\Omega = f(a) \tag{10.11.7}$$

can be found from relations (10.11.3) and (10.11.4), which imply

$$f(2a) = 4f(a) - f(a)^2 \tag{10.11.8}$$

The solution is

$$\Omega = f(a) = 2(1 - \cos ka) \tag{10.11.9}$$

where k is an arbitrary constraint, clearly recognizable as the wave vector of the usual approaches. Equation (10.11.9) is the usual full dispersion relation.

Equation (10.11.3) is the quadratic iterative map[97] and equation (10.11.8) defines a function which provides the general iterate. The special value 4 of the "control parameter" (the coefficient of Ω) makes it soluble here, and indeed gives rise to (fully developed) chaotic behavior under scaling. The next section shows that dynamic scaling on Sierpinsky gasket fractals again provides the quadratic map, but with more interesting values of the control parameter.

It can be shown that the sampling density of the variable Ω under iteration provides the density of states of the systems.[98] Further, the decimation procedure just illustrated can be applied also to the inhomogeneous equations satisfied by the Green function of the system to provide the scaling of the full dynamic response.[92]

10.11.3. Dynamic Properties of Fractals via Length Scaling (Decimation)

We consider the treatment, by the length-scaling procedure, of the dynamic properties of the Sierpinsky gasket illustrated in Figure 10.11. This is a non-random but nonuniform system, so it cannot be treated by k-space methods. Its (discrete) scale invariance suggests the application of scaling methods, and since decimation reverses its generation process it provides a direct approach to the problem.

We proceed exactly as in Sections 10.11.1 and 10.11.2, starting from equations of motion analogous to (10.11.1) and eliminating sites corresponding to the open circles in Figure 10.11. That leads[91,92] to the exact scalings $\Omega \to \Omega'$, $a \to a'$ where

$$\Omega' = \lambda\Omega - \Omega^2, \qquad a' = 2a \tag{10.11.10}$$

with $\lambda \equiv d + 3$ for the hypertetrahedral Sierpinsky gasket in general (integer) Euclidean dimension d.

For this system the general relationship [cf. (10.11.7)] between Ω and a cannot be found because the generalization of equation (10.11.8), in which 4 is replaced by $\lambda = d + 3$, is insoluble for $d > 1$.

However, the linearization steps [relations (10.11.5) and (10.11.6)] can obviously be carried out to provide the result $\Omega \propto a^z$ with dynamic exponent

$$z = \log_2 (d + 3) \tag{10.11.11}$$

The density of states for the fractal has been obtained[98] by the use of a "frequency sampling under iteration" method, and the Green function scaling has also been provided.[92] In addition, the scaling of a generating function[99] (analogous to the free energy in static scalings) has been used to obtain[100] the full dynamic response function $R(\Omega, k)$, which gives the intensity of scattering from the gasket at frequency and wave vector (transfer) Ω and k. This function is, for small Ω and k, of the form

$$R(\Omega, k) = k^{-(2-\eta+z)} F(\Omega/k^z, \log_2 k) \tag{10.11.12}$$

where η is a static transverse correlation exponent, z is the dynamic exponent given in equation (10.11.11), and the dependence on $\log_2 k$ is a periodic dependence arising from the discrete scale invariance of the fractal. Without it, equation (10.11.12) would be a conventional dynamic scaling form for a scale-invariant system ($\xi = \infty$) [cf. equation (10.8.6)].

The dynamic properties of other regular fractals, such as Berker lattices,[101] have also been investigated by similar length-scaling techniques, but we do not have space for details here.

10.12. Anomalous Dynamics in Random Scale-Invariant Systems

To conclude the discussion of critical dynamics, we briefly describe some real situations where scale invariance gives rise to anomalous features. These situations mostly occur in randomly self-similar systems. Examples are:

1. *Resins, glasses.* These are "quasi-scale invariant," in the sense of involving a large but finite range of lengths (from molecular size up to a size L_0, set by the size of macromolecules or the separation of cross links in the case of resins). So the dynamic behavior of, e.g., phonons should show anomalous properties for wavelength large but within the above length range, eventually crossing over to normal behavior for wavelengths exceeding L_0. Neutron-scattering studies of phonon density of states[102] have provided evidence for such "fracton" dynamics.[103]

2. *Diluted systems at the percolation threshold.* Such systems, especially dilute magnets,[24] can provide examples of almost perfect scale invariance: single-crystal substitutionally diluted magnets (such as the two- and three-dimensional Ising systems $Rb_2Co_pMg_{1-p}F_4$ and $RbCo_pMg_{1-p}F_3$, or related Heisenberg magnets (Mn in place of Co, etc.), can be prepared with percolation correlation lengths of hundreds of lattice spacings, and the dilution-induced static critical properties of such systems have already been very fully studied, both experimentally and theoretically.[24] Neutron scattering and other experiments are presently proceeding on the anomalous dynamics of such systems, particularly the spin dynamics on the percolation infinite cluster. Such experiments have already demonstrated anomalous Ising dynamics[104] due to the scale invariance of the infinite cluster at p_c. The theoretical description of that system,[77,105-107] and of spin wave dynamics on the percolation infinite cluster at p_c, are outlined below.[77,81,89,92,96]

While vibrational modes should also become anomalous at a percolation threshold, the system will fall apart at the same place so this effect is less ideal for experimental study, though it has been theoretically investigated. Theoretical treatments have been given of anomalous dynamic dielectric effects,[108] and these should be experimentally observable. Since the variety of spin dynamics is great enough to encompass most types [linear precessional (spin waves), relaxational, diffusive, and activated (Glauber[109] spin flip dynamics)] seen elsewhere, we limit our attention to the diluted spin systems.

10.12.1. Anomalous Dynamics in Diluted Spin Systems near the Percolation Threshold

(a) *Spin wave dynamics on the percolation infinite cluster at p_c.* This system, for which the Sierpinsky gasket calculation[91,92] of Section 10.11.3 provides a crude model,[110] can be more properly treated by a cluster decimation method of the type used there but applied to the spin wave equations of motion of the configurationally random system.[92] The technique requires the scaling of a probability distribution for a random variable[94] (exchange interaction divided by frequency) from which the joint scaling of a characteristic frequency variable Ω and a concentration can be obtained:

$$\Omega' = S(\Omega, p) \tag{10.12.1}$$

$$p' = R(p) \tag{10.12.2}$$

where $R(p)$ is the same function as appears in the corresponding treatment (cf. Section 10.9) of percolation.

The fixed point $(\Omega^*, p^*) = (0, p_c)$ controls the anomalous spin wave dynamics at the percolation threshold. Linearization of equations (10.12.1) and (10.12.2) about the fixed point yields the static percolation exponent ν (as in Section 10.9) together with a statement like transformation (10.11.5) from which the critical dynamic exponent z is obtained,[92] as in relation (10.11.6), but now for spin waves at the percolation threshold. This exponent is presently being measured for comparison with the theory.

An alternative, simpler, way of arriving at an equivalent value for the dynamic exponent z is by combining the Einstein relation [between conductivity, diffusion constant, and density, which here becomes a relation between percolation conductivity, spin wave stiffness, and percolation order parameter P(p)] with a crossover argument to obtain the relation[77,92,108]

$$z = 2 + (t - \beta)/\nu \tag{10.12.3}$$

Here t is the static percolation conductivity exponent, which can be calculated by scaling techniques relatively easily,[111] and β and ν are the standard percolation exponents.

(b) *Spin wave density of states on the percolation infinite cluster at p_c.* The low-frequency density of states of spin waves on the percolation infinite cluster at p_c has been numerically evaluated for $d = 2$ and shown to take the form[112]

$$\rho \equiv dN/d\Omega \propto \Omega^{-\mu}, \qquad \mu = 0.32 \pm 0.01 \tag{10.12.4}$$

This can be explained by the following argument. The number N of modes scales like the number of sites, i.e., like $N \propto a^{d_f}$ where d_f is the fractal dimensionality. However, $\Omega \propto a^z$, with z the dynamic exponent, so

$$\rho \propto a^{d_f - z} \propto \Omega^{(d_f/z)-1} \tag{10.12.5}$$

The exponent in relation (10.12.5) is related to the "spectral dimension"[90,91] and the value for μ in relation (10.12.4) is close to a conjecture of Alexander and Orbach[90] which would make $\mu = \frac{1}{3}$ and hence, using equation (10.7.6), $z \sim \frac{3}{2}(d - \beta/\nu)$. Though the conjecture is not exact (e.g., in first order in the ε-expansion from $d = 6$) this result for $d = 2$ is very close to other estimates, such as that from equation (10.12.3).

(c) *Spin waves on a dilute chain.* When the technique described in (a) above is applied to a dilute chain,[94] the scaling equation (10.12.2) is, of course, (10.10.4) and at its fixed points, $p^* = 0, 1$, the frequency scaling (10.12.1) becomes

$$\Omega' = \lambda(p^*)\Omega - \Omega^2 \tag{10.12.6}$$

where $\lambda(0) = 3$ and $\lambda(1) = 4$. These two forms of quadratic map are associated respectively with cyclic iterative behavior [related to the dynamics of spin waves on (small) finite clusters which have localized states and a discrete density of states] and with fully developed chaotic behavior (originating from the extended state dynamics of the pure chain, with its continuous density of states). The scaling behavior of the full dynamic response of the dilute chain has been obtained exactly[96,113] for arbitrary p.

(d) *Ising dynamics at the percolation threshold at low temperatures.*[77,105–107] This is perhaps the most interesting of all the anomalous effects, since owing to nonlinear dynamic effects the scale invariance here leads to a breakdown of dynamic scaling, according to which one expects

$$\tau \propto \xi^z \tag{10.12.7}$$

and instead its replacement by

$$\tau \propto \xi^{A \ln \xi + B} \tag{10.12.8}$$

where τ is here the characteristic time of the relaxational (Glauber) dynamics of domain wall motion, and ξ is the *thermal* correlation length (the percolation correlation length being infinite at p_c). The basic reason for the breakdown is that the relaxation (of the magnetization) involves the motion of the domain wall across the percolation infinite cluster at p_c which (see Figure 10.4) is a statistically self-similar object made up of chain segments branching into multiply connected parts in a hierarchical way. The motion of a domain wall on a chain is diffusive; and when it meets a branching, e.g., into two chains, the domain wall moves on in a time larger by a factor $e^{+2J/KT}$ due to the energetics of the activation process. This factor can be written as $(\xi/a)^{1/\nu}$ where ν is the usual percolation exponent and a the lattice constant. Owing to the self-similar nature of the infinite cluster ("branchings on branchings") this factor enters in the resulting scaling of the characteristic time τ, which is (keeping only the dominant terms)

$$\tau' = (\xi/a)^{1/\nu}\tau \tag{10.12.9}$$

while, since the correlation length is as usual held fixed in the scaling, which changes a by the dilatation factor b,

$$(\xi/a)' = (1/b)(\xi/a) \tag{10.12.10}$$

Comparison of equation (10.12.9) with (10.11.5) might suggest a behavior like relation (10.12.7) with dynamic exponent z related [by relation (10.11.6)] to the value of ξ at the transition (fixed point). However, the fixed point is at $p = p_c$ and $T = 0$, where ξ diverges. The proper anomalous dynamics has to be obtained by manipulating the joint scaling equations (10.12.9) and (10.12.10) in the vicinity of the fixed point. The result is the new form (10.12.8). This is an extreme form of critical slowing down ($\tau \to \infty$ as $\xi \to \infty$). An experiment[104] performed before the theory was completed did indeed show critical slowing down, but since it was done in a limited low-temperature range over which $\ln\xi$ changes very little the result could be fitted to relation (10.12.7) with an anomalously high effective exponent. Several Monte Carlo studies[114,115] of the dilute Ising dynamics at p_c have since confirmed the theoretical result (10.12.8), and further studies (experimental, theoretical, and computational) continue.

References

1. E. Ising, *Z. Phys.* **31**, 253 (1925).
2. G. H. Wannier, *Statistical Physics*, John Wiley and Sons, New York (1966).
3. R. E. Peierls, *Proc. Comb. Phil. Soc.* **32**, 477 (1936).
4. L. J. de Jongh and A. R. Miedema, *Experiments on Simple Model Magnetic Systems*, Taylor and Francis, London (1974).
5. L. J. de Jongh and A. R. Miedema, *Adv. Phys.* **23**, 1 (1974).
6. J. Ikeda and K. Hirakawa, *Solid State Commun.* **14**, 529 (1974).
7. H. Ikeda, I. Hatta, and M. Tanaka, *J. Phys. Soc. Jpn.* **40**, 334 (1976).
8. E. J. Samuelson, *Phys. Lett.* **31**, 936 (1973).
9. H. E. Stanley, *Introduction to Phase Transitions and Critical Phenomena*, Oxford University Press (1971).
10. M. E. Fisher, *Rep. Prog. Phys.* **30**, 615 (1967).
11. R. B. Stinchcombe, in: *Correlation Functions and Quasi-Particle Interactions in Condensed Matter* (J. Woods Halley, ed.), Plenum Press, New York (1978).
12. J. Als-Nielsen, O. W. Dietrich, and L. Passell, *Phys. Rev. B* **14**, 4908 (1977).
13. R. J. Birgeneau, R. A. Cowley, G. Shirane, H. Yoshizawa, D. P. Belanger, A. R. King, and V. Jaccarino, *Phys. Rev. B* **27**, 6747 (1983).
14. S. R. Broadbent and J. M. Hammersley, *Proc. Camb. Phil. Soc.* **53**, 629 (1957).

15. J. W Essam, in: *Phase Transitions and Critical Phenomena* (C. Domb and M. S. Green, eds.), Vol. 2, Academic Press, New York (1972).
16. D. J. Stauffer, *Phys. Rep.*, **54**, 1 (1979).
17. J. M. Kosterlitz and D. J. Thouless, *J. Phys. C* **6**, 1181 (1973).
18. J. M. Kosterlitz and D. J. Thouless, in: *Progress in Low Tempertaure Physics* (D. F. Brewer, ed.), Vol. VII B, North-Holland Publ. Co., Amsterdam (1978).
19. L. D. Landau and E. M. Lifshitz, *Statistical Physics*, Pergamon Press, Oxford (1959).
20. K. Huang, *Statistical Mechanics*, John Wiley and Sons, New York (1963).
21. M. Rasolt, M. J. Stephen, M. E. Fisher, and P. B. Weichman, *Phys. Rev. Lett.* **53**, 798 (1984).
22. V. K. S. Shante and S. Kirkpatrick, *Adv. Phys.* **20**, 325 (1971).
23. J. W. Essam, *Rep. Prog. Phys.* **43**, 833 (1980).
24. R. B Stinchcombe, in: *Phase Transitions and Critical Phenomena* (C. Domb and J. L. Lebowitz, eds.), Academic Press, New York (1983).
25. S. Kirkpatrick, in: *Ill-Condensed Matter* (R. Balian, R. Maynard, and G. Toulouse, eds.), North-Holland Publ. Co., Amsterdam (1979).
26. D. J. Thouless, in: *Ill-Condensed Matter* (R. Balian and G. Toulouse, eds.), North-Holland Publ. Co., Amsterdam (1979).
27. F. Y. Wu, *Rev. Mod. Phys.* **54**, 235 (1982).
28. C. M. Fortuin and P. W. Kasteleyn, *Physica* **57**, 536 (1972).
29. G. Toulouse, *Commun. Phys.* **2**, 115 (1977).
30. J. Villain, *J. Phys. C* **10**, 1717 (1977).
31. P. W. Anderson, in: *Ill-Condensed Matter* (R. Balian, R, Maynard, and G. Toulouse, eds.), North-Holland Publ. Co., Amsterdam (1979).
32. W. Selke and M. E. Fisher, *Phys. Rev. B* **20**, 257 (1979).
33. J. Smith, J. M. Yeomans, and V. Heine, in: *Modulated Structure Materials* (T. Tsakalakos, ed.), Nijhoff, Dordrecht (1984).
34. R. Pandit and M. Wortis, *Phys. Rev. B* **25**, 3226 (1982).
35. B. Mandelbrot, *Fractals*, Freeman, San Francisco (1977).
36. R. B. Stinchcombe, in: *Scaling Phenomena in Disordered Systems* (R. Pynn and A. Skjeltorp, eds.), Plenum Press, New York (1985).
37. L. Onsager, *Phys. Rev.* **65**, 117 (1944).
38. L. Onsager and B. Kaufman, *Phys. Rev.* **76**, 1232, 1244 (1949).
39. M. F. Sykes and J. W. Essam, *J. Math. Phys.* **5**, 1117 (1964).
40. R. Brout, *Phase Transitions*, Benjamin, New York (1965).
41. L. P. Kadanoff, *Physics* **2**, 263 (1966).
42. K. G. Wilson, *Phys. Rev. Lett.* **28**, 584 (1972).
43. K. G. Wilson and M. E. Fisher, *Phys. Rev. Lett.* **28**, 240 (1972).
44. K. G. Wilson and J. Kogut, *Phys. Rep.* **12C**, 75 (1974).
45. Th. Niemeyer and J. M. J. Van Leeuwen, *Physica* **71**, 17 (1984).
46. Th. Niemeyer and J. M. J. Van Leeuwen, in: *Phase Transitions and Critical Phenomena* (C. Domb and M. S. Green, eds.), Vol. 6, Academic Press, New York (1976).
47. L. P. Kadanoff and A. Houghton: *Phys. Rev. B* **11**, 377, (1975).
48. M. N. Barber: *J. Phys. C* **8**, L203 (1975).
49. A. P. Young and R. B. Stinchcombe, *J. Phys. C* **8**, L535 (1975); **9**, L643 (1976).
50. R. B. Stinchcombe, in: *Magnetic Phase Transitions* (R. J. Elliott and M. Ausloos, eds.), Springer-Verlag, Berlin (1983).
51. C. Domb, *Adv. Phys.* **9**, 149, 245 (1960).
52. K. Binder, *Applications of the Monte Carlo Method in Statistical Physics*, Springer-Verlag, Berlin (1984).
53. M. N. Barber, in: *Phase Transitions and Critical Phenomena* (C. Domb and J. L. Lebowitz, eds.), Vol. 8, Academic Press, New York (1983).
54. R. G. Priest and T. C. Lubensky, *Phys. Rev. B* **13**, 4159 (1976).
55. N. N. Bogoliubov and S. V. Tyablikov *Sov. Phys. Dokl.* **4**, 589 (1959).
56. F. Englert, *Phys. Rev. Lett.* **5**, 102 (1960).

57. R. B. Stinchcombe, in: *Polymers, Liquid Crystals, and Low-Dimensional Solids* (N. March and M. P. Tosi, eds.), Plenum Press, New York (1984).
58. R. B. Stinchcombe, in: *Electron-Phonon Interactions and Phase Transitions* (T. Riste, ed.), Plenum Press, New York (1977).
59. V. L. Ginzburg and L. D. Landau, *Zh. Eksp. Theor. Fiz.* **26**, 1064 (1950).
60. V. L. Ginzburg and L. P. Pitaevskii, *Zh. Eksp. Theor. Fiz.* **34**, 1240 (1958).
61. J. Bardeen, L. N. Cooper, and J. R. Schrieffer, *Phys. Rev.* **108**, 1175 (1957).
62. J. Goldstone, *Nuovo Cimento* **19**, 154 (1963).
63. C. N. Yang, *Phys. Rev.* **85**, 808 (1952.)
64. T. T. Wu: *Phys. Rev.* **149**, 380 (1966).
65. K. G. Wilson, *Physica* **73**, 119 (1974.)
66. S. K. Ma, *Modern Theory of Critical Phenomena*, Benjamin, New York (1976).
67. S. K. Ma, *Rev. Mod. Phys.* **45**, 589 (1973).
68. D. J. Wallace and R. K. P. Zia, *Rep. Prog. Phys.* **41**, 1 (1978).
69. D. J. Amit, *Field Theory, the Renormalization Group and Critical Phenomena*, McGraw-Hill, New York (1978).
70. E. Brézin and J. Zinn-Justin, *Phys. Rev. B* **14**, 3110 (1976).
71. T. W. Burkhardt and J. M. J. Van Leeuwen, *Real Space Renormalization*, Topics in Current Physics, Vol. 30, Springer-Verlag, Berlin (1982).
72. B. Widom, *J. Chem. Phys.* **43**, 3892, 3898 (1965).
73. R. B. Griffiths, *J. Chem. Phys.* **43**, 1958 (1965).
74. A. D. Bruce, in: *Nonlinear Phenomena at Phase Transitions and Instabilities* (T. Riste, ed.), Plenum Press, New York (1982).
75. P. C. Hohenberg and B. I. Halperin, *Rev. Mod. Phys.* **49**, 435 (1977).
76. R. J. Baxter, *Ann. Phys.* (N.Y.) **70**, 193 (1972); **76**, 1, 25, 48 (1973).
77. R. B. Stinchcombe, in: *Scaling Phenomena in Disordered Systems* (R. Pynn and A. Skjeltorp, eds.), Plenum Press, New York (1985).
78. A. A. Belavin, A. M. Polyakov, and A. B. Zamolodchikov, *Nucl. Phys. B* **241**, 333 (1984).
79. B. I. Halperin and P. C. Hohenberg, *Phys. Rev.* **117**, 952 (1969).
80. M. E. Fisher, *Rev. Mod. Phys.* **46**, 597 (1974).
81. R. B. Stinchcombe, in: *Highlights of Condensed Matter Physics* (F. Bassani, F. Fumi, and M. P. Tosi, eds.), Proceedings of the 1983 Varenna Summer School, North-Holland Publ. Co., Amsterdam (1985).
82. P. J. Reynolds, W. Klein, and H. E. Stanley, *J. Phys. C* **10**, L167 (1977).
83. R. H. Swendsen, *J. Appl. Phys.* **53**, 1920 (1982).
84. M. Nauenberg and B. Nienhuis, *Phys. Rev. Lett.* **33**, 1598 (1974).
85. A. N. Berker and S. Ostlund, *J. Phys. C* **12**, 4961 (1979).
86. A. A. Migdal, *Sov. Phys. JETP* **42**, 743 (1976).
87. L. P. Kadanoff; *Ann. Phys.* (*N.Y.*) **100**, 559 (1976).
88. R. B. Stinchcombe, in: *Fractals in Physics* (L. Pietronero and E. Tosatti, eds.), Elsevier, Amsterdam (1985).
89. R. B. Stinchcombe, in: *Static Critical Phenomena in Inhomogeneous Systems* (A. Pekalski and J. Sznajd, eds.), Springer-Verlag, New York (1984).
90. S. Alexander and R. Orbach, *J. Phys.* (*Paris*), *Lett.* **43**, L625 (1982).
91. R. Rammal and G. Toulouse, *Phys. Rev. Lett.* **49**, 1194 (1982).
92. C. K. Harris and R. B. Stinchcombe, *Phys. Rev. Lett.* **50**, 1399 (1983).
93. I. Ya Korenblit and E. F. Shender, *Usp. Fiz. Nauk* **126**, 233 (1978) [*Sov. Phys. Usp.* **21**, 832 (1978)].
94. R. B. Stinchcombe, *Phys. Rev. Lett.* **50**, 200 (1983).
95. L. G. Marland, D.Phil Thesis, University of Oxford (1978, unpublished).
96. R. B. Stinchcombe and C. K. Harris, *J. Phys. A* **16**, 4083 (1983).
97. H. G. Schuster, *Determinstic Chaos*, Physik Verlag-Weinheim (1984).
98. E. Domany, S. Alexander, D. Bensimon, and L. P. Kadanoff, *Phys. Rev. B* **28**, 3110 (1983).
99. A.-M. S. Tremblay and R. W. Southern, *J. Phys.* (*Paris*), *Lett.* **44**, 843, (1983).

100. A. C. Maggs and R. B. Stinchcombe, *J. Phys. A* **19** (1986), to appear.
101. J-M. Langlois, A.-M. S. Tremblay, and B. Southern, *Phys. Rev.* **28**, 218 (1983).
102. H. M. Rosenberg, *Phys. Rev. Lett.* **54**, 704 (1985).
103. R. Orbach, in: *Scaling Phenomena in Disordered Systems* (R. Pynn and A. Skjeltorp, eds.), Plenum Press, New York (1985).
104. G. Aeppli, H. Guggenheim, and Y. J. Uemura, *Phys. Rev. Lett.* **52**, 942 (1984).
105. C. K. Harris, D.Phil Thesis, University of Oxford (1983, unpublished)
106. C. Henley, *Phys. Rev. Lett.* **54**, 2030 (1985).
107. C. K. Harris and R. B. Stinchcombe, *Phys. Rev. Lett.* **56**, 896 (1986).
108. A. Aharony, in: *Scaling Phenomena in Disordered Systems* (R. Pynn and A. Skjeltorp, eds.), Plenum Press, New York (1985).
109. R. J. Glauber, *J. Math. Phys.* (*N.Y.*) **4**, 294 (1963); K. Kawasaki, in: *Phase Transitions and Critical Phenomena* (C. Domb and M. S. Green, eds.), Vol. 2, Academic Press, New York (1972).
110. Y. Gefen, A. Aharony, B. Mandelbrot, and S. Kirkpatrick, in: *Disordered Systems and Localization* [C. Castellani, C. di Castro, and L. Peliti, eds.), Springer-Verlag, Berlin (1981).
111. R. B. Stinchcombe and B. P. Watson, *J. Phys. C* **9**, 3221 (1976).
112. S. J. Lewis and R. B. Stinchcombe, *Phys. Rev. Lett.* **52**, 1021 (1984).
113. A. C. Maggs and R. B. Stinchcombe, *J. Phys. A* **17**, 1555 (1984).
114. S. Jain, *J. Phys. A* **19**, L57 (1986).
115. D. Chowdhury and D. Stauffer, *J. Phys. A* **19**, (1986), to appear.

11

Classical Chaos and Quantum Eigenvalues

M. V. Berry

The connection between classical and quantum mechanics (i.e., the semiclassical limiting asymptotics as $\hbar \to 0$) must be subtle and complicated, because classical mechanics itself (i.e., the classical limit $\hbar = 0$) is subtle and complicated: the orbits of systems governed by Hamilton's equations of motion may be predictable (regular) or unpredictable (irregular) depending on subtle details of the form of the Hamiltonian $H(\{q_i\}, \{p_i\})$.[1-3] A natural question is: how does the "chaology" of classical orbits reflect itself in the corresponding quantum system? Sometimes this question is put in the form: what is quantum chaos?

There are many approaches to this question. One is to study the *dynamics* of quantum systems which are classically chaotic, that is, to study nonstationary states. There have been many studies of mathematical models of such quantum evolution[4-10] which have found important recent application in interpreting experiments on the microwave ionization of hydrogen atoms.[11-13] Another approach is to look at *stationary states* and concentrate on the form of the wave functions: these are remarkably different for eigenstates corresponding to regular and chaotic systems.[14-19]

Here, however, we concentrate on the *energies* of stationary states, and ask how the *distribution of eigenvalues* $\{E_n\} = E_1, E_2, \ldots$ of a quantum Hamiltonian $\hat{H} = H(\{\hat{q}_i\}, \{\hat{p}_i\})$ reflects the chaology of the classical trajectories generated by the classical H, in which $\{q_i\}$ and $\{p_i\}$ are variables rather than operators. Of course the energies $\{E_n\}$ depend on $\hbar$. We consider only the nontrivial case where the number of freedoms N exceeds unity.

Ideally one would like an explicit asymptotic expression giving $\{E_n(\hbar)\}$ with an error that decreases as $\hbar \to 0$ faster than the mean level spacing. Such

---

M. V. Berry • H. H. Wills Physics Laboratory, Bristol BS8 1TL, England.

an expression has been found only for classically integrable (i.e., nonchaotic systems[20-22] and is a generalization of the familiar WKB theory for one dimension. For integrable systems with N freedoms, there are N constants of motion (including the energy) which confine motion to N-dimensional tori in the $2N$-dimensional phase space.[23] In lowest order, quantization selects the energies E of those tori whose N actions are separated by multiples of $\hbar$, i.e.,

$$E_{\{n_i\}} = H\{[I_i = (n_i + \alpha_i/4)\hbar]\} \tag{11.1}$$

where $\{n_i\} = n_1, \ldots, n_N$ are the quantum numbers, $\{I_i\}$ are the actions

$$I_i \equiv \oint_{\gamma_i} p_j \, dq_j \tag{11.2}$$

round the irreducible cycle γ_i of the torus, and the α_i are constants (Maslov indices[20]). Obviously equation (11.1) works only when tori exist. In the chaotic extreme, motion is ergodic and there are no constants of motion apart from E; therefore there are no tori and the semiclassical rule (11.1) cannot be applied: so far nobody has found a semiclassical quantization rule for chaotic systems.

In these circumstances one must seek less precise information, in the form of *average* properties of the distribution of energies. These spectral averages can be defined semiclassically, because as $\hbar \to 0$ infinitely many levels crowd into any fixed energy interval however small. The simplest spectral average is the mean *spectral density* $\langle d(E) \rangle$. This is the average of

$$d(E) \equiv \sum_n \delta(E - E_n) = \mathrm{Tr}\, \delta(E - \hat{H}) \tag{11.3}$$

and is given semiclassically by the "Weyl rule"[21]

$$\langle d(E) \rangle = \frac{d\Omega(E)/dE}{h^N} \tag{11.4}$$

where Ω is the phase volume given by

$$\Omega(E) = \int_{(H<E)} d^N q \int d^N p, \qquad \text{i.e.,}$$

$$d\Omega/dE = \int d^N q \int d^N p \delta[E - H(\{q_i\}, \{p_i\})] \tag{11.5}$$

The Weyl rule formalizes the old idea of "one quantum state per volume h^N of phase space."

The result (11.4) tells us nothing about quantum chaos, because the classical volume $\Omega(E)$ is insensitive to the regularity or chaos of the orbits. This is disappointing, but nevertheless two useful pieces of information can be obtained. First, the mean level spacing $\langle d\rangle^{-1}$ is of order $\hbar^N$, thus, for example, in a classically small energy range of size h there are many levels (of order $\hbar^{-(N-1)}$); this will be important later. And second, a rough quantization rule can be found by realizing that the integral of $d(E)$ is the *spectral staircase*

$$\mathcal{N}(E) \equiv \sum_n \theta(E - E_n) = \int_{-\infty}^{E} dE'\, d(E') \tag{11.6}$$

where θ is the unit step; the rule, expressing the idea that the smooth curve of the average staircase might intersect the steps halfway, on average, is then

$$(n + \tfrac{1}{2}) = \langle \mathcal{N}(E_n)\rangle = \Omega(E_n)/h^N \tag{11.7}$$

The above rule is rough because it fails to describe the fine-scale *fluctuations* in the levels (in graphs of the E_n as a function of a parameter on which H depends, these fluctuations appear as avoided crossings[24-26]). To describe these fluctuations it is necessary to employ statistics which (unlike $\langle d\rangle$ and $\langle \mathcal{N}\rangle$) involve correlations between nearby levels, that is, on scales $\hbar^N$. Such fluctuation measures have been devised in *random-matrix theory*[27] and applied to sequences of excited resonance levels of atomic nuclei. The fluctuation statistics depend not on the raw spectrum levels $\{E_n\}$ but on the "unfolded" spectrum of levels $\{x_n\}$ which have been scaled so as to have unit mean spacing. Thus

$$x_n + \tfrac{1}{2} \equiv \langle \mathcal{N}(E_n)\rangle \tag{11.8}$$

[without the fluctuations, equation (11.7) shows that x_n would be simply n]. Two particularly useful statistics are the probability distribution $P(S)$ of the *level spacings* $\{S_n \equiv x_{n+1} - x_n\}$, and the *spectral rigidity*[28-30]

$$\Delta(L) \equiv \left\langle \min \frac{1}{L} \int_{x-L/2}^{x+L/2} d\zeta [\mathcal{N}(\zeta) - A - B\zeta]^2 \right\rangle \tag{11.9}$$

(this is the least-squares deviation of the staircase from a straight line, over a range of L mean spacings, averaged over an interval of energies x that includes many levels). $P(S)$ is useful in describing spectral correlations on the finest scales—i.e., between neighboring levels—and $\Delta(L)$ is useful for describing how spectral correlations depend on range—i.e., large or small L.

When spectral statistics are computed for sequences of levels of Hamiltonian systems with classical limits, a remarkable "experimental" fact emerges: the statistics display *universality* and the spectral universality class depends on the chaology of the classical orbits. The universality classes are:

(a) *Classically integrable systems.* Here the spectral statistics are those of a *Poisson*—i.e., uncorrelated random—distribution of levels.[30,31] At first sight it is surprising that the quantum conditions (11.1) can give rise to a random sequence, but the surprise dissipates with the realization that neighboring levels E_n, E_{n+1} can have very different sets of quantum numbers $\{n_i\}$. For Poisson statistics,

$$P(S) = \exp(-S) \qquad \text{and} \qquad \Delta(L) = L/15 \tag{11.10}$$

(b) *Classically chaotic spinless (or integral-spin) systems with time-reversal symmetry.* Here the spectral statistics are those of the *Gaussian orthogonal ensemble* (GOE), which consists[27] of real symmetric matrices whose elements are Gauss-distributed so as to make the statistics of the ensemble invariant under orthogonal rotations. Only real symmetric matrices are involved because time-reversal symmetry implies that the wave functions are real. For the GOE, to a close approximation,

$$P(S) \approx \frac{\pi}{2} S \exp(-\pi S^2/4) \tag{11.11}$$

and

$$\begin{aligned} &\Delta(L)L/15 \qquad (L \ll 1) \\ &\to \frac{\ln L}{\pi^2} - 0.00695 \qquad (L \gg 1) \end{aligned} \tag{11.12}$$

(c) *Classically chaotic systems without time-reversal symmetry.* This is in a sense the generic case which best justifies the label "quantum chaos." Time-reversal symmetry (T) can be broken most simply by magnetic fields, either smoothly-varying[32] or consisting of a single (Aharonov-Bohm) flux line.[33] Here the spectral statistics are those of the Gaussian unitary ensemble (GUE),[27] of complex Hermitian matrices whose elements are Gauss-distributed so as to make the statistics of the ensemble invariant under unitary transformations. For the GUE, to a close approximation,

$$P(S) \approx \frac{32S^2}{\pi^2} \exp(-4S^2/\pi) \tag{11.13}$$

and

$$\Delta L \to L/15 \qquad (L \gg 1)$$
$$\to \frac{\ln L}{2\pi^2} + 0.05902 \qquad (L \gg 1) \tag{11.14}$$

Before anticipating that a system without T will have GUE statistics, care must be taken to determine whether it has any geometric symmetries, because these can act so as to mimic T-symmetry and generate levels with GOE statistics; in the theory of this "false time-reversal symmetry-breaking"[34] it is shown that for GUE the system must possess no antiunitary symmetry operator $\hat{A}$ [27,35] (commuting with H) and satisfying $\hat{A}^2 = 1$ or $\hat{A}\hat{A}^* = 1$ (T is represented in position representation by the operator A = complex conjugation). A set of numbers recently discovered to have GUE statistics is the imaginary parts of zeros of Riemann's zeta function[36]; this is surprising and suggestive.[37]

(d) *Classically chaotic systems with half-integer spin and with T* (or, more generally, chaotic systems with an antiunitary symmetry satisfying $\hat{A}^2 = -1$). Here there are so far no numerical experiments (I am planning one now) but the spectral statistics are expected to be those of the *Gaussian symplectic ensemble* (GSE),[27] of quaternion real Hermitian matrices whose elements are Gauss-distributed so as to make the ensemble invariant under symplectic transformations. For the GSE, to a close approximation,

$$P(S) \approx \frac{2^{18} S^4}{3^6 \pi^3} \exp\left(-64S^2/9\pi\right) \tag{11.15}$$

and

$$\Delta(L) \to L/15 \qquad (L \ll 1)$$
$$\to \frac{\ln L}{4\pi^2} + 0.07832 \qquad (L \gg 1) \tag{11.16}$$

That completes the list of universality classes. But I now reveal that life is really not so simple, and describe two ways in which universality is compromised. First, most classical systems are neither purely regular nor purely chaotic,[1] but exhibit mixed (or, in the jargon, "KAM") behavior in which some orbits are regular and some predictable, depending on initial conditions. Such cases are important in quantum mechanics because they correspond to the anharmonically coupled oscillators describing vibrating molecules and to atoms in strong magnetic fields occurring astrophysically. It is natural to expect that in lowest approximation the spectral statistics will interpolate between those of the Poisson and the appropriate random-matrix universality classes, to a degree which depends on the relative phase-space volumes of regions of regular and chaotic motion; a theory along these lines[38] is supported by numerical experiments.[39] Second, even for purely regular or purely chaotic systems the domain of universal behavior is limited to energy ranges not exceeding a quantity of order $\hbar$; for the rigidity $\Delta(L)$, this means that universality holds when $L < L_{\max} \sim \hbar^{-(N-1)}$, so that in the semiclassical limit the domain of universality shrinks to zero in energy but nevertheless extends over infinitely many levels. When $L > L_{\max}$ numerical experiments[40] (so far restricted to integrable systems) show, and a theory[30] (for both integrable and chaotic systems) explains, that $\Delta(L)$ does not continue to increase as in equations (11.10), (11.12), (11.14), and (11.16), but saturates at nonuniversal values characteristic of the particular system.

In spite of these caveats, the universality of semiclassical spectra is a remarkable phenomenon that demands explanation. One class of theories[41,42] considers the energies to depend on a parameter t which is regarded as akin to a time variable, and the "motion" of the eigenvalues $\{E_n(t)\}$ on the E axis is put into correspondence with the statistical mechanics of particles on a line. These theories can be made to generate random-matrix behavior but the derivations rest on statistical assumptions about the matrix elements of the t derivatives of $\hat{H}$ between different eigenstates. It is desirable to understand spectral statistics directly, without introducing parameters or extra statistical assumptions. Some progress has been made, as will now be described.

The behavior of $P(S)$ as $S \to 0$ can be related to the codimension K of degeneracies when the system is embedded in an ensemble of similar ones,[43,21-24] where K is the number of parameters that must be varied to produce a degeneracy: for separable systems $K = 1$, for real symmetric matrices $K = 2$, for complex Hermitian matrices $K = 3$, and for quaternion real matrices $K = 5$. The result is

$$P(S) \sim S^{K-1} \qquad \text{As } S \to 0 \tag{11.17}$$

and this agrees with equations (11.10), (11.11), (11.13), and (11.15). But K is only roughly related to the classical symmetries (subtleties arise from barrier penetration[44]) and a semiclassical understanding of $P(S)$ is still lacking.

The behavior of $\Delta(L)$, on the other hand, is rather well understood[30] in terms of a semiclassical theory. According to equation (11.9), $\Delta(L)$ is a quadratic functional of the spectral staircase (11.6). This can be expressed as its average $\langle \mathcal{N} \rangle$ [equation (11.7)] plus a series of correction terms which are oscillatory functions of E. Each such correction comes from a *closed orbit* of the classical system,[45-47] and gives an oscillation with energy period $\hbar/T$, where T is the time period of the orbit. So values of $L \ll L_{\max}$, corresponding to energy scales $\ll \hbar$, correspond to *very long orbits.* In particular, any fixed L corresponds as $\hbar \to 0$ to periods of order $\hbar^{-(N-1)}$. But for these very long orbits there exist *universal sum rules*[48] which depend on the classical chaology, and it is these that enable the theory[30] to reproduce the random-matrix results (11.10), (11.12), and (11.14) [but not yet relation (11.16)]. When $L \gg L_{\max}$, the previously-mentioned breakdown of universality occurs and is explained by $\Delta(L)$ depending only on short closed orbits which, or course, differ from system to system.

Much work remains to be done in understanding semiclassical spectra. The most pressing and also fundamental problem is to discover whether the semiclassical sum over the closed orbits or a chaotic system can be extended (or interpreted, or analytically continued[37]) so as to describe the finest spectral scales such as those embodied in $P(S)$ (or even—we are entitled to hope—a complete quantization formula). Then there are "crossover" phenomena associated with the breakdown of universality when $L \sim L_{\max}$. Finally, higher spectral statistics, depending more than quadratically on $\mathcal{N}(E)$, should be studied semiclassically. It is likely that from the program outlined in the last two sentences there might emerge a statistic which, for chaotic systems, depends on the Kolmogorov-Sinai[1] entropy which is so important in classical chaology.

References

1. A. J. Lichtenberg and M. A. Lieberman, *Regular and Stochastic Motion*, Springer-Verlag, New York (1983).
2. M. V. Berry, in: *Topics in Nonlinear Mechanics* (S. Jorna, ed.) Am. Inst. Phys. Conf. Proc. Vol. **46**, pp. 16-120 (1978).
3. H. G. Schuster, *Deterministic Chaos*, Physik-Verlag GMBH, Weinheim (1984).
4. G. Casati, B. V. Chirikov, J. Ford, and F. M. Izraelev in: *Stochastic Behaviour in Classical and Quantum Hamiltonian Systems* (G. Casati and J. Ford, eds), Springer Lecture Notes in Physics, Vol. 93, pp. 334-352 (1979).
5. M. V. Berry, N. L. Balazs, M. Tabor, and A. Voros, *Ann. Phys.* (*N.Y.*) **122**, 26-63 (1979).
6. H. J. Korsch and M. V. Berry, *Physica* **3D**, 627-636 (1981).
7. M. V. Berry, *Physica* **10D**, 369-378 (1984).
8. B. V. Chirikov, F. M. Izraelev, and D. L. Shepelyansky, *Sov. Sci. Rev.* **2C**, 209-267 (1981).
9. D. L. Shepelyansky, *Physica* **8D**, 208-222 (1983).
10. S. Fishman, D. R. Grempel, and R. E. Prange, *Phys. Rev. Lett.* **49**, 509-512 (1982).
11. R. V. Jensen, in: *Chaotic Behavior in Quantum Systems* (G. Casati, ed.), pp. 171-186, Plenum Press, New York (1985).
12. J. E. Bayfield, L. D. Gardner, and P. M. Koch, *Phys. Rev. Lett.* **39**, 76-79 (1977).
13. G. Casati, B. V. Chirikov, and D. L. Shepelyansky, *Phys. Rev. Lett.* **53**, 2525-2528 (1984).
14. M. V. Berry, *Phil. Trans. R. Soc. London, Ser. A* **287**, 237-271 (1977).
15. S. W. McDonald and A. N. Kaufman, *Phys. Rev. Lett.* **42**, 1189-1191 (1979).
16. M. V. Berry, *J. Phys. A* **10**, 2083-2091 (1977).
17. M. V. Berry, J. H. Hannay, and A. M. Ozorio de Almeida, *Physica* **8D**, 229-242 (1983).
18. E. J. Heller, *Phys. Rev. Lett.* **16**, 1515-1518 (1984).
19. M. V. Berry and M. Robnik, *J. Phys. A* 1365-1372 (1986).
20. V. P. Maslov and M. V. Fedoriuk, *Semiclassical Approximation in Quantum Mechanics*, D. Reidel, Dordrecht (1981).

21. V. M. Berry, in: *Chaotic Behavior of Deterministic Systems* (G. Iooss, R. H. G. Helleman, and R. Stora, eds.), pp. 171-271, Les Houches Lectures XXXVI, North-Holland, Amsterdam (1983).
22. I. C. Percival, *Adv. Chem. Phys.* **36**, 1-61 (1977).
23. V. I. Arnold, *Mathematical Methods of Classical Mechanics*, Springer-Verlag, New York (1978).
24. M. V. Berry, *Ann. Phys.* (*N.Y.*) **131**, 163-216 (1981).
25. R. Ramaswamy and R. A. Marcus, *J. Chem. Phys.* **74**, 1379-1384, 1385-1393 (1981).
26. M. V. Berry and M. Wilkinson, *Proc. R. Soc. London, Ser. A* **392**, 15-43 (1984).
27. C. E. Porter, *Statistical Theories of Spectra: Fluctuations*, Academic Press, New York (1965).
28. F. J. Dyson and M. L. Mehta, *J. Math. Phys.* **4**, 701-712 (1963).
29. O. Bohigas and M. J. Giannoni, in: *Mathematical and Computational Methods in Nuclear Physics* (J. S. Dehesa, J. M. G. Gomez, and A. Polls, eds.), Lecture Notes in Physics, Vol. 209, pp. 1-99, Springer-Verlag, New York (1984).
30. M. V. Berry, *Proc. R. Soc. London, Ser. A* **400**, 229-251 (1985).
31. M. V. Berry and M. Tabor, *Proc. R. Soc. London, Ser. A* **356**, 375-394 (1977).
32. T. H. Seligman and J. J. M. Verbaarschot, *Phys. Lett.* **108A**, 183-187 (1985).
33. M. V. Berry and M. Robnik, *J. Phys. A* **19**, 649-668 (1986).
34. M. Robnik and M. V. Berry, *J. Phys. A* **19**, 669-682 (1986).
35. J. J. Sakurai, *Modern Quantum Mechanics*, Benjamin, New York (1985).
36. A. M. Odlyzko, *Math. of. Comp.*, **48**, 273-308 (1987).
37. M. V. Berry, in: *Quantum Chaos and Statistical Nuclear Physics* (T. H. Seligman and H. Nishioka, eds.), Springer Lecture Notes in Physics No. 263, pp. 1-17 (1986).
38. M. V. Berry and M. Robnik, *J. Phys. A* **17**, 2413-2421 (1984).
39. H.-D. Meyer, E. Haller, H. Köppel, and L. S. Cederbaum, *J. Phys. A* **17**, L831-L836 (1984).
40. G. Casati, B. V. Chirikov, and I. Guarneri, *Phys. Rev. Lett.* **54**, 1350-1353 (1985).
41. P. Pechukas, *Phys. Rev. Lett.* **51**, 943-946 (1983).
42. T. Yukawa, *Phys. Rev. Lett.* **54**, 1883-1886 (1985).
43. M. V. Berry, in: *Chaotic Behavior in Quantum Systems* (G. Casati, ed.), pp. 123-140, Plenum Press, New York (1985).
44. M. Wilkinson, *J. Phys. A*, in press.
45. M. C. Gutzwiller, *J. Math. Phys.* **12**, 343-358 (1971).
46. M. C. Gutzwiller, in: *Path Integrals and their Applications in Quantum, Statistical and Solid-State Physics* (G. J. Papadopoulos and J. T. Devreese, eds.), pp. 163-200, Plenum Press, New York (1978).
47. R. Balian and C. Bloch, *Ann. Phys.* (*N.Y.*) **69**, 76-160 (1972).
48. J. H. Hannay and A. M. Ozorio de Almeida, *J. Phys. A* **17**, 3429-3440 (1984).

12

Renormalization Description of Transitions to Chaos

P. Cvitanović

12.1. Introduction

The aim of this chapter, and the following one, is to review the developments which have greatly increased our understanding of chaotic dynamics during the past decade, and given us new concepts and modes of thought that, we hope, will have far-reaching repercussions in many different fields.

Once it was believed that given the initial conditions, we knew what a deterministic system would do far into the future. That was immodest. Today it is widely appreciated that given infinitesimally different starting points, we often end up with wildly different outcomes. Even with the simplest conceivable equations, almost any nonlinear system will exhibit chaotic behavior.

Confronted today with a potentially turbulent nonlinear dynamical system, we analyze it through a sequence of three distinct steps. First, we determine the intrinsic *dimension* of the system—the minimum number of degrees of freedom necessary to capture its essential dynamics. If the system is very turbulent (its attractor is of high dimension) we are, at present, out of luck. We know only how to deal with the transitional regime between regular motions and weak turbulence. In this regime the attractor is of low dimension, the number of relevant parameters is small, and we can proceed to the second step; we classify all the motions of the system by a hierarchy whose successive layers require increased precision and patience on the part of the observer. We call this classification the *symbolic dynamics* of the system: the following chapter shows how to do this for the Hamiltonian systems, and in this chapter we take the period n-tuplings and circle maps as instructive examples.

P. Cvitanović • Institute of Theoretical Physics, Chalmers University of Technology, S-412 96 Göteborg, Sweden.

Though the dynamics might be complex beyond belief, it is still generated by a simple deterministic law and, with some luck and intelligence, our labeling of possible motions will reflect this simplicity. If the rule that gets us from one level of the classification hierarchy to the next does not depend on the level, the resulting hierarchy is self-similar, and we can proceed with the third step: investigate the *scaling structure* of the dynamical system. This is the subject of the present chapter.

The scaling structure of a dynamical system is encoded into the *scaling functions.* The purpose of scaling functions is twofold:

For an experimentalist, they are the theorist's prediction of the motions expected in a given parameter and phase-space range. Given the observed motions up to a given level, the symbolic dynamics predicts what motions should be seen next, and the scaling functions predict where they should be seen, and what precision is needed for their observation.

For a theorist, the scaling functions are the tool which resolves asymptotically the fine structure of a chaotic dynamical system and proves that the system is indeed chaotic, and not just a regular motion of period exhausting the endurance of an experimentalist. Furthermore (and that is theoretically very sweet), the scalings often tend to *universal* limits. In such cases, the finer the scale, the better the theorist's prediction! So what *a priori* appears to be an arbitrarily complex dynamics turns out to be something very simple, and common to many apparently unrelated phenomena.

This is the essence of the recent progress here: large classes of nonlinear systems exhibit transitions to chaos which are *universal* and *quantitatively* measurable in a variety of experiments. This advance can be compared to past advances in the theory of solid-state phase transitions (cf. Chapter 10); for the first time we can predict and measure "critical exponents" for turbulence. But the breakthrough consists not so much in discovering a new set of scaling numbers, as in developing a new way to do physics. Traditionally we use regular as zeroth-order approximations to physical systems, and account for weak nonlinearities perturbatively. We think of a dynamical system as a smooth system whose evolution we can follow by integrating a set of differential equations. The new insight is that the zeroth-order approximations to strongly nonlinear systems should be quite different. They show an amazingly rich structure which is not at all apparent in their formulation in terms of differential equations. However, these systems do show self-similar structures which can be encoded by universality equations of a type which we will develop here.

As there is already much good literature on this subject, there is no point in reproducing it here. Instead, we provide the references which cover the same ground. However, it must be realized that the essence of this subject is incommunicable in print; here the intuition is developed by computing.

An introduction to the subject is provided by either of the two general reprint collections: one edited by Cvitanović,[1] and the other by Hao Bai-Lin.[2] Even though the selections overlap in only a few articles, both collections contain introductory overviews, the fundamental theoretical papers, and the most important experimental observations.

The theory of period-doubling universal equations and scaling functions is most helpfully developed in Kenway's notes on Feigenbaum's lectures.[3] Another excellent text is the Collet and Eckmann monograph.[4]

The universality theory for complex period n-tuplings is developed by Cvitanović and Myrheim.[5]

A nice discussion of circle maps and their physical applications is given by Jensen *et al.*[6] and Bohr *et al.*[7] The universality theory for golden mean scalings was developed by Shenker,[8] Feigenbaum,[9] and Ostlund.[10] The scaling functions for circle maps are discussed by Cvitanović.[11] The above authors cover all of the theory discussed in this chapter. Experimental and theoretical advances can be swiftly appraised by scanning the *Physical Review Letters*, *Physics Today* "Search and Discovery" section, and the front pages of *The New York Times*!

12.2. Complex Universality

In this section (based on work in collaboration with Myrheim[5]), we develop the universality theory for period n-tuplings for complex maps.[12,13] This example is chosen for its beauty; here one should be able to visualize the renormalization transformations and the universal scalings as encodings of the self-similar patterns generated by the dynamics of the system.

We shall study metric properties of the asymptotic iterates of

$$z_{n+1} = f(z_n) \tag{12.2.1}$$

where $f(z)$ is a polynomial in the complex variable z with a quadratic critical point, i.e., with power-series expansion of the form

$$f(z) = a_0 + a_2(z - z_c)^2 + \cdots \tag{12.2.2}$$

Typical model mappings of this type are

$$f(z) = p + z^2 \tag{12.2.3}$$

$$f(z) = \lambda z(1 - z) \tag{12.2.4}$$

When such mappings are used to model dynamical systems with z a real variable and the "nonlinearity" parameter p real, the asymptotic attractor is conveniently represented by a "bifurcation tree," i.e., by a two-dimensional plot with p as one axis and values of the asymptotic iterates for given p plotted along the other axis.

It is not possible to describe asymptotics of complex iterations in this way, as their iteration space has two (real) dimensions, and period n-tuplings are induced by tuning a pair of (real) parameters. To describe the asymptotic iterates of complex maps we proceed in two steps.

First, we describe the *parameter* space by its *Mandelbrot set* M. The Mandelbrot set[14,15] is the set of all values of the mapping parameter [parameter p in the model mapping (12.2.3)] for which iterates of the critical point do not escape to infinity. [A *critical point* z_c is a value of z for which the mapping $f(z)$ has vanishing derivative, $f'(z_c) = 0$. In equation (12.2.3) $z = 0$ is the critical point.] The Mandelbrot set for the mapping (12.2.4) is plotted in Figure 12.1.

Second, we characterize the *asymptotic iterates* for a given value of the parameter either by their *basin of attraction*, or by their *attractor*. The basin of attraction K is the set of all values of z which are attracted toward the attractor under iteration by $f(z)$. A typical basin of attraction is plotted in Figure 12.2. The boundary of K, or the *Julia set* J, is the closure of all unstable fixed points of all iterates of $f(z)$.

Theorem. For parameter values within the Mandelbrot set M, the Julia set J is connected. If all critical points iterate to infinity, J is a Cantor set.

If the nth iterate of $f(z)$ equals z, the set of points $z_k = f^k(z_0)$, $k = 0, 1, 2, \ldots, n-1$ form a periodic orbit (or cycle) of length n. If

$$|df^n(z_k)/dz_k| < 1 \tag{12.2.5}$$

the orbit is *attractive*. The *attractor* L is the periodic orbit $z_0, z_1, \ldots, z_{n-1}$. If

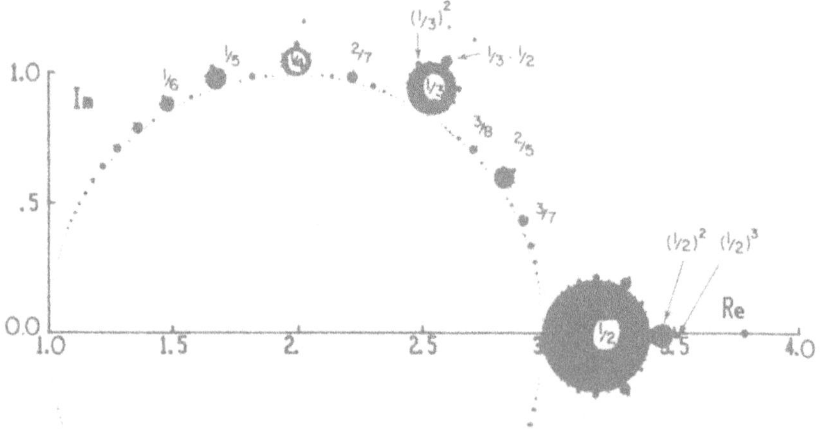

Figure 12.1. The Mandelbrot set M is the region in the complex parameter plane for which the critical point of the mapping (12.2.4) does not iterate away to infinity. Inside the big circle (left open for clarity) iterations converge to a fixed point. The full region has two symmetry axes, Re $\lambda = 1$ and Im $\lambda = 0$, so only one quarter is shown. The usual period-doubling sequence is on the real axis. The winding numbers of the periodic orbits corresponding to larger leafs of M are indicated. See Mandelbrot[15] for detailed scans of this set.

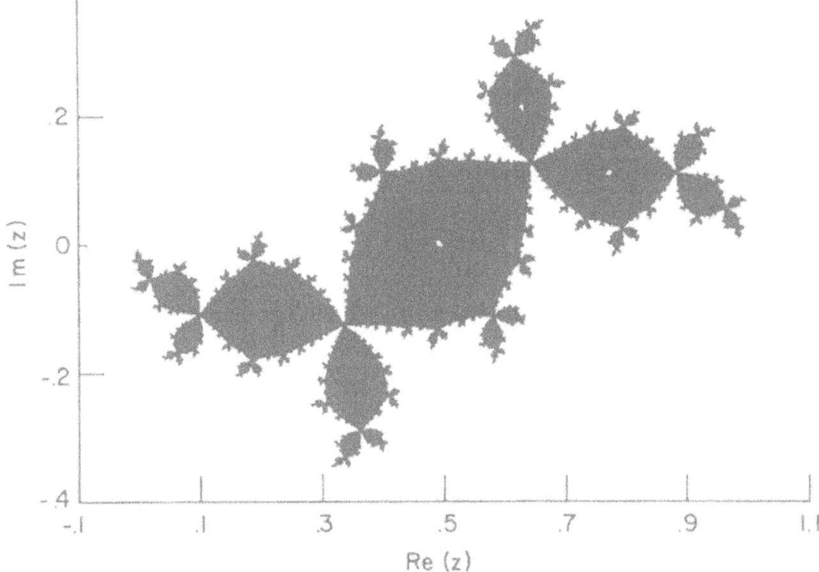

Figure 12.2. The basin of attraction for the superstable 3-cycle of mapping (12.2.4). Any initial z from the black region converges toward the superstable 3-cycle, denoted by the three white dots. The basin of attraction for mapping (12.2.3) superstable 3-cycle is the same, up to a coordinate shift and rescaling.

the derivative (12.2.5) is vanishing, the orbit is *superstable*, and (by the chain rule) a critical point is one of the cycle points. For polynomial mappings $z = \infty$ plays a special role; it is always a superstable fixed point. The following theorem eases attractor searches:

Theorem. The basin of attraction K contains at least one critical point.

The precise shape of the Mandelbrot set M depends on $f(z)$, but it always resembles a cactus; see Figure 12.1. Here we are not so much interested in the entire M, as in the *Mandelbrot cactus*, the set of connected components of M generated from a single fixed point by all possible sequences of all possible period n-tuplings.

To summarize, the parameter dependence of asymptotic iterates of mapping $f(z)$ is described by the Mandelbrot set M. For each point inside M, the asymptotic iterates are characterized by their basin of attraction K, the Julia set J, and the attractor L.

Now that the general setting is established, we can turn to a detailed study of the way in which a fixed point of the complex mapping (12.2.1) branches into an n-cycle. The fact that the same analysis applies to period n-tupling of any k-cycle into an nk-cycle will be seen to be the origin of the self-similarity of the Mandelbrot cactus.

The stability of a fixed point is given by

$$\rho = df(z)/dz \tag{12.2.6}$$

and we take, without loss of generality, the fixed point to be at $z = 0$, and $f(z)$ with a power-series expansion

$$f(z) = \rho z + a_J z^J \tag{12.2.7}$$

To bring the map into a standard form, we change the variable

$$w = z + b_J z^J \tag{12.2.8}$$

and "flatten" out the mapping close to the fixed point by choosing successively $b, b, \ldots$ in such a way that as many leading nonlinear terms as possible vanish in equation (12.2.7). If ρ is sufficiently close to the nth root of unity, $\omega = \exp(i2\pi m/n)$, and z is close to 0, the typical behavior of the new iteration function is the same as

$$f(z) = \rho z + z^{n+1} \tag{12.2.9}$$

This function has an n-cycle

$$z_J = \omega^J z_0, \qquad z_0^n = \omega - \rho \tag{12.2.10}$$

For $\rho = \omega$ this n-cycle coincides with the fixed point $z = 0$. In the neighborhood of $\rho = \omega$ we have

$$\begin{aligned} dz_n/dz_0 &= f'(z_0)f'(z_1)\cdots f'(z_{n-1}) \\ &= [\rho + (n+1)z_0^n]^n \\ &= 1 - (\rho - \omega)n^2/\omega + \cdots \end{aligned} \tag{12.2.11}$$

For $\rho = (1 + \varepsilon)\omega$ the n-cycle (12.2.10) of the mapping (12.2.9) is stable if

$$|1 - n^2\varepsilon| < 1 \tag{12.2.12}$$

while the fixed point is stable if

$$|1 + \varepsilon| < 1 \tag{12.2.13}$$

The mapping (12.2.9) is equivalent to equation (12.2.7) only for small z, so the above analysis of how a fixed point of equation (12.2.7) becomes unstable and branches into the n-cycle is valid only for infinitesimal $n\varepsilon$.

In conclusion, whenever a fixed point becomes unstable at $\rho = n$th root of unity, it branches into an n-cycle which immediately becomes stable. As any stable cycle becomes unstable in the same fashion, branching into a new stable cycle with a multiple of the original cycle length, and as any such cycle is stable inside a disk-like region in the complex parameter plane, the union of all these stability regions is a self-similar *Mandelbrot cactus.*

Next we turn to a study of infinite sequences of period n-tuplings, each branching characterized by the same ratio m/n.

As discussed above, a stable n^k-cycle becomes unstable and branches into an n^{k+1}-cycle when the parameter λ passes through a value such that the stability $\rho_k(\lambda)$ [as defined in equation (12.2.6.)] attains the critical value

$$\rho(\lambda) = \omega = \exp(i2\pi m/n) \qquad (12.2.14)$$

For ρ sufficiently close to this value the system is modeled by equation (12.2.9). From equation (12.2.11) it follows that near the transition from an n^k-cycle to an n^{k+1}-cycle

$$\rho_{k+1} = 1 - (\rho_k - \omega)n^2/\omega + \cdots \qquad (12.2.15)$$

hence

$$d\rho_{k+1}/d\lambda|_{\rho_{k+1}=1} = -(n^2/\omega)\, d\rho_k/d\lambda|_{\rho_k=\omega} \qquad (12.2.16)$$

and at the transition there is a scale change by the complex factor $-n^2/\omega$ which is independent of k.

Each leaf of the Mandelbrot cactus Figure 12.1 corresponds to an m/n cycle, and the parameter value for the superstable m/n cycle corresponds to the center of the leaf. The above argument suggests that the leaf is n times smaller than the cactus, and that it is rotated by a phase factor $-1/\omega$. The very geometry of the Mandelbrot cactus, Figure 12.1, suggests such scaling. This scaling is not exact, because the above analysis applies only to the infinitesimal neighborhood of the junction of a leaf to the cactus; however, the evaluation of the exact scaling numbers shows that this is a rather good approximation to the exact scaling. We conjecture that $\delta_{m/n} \to n^2/\omega$ as $m/n \to 0$, exactly. This conjecture is supported by the numerical evaluation of the value of δ.[5]

The exact scaling is obtained by comparing values of the parameter λ corresponding to successive $(m/n)^k$ superstable cycles, i.e., λ values such that $\rho_k(\lambda_k) = 0$. As each cactus leaf is similar to the entire cactus, the ratios of the sizes of the successive stability regions corresponding to successive $(m/n)^k$-cycles tend to a limit as $k \to \infty$:

$$\delta_{m/n} = \lim_{k \to \infty} (\lambda_k - \lambda_{k-1})/(\lambda_{k+1} - \lambda_k) \tag{12.2.17}$$

The scaling number δ tells us by how much we have to change the parameter λ in order to cause the next m/n period n-tupling. In particular, $\delta_{1/2} = 4.669\ldots$ is the Feigenbaum δ for the period doublings in the real one-dimensional mappings.

Scaling in the parameter space (generalized Feigenbaum δ) is a consequence of the self-similarity of the Mandelbrot cactuses. In the same way the self-similarity of the Julia sets (or the asymptotic attractors) suggests a scaling law in the iteration space z, which we discuss next. This law will characterize the scales of successive trajectory splittings (generalized Feigenbaum α).

The self-similarity we are alluding to can be seen by comparing the basin of attraction for the superstable three-cycle, Figure 12.2, and for the superstable nine-cycle, Figure 12.3. In the latter figure the three-cycle basin of attraction is visible in the center, rotated and scaled down by a factor whose asymptotic limit is the generalization of Feigenbaum α to period triplings.

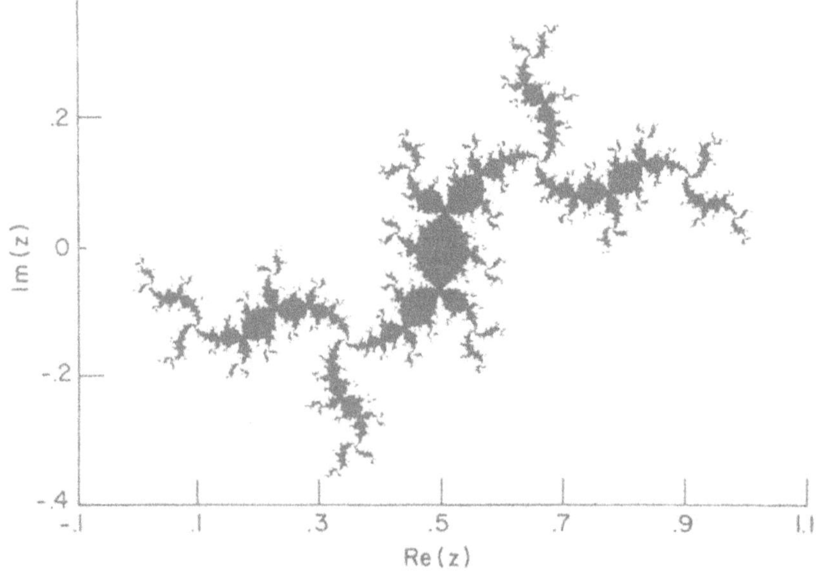

Figure 12.3. The basin of attraction for the superstable 9-cycle for iterates of the model mapping (12.2.4). The scaled down version of the 3-cycle basin of attraction, Figure 12.2, is visible in the center.

This scaling number α can be computed by comparing the successive superstable cycles at successive parameter values λ_k, λ_{k+1}. As $k \to \infty$, the sequence of values of λ converges to λ_∞, and the superstable n^k-cycles converge to an n^∞-cycle. The attractor is self-similar: the orbits on succeeding levels are related by rescaling and rotation by a complex number which asymptotically approaches

$$\alpha_{m/n} = \lim_{k \to \infty} (z_{n^k} - z_0)/(z_{n^{k+1}} - z_0) \tag{12.2.18}$$

α characterizes the scale of trajectory splitting at each period n-tupling. (For $m/n = \frac{1}{2}$ this is the Feigenbaum $\alpha = -2.5029\ldots$.)

So period n-tuplings are self-similar both in the iteration space and in the parameter space: not only does the asymptotic orbit resemble itself under rescaling and rotation by α, but also each leaf of the Mandelbrot cactus resembles the entire cactus under rescaling and rotation by δ.

These self-similarities can be described by means of the following three operations.

The first operation is a rescaling of the parameter and iteration spaces:

$$[Rf]_p(z) = af_{p/d}(z/a) \tag{12.2.19}$$

With the appropriate choice of complex numbers d (a), a leaf of the Mandelbrot cactus (a part of the attractor) can be rescaled and rotated to the size and the orientation of the entire cactus (entire attractor).

We fix the origin of p and z by requiring that $z = 0$ be a critical point of the mapping $f_p(z)$, and, for the parameter value $p = 0$, a superstable fixed point as well [equation (12.2.3) is an example of such mapping]. We fix the scale of p and z by requiring that the superstable m/n cycle occurs for the parameter value $p = 1$ and that

$$f_1(0) = 1 \tag{12.2.20}$$

The second operation shifts the origin of the parameter space to the center of the m/n-leaf of the Mandelbrot cactus (p corresponding to the superstable m/n cycle):

$$[Sf]_p(z) = f_{1+p}(z) \tag{12.2.21}$$

The third operation iterates $f_p(z)$ n times:

$$[Nf]_p(z) = f_p^n(z) \tag{12.2.22}$$

By definition, $[Sf]_0(z) = f_1(z)$ has a superstable m/n cycle, so its nth iterate has a superstable fixed point, $[NSf]_0(0) = 0$.

The parameter shift S overlies the Mandelbrot cactus over its m/n leaf, and the Julia set for $[Nf]_1(z)$ resembles the Julia set for the superstable fixed point $f_0(z)$ (see Figures 12.2 and 12.3, for example). Finally we adjust the scale of the new M, J sets by requiring that the scale factors a, d in equation (12.2.19) are such that $[RNSf]_p(z)$ satisfies the same normalization condition (12.2.20) as the initial function $f_p(z)$. This shifting and rescaling is illustrated in Figure 12.4.

The combined effect of the rescaling, parameter shift, and iteration is summarized by the operator $T^* = RNS$;

$$[T^*f]_p(z) = af^n_{1+p/d}(z/a) \tag{12.2.23}$$

If we take a polynomial $f_p(z)$ and act on it with T^*, the result will be a longer polynomial with similar M and J sets. For a finite number of T^* operations the scaling numbers d and a depend on the choice of the initial mapping $f_p(z)$. If we apply T^* infinitely many times, a and d converge to the universal numbers α and δ, and $T^*f_p(z)$ converges to a one-parameter family

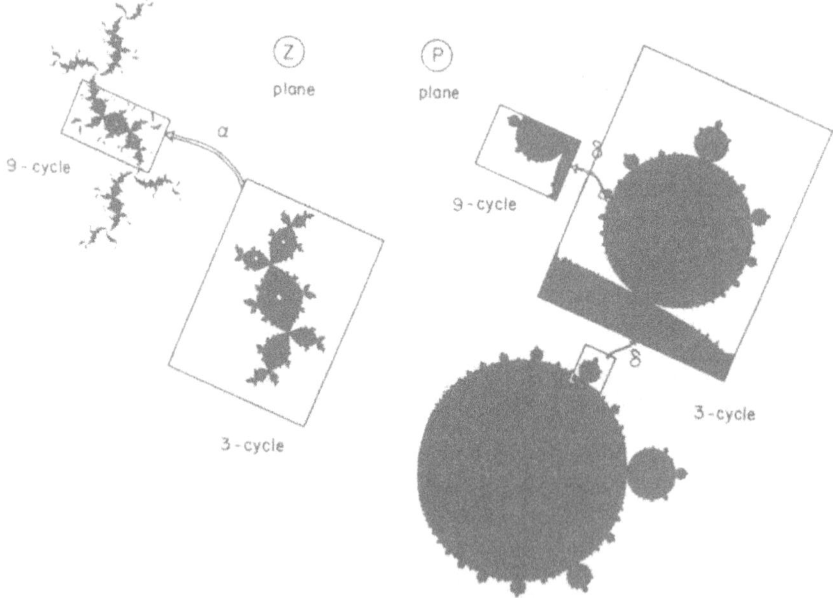

Figure 12.4. The unstable manifold method illustrated by period triplings. The parameter is shifted from the center of a cactus leaf to its $\frac{1}{3}$ leaf, the $\frac{1}{3}$ leaf is rescaled and rotated by δ, and the basin of attraction of third iterates is rescaled and rotated by α. The Mandelbrot cactus and the basin of attraction for the unstable manifold $g_p(z)$ is self-similar under such shifting and rescaling.

of universal functions which is the fixed point of the operator T^*:

$$\begin{aligned} g_p(z) &= [T^*g]_p(z) \\ &= \alpha g^n_{1+p/\delta}(z/\alpha) \end{aligned} \tag{12.2.24}$$

This universal equation determines both $g_p(z)$ and the universal numbers α and δ. The family of universal functions $g_p(z)$ is called the unstable manifold (the reason is explained in the introductory lectures of Cvitanović[(1)]).

To summarize, the T^* operation encodes simultaneously the self-similarity of the parameter space (Mandelbrot cactuses) and of the iteration space (Julia sets). Being no more than a redefinition of variables, it is exact, and it is an explicit implementation of the above self-similarities; T^* magnifies the nth iterate of the $(m/n)^{k+1}$-cycle and overlies it onto the $(m/n)^k$-cycle (see Figure 12.4). Asymptotically the self-similarities are exact, and the procedure converges to the unstable manifold, a one-dimensional line of universal functions g_p.

Not only are the N, S, R operations the natural encoding of the complex universality, but they are also useful computational tools.

The universal equation (12.2.24) can be solved numerically by approximating the unstable manifold by a truncation of the double power-series expansion

$$g_p(z) = \sum_{j,k\geq 0} c_{jk} z^{2j} p^k \tag{12.2.25}$$

We start with equation (12.2.3) as a two-term approximation to $g_p(z)$. Repeated application of the T^* operation (12.2.23) generate a longer and longer double polynomial in z and p; this procedure converges asymptotically to the unstable manifold $g_p(z)$. We implement the shifting and iteration operations S and N as numerical polynomial substitution routines, truncating all polynomials as in expansion (12.2.25). The T operation is completed by the rescaling operation R, in equation (12.2.19). The scaling numbers d and a are fixed by the normalization conditions (12.2.20). We use Newton's method to find the parameter value corresponding to the superstable m/n-cycle. This determines d, and a then follows directly from condition (12.2.20). The result is a new approximation to $g_p(z)$. Asymptotically, the values of d converge to δ and of a converge to α. We keep applying the truncated T^* operation until the coefficients in equation (12.2.25) stabilize to the desired accuracy.

The self-similar structure of the Mandelbrot cactus, Figure 12.1, suggests a systematic way of presenting the universal numbers that we have computed in the previous section. We observe that roughly halfway between any two large leafs on the periphery of a Mandelbrot cactus (such as $\frac{1}{2}$ and $\frac{1}{3}$) there is the next largest leaf (such as $\frac{2}{5}$). Furthermore, we know from equation (12.2.26) that the size of the "cactus leaf" corresponding to period n-tupling is of order n^{-2}. Hence the natural hierarchy is provided by an interpolation scheme, which organizes rational numbers m/n into self-similar levels of increasing period lengths n. Such a scheme is provided by Farey numbers.[16]

Implicit in the Farey numbers are scaling laws that relate the universal numbers. It turns out that the same Farey structure is a very useful tool for the study of mode-locking intervals for circle maps. We refer the reader to the circle-map references listed in the introduction.

References

1. P. Cvitanović (ed.), *Universality in Chaos*, Hilger, Bristol (1984).
2. B.-L. Hao, *Chaos*, World Scientific, Singapore (1984).
3. M. J. Feigenbaum and R. D. Kenway, in: *Statistical and Particle Physics* (K. C. Bowler and A. J. McKane, eds.), Scottish Universities Summer School in Physics, Dept. of Physics, University of Edinburgh (1984).
4. P. Collet and J.-P. Eckmann, *Iterated Maps on Interval as Dynamical Systems*, Birkhauser, Boston (1980).
5. P. Cvitanović and J. Myrheim, *Commun. Math. Phys.* (to appear).
6. M. H. Jensen, P. Bak, and T. Bohr, *Phys. Rev.* A **30**, 1960 (1984).
7. T. Bohr, M. H. Jensen, and P. Bak, *Phys. Rev. A* **30**, 1970 (1984).
8. S. J. Shenker, *Physica* **5D**, 405 (1982).
9. M. J. Feigenbaum, L. P. Kadanoff, and S. J. Shenker, *Physica* **5D**, 370 (1982).
10. S. Ostlund, D. Rand, J. Sethna, and E. D. Siggia, *Physica* **D8**, 303 (1983).
11. P. Cvitanović, B. Shraiman, and B. Soderberg, *Phys. Scr.* **32**, 263 (1985).
12. A. I. Golberg, Ya. G. Sinai, and K. M. Khanin, *Usp. Mat. Nauk* **38**, 159 (1983).
13. P. Cvitanović and J. Myrheim, *Phys. Lett.* **94A**, 329 (1983).
14. B. B. Mandelbrot, *Ann. N.Y. Acad. Sci.* **357**, 249 (1980).
15. B. B. Mandelbrot, *The Fractal Geometry of Nature*, Freeman, San Francisco (1982).
16. G. H. Hardy and E. M. Wright, *UX* (*Theory of Numbers*), Oxford Univ. Press, Oxford (1938).

13

Order and Chaos in Hamiltonian Systems

I. C. Percival

13.1. Survey of Modern Hamiltonian Dynamics

13.1.1. Regular and Chaotic Motion

Traditionally, in applications, Hamiltonian systems with a finite number of degrees of freedom have been divided into those with few degrees of freedom, which were supposed to exhibit some kind of regular ordered motions, and those with large numbers of degrees of freedom, for which the methods of statistical mechanics should be used.

The last few decades have seen a complete change of view. This change affects almost all the practical applications, particularly in mathematical physics, which has been dominated for many decades by linear mathematics, coming from quantum theory.

The motion of an actual Hamiltonian system is usually neither completely regular nor properly described by the methods of statistical mechanics. It exhibits both regular and irregular or chaotic motion for different initial conditions, and the transition between the two types of motion, as the initial conditions are varied, is complicated, subtle, and beautiful.

The nature of the regular motion of a conservative system of m degrees of freedom is the same as that of the traditional integrable systems; when bounded, it is quasi-periodic almost everywhere, with a discrete set of m frequencies, together with their integer linear combinations. The regular motion for a given initial condition is confined to a smooth m-dimensional region in the $2m$-dimensional phase space.

I. C. Percival • School of Mathematical Sciences, Queen Mary College, University of London, London E1 4NS, England.

By contrast, the nature of chaotic motion is still not fully understood. It is unstable in a strong exponential sense. For a conservative system it usually cannot be confined to any smooth region of dimension less than $2m - 1$, the confinement in the energy shell required by energy conservation. But it does not normally occupy the whole of the energy shell as required by the ergodic principle of traditional statistical mechanics. Far away from the regular regions of phase space, the chaotic motion resembles a simple diffusion process, but close to them it does not. Chaotic motion is common for conservative systems of two degrees of freedom.

Chaotic motion appears when local exponential divergence of trajectories is accompanied by global confinement in the phase space. The divergence produces a local stretching of the phase space, but, because of the confinement, this stretching cannot continue without folding (or sometimes chopping). Repeated folding and refolding produces very complicated behavior that is described as chaotic.

For Hamiltonian systems the stretching in one direction is exactly compensated by a shrinking in another direction, so that area in the phase space is conserved, while in dissipative systems there is no such compensation. There can be chaotic motion for both, but it differs in detail. Chaotic motion may appear to be ordered or regular over long periods of time, or may develop very rapidly.

Poincaré was interested in the application of Hamiltonian dynamics to the motion of the bodies of the solar system. He was the first to see something of this new view of dynamics. He recognized in particular that Hamiltonian systems can show qualitatively different behavior for arbitrarily small changes in a parameter. On the convergence of a particular form of perturbation theory for a Hamiltonian system with two degrees of freedom he wrote:

> I have shown that the irrational ratios of the periods can be separated into two categories: those for which the series converge, and those for which the series diverge, and that in any interval, however small it may be, there are values from the first category, and values from the second.

But Poincaré was unable to convey his magnificent vision to his contemporaries.

The recent advances, from the 1950s, have been more closely related to other applications. For Hamiltonian dynamics, these include the motion of the stars in the galaxies, of atoms in molecules and solids, of various fundamental particles in accelerators and storage rings, and the shape of magnetic field lines in plasma containment devices. Problems in the foundations of statistical mechanics have also played an important role. Only in the last few years have these advances helped to solve problems of the solar system, and only gradually was it realized that separate studies in these different fields of application had so much in common. Numerous books and reviews on modern Hamiltonian dynamics have appeared.[1-16]

13.1.2. Variational Principles

Variational principles are part of the traditional dynamics and they will be used to form a link with the new. A variational approach to Hamiltonian dynamics is valuable for a number of reasons:

1. Variational quantities are simpler than the corresponding equations of motion and so allow a simpler formulation.
2. A variational principle can be used as a basis for computation and a criterion for precision in approximations.
3. Variational principles can be used to prove the existence or non-existence of different types of motion.
4. Variational principles can be used to gain insight.

We concentrate on the insight, and start with some elementary theory.

For Hamiltonian or Lagrangian systems, there are variational principles for orbits and for other invariant sets of the phase space. Let

$$q = (q^1, q^2, \cdots, q^m) \tag{13.1.1}$$

be the m-dimensional coordinate of a system of m degrees of freedom, and let

$$L(q, \dot{q}, t) \tag{13.1.2}$$

be its Lagrangian, where the dot means differentiation with respect to time.

The variational principle then states that if the initial and final configurations and times are

$$(q_0, t_0) \qquad \text{and} \qquad (q_1, t_1) \tag{13.1.3}$$

then there will be an orbit $q(t)$ between these points if and only if

$$\Delta W \equiv \Delta \int_{t_0}^{t_1} dt\, L(q, \dot{q}, t) = O(\Delta q)^2 \tag{13.1.4}$$

for all variations Δq about the orbit in which the end points are fixed in configuration and time. By standard methods, Lagrange's equations

$$\frac{d}{dt}\left(\frac{\partial L}{\partial \dot{q}}\right) - \frac{\partial L}{\partial q} = 0 \tag{13.1.5}$$

follow, and by using the definitions

$$p = \frac{\partial L}{\partial \dot{q}}; \qquad H(q, p) = p \cdot \dot{q} - L \tag{13.1.6}$$

we get Hamilton's equations

$$\dot{p} = -\partial H/\partial q; \qquad \dot{q} = \partial H/\partial p \tag{13.1.7}$$

where the partial derivatives each represent m-dimensional gradient vectors

$$\partial/\partial q = \left(\frac{\partial}{\partial q^1}, \frac{\partial}{\partial q^2}, \ldots, \frac{\partial}{\partial q^m}\right); \qquad \text{etc} \tag{13.1.8}$$

For example, a particle of unit mass constrained to move along the x axis of a coordinate system in the potential $V(x)$ has the Lagrangian

$$L(x, \dot{x}) = \tfrac{1}{2}\dot{x}^2 - V(x) \tag{13.1.9}$$

Lagrange's equation is

$$\frac{d^2x}{dt^2} = -V'(x) = F(x) \tag{13.1.10}$$

which is the same as Newton's equation of motion, and Hamilton's equations are

$$\frac{dp}{dt} = -V'(x) = F(x)$$

$$\frac{dx}{dt} = p \tag{13.1.11}$$

13.1.3. Static Model

The same variational principle can be used to define the static equilibrium of a uniform elastic string of unit Hooke's constant hanging freely in the reversed potential $-V(x)$.

Suppose the points of the string are labeled by the distances from some fixed point on the string when it is under unit tension in zero potential. Let $x(s)$ be the position of the point s in the presence of the potential $-V(x)$. Then apart from an additive constant, the total energy E of the string is given by the sum of its elastic energy and potential energy:

$$E = \int ds\left[\tfrac{1}{2}\left(\frac{dx}{ds}\right)^2 - V(x)\right] \tag{13.1.12}$$

and the equilibrium is determined by the condition $\Delta E = O(\Delta x)^2$. The form of the variational principle is the same as for the Lagrangian dynamical system: the orbit of a particle and the configuration of a string, when V is the gravitational potential, are illustrated in Figure 13.1 for the case in which initial and final points are the same.

13.1.4. Discrete Time

If a continuous force is replaced by a sequence of impulses at integer times, then the generating function over unit time may be considered as the Lagrangian of a system with discrete time:

$$L(q_t, q_{t+1}, t) \tag{13.1.13}$$

The stationary functional is now an action sum

$$W = \sum_{t=t_0}^{t_1-1} L(q_t, q_{t+1}, t) \tag{13.1.14}$$

The orbit space is finite dimensional and the stationary principle is

$$\Delta W = O(\Delta q)^2 \quad \Leftrightarrow \quad \frac{\partial W}{\partial q_t} = 0 \qquad (t_0 < t < t_1) \tag{13.1.15}$$

giving Lagrange's equations

$$\frac{\partial}{\partial q_t} L(q_t, q_{t+1}, t) + \frac{\partial}{\partial q_t} L(q_{t-1}, q_t, t-1) = 0 \tag{13.1.16}$$

For a particle of unit subject to impulses or "kicks" $F(x_t)$, the Lagrangian is

$$L = \frac{(x_{t+1} - x_t)^2}{2} - V(x_t) \tag{13.1.17}$$

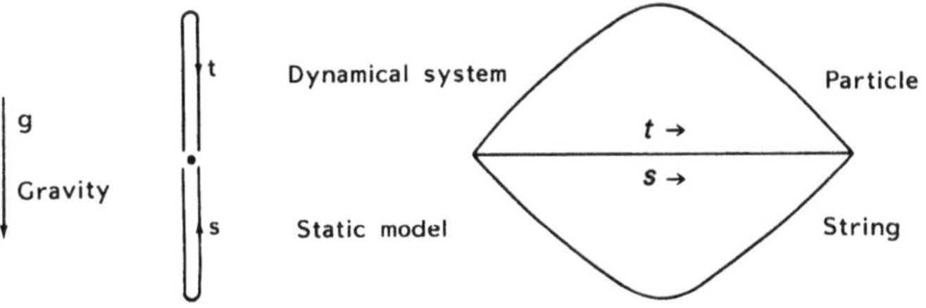

Figure 13.1. Uniform string hanging in uniform gravity, and upward particle trajectory.

and Lagrange's equations become

$$x_{t+1} - 2x_t + x_{t-1} \equiv \delta^2 x_t = \frac{-\partial V(x_t)}{\partial x_t} \equiv F(x_t) \tag{13.1.18}$$

This can be considered as the discrete time form of Newton's equation for impulses.

The corresponding Hamilton's equations are obtained by putting

$$p_t = x_t - x_{t-1}, \tag{13.1.19}$$

so that

$$p_{t+1} = p_t + F(q_t) \tag{13.1.20a}$$

$$x_{t+1} = x_t + p_{t+1} \tag{13.1.20b}$$

Each equation represents a shear, which is clearly area preserving, so Hamiltons's equations represent an area-preserving map.

The static model corresponding to the discrete time system is like the previous static model, but the elastic string is considered to be massless and unaffected by the potential. Unit masses s are distributed along the string, and they are spaced at unit intervals when the string is at unit tension; the masses are subject to a force $+V'(x_s)$, giving an energy

$$E = \frac{(x_{s+1} - x_s)^2}{2} - V(x_s) \tag{13.1.21}$$

The static and dynamical systems are illustrated in Figure 13.2.

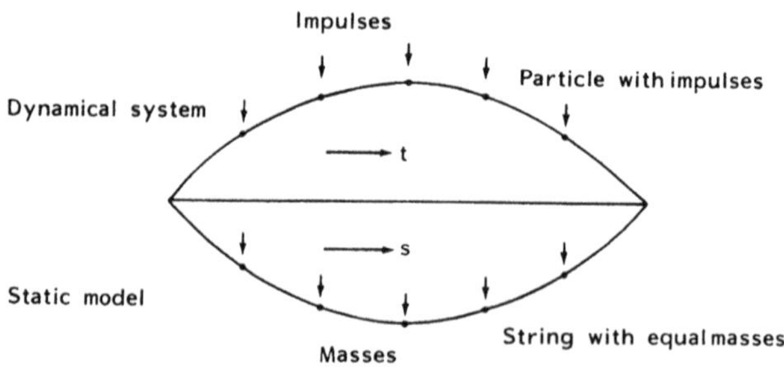

Figure 13.2. Light string with equal masses and particle subject to equal impulses.

13.1.5. Vertical Pendulum and Standard Map

For the vertical pendulum the configuration coordinate is the angle ψ, or, for convenience, the variable $x = \psi/2\pi$, that repeats at unit intervals. The Lagrangian for unit moment of inertia is

$$L(x, \dot{x}) = \tfrac{1}{2}\dot{x}^2 + \frac{k}{(2\pi)^2}\cos(2\pi x) \tag{13.1.22}$$

where the constant k has been chosen for later convenience. The equations of motion are

$$\frac{d^2x}{dt^2} = -(k/2\pi)\sin(2\pi x) \tag{13.1.23}$$

in Newtonian form and

$$\frac{dp}{dt} = -(k/2\pi)\sin(2\pi x) \tag{13.1.24a}$$

$$\frac{dx}{dt} = p \tag{13.1.24b}$$

in Hamiltonian form. This represents an area-preserving flow on a cylindrical phase space, as illustrated in Figure 13.3.

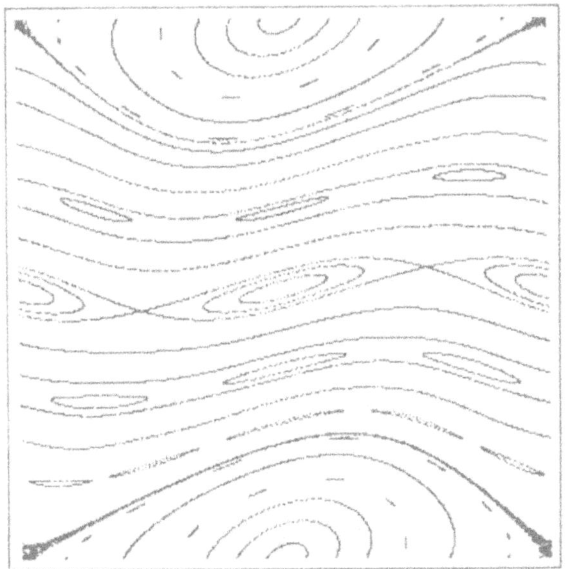

Figure 13.3. Standard map with different initial conditions, $k = 0.5$.

As for all conservative Hamiltonian systems of one degree of freedom, the system is integrable and the orbits are confined to contours of the Hamiltonian

$$H(x, p) = \frac{p^2}{2} - \frac{k}{(2\pi)^2} \cos(2\pi x) \tag{13.1.25}$$

The rotational and vibrational motion are both periodic and each is represented by a periodic function $x(t)$ of period $1/\nu$; for periodic motion there are no natural end points, and the orbit is better defined in terms of the period or the frequency ν. Let θ be an angle variable such that

$$\frac{dx}{dt} = \nu \frac{dx}{d\theta} \tag{13.1.26}$$

($2\pi\theta$ is an angle). Then for fixed ν we have a variational principle

$$\begin{aligned} \Delta W &= \Delta \oint d\theta \, L\Big(x(\theta), \nu dx/d\theta\Big) \\ &= O(\Delta x)^2 \end{aligned} \tag{13.1.27}$$

where $x(\theta)$ represents a closed curve, which is invariant when it satisfies the variational principle; following topological definitions such a closed curve is known as an invariant circle: periodic orbits are invariant circles.

The standard map is the discrete time version of the vertical pendulum. It represents a rotor subject to an angle-dependent impulse each unit time, and may be considered as a very simplified model of some types of particle accelerator: there are many other applications, and it is the most well-studied realistic area-preserving map. Its Lagrangian is

$$L(x_t, x_{t+1}) = \frac{(x_{t+1} - x_t)^2}{2} + \frac{k}{(2\pi)^2} \cos(2\pi x_t) \tag{13.1.28}$$

the Lagrange equations are

$$\delta^2 x_t \equiv x_{t+1} - 2x_t + x_{t-1} = -\frac{k}{2\pi} \sin(2\pi x_t) \qquad (\text{mod } 1) \tag{13.1.29}$$

or in standard Hamiltonian form

$$p_{t+1} = p_t - (k/2\pi) \sin(2\pi x_t) \qquad (\text{mod } 1) \tag{13.1.30}$$

$$x_{t+1} = x_t + p_{t+1} \qquad (\text{mod } 1) \tag{13.1.31}$$

The standard map is *not* integrable unless $k = 0$, and is used as a test case for nonintegrable systems of two degrees of freedom.

For k sufficiently small, it follows from the KAM theorem of Kolmogorov, Arnold, and Moser that a positive measure of invariant circles should exist,[17] but for the original forms of the theorem, sufficiently small meant less than 10^{-48} in appropriate units, which is smaller than the gravitation perturbation on a body falling under the Earth's gravity in London by the motion of a bacterium in Australia!

However, numerical calculations by Greene indicate that invariant circles of positive measure exist up to $k = 0.971635\ldots$[18] and the rigorous bounds are being improved.

Phase plots of points of orbits of the standard map are easy to produce using microcomputers, and are illustrated in Figures 13.3–13.5. There is a clear but complicated transition from orbits that appear to be regular and confined to invariant circles, to orbits that appear to be chaotic and seem to occupy regions of dimension two.

Figure 13.6, from Murray,[19] illustrates the chaotic regions in the 2:1 resonance with Jupiter, as a function of unperturbed semimajor axis and eccentricity. The asteroids clearly avoid the chaotic regions.

Now we extend the variational principles to area-preserving maps. Since periodic orbits consist of finite numbers of points they are not invariant circles, so we used a separate variational principle for each.

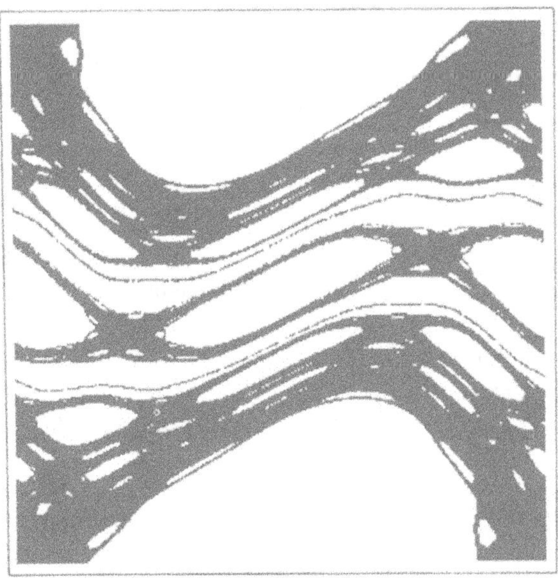

Figure 13.4. Standard map with $k = k_{\text{critical}}$.

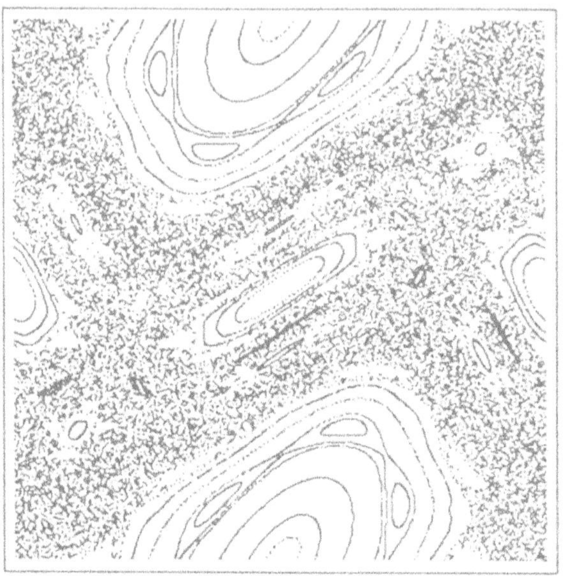

Figure 13.5. Standard map with $k = 1.3$.

For a periodic orbit of period n, we have

$$\Delta \sum_{t=0}^{n-1} L(x_t, x_{t+1}) = O(\Delta x)^2 \qquad (x_n = x_0) \tag{13.1.32}$$

while for the invariant circle $x(\theta)$ of frequency ν

$$\theta = \nu t \qquad (\text{mod } 1) \tag{13.1.33}$$

and

$$\Delta \oint d\theta\, L(x(\theta), x(\theta + \nu)) = O(\Delta x)^2 \tag{13.1.34}$$

But do invariant circles exist? For sufficiently small k they do, by the KAM theorem. There is a converse KAM theorem, which is much simpler than the KAM theorem itself, by which it has been proved that no rotational invariant circles exist for $k > 63/64$.[20] A rotational invariant circle wraps once around the cylindrical plane space, just like the rotational orbits of the vertical pendulum.

However, for any k, the Lagrangian integral (13.1.34) for the invariant circle is bounded below, so it must have at least one minimum. If the variational principle applies, then it must correspond to some invariant set. If it is not an invariant circle, what is it?

To find out, we consider the static models.

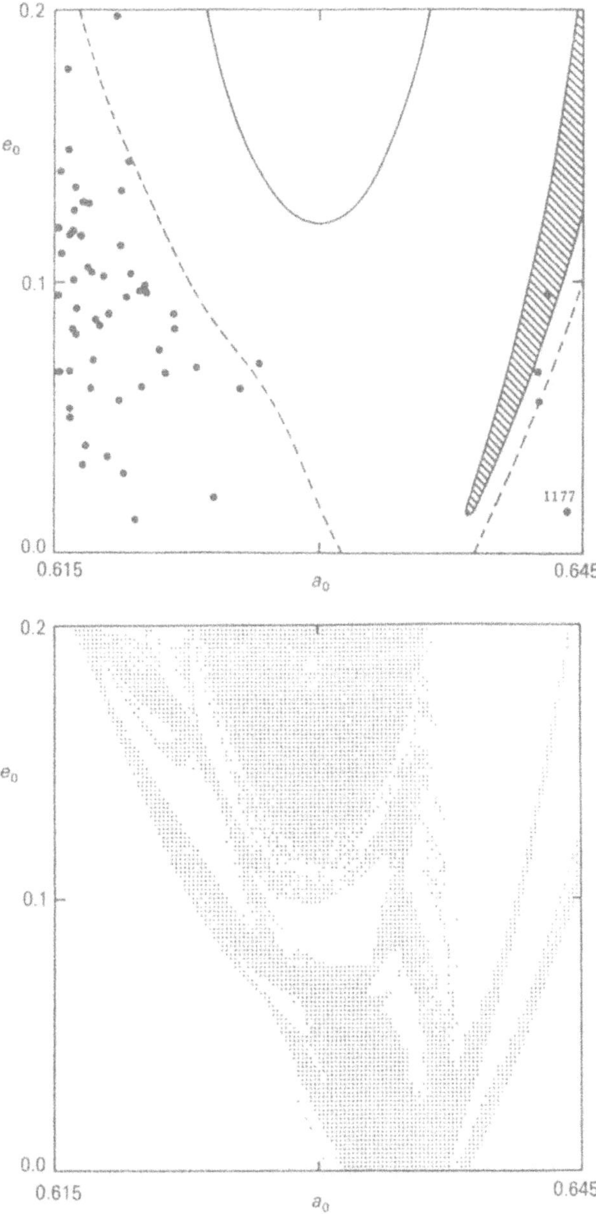

Figure 13.6. Irregular orbits near 2:1 resonance with Jupiter and location of observed asteroid orbits (after Murray[19]).

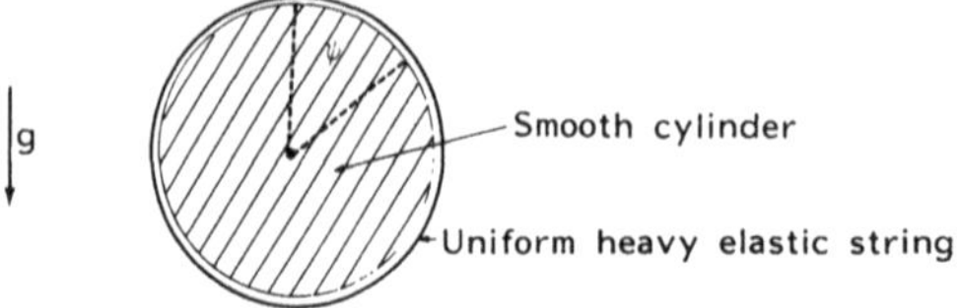

Figure 13.7. Static model for the vertical pendulum: a uniform heavy circular elastic string on a smooth horizontal cylinder.

13.1.6. Static Models and Cantori[21]

A rotational orbit of a vertical pendulum corresponds to a heavy elastic string wrapped once around a smooth cylinder with its axis situated horizontally in the Earth's gravitational field, as illustrated in Figure 13.7. In the absence of the field, $2\pi\theta$ is the angle that a point on the string would make with the upward vertical (as against the downward vertical for the pendulum). Keeping this labeling for the points on the string, $2\pi x(\theta)$ is the angle that such a point actually makes with the upward vertical in the Earth's gravitational field.

Clearly the string stretches at the top, and the resulting lower density corresponds to the greater velocity of the pendulum at the bottom of its swing.

The static model for the periodic orbits of the standard map is given in Figure 13.8, where two orbits of period 5 are illustrated.

The static model for an orbit of an invariant circle with irrational frequency ν is illustrated in Figure 13.9. The string is now an infinite helix, and the masses would be distributed uniformly along it in the absence of the field; in this unperturbed case, when viewed in projection along the axis, the masses are dense on a circle. By the KAM theorem, for most ν, a sufficiently small field preserves this dense structure: the invariant circle remains. If

$$\theta = \nu s + \theta_0 \qquad (\text{mod } 1) \tag{13.1.35}$$

is the angle variable, then $x(\theta)$ gives the orientation of the mass at position

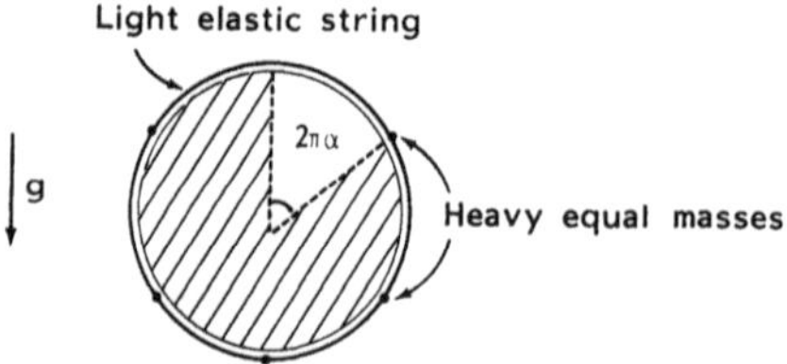

Figure 13.8. Static model for periodic orbit of standard map: equal heavy masses attached to a light circular elastic string on a smooth horizontal cylinder.

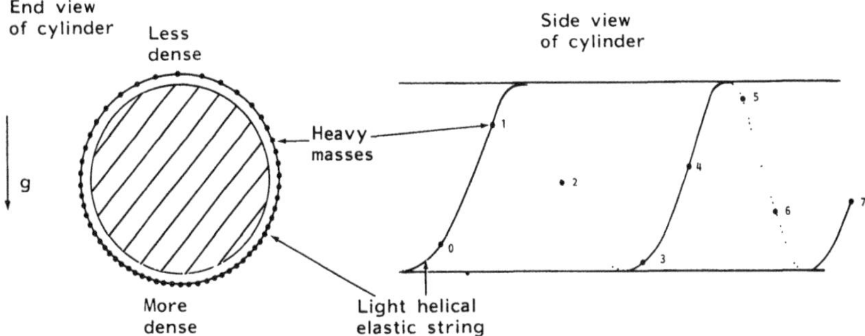

Figure 13.9. Static model for orbit on invariant circle: masses are on an infinite string wound helically on smooth horizonal cylinder. Viewed from the end the density of masses tends to infinity with the length of the string.

s, and the $x(\theta)$, which represents the distribution of masses around the cylinder, is the same $x(\theta)$ which represents the configuration of the standard map.

The corresponding momentum is given by

$$p(\theta) = x(\theta) - x(\theta - \nu) \tag{13.1.36}$$

Now suppose we increase the masses to 1 tonne. Clearly they will all bunch up near the bottom of the bottom of the cylinder, leaving a gap at the top, and the same must apply for any mass above some critical value. The gap will be iterated around the cylinder, and since the map is smooth, there must be a gap arbitrarily close to any mass. The invariant circle has become a Cantor set (cf. Chapter 1) as illustrated in Figure 13.10.

Since the variational principle and the equations are the same for the standard map as for the static model, there must be an invariant Cantor set or "cantorus" in place of an invariant circle of frequency ν for the standard map with k above some critical $k_c(\nu)$. Aubry and Le Daeron, and Mather[22] have proved that such cantori exist in place of any vanished invariant circle of the standard map. It has been shown that some of these cantori control the

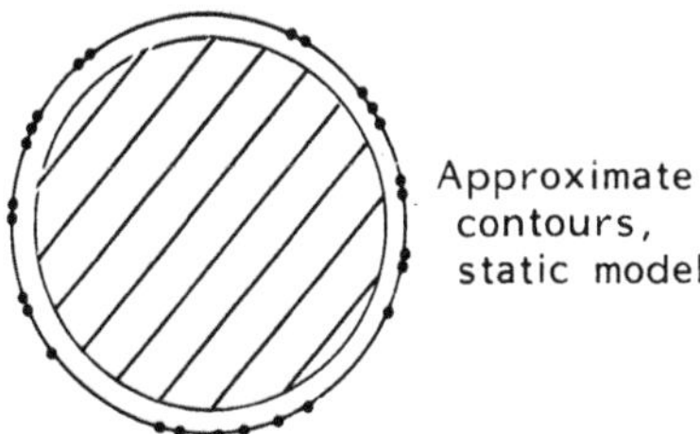

Figure 13.10. Static model for an orbit of a cantorus. There are a countable infinity of gaps.

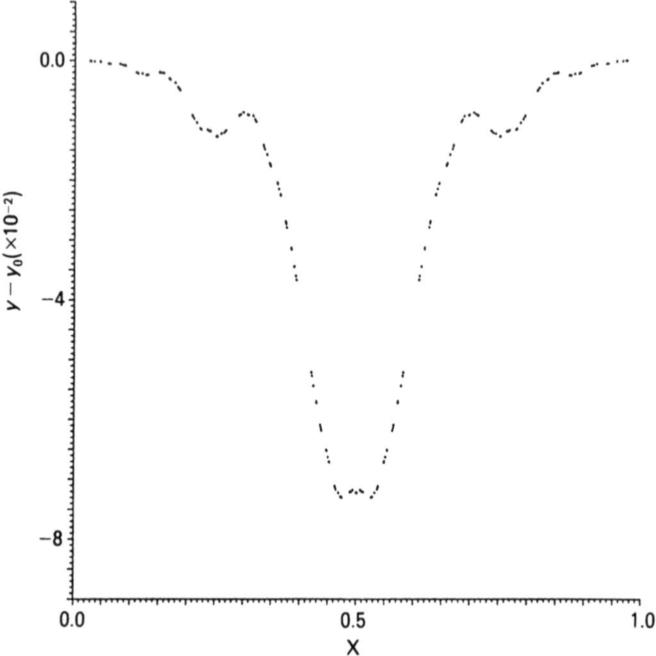

Figure 13.11. An approximate cantorus in the phase plane.

chaotic motion over an important range of k, as summarized in Section 13.5.
An approximate cantorus is illustrated in Figure 13.11.

13.2. Hamiltonian Systems with *m* Degrees of Freedom

In this section the general theory of systems of many degrees of freedom and the elementary theory of integrable systems will be briefly referred to. In the nineteenth century classical dynamics consisted of very little else. Resonances and nonintegrable systems were subsequently introduced. The reader is referred to the author's article, "Integrable and Nonintegrable Hamiltonian Systems," presented at the Joint US/CERN School on nonlinear aspects of particle accelerators in Sardinia in 1985.[23] There, Hamiltonian systems with m degrees of freedom are defined, and from Hamilton's equations, the area-preserving first Poincaré invariant introduced in geometrical form. This leads to canonical transformations, the Poisson bracket, and Noether's symmetry theory.

Invariant tori are defined[23] and the dimensions of invariant regions tabulated for integrable systems. This leads to resonances and nonintegrable systems, and Chirikov's overlap criterion[14] was introduced through resonance analysis.[23] After this brief summary, we will now discuss chaotic systems in detail.

13.3. Chaotic Systems and Symbolic Dynamics (1)

13.3.1. Introduction

According to one point of view, put forward by Laplace, dynamical systems like the solar system are completely deterministic, so probability theory can have no relevance. But this point of view requires a God-like omniscience to be able to determine initial conditions exactly. This requires an infinite number of digits, and is beyond the capacity of anybody or anything of finite size, including the observable universe.

In reality, measurement is only able to determine the state of a classical system to a finite number of digits, and even this determination is subject to probable errors, without quantum mechanics and irrespective of whether the determination is made by human or machine. Such measurements limit the known or recorded motion to a range of possible orbits. For most real systems the theory of such incompletely known conditions is complicated, but for some model systems and model measurements a theory does exist.

The relevant theory is known as symbolic dynamics, which relates three things:

1. Sequences of symbols
2. Orbits of a dynamical system
3. A partition of its phase space

The phase space is partitioned into pieces or "atoms," in the terminology of ergodic theory. The partitions are determined from an elementary partition, which is open to choice, and the rest are obtained from the dynamics. Each piece of phase space has a different code. The code is complete if the pieces are points.

To show how this works, we start with two very simple cases, the area-preserving map known as the baker's transformation, which is approached through the binary map. The general theory is more difficult, and the relevant parts are largely contributions of Kolmogorov, Sinai, and Bowen; we discuss them only briefly.

13.3.2. The Binary Map

We consider the binary representation of a real number x_0 in a unit interval. It can be obtained by an iteration procedure:

$$x_1 = 2x_0 \pmod 1; \qquad b_0 = 2x_0 - x_1$$

$$\vdots$$

$$x_t = 2x_{t-1} \pmod 1; \qquad b_{t-1} = 2x_{t-1} - x_t \tag{13.3.1}$$

Then

$$x_0 = 0.b_0 b_1 b_2 \cdots b_t \cdots \tag{13.3.2}$$

in binary representation.

The basic operation $x_t = 2x_{t-1}$ (mod 1) is a noninvertable map from the unit interval onto itself and may be considered as an abstract irreversible dynamical system, with orbits $x_0, x_1, x_2, \ldots, x_t, \ldots$ and discrete integer time t.

The binary sequence (13.3.2) for x_0 is a code representing x_0. The codes representing successive points on the orbit are obtained by shifting the whole sequence $b_0, b_1, \ldots, b_t, \ldots$ to the left and truncating the leading terms.

For general dynamical systems a phase point is represented by a symbol sequence, and the orbit for which it is the initial point by a shift on the sequence, labeled by an integer time. There is not always truncation; in particular there is none for area-preserving maps.

We note that for the binary map, the phase space is stretched by a factor of 2 at each stretching, and that the entropy of the sequence of symbols is one bit per unit time. The entropy appears as the logarithm of the stretching factor. This property is general.

Symbolic dynamics provides a correspondence between sequences of symbols and orbits with given initial conditions. Some things are very easy to do with the symbols. For example, for the binary map, periodic orbits of period n correspond to periodic binary sequences of period n. For period 1, .000000 is the only fixed point (0.111111 . . . is not in the phase space). There is one orbit of period 2, with points at $0.010101\ldots = \frac{1}{3}$ and $0.101010\ldots = \frac{2}{3}$, and so on. We can also find orbits that finish up on a periodic orbit: they have codes that start with any finite sequence of 0s and 1s, with the rest of the code being the same as for the periodic orbit.

What about partitions? Look at the parts of the phase space represented by the condition $b_t = 0$.

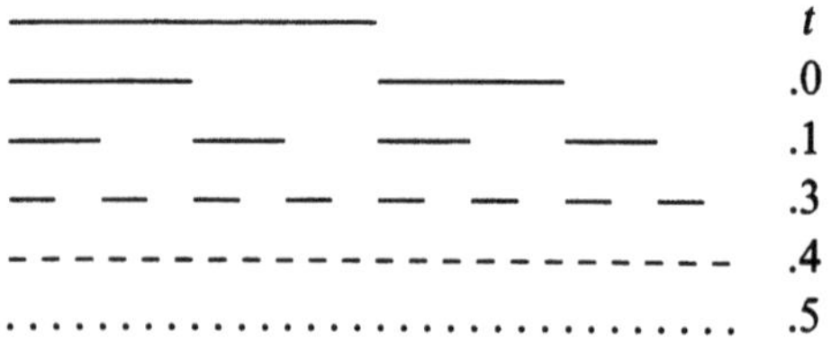

```
_______________                 t
_______        _______          .0
___   ___   ___   ___           .1
_ _ _ _ _ _ _ _                 .3
- - - - - - - - - - - - - - -   .4
. . . . . . . . . . . . . . .   .5
```
(13.3.3)

It is easy to see how the codes, by choosing a binary digit for each time, define smaller and smaller intervals as t increases, with the infinite sequence defining a point, so that the partition is complete. Each digit is determined by locating the t-iterate of the initial point to the left or not to the left of $x = \frac{1}{2}$.

The partitions of our figure are given by the successive inverse iterates of the point $x = 0$.

13.3.3. The Baker's Transformation

Hamiltonian systems of one degree of freedom subject to periodic forces, impulsive or smooth, can be represented by area-preserving maps. Normally the unit time is chosen so that the period is 1. Conservative Hamiltonian systems of two degrees of freedom can also be represented by area-preserving maps, using Poincaré surfaces of section. The study of these maps has become a focal point and a testing ground for the new theories of dynamics.

The simplest area-preserving map for our purposes is the baker's transformation, obtained as a simple extension of the binary map. Denoting the phase points by (x, y), the phase space is the unit square $[0, 1), [0, 1)$. The phase point is represented by the doubly infinite sequence of binary digits

$$\ldots, b_{-1}, b_0, b_1, \ldots \tag{13.3.4}$$

in which, using binary expansion of the real numbers x_0 and y_0,

$$(x_0, y_0) = (0 \cdot b_0 b_1 \ldots, 0 \cdot b_{-1} b_{-2} \ldots) \tag{13.3.5}$$

If we represent the map by T, then it is given, as before, by shifting the sequence to the left by one place, in this case without truncation, so

$$T(x_0, y_0) = (x_1, y_1) = (0 \cdot b_1 b_2 \ldots, 0 \cdot b_0 b_{-1} \ldots) \tag{13.3.6}$$

e.g., since

$$(\tfrac{1}{4}, \tfrac{3}{4}) = (0.0100\ldots, 0.1100\ldots)$$

then

$$T(\tfrac{1}{4}, \tfrac{3}{4}) = (0.1000\ldots, 0.0110\ldots) = (\tfrac{1}{2}, \tfrac{3}{8}) \tag{13.3.7}$$

In general, if (x, y) is the phase point at time t then the point at time $t + 1$ is

$$(x', y') = T(x, y) \tag{13.3.8}$$

where

$$\begin{aligned} x' &= 2x \qquad (\text{mod } 1) \\ y' &= \tfrac{1}{2}(2x - x' + y) \\ &= \tfrac{1}{2}(b_t + y) \end{aligned} \tag{13.3.9}$$

and b_t is the first bit in the binary expansion of x_t.

Figure 13.12 gives a picture of the effect of the map on the points of the phase space.

The baker's transformation would be very effective for mixing dough, which is the reason for its name. It is a particularly simple example of a "mixing" dynamical system, in the mathematical sense.

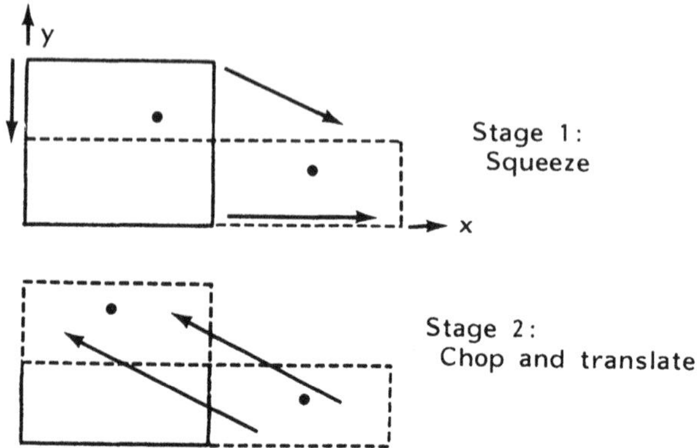

Figure 13.12. Picture of the baker's transformation.

The elementary partition is into a congruent pair of left and right rectangles, and the complete partition is given by the forward and backward iterates of these. For $t \geq 0$ the bit b_t divides the phase space into 2^{t+1} equal strips. Clearly the partition separates points, and is complete.

Again periodic points represent periodic orbits. A code of zeros represents the only fixed point; other codes represent the orbits as illustrated in Figures 13.13–13.16.

A set S of points that remains unchanged under a transformation T is said to be invariant under T. An orbit that is not in S, but is asymptotic to S in both directions of time, is said to be "homoclinic" to S. If the orbit is asymptotic to S_- as $t \to \ldots \infty$ and S_+ as $t \to +\infty$, then it may be described as "heteroclinic" orbits S_- to S_+. Periodic orbits, homoclinic orbits, and heteroclinic orbits are easy to construct from symbol sequences: periodic orbits have periodic sequences, homoclinic and heteroclinic orbits can be constructed by taking arbitrary codes and sandwiching them between the appropriate infinite codes for the invariant sets. How would you construct an orbit that was heteroclinic from one periodic orbit S_- to another S_+, yet spent a long time in the close neighborhood of yet a third, S_0?

With area-preserving maps, or any measure-preserving maps on a compact phase space, the measure can be normalized to unity for the whole phase space, and then provides a natural basis for defining probabilities for the symbol sequences of the code. The probability of any finite sequence of symbols is given by the measure of the orbits which contain that finite sequence—beginning with a specified time, such as $t = 0$.

It is then natural to define the entropy in terms of these probabilities in the same way as for an information source in communication theory. Such an entropy is known as a metric entropy, although strictly it should be called a measure entropy. But some maps have no single natural measure, so such an

Figure 13.13. Period-2 orbit of the baker's transformation, with a finite part of its symbol sequence above.

entropy cannot be uniquely defined for them. For these maps it is convenient to use the "topological entropy," defined as the long-time limit of the mean over time of the logarithm of the number of distinct symbol sequences. This is analogous to the capacity of a noiseless channel. The topological entropy is defined for area-preserving maps and it follows from the basic inequality for entropies that maximum entropy is given by equal probabilities, that the topological entropy is an upper bound on the metric entropy.

For the baker's transformation the topological and metric entropy are both 1 per unit time.

Symbol sequences are not unique: different partitions give different sequences. Any coding of symbol sequences giving a new symbol sequence can be used to represent the orbits. The fundamental theorem for a noiseless channel tells us that if the sequences are ergodic, then any two sequences of the same entropy rate can be coded into one another, with arbitrarily small error. In this sense any two ergodic Markov processes, including Bernoulli sequences, are equivalent to one another, provided they have the same entropy

Figure 13.14. Period-3 orbit.

rate, and the same goes for dynamical systems. Providing such equivalence is one of the main aims of ergodic theory. For some interesting problems, long-time correlations which do not decay exponentially are a challenge in this field.

A new code for a dynamical system corresponds to a new partition. However, these partitions can be fantastically singular and complicated. So although in some abstract sense any two ergodic dynamical systems with the same entropy rate are equivalent, this does not imply equivalence in any practical sense, and we really have a wide variety of dynamical systems: even those with the same entropy rate are worth independent study.

13.3.4. Bernoulli Maps and Liapunov Exponents

The baker's transformation admits many generalizations. A useful one is an area-preserving map which represents an arbitrary Bernoulli sequence with

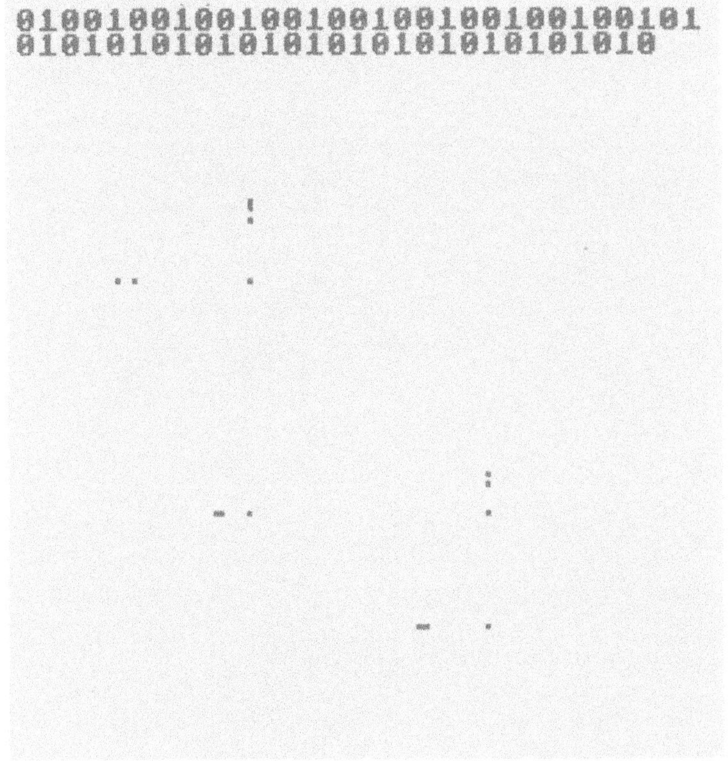

Figure 13.15. Orbit heteroclinic from period 3 to period 2. Compare the symbol sequence with the previous two.

a finite number of symbols. Figure 13.17 represents the map for three symbols A, B, C with probabilities

$$p(A) = \tfrac{1}{7}, \qquad p(B) = \tfrac{4}{7}, \qquad p(C) = \tfrac{2}{7} \tag{13.3.10}$$

The elementary partition is given by the regions defined by the solid vertical lines of Stage 1, and the complete partition by its forward and backward iterates under the map.

The analytic expression for the Bernoulli map depends on the location of (x, y) in the regions of the elementary partition, in other words, on the value of the current symbol. For symbol I it is a squeeze with eigenvalues

$$\lambda_1^+ = 1/p(I) = e^{\alpha(I)} \qquad [\text{defining } \alpha(I)] \tag{13.3.11}$$

$$\lambda_1^- = p(I) \tag{13.3.12}$$

Figure 13.16. Chaotic orbit with "random" symbol sequence.

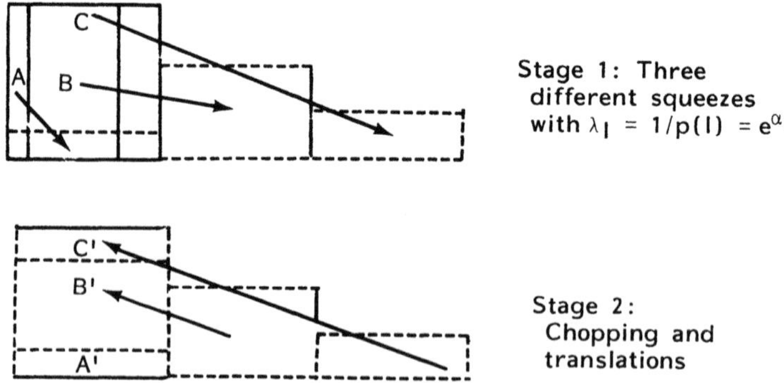

Figure 13.17. Picture of Bernoulli maps with probabilities 1/7, 4/7, 2/7.

For a given orbit the expansion in the x direction from time s to time $u-1$ inclusive is given by

$$\prod_{t=s}^{u-1} e^{\alpha(t)} \tag{13.3.13}$$

where we have written $\alpha(t)$ for $\alpha[(I(t)]$. The Liapunov exponent is defined as the infinite time limit of the arithmetic-mean exponent of expansion. Such a limit will be referred to as the time mean. Thus the time mean of a function $f(t)$ is denoted

$$\langle f(t)\rangle_t = \lim_{s\to\infty,\, u\to\infty} \frac{1}{u-s} \sum_{t=s}^{u-1} f(t) \tag{13.3.14}$$

and the Liapunov exponent for an orbit is

$$\alpha = \langle \alpha(t)\rangle_t = -\langle \ln p(I(t))\rangle_t \tag{13.3.15}$$

where we are now using natural logarithms, as this is dynamics.

But if the system is ergodic, which it is for Bernoulli systems, time means are equal to phase-space means for all orbits except a set of measure zero. So the probability of symbol I in the sequence is $p(I)$, and the time mean of functions $f(x(t), y(t))$ is the same as the phase-space mean of $f(x, y)$ which is given by

$$\langle f(x, y)\rangle_{x,y} = \int dx\, dy f(x, y) \tag{13.3.16}$$

The Liapunov exponent for almost all orbits is therefore

$$\alpha = -\langle \ln p(I)\rangle_{x,y} = -\sum_I p(I) \ln p(I) \tag{13.3.17}$$

which is the metric entropy of the system. Since almost all exponents are the same, this exponent is considered to be the Liapunov exponent of the system.

This result holds for ergodic systems with Bernoulli symbol sequences. If the symbol sequences are only Markov, then a further limiting process is required for sequences of symbols.

In practice, this result is often used in the reverse sense. It is easier, or at least more familiar, to calculate mean exponents of an orbit, and this is used to obtain a numerical estimate of the metric entropy, without using any partition. This is even done for nonergodic systems, in which case the entropy usually refers to a connected invariant region occupied by chaotic motion. Such cases often have long-time correlations, which lead to difficulties.

Symbolic dynamics is valuable for mixing systems for which the motion is chaotic everywhere except on a set of measure zero. The entropy or the Liapunov exponent measures the degree of chaos in the motion. So what happens when the motion is regular? We should expect the entropy rate to be zero, and so it is.

There is an extensive literature on symbolic dynamics and ergodic theory. Arnold and Avez[1] introduces some of the ideas, as does Moser.[2] Much of the literature is mathematical rather than physical in its presentation.[24–27]

13.4. Chaotic Systems and Symbolic Dynamics (2)

We merely record here one application of symbolic dynamics to the discontinuous area-preserving maps known as the sawtooth maps, which are a convenient model for chaotic systems. Percival and Vivaldi[28] have shown how these can be derived from a practical problem of stabilization, and how the practical problem leads naturally to a code for the symbol sequences.

A connection between the so-called cat maps and number theory is the subject of further research.[29]

13.5. Transport

Computers and visual display allow us to watch chaotic motion in mixed systems which have a phase space divided into regular and chaotic regions. The motion is subtle and does not always resemble a diffusion process. For systems of two degrees of freedom with continuous time, or for area-preserving maps, which represent systems of one degree of freedom with discrete time, the transport of phase points from one part of phase space to another is controlled by partial barriers, which are formed on a framework of the cantori introduced in earlier sections.

The flux of phase points through the partial barriers can be computed from action sums, resulting in a theory of transport in Hamiltonian systems, which has been published by MacKay, Meiss, and Percival.[30]

13.6. Conclusion—The Shift in Viewpoint

We have only been able to reveal a very small part of modern dynamics. This part, like many others, has been strongly influenced by electronic computers. Not only do they allow us to calculate what could not be calculated before, but they can present us with moving pictures of dynamical processes, which are a challenge to our understanding. They enable many to share a picture which was at one time the preserve of very few, like Poincaré.

Of course computation on its own is not enough. The computer is like a laboratory in which we can carry out an enormous variety of experiments at relatively little cost. The design of significant experiments is no easier in this field than in others, and neither is their interpretation. Without proper design and interpretation, computer experiments are worse than useless.

But although the computer has played an important role, there are many other important strands to the recent development of the subject. The recognition that problems in different fields of application have much in common has provided a unification and an enormous stimulus to research.

Furthermore, there has been a great increase in the interaction between the mathematics of dynamics and its application. It took a long time for applied dynamics to absorb the significance of the theorem of Kolmogorov, Arnold, and Moser, but its influence has been profound. In the other direction, variational principles that were originally formulated for purposes of practical computation have been used for predicting and then for proving the existence of new invariant structures in phase space. Mathematical problems in ergodic theory of long standing have recently been solved.

The mathematics of dynamics itself has changed. It has become much more geometrical than it was. Few would have dared to predict a few decades ago that it was necessary to understand Cantor sets in order to develop a satisfactory theory of transport in Hamiltonian systems, and that these Cantor sets need to be approximated numerically in order to determine whether fast particles can be confined effectively by magnetic fields, but that is the case.

What of the future? Number theory and fractals are already important for dynamics; their role is likely to increase considerably.

References

Books

1. V. I. Arnold and A. Avez, *Ergodic Problems of Classical Mechanics*, W. A. Benjamin, New York (1968).
2. J. Moser, *Stable and Random Motions in Dynamical Systems*, Annals of Mathematical Studies No. 77, Princeton University Press, Princeton (1973).
3. V. I. Arnold, *Mathematical Methods of Classical Mechanics*, Springer-Verlag, New York (1978).
4. V. I. Arnold, *Geometrical Methods in the Theory of Ordinary Differential Equations*, Springer-Verlag, New York (1983).
5. I. C. Percival and D. Richards, *Introduction to Dynamics*, Cambridge University Press (1982) (Elementary introduction).
6. A. J. Lichtenberg and M. A. Lieberman, *Regular and Stochastic Motion*, Applied Mathematical Sciences 38, Springer-Verlag, New York (1983).
7. J. Guckenheimer and P. Holmes, *Nonlinear Oscillations, Dynamical Systems, and Bifurcations of Vector Fields*, Applied Mathematical Sciences 42, Springer-Verlag, New York (1983).

Reviews

8. J. Ford, Stochastic behaviour in nonlinear oscillator systems, in: *Lectures in Statistical Physics* (W. L. Schieve, ed.), Springer-Verlag, New York (1972).
9. J. Ford, in: *Fundamental Problems in Statistical Mechanics III* (Cohen, ed.), North-Holland, Amsterdam (1975).
10. K. J. Whiteman, *Rep. Prog. Phys.* **40**, 1033 (1977).

11. M. J. Berry, in: *Am. Inst. Phys. Conf. Proc.* **46**, 16 (1978).
12. Y. M. Treve, in: *Am. Inst. Phys. Conf. Proc.* **46**, 147 (1978).
13. G. Casati and J. Ford (eds.), *Stochastic Behaviour in Classical and Quantum Hamiltonian Systems*, Springer-Verlag, Berlin (1979).
14. B. V. Chirikov, Universal instability of many-dimensional oscillator systems, *Phys. Rep.* **52**, 263-379 (1979).
15. D. F. Escande, Large scale stochasticity in Hamiltonian systems, *Phys. Scr.* **1/2**, 126-141 (1982).
16. L. Garrido (ed.), *Dynamical Systems & Chaos: Proc. Sitges 1982*, Lecture Notes in Physics No. 179, Springer-Verlag, Berlin (1983).
17. A. N. Kolmogorov, Dokl. Akad. Nauk **98**, 527 (1954); A. N. Kolmogorov, in: *Proceedings of the International Congress of Mathematicians Ser. II, 7th Congress* (Gerretson and de Groot, eds.), North-Holland, Amsterdam (1954), Vol. 1, p. 315 (1957) (in Russian). An English translation of this article forms Appendix D of R. Abraham, *Foundations of Mechanics*, W. A. Benjamin, New York (1967). V. I. Arnold, *Izv. Akad. Nauk, Ser. Matem.* **25**, 21 (1961); V. I. Arnold, *Usp. Matem. Nauk* **18**, No. 5, 13 (1963) [English translation; *Russ. Math. Surv.* **18**, No. 5, 9 (1963)]; V. I. Arnold, *Usp. Matem. Nauk* **18**, No. 6, 91 (1963) [English translation: *Russ. Math. Surv.* **18**, No. 6, 85 (1963)]; J. Moser, *Nachr. Akad. Wiss. Göttingen*, No. 1, 1 (1962).
18. J. M. Greene, Method for determining a stochastic transition, *J. Math. Phys.* **20**, 1183-1201 (1979).
19. C. D. Murray, Structure of the 2:1 and 3:1 Jovian resonances, *Icarus* **65**, 70-82 (1986).
20. R. S. MacKay and I. C. Percival, Converse KAM: Theory and practice, *Commun. Math. Phys.* **98**, 469-512 (1985).
21. I. C. Percival, Variational principles for invariant tori and cantori, in: *AIP Conf. Proc.* No. 57, 302-310 (1980).
22. S. Aubry and P. Y. Le Daeron, The discrete Frenkel-Kontorov model and its extensions, Physica **8D**, 381 (1983). J. N. Mather, Existence of quasi-periodic orbits, *Topology* **21**, 457-467 (1982).
23. I. C. Percival, Integrable and nonintegrable Hamiltonian systems, in: *Nonlinear Dynamics Aspects of Particle Accelerators. Proceedings, Sardinia 1985*, Lecture Notes in Physics No. 247 (J. M. Jowett *et al.*, eds.), Springer-Verlag, Berlin (1986).
24. D. S. Ornstein, *Ergodic Theory, Randomness and Dynamical Systems*, Yale University Press Math. Monographs (1974).
25. P. Shields, *Theory of Bernoulli Shifts*, University of Chicago (1973).
26. I. P. Cornfeld, S. V. Fomin, and Ya. G. Sinai, *Ergodic Theory*, Springer Math. Series 245 (1982).
27. Ya. G. Sinai, *Introduction to Ergodic Theory*, Princeton University (1976).
28. I. C. Percival and F. Vivaldi, A linear code for the sawtooth and cat maps, Queen Mary College Dynamics preprint QMC DYN 86-3; *Physica* **25D** 373 (1987).
29. I. C. Percival and F. Vivaldi, Arithmetical properties of strongly chaotic motions, Queen Mary College Dynamics preprint QMC DYN 86-2; *Physica* **25D** 105 (1987).
30. R. S. MacKay, J. D. Meiss, and I. C. Percival, Transport in Hamiltonian systems, *Physica* **13D**, 55-81 (1984); Resonances in area-preserving maps, submitted to *Physica D*.

14

Elementary Symbolic Dynamics

Hao Bai-Lin

In a sense the method of symbolic dynamics[1] is the only mathematically rigorous approach to the study of chaotic behavior in dynamical systems. However, having its roots in the topological theory of dynamical systems, symbolic dynamics remains a rather abstract chapter of mathematics and still seems to be useless for a practical physicist. The aim of this chapter is to show that there exists an elementary way to use at least a small "subset" of the method of symbolic dynamics to help the practical physicist in dealing with his data from computer or laboratory experiments. The adjective "elementary" means we shall use elementary mathematics only.

14.1. One-Dimensional Mappings and Their Bifurcation Diagrams

In physics there have been a few simple but by far nontrivial model problems or paradigms that have served as touchstones for many important theories that have developed over decades. In the first place we have in mind the two-body Kepler or hydrogen-atom problem in classical and quantum mechanics, both nonrelativistic and relativistic. As a second example we may recall Brownian motion, which has been the source of inspiration for the whole stochastic approach in physical sciences. In studying chaotic behavior in nonlinear physical systems we are lucky enough to have another such paradigm, namely, one-dimensional mappings or endomorphisms of the interval into itself.

Unlike many one-dimensional models in physics which are either too specific or too trivial to have serious practical meaning other than pedagogical, one-dimensional mappings are very rich in content and, at the same time, are simple enough to be accessible by some analytical tools and not very time-

Hao Bai-Lin • Institute of Theoretical Physics, Academia Sinica, Beijing, China.

consuming in numerical studies. Although one-dimensional mappings cannot cover all properties of higher-dimensional nonlinear systems, they do have many properties in common. Therefore, one-dimensional mappings may serve us as good beachheads to enter the vast Chaosland and it is quite rewarding to return to them when studying more complicated nonlinear systems.

In this chapter we shall consider one-dimensional mappings

$$x_{n+1} = f(A, x_n), \qquad x_n \in I, \qquad n = 0, 1, 2, \ldots \tag{14.1.1}$$

of the interval I into itself. In equation (14.1.1) $f(A, x)$ is a nonlinear function, depending on one or more parameters, denoted simply by A. We consider smooth, at least once differentiable mappings. To be more specific, we are going to employ the following mappings:

1. The logistic map (Figure 14.1)

$$x_{n+1} = 1 - Ax_n^2, \qquad x_n \in (-1, 1) \tag{14.1.2}$$

which describes, among other things, the population dynamics of certain insects without generation overlap. Equation (14.1.2) is a symmetric function with only one maximum at $x = 0$, hence the name "unimodal" for this kind of mapping.

2. The antisymmetric cubic map (Figure 14.2)[2]

$$x_{n+1} = Ax_n^3 + (1 - A)x_n, \qquad x_n \in (-1, 1) \tag{14.1.3}$$

As we shall see, this map has a close relation[3] to the celebrated Lorenz model[4] in which one of the first strange attractors was observed.

3. The sine-square map (Figure 14.3)[5]

$$x_{n+1} = A \sin^2 (x_n - B), \qquad |x_n - B| \leq \pi \tag{14.1.4}$$

which has been used to model an optical bistable device using liquid

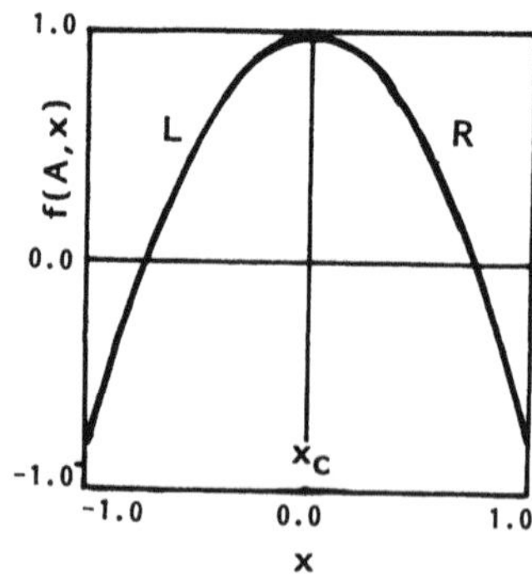

Figure 14.1. The logistic map (14.1.2).

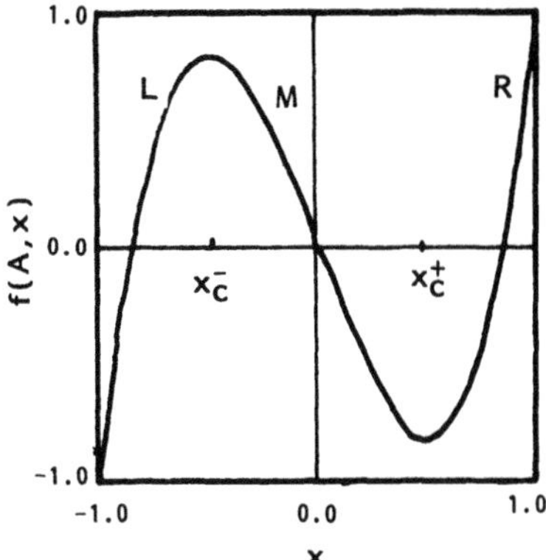

Figure 14.2. The antisymmetric cubic map (14.1.3).

crystal as the nonlinear medium. The map (14.1.4) has two maxima and two minima.

We shall call a point $x_c \in I$ critical when the map has a maximum or minimum at x_c. Thus the logistic map has only one critical point

$$x_c = 0 \tag{14.1.5a}$$

the cubic map has two critical points

$$x_c^{\pm} = \pm[(1 - 1/A)/3]^{1/2} \tag{14.1.5b}$$

while the sine-square map has four,

$$x_{c1} = B - \pi/2, \qquad x_{c2} = B, \qquad x_{c3} = B + \pi/2, \qquad x_{c4} = B + \pi \tag{14.1.5c}$$

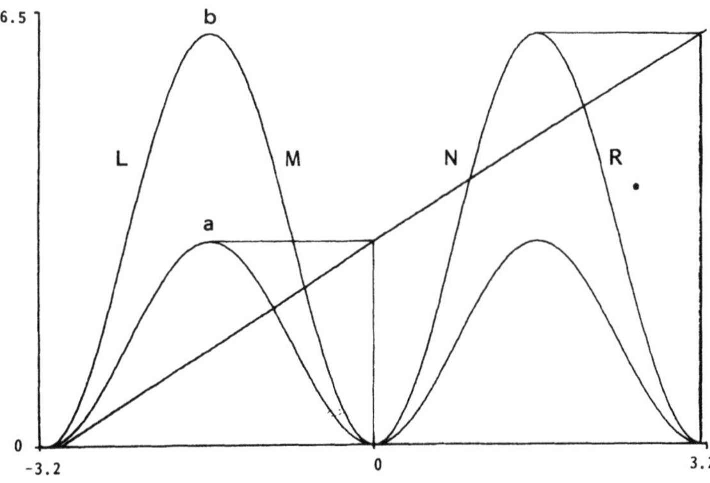

Figure 14.3. The sine-square map (14.1.4) for $A = 3.0$ and 6.14. The abscissa is $x - B$, $B = 3.0$.

In Figures 14.1–14.3 letters R, L, M, ... denote the monotonic segments of the curve. These are the "symbols" in our game. More about them later.

We are concerned with the long-time (i.e., $n \to \infty$) behavior of these mappings. The simplest possibility is a fixed point, that is, $x_n \to x^*$ when n goes to infinity:

$$x^* = f(A, x^*) \tag{14.1.6}$$

There may be periodic points too, i.e., when n gets large enough one may encounter the reappearance of K points:

$$\begin{aligned} x_1 &= f(A, x_0) \\ x_2 &= f(A, x_1) \\ &\cdots \\ x_{K-1} &= f(A, x_{K-2}) \\ x_0 &= f(A, x_{K-1}) \end{aligned} \tag{14.1.7}$$

We say this is a period-K orbit or a K-cycle.

A third possibility consists in having aperiodic output without recycling and ending. This is the chaotic behavior in which we are interested. In some other kinds of mappings, such as the mapping of a circle into itself or higher-dimensional mappings, one may have a quasi-periodic regime as another form of aperiodic behavior. It requires a little additional effort to distinguish it from chaos.

The simplest way to study mappings (14.1.2)–(14.1.4) is to visualize them on a personal computer with graphics. One covers the parameter axis A by small steps. At each parameter value one calculates the iterates of equation (14.1.1), throws away a few hundred points as "transient," and then displays the remaining points on the screen. In this way we get the bifurcation diagrams in x–A coordinates. At every given parameter value A, plotted along the x direction are the "limiting sets" of the mapping, i.e., fixed points, periodic cycles, or chaotic outputs. One can "zoom" into the details of these diagrams by changing to smaller and smaller scales both in x and in A. Figure 14.4 is a blow-up of the bifurcation diagram for the logistic map (14.1.2) for $A =$ 1.78632 to 1.78650. Figures 14.5a and b are bifurcation diagrams for the sine-square map at fixed B and A, respectively.

Two characteristic features are immediately seen when inspecting these diagrams. First, there are many dark lines through the chaotic regions or becoming the boundaries of the chaotic bands. In a sense, they form the skeleton of the bifurcation diagram. Second, there are many, in fact, countable infinite, windows in the chaotic region. These are the locations of periodic orbits, embedded in the chaotic regime.

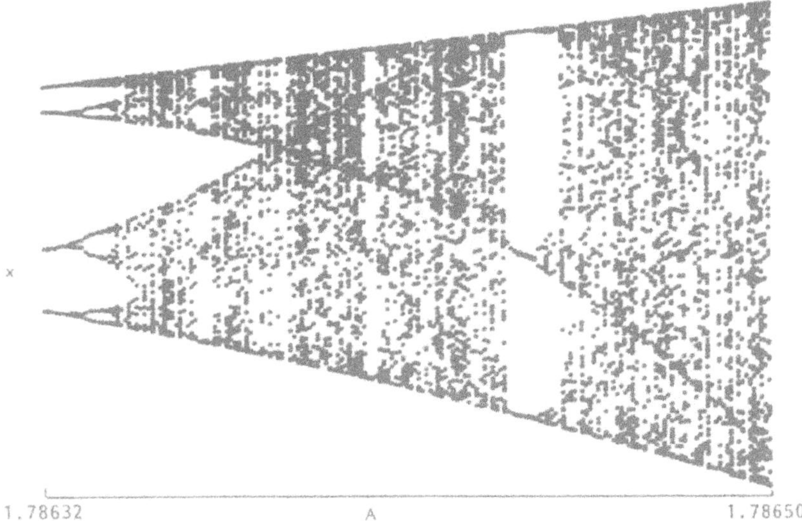

Figure 14.4. A blow-up of the bifurcation diagram for the logistic map (14.1.2); $A = 1.78632$ to 1.78650, x near zero.

Can one write down the equations for all the dark lines seen in the bifurcation diagrams? Is there a procedure to enumerate the periodic windows and to tell their location on the parameter axis? The answer to both questions is in the affirmative. In the following sections we shall show how to do this using elementary mathematics only, provided the original map (14.1.1) is given in terms of elementary functions.

14.2. The Skeleton of the Bifurcation Diagrams

Let us first "lift" the iteration of numbers, given by equation (14.1.1), to a recursive definition of functions using the same nonlinear transformation $f(A, x)$.

Starting from a critical point x_c, we define a set of functions $\{P_n(A)\}$ recursively:[6]

$$
\begin{aligned}
&P_0(A) = x_c \\
&P_{n+1}(A) = f(A, P_n(A)), \qquad n = 0, 1, \ldots
\end{aligned}
\tag{14.2.1}
$$

If there is more than one critical point, one has to define different sets of functions by adding a superscript i in the definition:

$$
\begin{aligned}
&P_0^i = x_{ci} \\
&P_{n+1}^i = f(A, P_n^i), \qquad n = 0, 1, \ldots
\end{aligned}
\tag{14.2.2}
$$

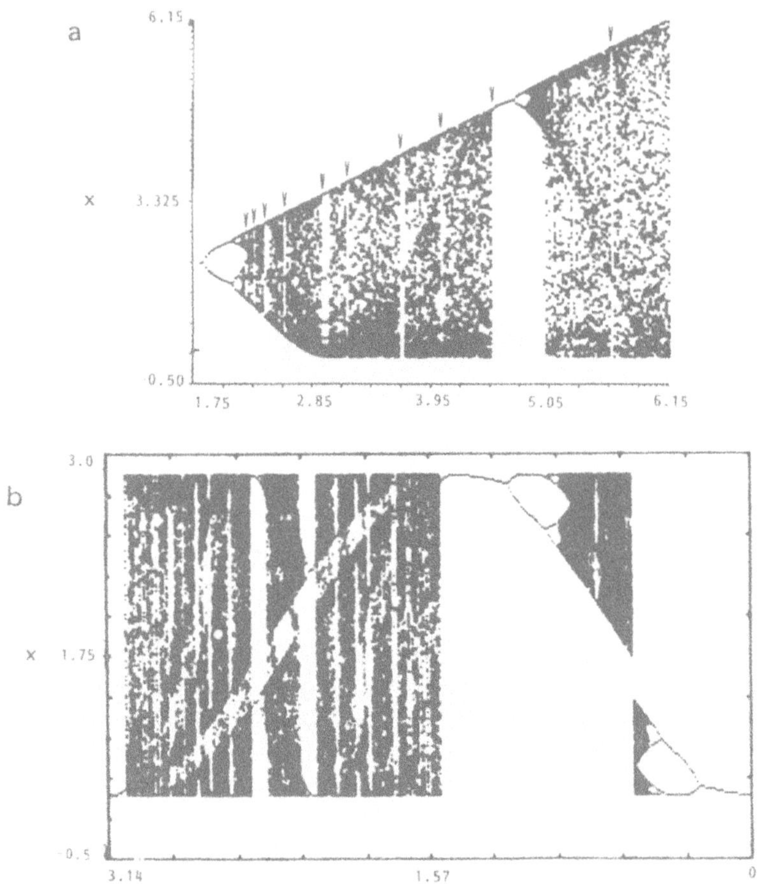

Figure 14.5. Bifurcation diagrams for the sine-square map (14.1.4): (a) x vs. A, $B = 3.0$; (b) x vs. B, $A = 3.0$.

Written down explicitly, they are polynomials of A in the case of the logistic map:

$$
\begin{aligned}
&P_0(A) = 0 \\
&P_1(A) = 1 \\
&P_2(A) = 1 - A \qquad\qquad (14.2.3) \\
&P_3(A) = 1 - A(1 - A)^2 \\
&\ldots \qquad \ldots
\end{aligned}
$$

and two sets of composite trigonometrical functions in the case of the sine-square map (14.1.4):[5]

$$P_0(A, B) = B + \pi/2$$

$$P_{n+1}(A, B) = A \sin^2 (P_n(A, B) - B) \tag{14.2.4}$$

$$Q_0(A, B) = B \text{ or } B + \pi$$

$$Q_{n+1}(A, B) = A \sin^2 (Q_n(A, B) - B) \tag{14.2.5}$$

These functions give all the dark lines or band boundaries in the corresponding bifurcation diagrams. For instance, the first few functions $P_n(A, B)$ and $Q_n(A, B)$ are shown in Figures 14.6a and b. The latter two figures are to be compared with Figures 14.5a and b. In particular, the almost imperceptible dark lines in the lower part of Figure 14.5a are just Q_2, Q_3, and Q_4 shown in Figure 14.6a.

These functions $P_n(A)$ have some nice properties. For example, the real roots of $P_n(a) - x_c = 0$ in suitable range give the superstable parameter values for all n-cycles (see the next section). The band-merging points are given by the root of a certain equation obtained as a combination of different functions $P_n(A)$,[6] etc.

These functions $P_n(A)$ have some nice properties. For example, the real roots of $P_n(a) - x_c = 0$ in suitable range give the superstable parameter values for all n-cycles (see the next section). The band-merging points are given by the root of a certain equation obtained as a combination of different functions $P_n(A)$,[6] etc.

To understand why these functions describe all the dark lines and band boundaries in the chaotic regime, we digress a little to recollect a middle-school physics problem: the rainbow. If asked to explain the origin of a rainbow, a good student would draw a picture like Figure 14.7 and then point out how the droplet works as a prism owing to two refractions and one reflection of the light ray at the water–air interface.

This might be an excellent answer for a middle-school student, but certainly not a full answer for a graduate student in physics. The angle of deflection θ of the outgoing ray depends not only on the refraction index n of the droplet, but also on the sight distance δ of the incoming ray. Since all sight distance from zero to R (the radius of the droplet) are present, the outgoing rays of different colors may overlap and mix up again. Therefore, it remains to be explained why one can see the rainbow at all.

The correct answer comes from inspecting the quantitative dependence of θ on n and δ. It is an elementary exercise to obtain the following expression:

$$\theta = 2 \sin^{-1}\left\{x\left[\frac{2}{n(1-x^2)^{1/2}}\left(\frac{1-x^2}{n^2}\right)^{1/2} - 1 + \frac{2x^2}{n^2}\right]\right\} \tag{14.2.6}$$

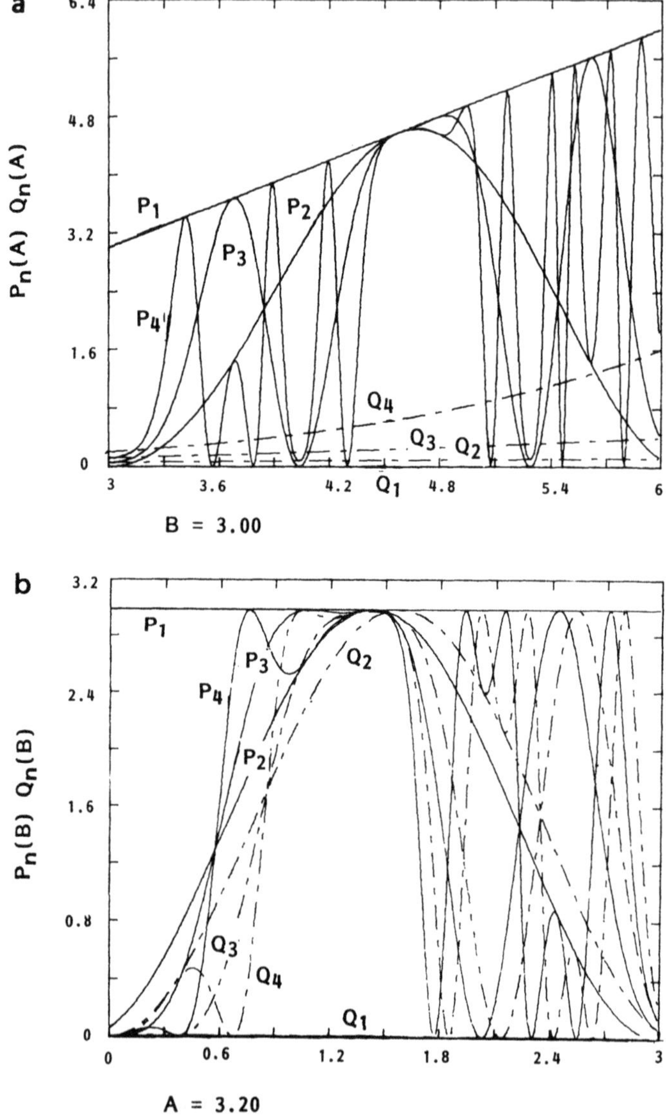

Figure 14.6. Skeleton of the bifurcation diagram for the sine-square map (14.1.4): (a) $P_n(A, 3.0)$ and $Q_n(A, 3.0)$, $n = 1$ to 4, to be compared with Figure 5a; (b) $P_n(3.0, B)$ and $Q_n(3.0, B)$, $n = 1$ to 4, to be compared with Figure 5b.

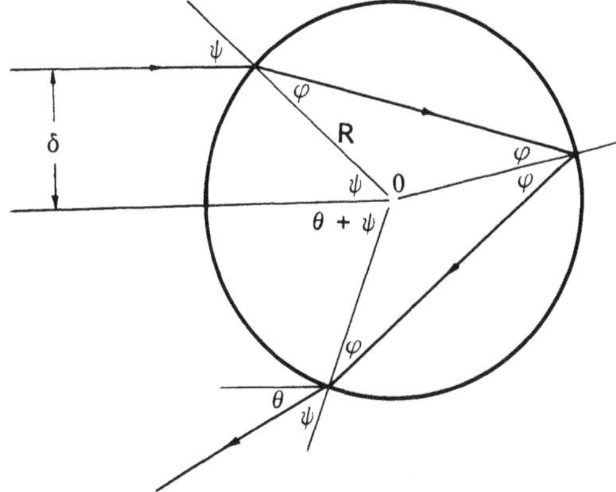

Figure 14.7. Refraction and reflection of a light ray in a water droplet.

where $x = \delta/R$. The $\theta - x$ dependence at fixed n happens to be a unimodal function (Figure 14.8). If we take incoming rays at equally spaced discrete values of the sight distance, then the resulting rays will come out more densely at angles closer to that corresponding to the critical x_c. If the incoming rays are homogeneously distributed in the interval $x(0, 1)$, then the outcoming distribution will have a singularity at the maximum of expression (14.2.6). This gives the rainbow we see.

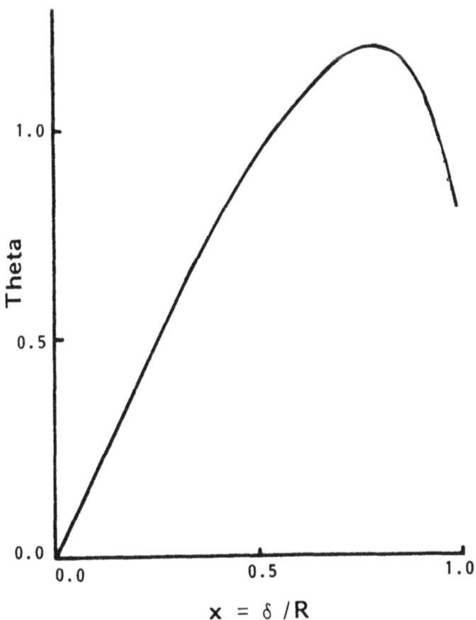

Figure 14.8. The θ-x dependence at $b = 1.3$. See equation (14.2.6) in the text.

We note that the amplification effect of a function near its extremes is a common phenomenon, e.g., the density of states in the energy band of a one-dimensional crystal may be explained in the same way (cf. Jose[7]).

14.3. Symbolic Description of Periods and the Determination of Superstable Parameter Values

A one-dimensional mapping like (14.1.1) may be viewed as a discrete time evolution equation. The most complete description of its dynamics requires knowledge of the full set of iterates $\{x_n, n = 0 \text{ to } \infty\}$ starting from a given x_0. However, one can design a certain coarse-grained description by ignoring the concrete numbers in the set, while retaining the essential feature of the evolution. To do so we divide the one-dimensional phase space (the interval I) into several segments, each corresponding to a monotonic branch of the map and separated by critical points of the map (and by end points of the map, if those do not belong to critical ones). We label each segment or, equivalently, each monotonic branch of the map by a letter, shown in Figures 14.1–14.3. Then every set $\{x_n\}$ may be replaced by a sequence of letters (if necessary, critical points may require additional symbols). Of course, different sets $\{x_n\}$ may correspond to one and the same symbolic sequence, thus opening the possibility of introducing a classification scheme for the numerical sequences $\{x_n\}$. All numerical sequences corresponding to one and the same symbolic sequence will be considered as equivalent. It is clear that we may lose some numerical details of the iterations, but essential features of the evolution, such as periodicity, will be preserved.

To show the usefulness of this symbolic description we need the notion of superstable periods. It is the ABC of nonlinear mathematics that having a solution of a nonlinear equation, the next step is to examine the stability of the solution by performing linear stability analysis. In the case of a fixed point x^* for the map (14.1.6) the analysis is extremely simple. One tests the iterations near x^*, i.e., let

$$x_n = x^* + \varepsilon_n$$
$$x_{n+1} = x^* + \varepsilon_{n+1} \tag{14.3.1}$$

and expand (14.1.1) to the first order in ε to get

$$|\varepsilon_{n+1}/\varepsilon_n| = |f'(A, x^*)| < 1 \tag{14.3.2}$$

for the iteration (14.3.1) to converge.

The stability condition (14.3.2) may be easily generalized to a K-cycle, since a K-cycle [see equations (14.1.7)] gives K fixed points of the Kth iterate of the map (14.1.1):

$$x_i^* = f^{(K)}(A, x_i^*), \qquad i = 0, 1, \ldots, K-1 \tag{14.3.3}$$

where x_i^* are points of the K-cycle. The functional composition, i.e., the K-times nested function, is denoted by a superscript (K):

$$f^{(K)} = \underbrace{f \circ f \circ f \circ \cdots \circ f}_{K \text{ times}}(x) = \underbrace{f(f(\ldots f}_{K \text{ times}}(x)\ldots)) \tag{14.3.4}$$

Do not confuse $f^{(K)}$ with the symbol of taking derivatives, of which we use only $f'(x)$ for the first derivative in this chapter.

The stability condition (14.3.2) is applied by using the chain rule of differentiation. This yields

$$|df^{(K)}(A, x)/dx|_{x^*} = \left| \prod_{i=0}^{K-1} f'(A, x_i^*) \right| < 1 \tag{14.3.5}$$

for the K-cycle to be stable.

Among all the K-cycles satisfying relation (14.3.5) there is a most favorable one, which corresponds to

$$\prod_{i=0}^{K-1} f'(\tilde{A}, x_i^*) = 0 \tag{14.3.6}$$

and occurs only for some specific $\tilde{A}$. It is called a superstable period (or cycle). In fact, all stable cycles sufficiently close to $\tilde{A}$ on the A-axis are the period K orbits. They form an interval on the A-axis with $\tilde{A}$ located somewhere near the center of the interval. Together with their "period-doubled" partners they form the windows in the bifurcation diagrams (e.g., Figures 14.4 and 14.5). The superstable K-cycle at A acts as a representative of the whole interval. To determine the superstable parameter value $\tilde{A}$ one can, in principle, to solve the $K+1$ simultaneous equations (14.1.7) and (14.3.6) for $K+1$ unknowns x_i^* and A. This is not a practical approach when K gets large, but there exists a very simple procedure to obtain A from the symbolic description of periods.

In the case of smooth differential mappings such as equations (14.1.2)–(14.1.4) any superstable cycle must contain at least one of the critical points x_{ci}, because the criticality condition $f'(A, x_{ci}) = 0$ provides a zero factor in the superstability condition (14.3.6). Therefore, we can define a superstable period-K orbit as a K-cycle containing at least one of the critical points. Now let us write down what was just said. For simplicity we discuss only the logistic map (14.1.2). Take a superstable period K and write it down explicitly as a

K-cycle starting from and ending at the critical point x_c:

$$\underbrace{f \circ f \circ \cdots \circ f}_{K \text{ times}}(A, x_c) = x_c \tag{14.3.7}$$

Now let us move all but one of the functions f from the left-hand to the right-hand side of equation (14.3.7). However, since the inverse map is multi-valued, we must attach a subscript to each f^{-1} to indicate which single-valued branch has been used in taking the inverse:

$$f(A, x_c) = f_{\alpha_1}^{-1} \circ f_{\alpha_2}^{-1} \circ \cdots \circ f_{\alpha_{k-1}}^{-1}(A, x_c) \tag{14.3.8}$$

The subscripts α_i equal either R or L in our case. Put together all the subscripts; we get a word made of $K - 1$ letters:

$$W = \alpha_1 \alpha_2 \cdots \alpha_{K-1} \tag{14.3.9}$$

We note that a K-cycle is described by a word made of $K - 1$ letters. Let us introduce a new notation by naming each single-valued branch of the inverse function by its subscript, i.e.,

$$\alpha_i(y) \equiv f_{\alpha_i}^{-1}(y) \tag{14.3.10}$$

Using these new notations, we can "lift" the word W to a composite function $W(y)$ (depending also on parameter A):

$$W(y) = \alpha_1 \circ \alpha_2 \circ \cdots \circ \alpha_{K-1}(y) \tag{14.3.11}$$

Now equation (14.3.8) takes the very simple form

$$f(A, x_c) = W(x_c) \tag{14.3.12}$$

Since x_c, f, and f^{-1} are known, relation (14.3.12) is an equation to determine the superstable parameter A for the K-cycle described by the word W. For the logistic map $x_c = 0$, $f(A, 0) = 1$, equation (14.3.12) looks even simpler:

$$W(0) = 1 \tag{14.3.13}$$

We shall see in the next section that the only 2-cycle and 3-cycle for the logistic map are described by the words R and RL, so their superstable parameter values may be calculated by solving

$$R(0) = 1$$

and

$$R \circ L(0) = 1$$

where

$$\begin{aligned} R(y) &= +[(1-y)/A]^{1/2} \\ L(y) &= -[(1-y)/A]^{1/2} \end{aligned} \tag{14.3.14}$$

To show how this kind of equation is solved, we calculate the superstable parameter values for all the 5-cycles in the logistic map. There are only three of them: RLR^2, RL^2R, and RL^3 (see the next section). Therefore, we have to solve the following three equations:

$$\begin{aligned} R \circ L \circ R \circ R(0) &= 1 \\ R \circ L \circ L \circ R(0) &= 1 \\ R \circ L \circ L \circ L(0) &= 1 \end{aligned} \tag{14.3.15}$$

Written explicitly with both sides multiplied by A, they read

$$\begin{aligned} RLR^2: &\quad A = \{A + [A - (A - A^{1/2})^{1/2}]^{1/2}\}^{1/2} \\ RL^2R: &\quad A = \{A + [A + (A - A^{1/2})^{1/2}]^{1/2}\}^{1/2} \\ RL^3: &\quad A = \{A + [A + (A + A^{1/2})^{1/2}]^{1/2}\}^{1/2} \end{aligned} \tag{14.3.16}$$

If one tries to eliminate the square roots by squaring them several times, all these three equations will lead to one and the same polynomial equation $P_5(A) = 0$ in our notation of the previous section. As we have mentioned in Section 14.2, real zeros of $P_n(A) = 0$ in a suitable range indeed give the superstable parameter values for all period-n orbits. Since almost all of these roots are located in the finite interval (1.4, 2), it would be extremely difficult to calculate them using any existing numerical algorithm when n becomes large enough.

A better way consists in solving equations (14.3.16) separately by converting them into iterations,[8] e.g., let

$$RLR^2: \qquad A_{n+1} = \{A_n + [A_n - (A_n - A_n^{1/2})^{1/2}]^{1/2}\}^{1/2}$$

The iteration converges very quickly and we have good estimates for the initial value A_0: any number between 1.4 and 2 will do. In this way we get

$$A_{RLR^2} = 1.625413725$$

$$A_{RL^2R} = 1.860782522$$

$$A_{RL} = 1.985424253$$

It is very easy to read out the rule to write down these iteration equations and to have the procedure to solve them programmed. The question we are facing now is how to find all the admissible words and how to order them along the parameter axis. The answer is given by using the method of symbolic dynamics.

14.4. Symbolic Dynamics of Two Letters

We have already pointed out that for a physicist symbolic dynamics may be understood as a coarse-grained description of the time evolution of dynamical systems. The symbolic dynamics of two letters for unimodal mappings of the interval was constructed before and concurrently with Feigenbaum's discovery of scaling properties of these mappings by Metropolis *et al.*[9] and by Derrida *et al.*[10] It has been described in detail in the book of Collet and Eckmann[11] on mappings of the interval.

In this section we shall take a pragmatic point of view, i.e., we formulate the related results without going into detailed proof or derivation and only show how the method works in practice. Having mastered the practical rules, a reader may wish to learn more mathematical details from the original papers.[9,10]

14.4.1. The Ordering of Words

Admissible Word. A word W is said to be admissible if it corresponds to a periodic orbit of the unimodal map [see equation (14.3.9)]. A word comprising $n-1$ letters is said to be of length n.

Parity of a Word. A word W made of the two letters R and L is said to be odd (even) if it contains an odd (even) number of Rs. Thus the word RL is odd, but RLR is even. To anticipate a little we indicate that the parity of a word is determined by the number of descending segments of the map in the orbit it represents. This is the hint to generalize the definition of parity to dynamics of more letters.

Harmonic of a Word. Given a word W one can construct another word $H(W)$, called the harmonic of W, according to the following rule:

$$H(W) = W\sigma W \qquad \text{where } \sigma = R(L) \text{ if } W \text{ is even (odd)} \tag{14.4.1}$$

Thus, $H(R) = RLR$, $H(RLR) = RLR^3LR$, etc. If W contains $n-1$ letters, then there are $2n-1$ letters in $H(W)$. If W is an admissible word, then $H(W)$ is admissible too and corresponds to the period-doubled cycle of W.

The Order of Two Admissible Words. Two admissible words W_1 and W_2 can be ordered in the following way. Comparing these two words letter by letter from left to right and denoting the largest common part by W^*, i.e.,

$$\begin{aligned} W_1 &= W^*\sigma_1\sigma_2\cdots \\ W_2 &= W^*\tau_1\tau_2\cdots \end{aligned} \tag{14.4.2}$$

where $\sigma_1 \neq \tau_1$. If W^* exhausts one of these words, say W_1, then we put a letter C after it, i.e., take $\sigma_1 = C$. C means critical [or central in the case of the logistic map (14.1.2)]. Now an order is introduced on the basis of the natural order on the interval I:

$$L < C < R \tag{14.4.3}$$

If W^* is even, then the order of W_1 and W_2 is determined by the order of σ_1 and τ_1, i.e., $W_1 > W_2$ if $\sigma_1 > \tau_1$ in the sense of (14.4.3), and vice versa. If W^* is odd, then the order of W_1 and W_2 is the antiorder of σ_1 and τ_1, i.e., $W_1 > W_2$ if $\sigma_1 < \tau_1$ in the sense of relation (14.4.3).

Example 1. A blank word (denoted by b if necessary) corresponds to period 1 and the letter C alone. Thus we have

$$b < R < RLR < RLR^3LR < \cdots \tag{14.4.4}$$

This determines the direction of bifurcations for the main period-doubling sequence of Feigenbaum.[12]

Example 2. The chain of inequalities

$$R < RL < RL^2 < \cdots < RL^n < RL^{n+1} < \cdots \tag{14.4.5}$$

determines the most clearly seen periodic windows in many mappings and ODEs (see Section 14.8 below). When n gets large and other windows in between RL^n and RL^{n+1} cannot be resolved, the sequence (14.4.5) has a close relation to the period-adding sequence, studied by Kaneko.[13]

For those unimodal mappings whose height (maximum) varies monotonically with the parameter A, e.g., the logistic map (14.1.2), the order we just defined is identical to the order of appearance of the corresponding periods along the parameter axis A. If $W_1 > W_2$, then the parameter values for these periods must satisfy $A_{W_1} > A_{W_2}$. This is no longer true for unimodal mappings with an arbitrary dependence on the parameter A. We shall return to this point later.

Antiharmonic of a Word. The antiharmonic $A(W)$ of a word W is constructed according to a rule opposite to (14.4.1), i.e.,

$$A(W) = W\sigma W \qquad \text{where } \sigma = R(L) \text{ if } W \text{ is odd (even)} \tag{14.4.6}$$

The antiharmonic is an artificial construct needed in the intermediate stage of discussion. It is not an admissible word and does not correspond to a real period of the map.

High-Order Harmonics and Antiharmonics. The definition of harmonic (antiharmonic) can be applied repeatedly to generate high-order harmonics and antiharmonics, e.g.,

$$H^2(R) = H(H(R)) = H(RLR) = RLR^3LR$$

Finding the Lowest Period in between Two Admissible Words. Given two admissible words $W_1 < W_2$ where W_1 is of length n_1, one can find another admissible word W, if any, which corresponds to the lowest period located in between W_1 and W_2, i.e.,

$$W_1 < W < W_2$$

One first constructs $H(W_1)$ and $A(W_2)$, and then denotes their leading common part of length n^* by W^*, i.e.,

$$\begin{aligned} H(W_1) &= W^*\sigma_1\sigma_2\cdots \\ A(W_2) &= W^*\tau_1\tau_2\cdots \end{aligned} \tag{14.4.7}$$

Then one must distinguish two cases:

1. $n^* > 2n_1$, then the period of lowest order is given by $H(W_1)$.
2. $n^* < 2n_1$, then the period of the lowest order is given by W^* itself.

In so doing one may encounter cases where the above rules cannot be readily applied. Then one must go to higher-order harmonics (and/or antiharmonics) instead of using the first order ones only.

Example 3. $W_1 = RLR^2$, $W_2 = RL$, $H(W_1) = RLR^2LRLR^2$, $A(W_2) = RLR^2L$. Since formally $W^* = A(W_2)$, but an antiharmonic cannot correspond to any admissible period, we have to construct the second-order antiharmonic

$$A^2(W_2) = A(A(W_2)) = RLR^2LR^2LR^2L$$

This is a period 7, in fact, the shortest period between period 5 (RLR^2) and period 3 (RL).

The $*$-composition. The harmonics and antiharmonics defined above happened to be particular cases of a more general composition rule, introduced and denoted by $*$ by Derrida *et al.*[10] The composition of two words P and $Q = \sigma_1\sigma_2\cdots\sigma_{k-1}$ is defined as

$$P * Q = P\bar{\sigma}_1 P\bar{\sigma}_2 P\cdots P\bar{\sigma}_{k-1}P \tag{14.4.8}$$

where $\bar{\sigma} = \sigma_i$, if P is even, and $\bar{\sigma}_i$ equals the conjugate of σ_i, i.e., with R and L interchanged, if P is odd. Therefore, a harmonic (antiharmonic) is nothing but the $*$-composition with $Q = R(L)$:

$$H(W) = W * R, \qquad A(W) = W * L \tag{14.4.9}$$

It is easy to verify that the $*$-composition is associative but noncommutative. The existence of the $*$-composition is the mathematical manifestation of the self-similar structure of the bifurcation diagram for the unimodal mappings.[10]

If a word cannot be decomposed using the $*$ operation, it is called a prime word. The prime components of a nonprime word, say P, Q, and S in $W = P * Q * S$, describe periodic motion at different scales, P corresponding to the largest scale, while S corresponds to the smallest one. Reflected in the power spectrum, these different scales give rise to fine structures of different order. Take, for example, the two period-6 orbits $W_1 = R * RL = RLR^3$ and $W_2 = RL * R = RL^2RL$, the former being the period-3 window embedded in the two-band chaotic zone and the latter the period-doubled regime of the period 3 in the one-band region. Their power spectra are shown schematically in Figure 14.9. The fine structure of power spectra is very useful in experiments to tell the whereabouts in the parameter space.

14.4.2. The λ-Expansion

In the last subsection we have learned how to order the periods according to their words and how to find the lowest period included in between two

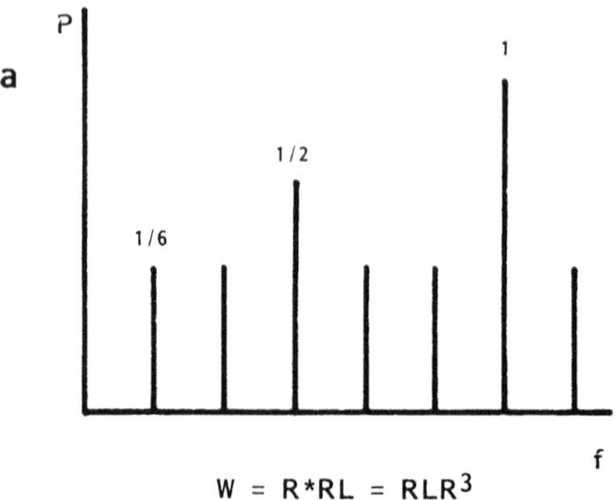

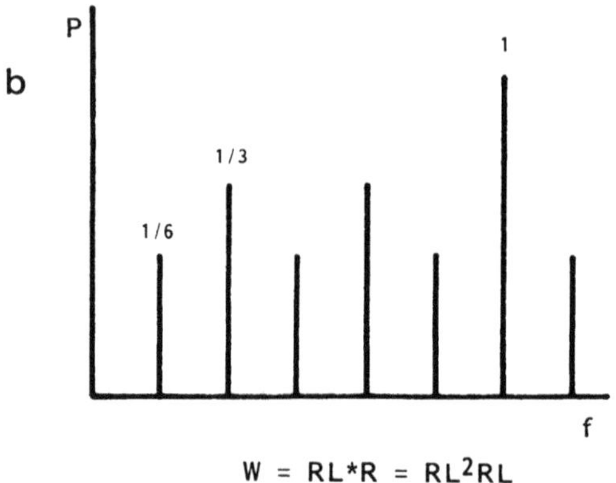

Figure 14.9. Fine structure of power spectra for two different 6-cycles (schematic): (a) $W = R*RL = RLR^3$; (b) $W = RL*R = RL^2RL$.

given words. We address this subsection to another question, namely, given the period n, how to generate all the admissible words comprising $n-1$ letters. This is accomplished by invoking the so-called λ-expansion of real numbers.[10]

To introduce the λ-expansion we depart temporarily from our original convention of studying smooth maps only. Let us consider the piecewise linear map

$$
\begin{aligned}
x_{n+1} = T(\lambda, x_n) &= \lambda x_n \qquad &\text{for } 0 < x_n < 1 \\
&= \lambda(2 - x_n) \qquad &\text{for } 1 < x_n < 2
\end{aligned}
\tag{14.4.10}
$$

Function $T(\lambda, x)$ maps the interval $(0, 2)$ into itself when the parameter λ is kept between 1 and 2; T is called the tent map according to its shape (see Figure 14.10). Since $|T'(\lambda, x)| < 1$ never holds, the tent map does not have any stable periods, but still one can use the method of Section 14.3 to determine the parameter value for a period containing the "critical point" $x_c = 1$. (In fact, the derivation of equation (14.3.12) did not require the stability of the period.) Now the inverse branches are given by

$$R(y) = 2 - y/\lambda$$
$$L(y) = y/\lambda \tag{14.4.11}$$

Consider again the three different orbits of period 5, i.e., write down the analogue of equation (14.3.15) using the functions (14.4.11); we have:

$$\begin{aligned} RLR^2&: \quad \lambda = 2 - 2/\lambda^2 + 2/\lambda^3 - 1/\lambda^4 \\ RL^2R&: \quad \lambda = 2 - 2/\lambda^3 + 1/\lambda^4 \\ RL^3&: \quad \lambda = 2 - 1/\lambda^4 \end{aligned} \tag{14.4.12}$$

Since the degenerated value $\lambda = 1$ provides a solution to each of the equations $T(\lambda, 1) = W(1)$, we cancel a factor $\lambda - 1$ in equations (14.4.12) and rewrite them as

$$\begin{aligned} RLR^2&: \quad \lambda = 1 + 1/\lambda - 1/\lambda^2 + 1/\lambda^3 \\ RL^2R&: \quad \lambda = 1 + 1/\lambda + 1/\lambda^2 - 1/\lambda^3 \\ RL^3&: \quad \lambda = 1 + 1/\lambda + 1/\lambda^2 + 1/\lambda^3 \end{aligned} \tag{14.4.13}$$

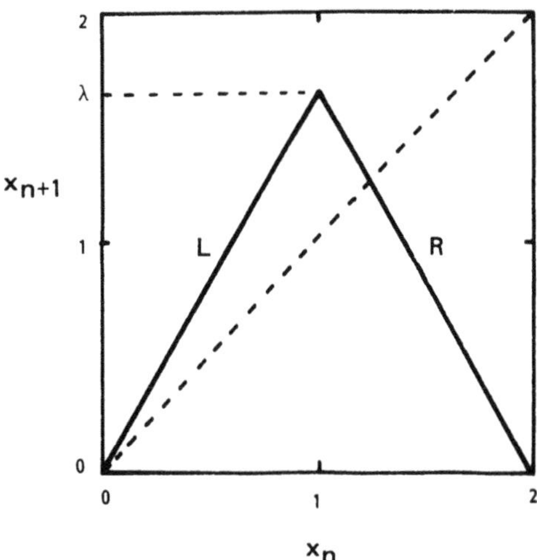

Figure 14.10. The tent map (14.4.10).

These are the equations to be transformed into iterations in λ_n. It is not incidental that equations (14.4.13) differ only in the signs of some of the coefficients. Actually, Derrida *et al.*[10] have shown that each admissible word of length n for the tent map corresponds to an expansion of λ in terms of $1/\lambda$, i.e.,

$$\lambda = \sum_{i=0}^{n-2} a_i/\lambda^i \tag{14.4.14}$$

where $a_0 = a_1 = 1$, $a_i = +1$ or -1 for $i > 1$. However, not every expansion of the form (14.4.14) with arbitrary $|a_i| = 1$ corresponds to an admissible word. In order to be admissible, the coefficients a_i must satisfy all the inequalities

$$\pm(a_n, a_{n+1}, \ldots) < (a_0, a_1, \ldots), \qquad n > 1 \tag{14.4.15}$$

where the comparison is made on a term-by-term basis. The < relation holds for two sequences of coefficients

$$(c_1, c_2, \ldots, c_{k-1}, c_k, \ldots) < (d_1, d_2, \ldots, d_{k-1}, d_k, \ldots)$$

as far as $c_i = d_i$ for $i = 1, 2, \ldots, k-1$, and $c_k < d_k$ in the sense of $-1 < 0 < 1$. Put in other words, the first occurrence of an unequal relation determines the inequality. Furthermore, the coefficients for an admissible expansion may be written as products of $\pm$ 1s:

$$a_i = \beta_1\beta_2 \cdots \beta_i, \qquad i = 1, 2, \ldots, n-2 \tag{14.4.16}$$

($A_0 = -\beta_0 = 1$ by definition). These β_i are in one-to-one correspondence to the letters in the admissible words: -1 means R and $+1$ means L.

What has just been said may be reverted to yield a rule to generate all the admissible words of given length. We show how to do this by the example of period-6 orbits. First, the β_i are given all possible ± 1 values. Then the a_i are computed by using equation (14.4.16):

β_0	β_1	β_2	β_3	β_4		a_0	a_1	a_2	a_3	a_4
−1	1	1	1	1	RL^4	1	1	1	1	1
−1	1	1	1	−1	RL^3R	1	1	1	1	−1
−1	1	1	−1	1	RL^2RL	1	1	1	−1	−1
−1	1	1	−1	−1	RL^2R^2	1	1	1	−1	1
−1	1	−1	1	1	inadmis.	1	1	−1	−1	−1
−1	1	−1	1	−1	inadmis.	1	1	−1	−1	1
−1	1	−1	−1	1	inadmis.	1	1	−1	1	1
−1	1	−1	−1	−1	RLR^3	1	1	−1	1	−1

After checking the inequalities (14.4.15) only 5 of these 8 combinations remain valid. Obviously, the procedure to generate all admissible words can be easily programmed. A list of all admissible words with $n \leq 11$ was given in the Appendix of the paper by Metropolis *et al.*[9]

We emphasize once more that there are no stable periods at all in the tent map, but it does have unstable periods corresponding to all prime admissible words in the symbolic dynamics of two letters. Therefore, one can use the tent map to generate the words needed in studying other unimodal maps. We give an example in the next section.

A question related to the symbolic dynamics of two letters is the number of admissible words for given period n. It was answered elsewhere[9] by borrowing some known mathematical results on the number of symmetry types of primitive sequences of two letters under the cyclic group Z_n. We mention in passing that there exists a very simple recursion formula for this number $M(n)$, namely,[18]

$$M(1) = 1$$

$$M(n) = 1/n\left[2^{n-1} - 1 - \sum_{i=1}^{p} 2^{i-1} - \sum_{1<d<n} {}_{d/n}\, dM(d)S(d)\right] \qquad (14.4.17)$$

where the sum over i exists if $n = k\,2^p$ with k odd and $p > 1$, $S(d) = 2^l$ if $n = d\,2^l$, and the last sum is over such d which divides n.

14.5. Period-n-Tupling Sequences in Unimodal Mappings

Now we are well equipped to study the scaling properties of period-n-tupling sequences in unimodal mappings.[6] Using the $*$-composition rule, explained in the previous section, we may represent the main Feigenbaum period-doubling bifurcation sequence by the symbolic sequence

$$R^{*n} = R * R * \cdots * R \ (n \text{ times}), \qquad n = 0, 1, 2, \ldots$$

In general, every prime word W gives birth to a period-doubling bifurcation sequence $W * R^{*n}$, $n = 0, 1, 2, \ldots$, the simplest case being $W = RL$, corresponding to the period-doubling sequence seen in the period-3 window starting at $A = 1.75$.

However, period-doubling sequences are not the only ones that can be selected from the infinitely many periodic windows embedded in the chaotic regime. In fact, there are infinite many ways to select other sequences. In particular, any prime word W leads to a sequence W^{*n}, $n = 1, 2, \ldots$. For example, $W = RL$ leads to a period-tripling sequence with periods 3^n, $W = RL^2$ leads to a period-quadrupling sequence, and $W = RLR^2$, RL^2R, RL^3 lead to three different kinds of 5^n sequences. These are the period-n-tupling sequences we study in this section. Unlike period-doubling sequences, orbits in the period-n-tupling sequences are not adjacent on the parameter axis, but they do enjoy scaling properties characterized by various exponents: the scaling factor α_W, convergence rate δ_W, noise exponent κ_W, etc., all depending on the particular prime word W. These scaling properties can be checked numerically or studied almost analytically, starting from a generalized renormalization group (RG) equation.[6]

To generalize the Feigenbaum-Cvitanović RG equation[12]

$$g(x) = \alpha g[g(x/\alpha)] \tag{14.5.1}$$

to the period-n-tupling case, it is better to rewrite this equation in the form

$$\alpha^{-1}g(\alpha x) = g^{(2)}(x) \tag{14.5.2}$$

which can be readily generalized to the period-n-tupling case:

$$\alpha^{-1}g(\alpha x) = g^{(n)}(x) \tag{14.5.3}$$

To fix the solution it is insufficient to impose the normalization condition $g(0) = 1$ and the superstability condition $g'(0) = 0$. Even in the period-doubling case one has to indicate the behavior of $g(x)$ near its maximum, i.e., indicate the power z in the expansion

$$g(x) = g(0) + ax^z + \cdots$$

The well-known Feigenbaum constants $\delta = 4.6692$ and $\alpha = 2.5029$ were obtained for the particular case $W = R$ and $z = 2$. However, in the general period-n-tupling case equation (14.5.3) may still have several solutions, each corresponding to a prime word W of the same length. For example, when $n = 5$, equation (14.5.3) has three different solutions, corresponding to the prime words RLR^2, RL^2R, and RL^3, respectively. One has to choose the initial value for α carefully in order to pick up the desired solution.

A few words about the numerical method for solving the RG equation (14.5.3) and the choice of initial value of α. Since equation (14.5.3) determines a saddle-type fixed point $g(x)$ in the functional space, precautions must be taken in dealing with any systematic approximation to it, because an initially convergent sequence may eventually get away from the true solution. However, if an algebraic manipulation language such as REDUCE or MACSYMA is available, one can construct a systematic approximation to equation (14.5.3) by first expanding $g(x)$ to a given order:

$$g(x) = 1 + ax^z + bx^{2z} + \cdots + nx^{nz} \tag{14.5.4}$$

and then substitute it in both sides of equation (14.5.3). By equating coefficients at equal powers one gets a system of algebraic equations for α, a, b, ... At the lowest order the equations for α and a can be solved by hand, then these α and a are used as initial values to solve the next system of equations for α, a, and b, etc. This procedure happens to have good convergence and is very easily programmed.

To estimate the initial value for α we replace the yet unknown universal function $g(x)$ in equation (14.5.3) by $f(A_\infty, x)$ and then take $x = 0$. The limiting parameter A_∞ may be estimated from the relation

$$A_\infty = (\delta A_n - A_{n-1})/(\delta - 1)$$

where δ as well as A_n may be estimated from the first few bifurcation points using the method of Section 14.3. Therefore, we have

$$\alpha = 1/g^{(n)}(0) \cong 1/P_n(A_\infty) \tag{14.5.5}$$

where P_n is just the composite function introduced in Section 14.2.

Once equation (14.5.3) has been solved, we turn to the linearized functional equation near the fixed point $g(x)$. To do this we first introduce an n-tupling operator O_n which, when acting on any arbitrary unimodal function $f(x)$, defines the scaling transformation of period-n-tupling, i.e.,

$$O_n f(x) = \alpha f^{(n)}(x/\alpha) \tag{14.5.6}$$

The RG equation (14.5.3) gives clearly the fixed point of O_n in the space of $\{f(x)\}$:

$$g(x) = O_n g(x) \tag{14.5.7}$$

Now take a function $f(x)$, sufficiently close to $g(x)$, i.e.,

$$f(x) = g(x) + h(x) \tag{14.5.8}$$

where $h(x)$ is a "small" function. Substituting function (14.5.8) into the right-hand side of equation (14.5.6) and expanding it to the first order in $h(x)$ yields

$$O_n f(x) = g(x) + L_n h(x) \tag{14.5.9}$$

by using the fixed point equation (14.5.7). The linearized operator L_n in the latter equation is given by

$$L_n h(x) = \alpha\{h[g^{(n-1)}(x/\alpha)] + g'[g^{(n-1)}(x/\alpha)]h[g^{(n-2)}(x/\alpha)] + g'[g^{(n-1)}(x/\alpha)]g'[g^{(n-2)}(x/\alpha)h(g^{(n-3)}(x/\alpha)] + \cdots + g'(g^{(n-1)}(x/\alpha)) \cdots g'(g(x/\alpha))h(x/\alpha)\} \quad (14.5.10)$$

Among the eigenvalues of the linearized equation

$$L_n h(x) = \delta h(x) \quad (14.5.11)$$

there is one $\delta > 1$ which determines the convergence rate of the bifurcation parameters A_n.[12,14]

To take into account the influence of external noise, one adds a random source to the map (14.1.1), transforming it into a discrete Langevin equation[15]

$$x_{n+1} = f(A, x_n) + \xi_n \quad (14.5.12)$$

where ξ_n is taken to be a Gaussian random variable.

As we know the scaling property of high-order iterates of any unimodal function near its maximum at $x = 0$ may be represented by the universal function $g(x)$ (with the same z), i.e., we have the chain [cf. equation (14.5.3)]

$$\begin{aligned} &f(x) \to g(x) \\ &f^{(n)}(x) \to g^{(n)}(x) = \alpha^{-1}g(\alpha x) \\ &\cdots \qquad \cdots \\ &f^{(n^k)}(x) \to \alpha^{-k}g(\alpha^k x) \end{aligned} \quad (14.5.13)$$

At the presence of small external noise if the scaling property is still to be expected, one might replace system (14.5.13) by

$$\begin{aligned} &f(x) + \xi \to g(x) + \xi D(x) \\ &(f+\xi)^{(n)}(x) \to \alpha^{-1}[g(\alpha x) + \xi\kappa D(\alpha x)] \\ &\cdots \qquad \cdots \\ &(f+\xi)^{(n^k)}(x) \to \alpha^{-k}[g(\alpha^k x) + \xi\kappa^k D(\alpha^k x)] \end{aligned} \quad (14.5.14)$$

where a new universal constant κ and a new universal dispersion function $D(x)$ have been introduced. By analogy with equation (14.5.3) we require the

Table 14.1. Universal Exponents for Period-n-Tupling Sequences W^{*n} ($z = 2$) (after Zeng *et al.*[6] and Wang and Chen[14])

W	δ	α	κ
R	4.669	−2.5029	6.6213
RL	55.247	−9.2773	89.522
RL^2	981.595	−38.819	1558.7
RLR^2	255.54	−20.128	431.91
RL^2R	1 287.0	45.804	2 182.6
RL^3	16 930	−160	26 458

relation

$$(g + \xi D)^{(n)}(x) = \alpha^{-1}[g(\alpha x) + \xi\kappa D(\alpha x)] \tag{14.5.15}$$

to hold to first order in $D(x)$. This leads to an eigenvalue problem

$$L_n^{(2)}[D(x)]^2 = \kappa^2[D(x)]^2 \tag{14.5.16}$$

where we have introduced another linearized operator $L_n^{(2)}$, which reads

$$\begin{aligned} L_n^{(2)}f(x) = \alpha^2\Big\{&f[g^{(n-1)}(x/\alpha)] + \{g'[g^{(n-1)}(x/\alpha)]\}^2 f[g^{(n-2)}(x/\alpha)] \\ &+ \{g'[g^{(n-1)}(x/\alpha)]g'[g^{(n-2)}(x/\alpha)]\}^2 f[g^{(n-3)}(x/\alpha)] \\ &+ \cdots + \{g'[g^{(n-1)}(x/\alpha)]\cdots g'[g(x/\alpha)]\}^2 f(x/\alpha)\Big\} \end{aligned} \tag{14.5.17}$$

The squares appear in this equation owing to the additivity theorem for the dispersion of Gaussian random variables. The similarity between equations (14.5.10) and (14.5.17) enables one to solve the eigenvalue problems (14.5.11) and (14.5.16) using the same subroutine. We omit the computational details and summarize the numerical results for exponents δ, α, and κ in Table 14.1.[14]

Other universal properties, e.g., the qth-order information dimension d_q of the limiting sets for the period-n-tupling sequences, in fact, the D_q versus q curve, may be calculated as well.[16]

14.6. The Antisymmetric Cubic Map and Symbolic Dynamics of Three Letters[2,17]

The antisymmetric cubic map (14.1.3) may be studied by using the method of Sections 14.2 and 14.3 and by extending some of the results we referred to in Section 14.4. This map is interesting for two reasons at least. First, it is the next step toward more complicated one-dimensional mappings. Second, it enjoys the same discrete symmetry as the Lorenz model[4] (see Section 14.8).

The antisymmetric cubic map has two symmetrically located critical points [see equation (14.1.5b)] and we proceed from the natural order

$$L < x_c^- < M < x_c^+ < R \tag{14.6.1}$$

Periodic orbits are of two kinds: symmetric and asymmetric. The asymmetric orbits always appear in pairs located symmetrically with respect to the origin and can be reached from two different basins. A stable symmetric orbit will first undergo symmetry-breaking bifurcation to a pair of asymmetric orbits of the same period, and then undergo period-doubling and symmetry restoration. A superstable symmetric orbit must contain both critical points and has the form

$$x_c^- P x_c^+ \bar{P} x_c^- \tag{14.6.2}$$

where P is a pattern made of the letters R, L, M, and $\bar{P}$ is the conjugate of P obtained by interchanging L and R with M unchanged. Similarly, the pair of superstable asymmetric orbits looks like

$$x_c^- W x_c^- \qquad \text{and} \qquad x_c^+ \bar{W} x_c^+$$

In what follows we shall consider only superstable orbits starting from x_c^- and omit x_c^- as understood, e.g., there are exactly two period-3 orbits RM and RR.

Between any two given words W_1 and W_2 an order can be introduced by analogy with Section 14.4 with the natural order (14.4.3) replaced by (14.6.2) and the parity defined by the number of Ms contained in W^*. A rule similar to that associated with system (14.4.7) can be derived to identify the shortest period included in between two given periods. We omit the related definitions and theorems as well as the list of all the admissible words (starting from x_c^-) with period less than 7 (see Zeng *et al.*[2] for details).

To generate all the admissible words of a given period one extends the λ-expansion, defined in Section 14.4, to the range $2 < \lambda < 3$.[17] The resulting working rule may be formulated very simply: one associates a number sequence $(a_1, a_2, a_3, \ldots, a_n, C)$ comprising ± 1, 0 (and $\pm\frac{1}{2}$ for C only) to the pattern

$$Px_c^- = \sigma_1\sigma_2 \cdots \sigma_n x_c^-$$

in the following way:

1. $a_i = 0$ if $\sigma_i = M$.
2. If the number of Ms before σ_i is even, then $a_i = 1$ (-1) when $\sigma_i = R$ (L).
3. If the number of Ms before σ_i is odd, then $a_i = 1$ (-1) when $\sigma_i = L$ (R).
4. $C = -\frac{1}{2}$ $(+\frac{1}{2})$ if P is even (odd).

To ensure the admissibility of the pattern the following inequalities must be satisfied:

$$\pm(a_i, a_{i+1}, \ldots, a_n, C, 0, 0, \ldots) < (a_1, a_2, \ldots, a_n, C), \qquad i = 1, 2, \ldots, n \tag{14.6.3}$$

Comparison between the two sides of the inequalities (14.6.3) must be made on a term-by-term basis; the first unequal relation determines the direction of the inequality, just like it was defined for the case of (14.4.15).

Example 1. *RMLMR* is an admissible word because the number sequence $(1\,0\,1\,0\,1\,-\frac{1}{2})$ satisfies all the inequalities (14.6.3).

Example 2. *RMMRL* is not allowed, because one of the inequalities, namely,

$$-(-1\,-\tfrac{1}{2}\,0\,0\,0\,0) < (1\,0\,0\,1\,-1\,-\tfrac{1}{2})$$

fails when comparing $\frac{1}{2}$ with 0.

We note that the number of stable periodic orbits, i.e., the number of admissible words of fixed length n, may be calculated from the following recursion formulas:[18]

$$M(1) = 1$$

$$M(n) = 1/4n\left[3^n - 3 - \sum_{j=0}^{p} 2^{j+1} - \sum_{1<d<n} {}_{d/n}\, 4dM(d)S(d)\right] \tag{14.6.4}$$

where the sum over j exists only if $n = k\,2^p$ with k odd and $p > 1$, $S(d) = 2^l$ if $n = d\,2^l$; $\sum_{d/n}$ means summation over such d which divides n. To get the total number of cycles one must add to $M(n)$ the period-doubled orbits that originated from lower periods.

14.7. The Sine-Square Map and Optical Bistability[5]

In an abstract dimensionless form optical bistable devices are described by time-delayed differential equations of the following type:

$$\tau\, dx(t)/dt + x(t) = f[A, x(t - t_R)] \tag{14.7.1}$$

where, as before, $f(A, x)$ is a nonlinear function depending on parameter(s) A, τ is the relaxation time of the system, and t_R the time delay introduced by a feedback loop. In the limit of very long delay one can neglect the time derivative and measure the time in units of t_R, reducing equation (14.7.1) to a map

$$x_{n+1} = f(A, x_n) \tag{14.7.2}$$

Several nonlinear functions f have been studied in connection with various experimental setups. We dealt with the case of an optical cavity filled with liquid crystal. Since the transmittance of the liquid crystal can be well approximated by a bimodal sine-square function, we were led to the map (14.1.4), i.e.,

$$x_{n+1} = A \sin^2 (x_n - B), \qquad |x_n - B| \leq \pi \tag{14.7.3}$$

It falls into the category of maps that can be studied in detail using the analytical approach outlined in Sections 14.2 and 14.3. The only complication comes from the necessity of using four letters and the presence of the second parameter B. Having in mind the definition of proper branches of inverse trigonometric functions, we get the following explicit expressions for the inverse functions:

$$\begin{aligned} N(y) &= f_N^{-1}(y) = B + \tfrac{1}{2} \cos^{-1} (1 - 2y/A) \\ L(y) &= f_L^{-1}(y) = N(y) - \pi \\ M(y) &= f_M^{-1}(y) = 2B - N(y) \\ R(y) &= f_R^{-1}(y) = M(y) + \pi \end{aligned} \tag{14.7.4}$$

The arguments of $\cos^{-1}$ must satisfy the condition

$$|1 - 2y/A| \leq 1 \tag{14.7.5}$$

The ordering of words in the symmetric $B = 0$ case can be treated by extending the method of Section 14.6 (see a simpler example given by Zeng *et al.*[(2)]). In the general case of nonmonotonic dependence on the second parameter B one can write down the words, at least for short periods, by inspecting the map.

Period-1 orbits correspond to blank words, which we denote as $W = b = 1$. Thus, the equation to determine the period-1 superstable parameter values reads in our case

$$f(x_{ci}) = x_{ci}, \qquad i = 1, 2, 3, 4 \tag{14.7.6}$$

This is, of course, a trivial result, yielding four straight lines in the A–B plane:

$$A = B - \pi/2, \qquad B = 0, \qquad A = B + \pi/2, \qquad B = \pi \tag{14.7.7}$$

for $i = 1, 2, 3, 4$, respectively (see Figure 14.11, where these lines are labeled b_i, while the line $B = -\pi$ is labeled b_0).

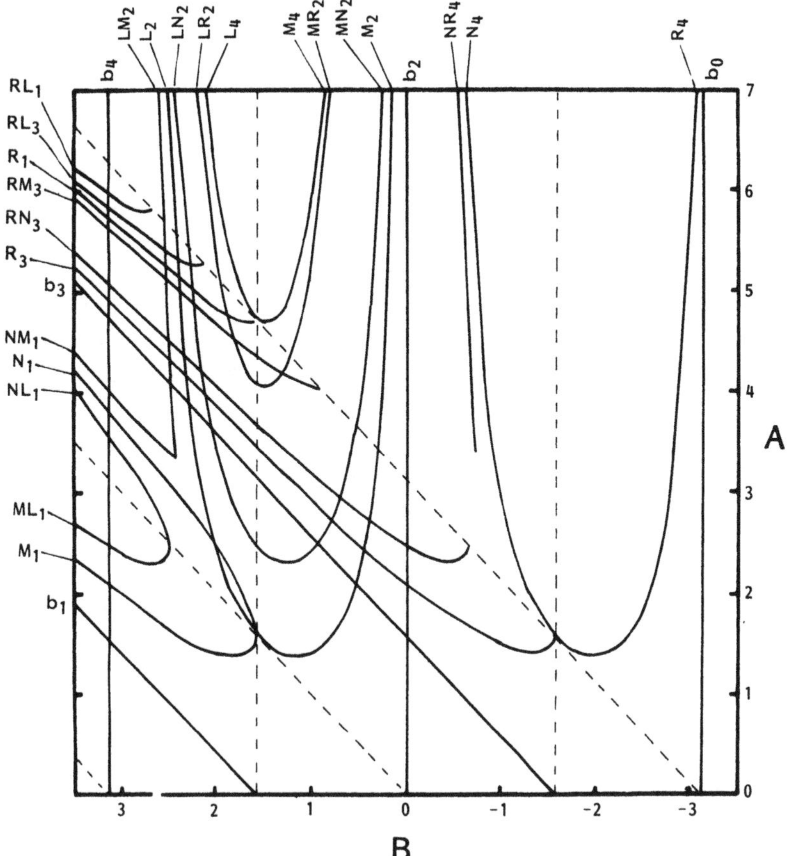

Figure 14.11. The A-B phase diagram for the sine-square map (14.1.4).

There are ten period-2 orbits: M_1, N_1, R_1, M_2, L_2, R_3, L_4, M_4, N_4, and R_4 (the subscript indicates which critical point the orbit contains). Take, for example, the orbit M_1. Equation (14.3.12) for this case reads

$$A = B - \tfrac{1}{2}\cos^{-1}(1 - (2B - \pi)/A) \tag{14.7.8}$$

which can be solved easily by transforming into an iteration:

$$A_{n+1} = B - \tfrac{1}{2}\cos^{-1}(1 - 2B - \pi)/A_n) \tag{14.7.9}$$

Actually, one can present equation (14.7.8) as a curve in the A-B plane (see Figure 14.11).

The equations for M_2 and L_2, as well as that for L_4, M_4, N_4, R_4, can be solved explicitly to yield

$$A = 2B/[1 - \cos(2B)] \qquad (\text{for } L_2\text{-}M_2) \tag{14.7.10}$$

and

$$A = 2(B + \pi)/[1 - \cos(2B)] \qquad (\text{for } L_4\text{-}M_4 \text{ and } N_4\text{-}R_4) \qquad (14.7.11)$$

Among the infinite segments of these functions one should retain only those satisfying condition (14.7.5).

As regards the period-3 orbits, there exist at least 17 words: ten words which are analogues of RL in the unimodal map, i.e., ML_1, NL_1, RL_1, MN_2, LN_2, MR_2, LR_2, RL_3, RM_3, RN_3, and seven orbits similar to those specific for the cubic map, i.e., RM_1, RN_1, NM_1, LM_2, MR_4, LR_4, and NR_4.

Figure 14.11 is a "phase diagram" of the optical bistable device, i.e., the loci of superstable periods in the parameter plane A–B. This kind of diagram is useful for the experimentalists to look for bistable and even multistable regimes. Drawn in Figure 14.11 are period-1, period-2, and some of period-3 orbits. The loci for period-3 orbits may serve as indicators of chaotic regions, because they are periodic windows embedded in the chaotic bands. In principle, each of the periodic windows in, say, Figures 14.5a and b can be found from phase diagrams like Figure 14.11 and be assigned a specific word. We note that Figure 14.11 was computed using only a programmable desk calculator.

Our method to determine the superstable parameter values outlined in Sections 14.3 and 14.7 may be applied to any one-dimensional map, provided the inverse branches $f_{\alpha_i}^{-1}$ are known explicitly. In particular, one may construct C^1 multimodal maps of any complexity by matching appropriately scaled segments of sines or cosines and study their phase diagram. The method of Section 14.2 is valid for any explicitly known maps.

14.8. Symbolic Description of Periodic Windows in Ordinary Differential Equations

Many physical systems are modeled by ordinary differential equations. We are concerned in the first place with bifurcation and chaos in dissipative systems, where dissipation plays a globally stabilizing role against local orbital instability and leads to the formation of strange attractors. In principle, one can have autonomous systems with three or more variables, nonautonomous systems with two or more variables, or time-delayed systems with one or more variables. Technically, however, nonautonomous periodically driven systems are preferred to autonomous ones, since it is easier to reach higher-frequency resolution using the driving frequency at our disposal. Time-delayed systems are too complicated to get a more or less complete picture for the time being.[19]

In principle, there might be serious objection against the feasibility of studying chaotic phenomena by numerical methods, because within finite computing time one cannot even realize a true quasi-periodic orbit, not to mention chaotic behavior. However, recent progress in studying chaotic transitions seems unthinkable without using computers. The point is that most conclusions about the bifurcation and chaos structure in the parameter space are based on the presence of various periodic orbits, just as mathematicians often rely on the densely existing periodic orbits or rational numbers in their analytical proofs. A practical prescription for studying bifurcation and chaos in computer experiments (and, to a certain extent, in laboratory experiments) consists of two steps. First, one should identify periodic and quasi-periodic regimes in the space of control parameters with as high as possible frequency resolution. This can be done by subharmonic stroboscopic sampling[20] in periodically driven systems, and by extensive power spectrum analysis as well as by calculating the Poincaré maps with high enough precision in autonomous cases. What remains in the parameter space after separating regions of periodic and quasi-periodic regimes may correspond to chaotic motion. Then comes the second step, i.e., characterization of the chaotic attractors. This can be accomplished by studying various dimensions, entropies, and characteristic exponents.

We have studied in detail the periodically forced Brusselator (for a recent review see B.-L. Hao[21])

$$\begin{aligned} \dot{X} &= A - (B+1)X + X^2Y + \alpha\cos(\omega t) \\ \dot{Y} &= BX - X^2Y \end{aligned} \tag{14.8.1}$$

and discovered the ordering of periodic windows according to the *MSS* sequence comprising two letters, i.e., the U-sequence described in Section 14.4.

To present the results we use another two parameters instead of A and ω. After carrying out the linear stability analysis of the free Brusselator, one gets two eigenvalues for the linearized problem, namely,

$$\lambda_{\pm} = (B - A^2 - 1)/2 \pm i[A^2 - (B - A^2 - 1)^2/4]^{1/2}$$

Now let us write

$$\lambda_{\pm} = \gamma\omega \pm i\beta\omega \tag{14.8.2}$$

where the frequency ω of the external force has been introduced into equation (14.8.2). Then γ and β will be measured in units of ω. In particular, β is the ratio of the limit cycle frequency in the linearized regime to the driving frequency. It happens that (γ, β) is a better plane to see mode-lockings, quasi-periodicity-to-chaos transitions, and the relation between the U-sequence and the locking frequencies. Figure 14.12 is the γ–β phase diagram. We shall not enter into detailed explanation of this diagram but only indicate that near and above the $\gamma = 0.05$ line the periods are given by the U-sequence up to

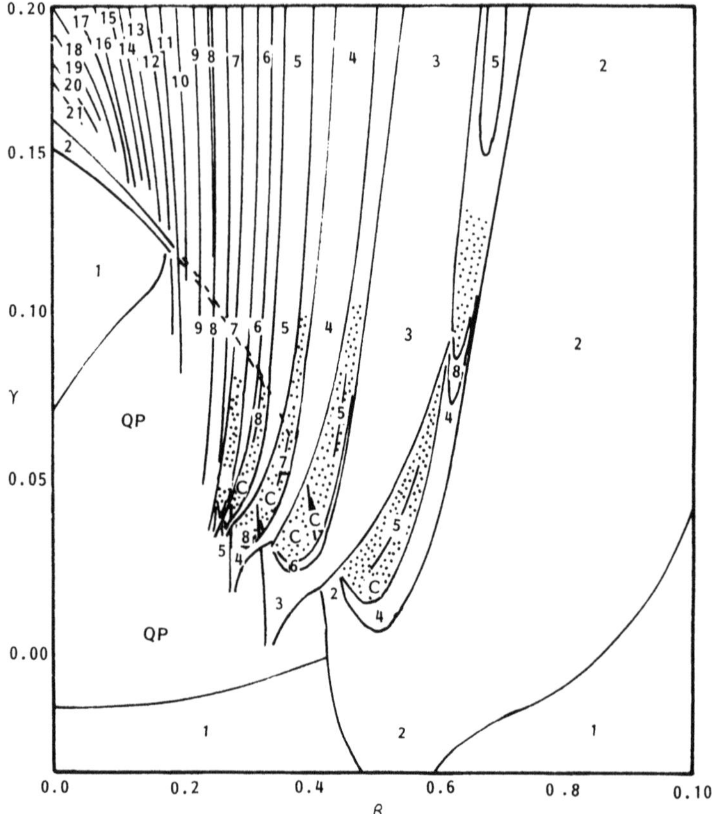

Figure 14.12. The γ-β phase diagram for the periodically forced Brusselator (14.8.1) with $B = 1.2$, $\alpha = 0.05$. Integers indicate periods, C chaos, and QP quasi-periodicity.

period 6. As a matter of fact, we could assign letters to the observed periods and the words thus obtained are ordered strictly in accordance with the U-sequence.

Figure 14.12 shows clearly the global nature important for understanding the transition from quasi-periodicity to chaos. In one region of the parameter space one sees chaotic regimes separated by periodic zones described by the U-sequence (not necessarily made of two letters—in principle, symbolic dynamics of more letters can be involved as well), while in another region we have mode-locking tongues, ordered according to the Farey sequences and embedded in the quasi-periodic sea. Since these two regions coexist in one and the same model, they must be somehow interrelated. However, the U-sequence and the Farey sequence are so different in nature that they cannot extend smoothly from one to another when the parameters are suitably varied.

Exceptions can occur only for a few short periods, where uniqueness guarantees a one-to-one correspondence between them. For example, in the

U-sequence the only period 2 and period 3 are described by the words R and RL, respectively, while in the Farey sequences there exists a unique period 2 (described by winding number $\frac{1}{2}$) and a unique period 3 ($\frac{1}{3}$ and $\frac{2}{3}$ only differ in the sense of rotation). Therefore, these periods can develop smoothly from mode-locking tongues in the quasi-periodic sea into periodic windows interspersed in the chaotic bands, as Figure 14.12 actually shows. It is clear that this cannot be the general case for higher periods, e.g., for period 11 there are 93 different prime words in the U-sequence but only 5 different rational fractions in the Farey sequence. Most of the U-sequence periods must close in themselves or terminate somewhere. What has just been said provides the physical reason for the folding and bending of the periodic sequences in Figure 14.12.

On the other hand, Figure 14.12 can be viewed as a distorted picture of the corresponding phase diagram for the circle mapping, studied in detail by, e.g., Balair and Glass[22] recently. It is interesting to note that if one keeps staying in the winding number $\frac{1}{2}$ region, one can enter the main period-doubling sequence, passing successfully the winding numbers $\frac{2}{4}$, $\frac{4}{8}$, $\frac{8}{16}$, etc., as has been checked in our computer experiment. However, one can never form rational fractions with odd denominators (i.e., for periods 3, 5, 7, ...), keeping the value of the winding number equal to $\frac{1}{2}$. This is why the complete U-sequence as seen in Figure 14.12 is formed by involving bent periodic sequences originating from other winding numbers.

To the best of our knowledge, there exist at least two other examples of similar interrelation between the U-sequence and the Farey sequence in systems of ODEs. The first is a driven nonlinear oscillator with exact solution, studied by Gonzalez and Piro.[23] In their Figure 1 the U-sequence could be seen clearly, though it was not stated explicitly in the paper. Another example comes from an ODE model describing the yeast glycolysis.[24] The dependence of glycolyzing yeast extract on the sugar input flux frequency ω_e and amplitude A is shown in Figure 14.13. It is quite interesting that Figure 14.13 resembles the experimental phase diagram obtained by the same authors. In both cases a little more effort would reveal the nature of the ordering of the periodic windows. It seems to be dynamics of two letters too. These examples show that this kind of global picture may be a universal property and certainly requires further study.

The systematics of periodic windows in the Lorentz model[4]

$$\begin{aligned} \dot{x} &= \sigma(y - x) \\ \dot{y} &= rx - xz - y \\ \dot{z} &= xy - bz \end{aligned} \tag{14.8.3}$$

requires at least three letters owing to the discrete antisymmetry of system

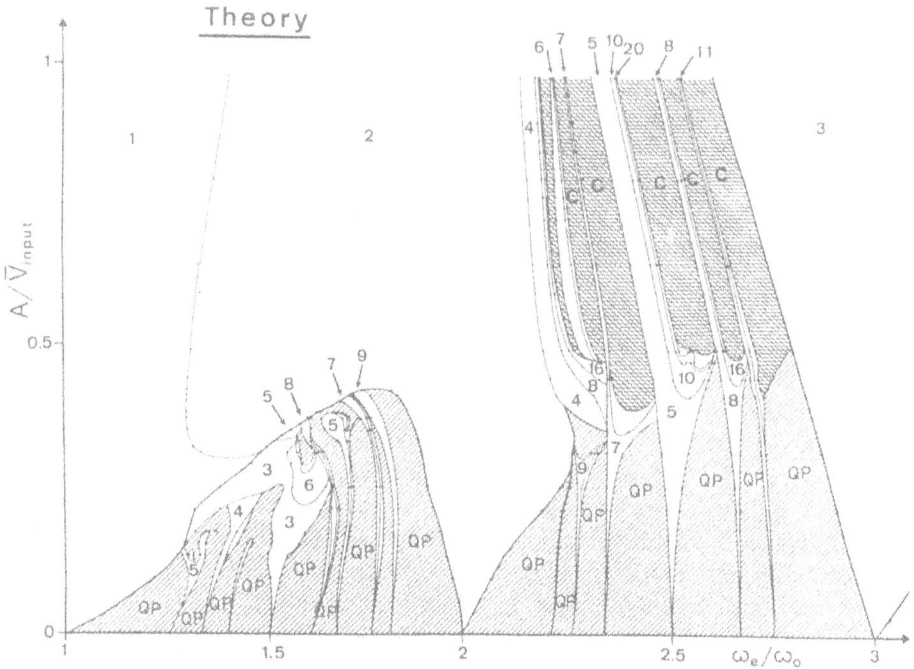

Figure 14.13. Dependence of glycolyzing yeast extract on the sugar input flux frequency ω_e and amplitude A in an ODE model (after Markus *et al.*[24]).

(14.8.3) with respect to $x \to -x$, $y \to -y$, which makes it closer to the antisymmetric cubic map (14.1.3).

In the case of periodically driven systems, the external forcing frequency is at our disposal and serves as the reference frequency to measure all other periodic components occurring in the motion. However, in an autonomous system like (14.8.3) the fundamental frequency of the motion may drift as the parameter varies and this seems to prevent us from associating each encountered period with an absolute designation like period 5. Nevertheless, by carrying out extensive power spectrum analysis for all available periodic orbits, changing parameter r with σ and b fixed at the usual values, we have obtained a calibration curve for the fundamental frequency.[25] All the windows known before and newly discovered by us have been assigned a period referring to this curve. To indicate the relation between different periods we have devised a way to assign letters to the numerically computed Poincaré map. It is a surprising fact that 47 out of 53 prime periods fit into the "cubic" scheme, i.e., appear in the same order as they appear in the antisymmetric cubic map (14.1.3), (see Ding and Hao[3] for the words thus assigned and the parameter values). Anyway, apart from a few exceptions the skeleton of the systematics for the occurrence of periodic windows in the Lorenz model obeys the "cubic" law. Or, put in another way, the symbolic dynamics underlying the global periodic structure appears to be the same as that in the cubic map. Sparrow[26] used a symbolic system of two letters X and Y to describe periodic orbits in the Lorenz model. Ours is a more detailed description, but is restricted to stable periods only.

To conclude this chapter we note that in the symbolic description of periods we have used only a subset of "measure zero" in the space of symbolic sequences used by mathematicians,[1] namely, the set of finite periodic sequences. Nevertheless, it has provided us with an analytical tool to treat many one-dimensional mappings.

Acknowledgments

I would like to express my gratitude to my coworkers S.-G. Chen, G.-R. Wang, S.-Y. Zhang, W.-Z. Zeng, M.-Z. Ding, and J.-N. Li for collaboration and discussion on various topics summarized in this chapter. My thanks go to Prof. B. Hess for allowing us to use Fig. 14.13 prior to publication. This work was supported in part by the Science Fund of the Chinese Academy of Sciences.

References

1. V. M. Alekseev and M. V. Yakobson, *Phys. Rep.* **75**, 287 (1981); J. Guckenheimer and P. Holmes, *Nonlinear Oscillations, Dynamical Systems, and Bifurcations of Vector Fields*, Chapter 5, Springer-Verlag, Berlin (1983).
2. W.-Z. Zeng, M.-Z. Ding, and J.-N. Li, *Chinese Phys. Lett.* **2**, 293 (1985); Symbolic dynamics for one-dimensional mappings with multiple critical points. The Academia Sinica Institute of Theoretical Physics Preprint ASITP-86-013, submitted to *Commun. Theor. Phys.*
3. M.-Z. Ding and B.-L. Hao, Systematics of the periodic windows in the Lorentz model and its relation with the antisymmetric cubic map, ASITP-86-015, submitted to *Commun. Theor. Phys.*
4. E. N. Lorenz, *J. Atmos. Sci.* **20**, 130 (1963).
5. H.-J. Zhang, J.-H. Dai, P.-Y. Wang, C.-D. Jin, and B.-L. Hao, *Chinese Phys. Lett.* **2**, 5 (1985); *Commun. Theor. Phys.* **8**, 281 (1987).
6. W.-Z. Zeng, B.-L. Hao, G.-R. Wang, and S.-G. Chen, *Commun. Theor. Phys.* **3**, 283 (1984).
7. J. V. Jose, *J. Phys. A* **16**, L205 (1983).
8. H. Kaplan, *Phys. Lett.* **97A**, 365 (1983).
9. N. Metropolis, M. L. Stein, and P. R. Stein, *J. Comb. Theory, Ser. A* **15**, 25 (1973).
10. B. Derrida, A. Gervois, and Y. Pomeau, *Ann. Inst. Henri Poincaré, Sect. A* **29**, 305 (1978).
11. P. Collet and J.-P. Eckmann, *Iterated Maps on the Interval as Dynamical Systems*, Birkhauser, Boston (1980).
12. M. J. Feigenbaum, *J. Stat. Phys.* **19**, 25 (1978); **21**, 669 (1979).
13. K. Kaneko, *Prog. Theor. Phys.* **69**, 403 (1983).
14. G.-R. Wang and S.-G. Chen, *Acta Physica Sinica* **35**, 58 (1986).
15. B. Shraiman, C. E. Wayne, and P. C. Martin, *Phys. Rev. Lett.* **46**, 935 (1981).

16. W.-Z. Zeng and B.-L. Hao, Dimensions of the limiting sets of period-n-tupling sequences, *Chinese Phys. Lett.* **3**, 285 (1986).
17. W.-Z. Zeng, On the number of stable cycles in the cubic map, *Commun. Theor. Phys.* **8**, 273 (1987).
18. W.-Z. Zeng, *Chinese Phys. Lett.* **2**, 429 (1985).
19. J.-N. Li and B.-L. Hao, Bifurcation spectrum in a delay-differential system related to optical bistability, ASITP-85-021, to appear in *Commun. Theor. Phys.*
20. B.-L. Hao and S.-Y. Zhang, *Phys. Lett.* **87A**, 267 (1982).
21. B.-L. Hao, Bifurcations and chaos in the periodically forced Brusselator, ASITP-86-011, to appear in a book dedicated to the occasion of the retirement of Professor K. Tomita.
22. J. Balair and L. Glass, *Physica* **16D**, 143 (1985).
23. D. L. Gonzalez and O. Piro, *Phys. Rev. Lett.* **50**, 870 (1983).
24. M. Markus, S. C. Muller, and B. Hess, personal communication, in preparation for publication.
25. M.-Z. Ding, B.-L. Hao, and X. Hao, *Chinese Phys. Lett.* **2**, 1 (1985).
26. C. Sparrow, *The Lorenz Equations: Bifurcations, Chaos, and Strange Attractors*, Springer-Verlag, Berlin (1982).

15

Nonlinear Mechanical Properties

G. Caglioti and C. E. Bottani

The present chapter has been written in consideration of the tremendous practical importance of the mechanical properties of materials, and is divided into two sections. The first section in the form of an extended abstract, contains a brief review of the irreversible thermodynamics of stressed materials. In the second section, an overview is presented of a dislocation field theory of the elastic-plastic instability in metals.

15.1. Irreversible Thermodynamics of Materials under Stress (by G. Caglioti)

This brief survey outlines the experience gained in the Materials Laboratory of Milano Polytechnic and is not intended to be a general review. Attention has been focused on two topics: the thermoelastic effect,[1-3] and the thermo-elastic-plastic instability (yield),[4-8] leading to plastic flow. The objective of this section is to provide experimental evidence for the proposition that the mechanical behavior of materials under stress cannot be adequately understood in purely mechanical terms, but should be framed instead within the thermodynamics of nonlinear irreversible processes.[9-13]

In fact, consider a metallic sample undergoing a tensile test: as the linear (and only slightly) irreversible thermodynamic branch describing thermoelastic cooling approaches its thermoelastic plastic limit, the mechanical variables, stress σ and strain ε, might become no longer sufficient to model the evolution of the state of the system. In particular, around and beyond yield, the state

G. Caglioti and C. E. Bottani • Istituto di Ingegneria Nucleare, CESNEF, Polytecnico di Milano, 20133 Milan, Italy; Gruppo Nazionale di Struttura della Materia del CNR, Unita di Ricerca 7, Milan, Italy; Centro Interuniversitario Struttura della Materia (CISM) del Ministero Pubblica Istruzione, Milan, Italy.

of the system depends also on hidden variables, such as dislocation density, as well as on the previous history of the sample. Furthermore, as plastic flow proceeds, the irreversible dissipation of plastic mechanical work into thermal energy and the associated entropy production are accompanied by a conspicuous increase in the materials temperature. Therefore, temperature changes are indices of fundamental processes occurring in the bulk of the materials.[9]

Results of several experiments of deformation calorimetry, performed using suitable thermistors, have been reported and discussed. Measurement of the evolution of the temperature field on the surface of metallic samples undergoing approximately isoentropic elastic elongations and volume increases yields the Grüneisen parameter[14] and the thermal diffusivity.[15,16]

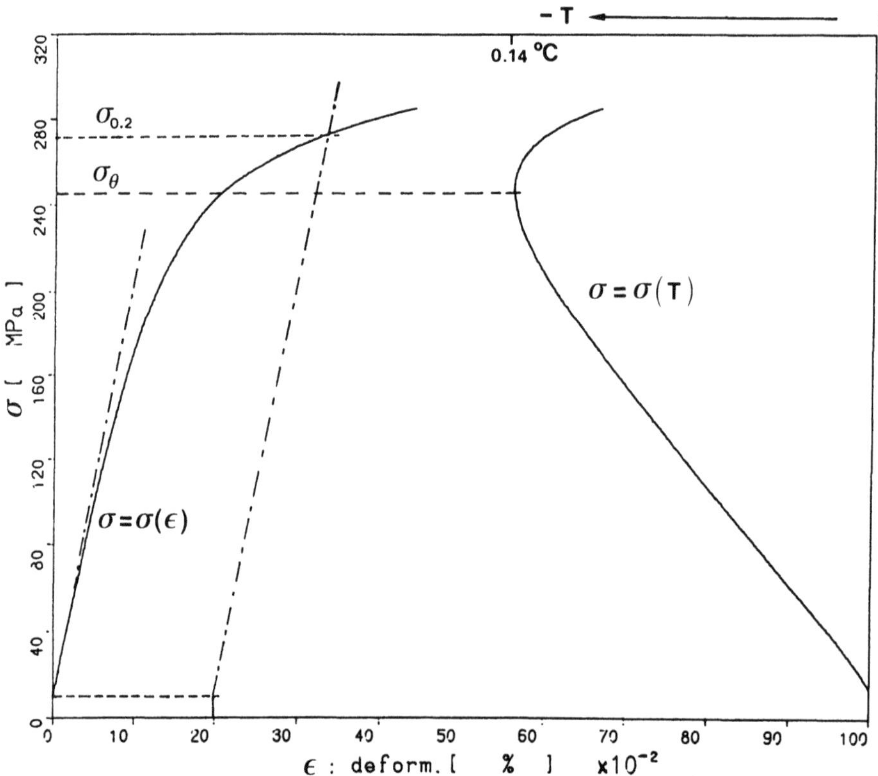

Figure 15.1. Stress-temperature characteristic (rhs) and stress-strain characteristic (1 hs) for a sample of stainless steel AISI 316 undergoing a tensile test, as obtained by eliminating time between the time behavior of stress and temperature, and stress and strain, respectively. The thermoelastic-plastic limit stress σ_θ is obtained from the stress-temperature characteristic as the stress at which the rate of the (adiabatic) thermoelastic cooling is exactly compensated by the rate of the thermoplastic heating. The graphical determination of the offset yield stress $\sigma_{0.2}$ is also indicated. The offset stress corresponds to the boundary between "small" and "large" plastic deformations, and is still adopted as a standard by mechanical engineers.

Similar procedures on concrete allow one to obtain the thermal expansion coefficient.[17]

As the thermoelastic plastic limit is approached, microplasticity sets in, and in near-adiabatic conditions thermoplastic heating becomes increasingly important.[18] Physical yield is defined to occur at the bifurcation of the thermodynamic branch, where the rate of thermoelastic cooling is exactly compensated by the rate of thermoplastic heating, so that the stress vs. temperature characteristic, $\sigma = \sigma(T)$, exhibits a vertical slope.[19] Beyond the bifurcation the thermodynamic branch is no longer stable. Physical yield is characterized by the thermoelastic plastic limit stress or critical stress σ_θ (Figure 15.1).[20] The value of σ_θ turns out to be generally smaller than the yield stress $\sigma_{0.2}$ defined by traditional engineering standards. In contrast with the empirical nature of the widely offset used parameter $\sigma_{0.2}$, the definition of σ_θ in terms of a dynamical instability marking the onset of (thermodynamically and mechanically) irreversible processes is unambiguous and physically sound.

As the applied stress increases beyond the limit σ_θ, thermoplastic heating definitely dominates over thermoelastic cooling. From the evolution of the temperature field, the fraction f of the mechanical plastic work converted immediately into heat can be obtained experimentally. In typical structural steels $f \simeq 0.5$. Should f vanish, the material would be ideally brittle.[13]

A correlation has been attempted between the evolution of the temperature field during plastic flow and the dislocation dynamics. A model based on a viscous force acting on dislocations, linearly dependent on velocity, indicates that σ_θ marks the value of the stress corresponding to the threshold for a copious production of mobile dislocations.[12,13]

Results on fatigue and fracture experiments can be found in the literature, where the nonlinear nature of most of the investigated processes has been emphasized (see especially Beghi *et al.*[21]). Unfortunately, a major departure from some other nonlinear physical systems investigated in this volume for materials undergoing plastic flow, is that the definition of even the reference state is practically inaccessible, since this state depends strongly on the defective microstructure of the nonhomogeneous sample.[12]

15.2. An Overview of a Dislocation Field Theory of the Elastic–Plastic Instability in Metals (by C. Bottani)

The fundamentals of the dynamic theory of continuously distributed dislocations are briefly summarized for the sake of self-consistency. The theory is then generalized to take into account dislocation reactions, crucial to the understanding of macroscopic plastic flow. The peculiarity of the nonlinear equations of the model is stressed. A case study (the breakaway occurring in a population of dislocations trapped at an obstacle under the action of an external shear) is fully treated solving, in the quasi-static case, the underlying singular integral equation with unknown integration bounds. The theory, though highly idealized, explains some intriguing memory effects of yield instability.

The macroscopic elastic–plastic transition in crystals is controlled at the mesoscopic level by dislocation motions. Some basic properties of dislocations are synthesized in Section 15.2.1. For a comprehensive treatment of the subject the reader is referred elsewhere.[22] Various kinds of instabilities of both static and dynamic natures, depending on the type of external excitation, material properties, and geometry, may occur, examples being creep, yield, jerky flow, thermoplastic strain avalanches, fatigue crack growth (with the related microstructural patterning), and elastic–plastic fracture.

Dislocation theory is not mature enough to explain all the related phenomenology in a quantitative way, but important qualitative results have been obtained and guidelines are available for the future. In our opinion, the dynamic theory of continuously distributed dislocations as formulated, e.g., by Kosevich[23] is the natural theoretical frame to start a field theory of small plastic deformations in crystals. The theory, improved to take into account dislocation reactions (such as multiplication and annihilation) playing a major role in determining the macroscopic plastic properties of a material, is summarized in Section (15.2.2). The inclusion of dislocation sources in the three-dimensional model is a delicate theoretical problem[24] and must be carried out very carefully to preserve compatibility with the basic geometrical statements of dislocation theory (see below). In particular, the connection between dislocation production and plastic strain rates turns out not to be the "familiar" Orowan's equation but a more complex differential equation containing the latter equation as a special case.

All the above material is treated briefly assuming that the reader possesses some elementary knowledge of the theory of discrete dislocation loops. Section 15.2.3 may be considered as either an application of the general theory to a treatable case of physical interest, or an example of rather general relevance of the phenomena that may arise in a dynamic nonlinear system of "particles" in the presence of long-range interactions not reducible to any local approximation. The problem treated is that of a one-dimensional distribution of parallel interacting dislocations moving in a common slip plane under the action of an external shear and in the presence of an internal finite obstacle. Regions of two-component confinement (a compact distribution of dislocations of both signs), stable single-component confinement, and breakaway are found as functions of the applied stress and total populations. Stability properties are summarized and critical lines on the bifurcation surface are identified. Open problems are discussed in the conclusion.

15.2.1. Dislocations

A dislocation is a linear topological defect of a crystal lattice (see Figures 15.2 and 15.3). Its geometrical definition is given in the next section in the

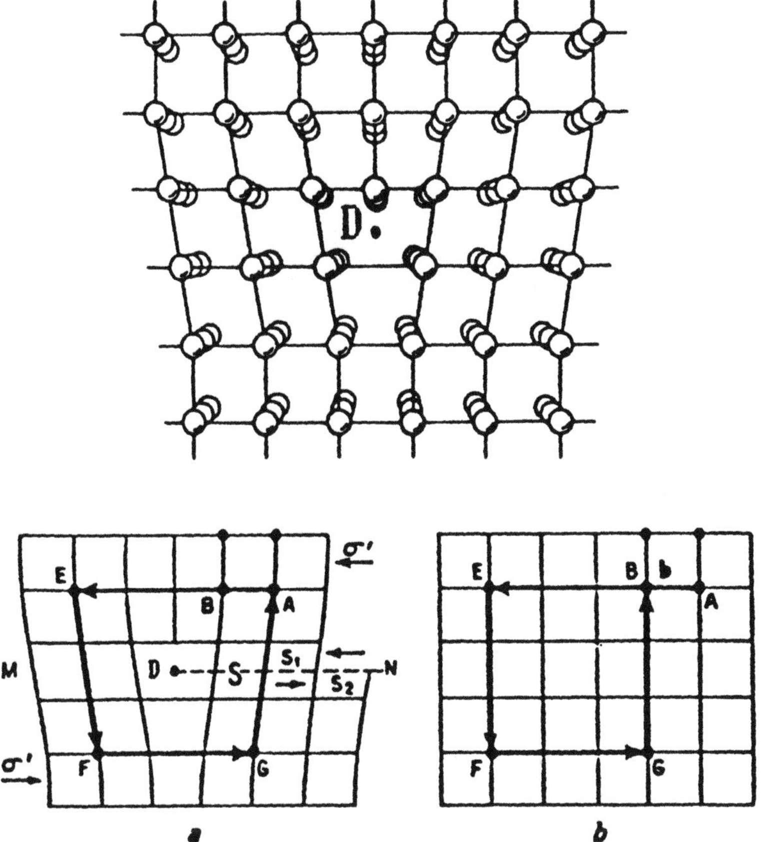

Figure 15.2. An edge dislocation in a simple cubic crystal with the associated Burgers circuit. The corresponding perfect crystal is shown for comparison. The Burgers vector is normal to the dislocation line. Edge dislocations are represented by the off-diagonal components of the dislocation density tensor (see Section 15.2.2). If the theory of Section 15.2.3 is applied to edge dislocations, the density ρ corresponds to the component $\alpha_{31} = b_1\rho(x_1)\delta(x_2)$.

case of the continuum approximation. Dislocations may be either sessile (in static equilibrium) or mobile. It is believed, upon experimental evidence, that mobile dislocations are the carriers of plastic deformation in crystalline bodies. The motion of dislocations at or below room temperature occurs mainly without transport of matter (conservative motion or glide). The defect transports (and partly radiates) its own elastic strain field and carries an incompatible "quantized" shear relative displacement (of amount b: the Burgers vector; see below and Figure 15.4) of the two parts of the crystal above and below the glide plane behind the dislocation line.

Dislocation lines are either close loops or terminate onto other crystal defects like free surfaces, grain boundaries, or different dislocation lines. Frequently dislocations in equilibrium form a three-dimensional network. Only dislocations with particular Burgers vectors (the Burgers vector of normal

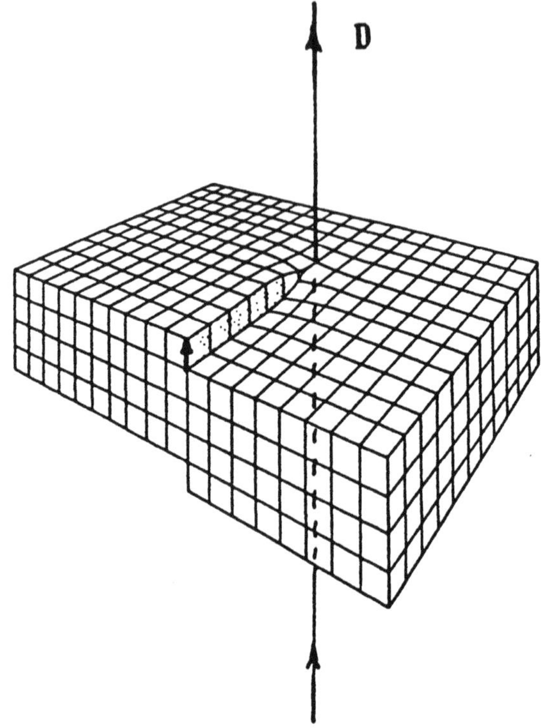

Figure 15.3. A screw dislocation in a simple cubic crystal. The Burgers vector is parallel to the dislocation line. Screw dislocations are represented by the diagonal components of the dislocation density tensor (see Section 15.2.2). If the theory of Section 15.2.3 is applied to screw dislocations, the density ρ corresponds to the component $\alpha_{33} = b_3\rho(x_1)\delta(x_2)$.

dislocations coincides with a lattice vector) can move on specific crystal planes $\{hkl\}$ along specific crystal directions $\langle HKL\rangle$ (slip systems; see Nabarro[1]).

The elastic energy stored in the self-field of a dislocation is so large (of the order of 1 eV per atomic unit length of dislocation line) that a crystal containing dislocations may be in mechanical equilibrium but never in true thermodynamic equilibrium. This happens because the dislocation entropic contribution to the free energy of the crystal is not high enough to compensate the corresponding dislocation internal energy contribution. This makes plastic deformations strongly path-dependent and dislocation density (see below) not a good state variable.

15.2.2. The Dynamic Theory of Continuously Distributed Dislocations

"There exists a complete system of equations which relates, in the general case, the elastic fields with the distributions and fluxes of the dislocations. There exist rigorous equations for obtaining the plastic deformation of a body from the known dislocation motion."[23] The theory to which the above citation refers is the elastodynamic theory of continuously distributed dislocations, due to several sources.[25-27] Here we follow mainly Kosevich's point of view[23] and notation as the more "physical" in our opinion. The main limitations of the theory are: the dislocation motion must be known, inelastic forces acting on dislocation loops cannot be derived from the theory (but can be introduced "from outside" in a compatible manner), and dislocation reactions cannot be accounted for and hence conservation of the total Burgers vector is implied.

There exist several atomistic models of a dislocation in a crystal but, from the standpoint of continuum physics, only two alternative definitions are commonly used. A dislocation line can be either conceived as a one-dimensional continuous set of singular points of the elastic displacement vector field (not a single-valued function of the coordinates) defined by the Burgers circuit (see below), or as the boundary of a singular surface on which the single-valued total displacement vector field undergoes a uniform discontinuity. Both definitions are in fact employed in the theory in a compatible way: the first is used to introduce the dislocation density tensor and relate it with the elastic strains, while the second is particularly useful in obtaining the plastic strains produced by a prescribed dislocation motion.[23]

We start by defining the dislocation density tensor of a single dislocation loop and then generalize the concept to the case of continuously distributed dislocations. With reference to Figure 15.4, we state that a dislocation line D is a singular line defined by the property that the closed line integral of $d\mathbf{u}^e$ (where $\mathbf{u}^e$ is the elastic displacement vector) around any contour L that encircles the dislocation line, is equal to a constant vector $\mathbf{b}$, called the Burgers vector. If the contour L is not coupled to the line D, the integral of $d\mathbf{u}^e$ is zero. The

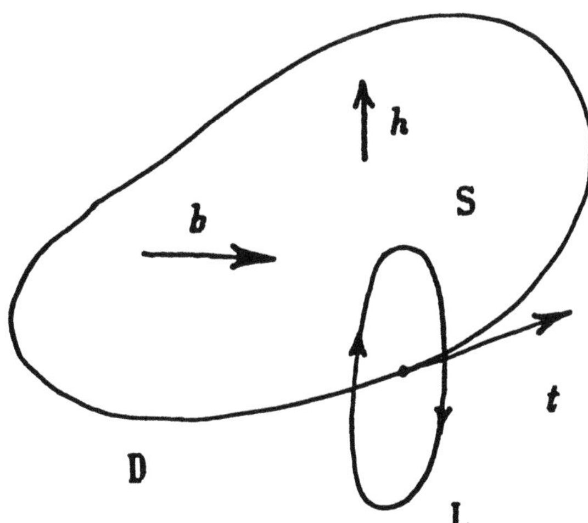

Figure 15.4. The Burgers definition of a dislocation loop (see text).

above property is written as

$$\oint_L d\mathbf{u}^e = -\mathbf{b} \qquad (15.2.1)$$

Alternatively, on introducing the elastic distortion tensor $\mathbf{W}^e$, gradient of the elastic displacement vector, given by $W^e_{ij} = \partial u^e_j/\partial x_i$, equation (15.2.1) can be expressed in the scalar form

$$\oint_L W^e_{ji}\, dx_j = -b_i \qquad (15.2.2)$$

Thus, in the presence of a dislocation loop, the elastic displacement vector is an ambiguous function of the coordinates. Equation (15.2.2) is our basic equation which states that the circulation of the elastic distortion tensor is equal to minus the Burgers vector if computed along a contour chained to a dislocation loop. The distortion tensor is the primordial strain measure. In the presence of discrete dislocation loops the definition of the elastic strain tensor in the form

$$\varepsilon^e_{ij} = \text{Sym}\left\{\frac{\partial u^e_i}{\partial x_j}\right\} = \text{Sym}\,\{W^e_{ji}\} \qquad (15.2.3)$$

is meaningful everywhere but on singular dislocation lines. In dislocation theory the elastic distortion tensor is assumed to be a one-valued function of the coordinates, continuous, and differentiable over all of space.

We now apply Stokes theorem to the left-hand side of equation (15.2.2):

$$\oint_L W^e_{ik}\, dx_i = \iint_S \varepsilon_{ilm} \frac{\partial W^e_{mk}}{\partial x_l}\, n_i\, dS \qquad (15.2.4)$$

Here ε_{ilm} is the Ricci tensor,[34] or permutation symbol, equal to +1 if the values of (ilm) are an even permutation of (123), −1 if the values of (ilm) are an odd permutation of (123), and 0 otherwise. Equation (15.2.4) may be read as: the circulation of the elastic distortion tensor along any contour L is equal to the flux of the curl of the same tensor through a surface S spanning this contour; $\mathbf{n}$ is the unit vector normal to a generic element dS of the surface. The differential form of the Burgers definition (15.2.1) can be obtained by writing also the vector $\mathbf{b}$ as a surface integral using the properties of the two-dimensional Dirac delta function,

$$b_k = \iint_S t_i b_k \delta(\boldsymbol{\xi}) n_i\, dS \qquad (15.2.5)$$

where $\boldsymbol{\xi}$ is the two-dimensional radius vector taken from the axis of the dislocation in the plane perpendicular to the tangent unit vector $\mathbf{t}$ at the given point. The contour L is arbitrary, so we obtain the required differential form:

$$\varepsilon_{ilm} \frac{\partial W^e_{mk}}{\partial x_l} = -t_i b_k \delta(\boldsymbol{\xi}) \tag{15.2.6}$$

where the dyadic tensor **tb** is naturally introduced.

Equations (15.2.5) and (15.2.6) may be generalized to represent a "coarse-grained" averaged description of practical interest. Yet, the averaging procedure is not unique owing to the line nature of dislocation defects. For a thorough analysis of the problem the reader is referred elsewhere.[23] We regard here as infinitesimal those volume elements that are "macroscopic" with respect to the average distance between dislocation segments, but "microscopic" with respect to the average loop dimensions. In this case we can define the *total Burgers* vector **b** coupled with a surface S as the sum of the Burgers vectors of all the dislocation lines embraced by the contour spanned by S. Then, going to the continuum limit, a surface density of the vector **b**, having a typical tensor character, is introduced. The formal procedure is the following: we define a macroscopic dislocation density tensor $\boldsymbol{\alpha}$ generalizing equation (15.2.5):

$$b_k = \iint_S \alpha_{ik} n_i \, dS \tag{15.2.7}$$

The surface density of a Burgers vector relative to any surface element $dS\,\mathbf{n}$ is then given by the relation

$$\frac{db_k}{dS_\mathbf{n}} = \alpha_{ik} n_i \tag{15.2.8}$$

strictly analogous to the Cauchy relation connecting the internal surface tensions to the stress tensor. At this point the fundamental equation (15.2.6) becomes

$$\varepsilon_{ilm} \frac{\partial W^e_{mk}}{\partial x_l} = -\alpha_{ik} \tag{15.2.9}$$

The averaging procedure to compute the tensor $\boldsymbol{\alpha}$ can now be specialized to different physical situations, introducing suitable scalar statistical joint distribution functions for the vectors **t** and **b**. In the case of a single crystal, for example, the vectors **b** belong to the discrete finite set of lattice vectors of minimal length and the dislocation density tensor can be written as

$$\boldsymbol{\alpha} = \sum_\beta \int d\Omega_t \mathbf{t}\mathbf{b}^\beta \rho^\beta(\mathbf{t}; \mathbf{r}) \tag{15.2.10}$$

the index β running over the possible directions of the vector **b**; $\rho^\beta(\mathbf{t}; \mathbf{r})$ is the scalar density of the distribution of the vectors **b** and **t** over the possible directions, $\rho^\beta(\mathbf{t}, \mathbf{r})\, d\Omega_t$ being the number of dislocations having a Burgers vector direction β passing through a unit area perpendicular to the vector **t**, and located inside a solid angle $d\Omega_t$ around the direction of **t**. The integration is carried out over the complete solid angle.

The overall density of dislocations ρ traditionally used by metallurgists is related to $\rho^\beta(\mathbf{t}, \mathbf{r})$ in the obvious way:[35]

$$\rho = \sum_\beta \int d\Omega_t \rho^\beta(\mathbf{t}; \mathbf{r}) \tag{15.2.11}$$

and represents the summary length of dislocation lines per unit volume. It is noteworthy that this latter density *is not directly connected with the strains.*

The dislocation density tensor is the curl of another tensor, so its divergence must be zero:

$$\operatorname{div} \boldsymbol{\alpha} = 0 \tag{15.2.12}$$

In the case of continuously distributed dislocations the elastic distortion tensor entering equation (15.2.9) is no longer related to any elastic displacement vector; in fact, if we assumed that $\mathbf{W}^e$ was the gradient of some vector potential, the left-hand side of equation (15.2.9) would vanish identically ($\widehat{\text{rot}}\, \widehat{\text{grad}} \equiv 0$). Hence in dislocation theory we must consider the elastic distortion tensor as the primary quantity defining the dislocation elastic strains via equation (15.2.9) and connected to the elastic strain tensor by the second part of equation (15.2.3) only (the first part of this equation being now meaningless). Of course, a total geometrical displacement vector field continues to exist even in the presence of continuously distributed dislocations. We assume that its gradient, the total geometrical distortion tensor $\mathbf{W}^T$, is simply the sum of the elastic $\mathbf{W}^e$ plus a plastic tensor $\mathbf{W}^p$. In this way allowance is made for a simple superposition of elastic and plastic deformations, while the notion of separate elastic and plastic displacements loses any meaning. The above *additive decomposition* of the total strains in an elastic plus a plastic part is valid as long as the total strains are small ($|\varepsilon_{ij}| \ll 1$). Consideration of finite deformations introduces formidable geometrical and physical difficulties in the theory and will not be further considered.[34] It is just worth noting that our aim is to elucidate some aspects of incipient plastic flow for which the small-strains approximation suffices. Owing to its definition, the curl of $\mathbf{W}^T$ is zero. Taking this fact into account in equation (15.2.9), we see that the dislocation density tensor is also directly related to the curl of plastic distortions:

$$\operatorname{rot} \mathbf{W}^p = \boldsymbol{\alpha} \tag{15.2.13}$$

So far we have not specified whether the dislocations were in motion or stationary. We complete the kinematic description by introducing the velocity

$\mathbf{v}(\mathbf{r}, t)$ of the total geometrical displacement. Its gradient is connected in an obvious way with the rates of elastic and plastic distortions:

$$\frac{\partial v_i}{\partial x_k} = \frac{\partial W^{e}_{ki}}{\partial t} + \frac{\partial W^{p}_{ki}}{\partial t} \tag{15.2.14}$$

In accordance with Kosevich,[23] we first write the plastic distortion rate as a purely flux term, ignoring dislocation reactions:

$$-\frac{\partial W^{p}_{ki}}{\partial t} = J_{ik} = \varepsilon_{ilm} \sum_{\beta} \int d\Omega_{t}\, t_l b^{\beta}_{k} V^{\beta}_{m}(\mathbf{t}; \mathbf{r}, t)\rho^{\beta}(\mathbf{t}; \mathbf{r}, t) \tag{15.2.15}$$

This equation introduces a new independent macroscopic quantity, the dislocation flux density tensor $\mathbf{J}$, relating it to the above scalar distributions ρ^{β} and the mean velocities of groups of dislocations $\mathbf{V}^{\beta}(\mathbf{t}; \mathbf{r}, t)$. It is clear from the definition that minus the symmetric part of the dislocation flux density tensor gives, in this case, the plastic strain rate tensor. Equation (15.2.15) is nothing else than a generalization of the popular Orowan's equation in metallurgy.[23] If we now take the time derivative of equation (15.2.9) and substitute equations (15.2.14) and (15.2.15) in it, we obtain the law of conservation of Burgers vector in the medium in differential form:

$$\frac{\partial \boldsymbol{\alpha}}{\partial t} + \operatorname{rot} \mathbf{J} = 0 \tag{15.2.16}$$

The statistical definitions of $\boldsymbol{\alpha}$ and $\mathbf{J}$ introduced in equation (15.2.16) yield the explicit form of this conservation equation in terms of the scalar distribution $\rho^{\beta}(\mathbf{t}; \mathbf{r}, t)$:

$$\sum_{\beta} \int d\Omega_{t} \left[t_i b^{\beta}_{k} \frac{\partial \rho^{\beta}}{\partial t} + t_i b^{\beta}_{k} \operatorname{div}(\rho^{\beta}\mathbf{V}^{\beta}) - b^{\beta}_{k}\mathbf{t} \cdot \mathbf{grad}\,(\rho^{\beta} V^{\beta}_{i}) \right] = 0 \tag{15.2.16'}$$

where the tensorial identity

$$\varepsilon_{ipq}\varepsilon_{qlm} = \delta_{il}\delta_{pm} - \delta_{im}\delta_{pl} \tag{15.2.16''}$$

has been used. At this point two facts should be noted. First, a realistic theory of yield and work hardening must take dislocation reactions into account and find precise relations between dislocation production and plastic strains. Second, as opposed to more common phenomena (like mass diffusion), in dealing with the volume density of particles the usual flow term $t_i b^{\beta}_{k} \operatorname{div}(\rho^{\beta}\mathbf{V}^{\beta})$ in equation (15.2.16') precedes another flow term $-b^{\beta}_{k}\mathbf{t} \cdot \mathbf{grad}\,(\rho^{\beta} V^{\beta}_{i})$ that may vanish only in particular cases (e.g., in the case of equal infinite straight edge dislocations with velocity perpendicular to the line direction). These last important points will be discussed briefly at the end of this section.

We further assume that Hooke's law still connects the stress tensor with the elastic strain tensor. For example, in the case of isotropic solids, this law reads

$$\sigma_{ik} = 2\mu W^{e}_{ik} + \lambda \delta_{ik} W^{e}_{ll} \tag{15.2.17}$$

where λ and μ are the two Lamé constants (μ, in particular, is the shear modulus). The linearity of this law allows one to simply superimpose the elastic stresses due to external loads (or external fields) on those produced by the incompatible strains associated with the presence of dislocations [equation (15.2.9)], to which no volume forces correspond.[23]

Finally, we write the linear momentum balance

$$\rho_{\mathrm{m}} \frac{d\mathbf{v}}{dt} = \operatorname{div} \boldsymbol{\sigma} \tag{15.2.18}$$

where ρ_{m} is the mass density.

Provided dislocation densities and fluxes are known, equations (15.2.9), (15.2.14), (15.2.17), and (15.2.18) together with definitions (15.2.10) and (15.2.15) constitute a closed system of equations.[23]

It is clear from the preceding that to close the latter system of equations one requires, in principle, the equation of motion of every dislocation segment $(\mathbf{t}\, dl)^{\beta}$ under the influence of the total stress field and then its averaged version in the case of continuously distributed loops. Unfortunately, there are several limitations to this approach. First, we stated that the dislocation density and dislocation flux density tensors are statistical quantities, so that one should instead formulate kinetic equations relating $\boldsymbol{\alpha}$ and $\mathbf{J}$, including also dislocation reactions. Second, it is impossible to derive forces related to the discrete atomic nature of the dislocation core in a continuum elastic theory.[29] Similar difficulties exist for all other forces of inelastic origin acting on dislocations (such as the velocity-dependent viscous drag damping the glide motion).[30,31]

Yet, the derivation of the elastic equation of motion of an individual dislocation loop[32] attracts attention as a basic procedure in understanding the physical origin of the long-range static and dynamic elastic interactions among different dislocation segments and with the external stress field. These interactions constitute the main driving force of dislocation motion and play a crucial role in the activation of dislocation sources, interaction with point defects and surfaces, formation of sessile dislocation networks, work hardening processes, and breakaways of pileups[22] (see Section 15.2.3 and Bottani[36]).

The specific problem of the field nature of the nonlocal mass tensor of a dislocation loop will not be discussed below. We merely use Kosevich's final result while stressing that it holds as long as the dislocation velocity is much less than the velocity of sound in the host material.

A Lagrangian formalism[32] can be employed to derive the elastic equation of motion of a segment of a dislocation loop in implicit Eulerian form:

$$\mathbf{t} \times [\boldsymbol{\sigma}(\mathbf{r}_\mathrm{D}(\mathbf{R}_\mathrm{D}, t)\, t) \colon \mathbf{b}^\beta] = 0 \tag{15.2.19}$$

The left-hand side is formally identical to the Peach–Koehler[33] static expression for the force per unit length acting on a dislocation in an external stress field. However, the stress tensor components σ_{ij} appearing in it do not denote only the "external" stress, but rather the total self-consistent stress, including also the dislocation self-action (both the "line tension"[30] and the "inertial" parts) and must be regarded as taken at the same point of space at which the dislocation segment under consideration is located at the given instant of time. Quantity $\mathbf{r}_\mathrm{D}(\mathbf{R}_\mathrm{D}, t)$ is the "material" coordinate of a dislocation segment, of unit tangent $\mathbf{t}$ and Burgers vector $\mathbf{b}^\beta$, that was initially (at $t = 0$) at position $\mathbf{R}_\mathrm{D}$. It can be shown[30] that the averaging procedure needed to adapt equation (15.2.19) to the case of a continuous distribution of dislocations conserves the form of the aforementioned equation, provided the stress in it is considered as an average quantity related to the dislocation density tensor and to the dislocation flux density tensor by the system of equations (15.2.9)–(15.2.18).

In Section (15.2.3), equation (15.2.19) will be given an explicit Newtonian form (in the quasi-static approximation).

Now the above theory can be generalized to account for dislocation production in the following formal way.[36] Equation (15.2.15) is rewritten in the form

$$-\frac{\partial W^\mathrm{p}_{ik}}{\partial t} = J_{ki} + s_{ki} \tag{15.2.15 bis}$$

where the new tensor quantity $\mathbf{s}$ is associated with plastic distortion rates produced by dislocation reactions (generation and annihilation) while $\mathbf{J}$ still represents the contribution of dislocation motion (flux). Equation (15.2.16) consequently assumes the form

$$\frac{\partial \boldsymbol{\alpha}}{\partial t} + \mathrm{rot}\, \mathbf{J} = \mathbf{P} \tag{15.2.16 bis}$$

where the tensor $\mathbf{P}$, the "dislocation production tensor," must be related to $\mathbf{s}$ by the differential condition

$$-\mathrm{rot}\, \mathbf{s} = \mathbf{P} \tag{15.2.20}$$

By analogy with equation (15.2.10), $\mathbf{P}$ may be expressed as

$$\mathbf{P} = \sum_\beta \int d\Omega_\mathrm{t}\, \mathbf{t}\mathbf{b}^\beta P^\beta(\mathbf{t}; \mathbf{r}, t) \tag{15.2.21}$$

where the scalar function $P^{\beta}(\mathbf{t}; \mathbf{r}; t)$ has the obvious meaning of production rate of dislocation segments of "species" $(\mathbf{t}, \mathbf{b}^{\beta})$. We present (and discuss the physical implications of) the explicit form of the above generalized equations for a special case elsewhere.[24] Here we just note that dislocation production necessarily introduces spatial inhomogeneities in the plastic strain field and may lead to breaking of the continuity of the host medium.

15.2.3. Confinement–Deconfinement Transitions in a Population of Dislocations

The problem of the motion of a variable number of dislocations under the combined action of an applied shear, the internal stresses due to fixed obstacles (such as locked dislocations, pinning centers, or other static arrays), and dislocation-dislocation reactions (such as production) is basic to the understanding of plastic deformation, including phenomena such as pileups, breakaways, shear bands, and so on.

For the case in which infinite, straight, parallel dislocations of the same type move on a single slip plane, with a velocity much lower than the velocity of sound, it may be shown[22] that the problem decouples into two separate problems: finding the dislocation motion (and hence the plastic strains) in a given "self-consistent" stress field, and finding the elastic stresses produced by a given dislocation motion. The assumption that the decoupling remains valid even in the presence of dislocation sources should be proved. Yet, in this simplified model the exact form of the most crucial feature of the physical situation, the long-range Peach–Köhler force[33] between dislocations, is retained. The model is specified by a set of coupled, nonlinear, singular integrodifferential equations for the dislocation density and velocity fields. These equations describe a richly-structured dynamical system. Even the static configurations, in the absence of dislocation reactions, are quite intricate and the dynamical details of transitions among them remain largely unexplored.

15.2.3.1. The Model Equations

If only one component of the dislocation density tensor $\boldsymbol{\alpha}$ exists with only one population of (positive and negative) dislocations of the same type moving in a single slip plane, the basic equations (15.2.16 bis) and (15.2.19) assume the form

$$\frac{\partial \rho}{\partial t} = -\frac{\partial}{\partial x}(\rho V) + P \qquad (15.2.22)$$

and

$$m^* \frac{\partial V}{\partial t} = b\sigma^{e}(x, t) - \frac{\mu b^2}{2\pi f} \fint_{L(t)} dx' \frac{\rho(x', t)}{x' - x} - BV \qquad (15.2.23)$$

where ρ is a linear dislocation number density, **V** the Eulerian dislocation velocity field, m^* the dislocation mass linear density,[32] f is equal to 1 for screw dislocations and $(1-\nu)$ for edge dislocations,[30] σ^e is the external shear stress field, L is the integration domain where a non-zero density of dislocations does exist (see Muskhelishvili[38] and Section 15.2.3.2), and B is a friction coefficient of inelastic origin[30] describing dislocation–phonon and dislocation–electron interactions. The derivation of the approximate equation of motion (15.2.23) from the exact equation (15.2.19) can be found elsewhere. It is worth noting the presence of the long-range dislocation–dislocation interaction expressed by

$$-\frac{\mu b^2}{2\pi f}\int_{L(t)} dx'\,\frac{\rho(x',t)}{x'-x}$$

where the integral over L must be understood as a Cauchy principal value[38,39] to exclude the nonphysical interaction of a dislocation with itself.

Below we shall consider the simplified problem with no production and negligible inertia ($m^*=0$), but it is important to note that the production term P cannot be taken arbitrarily if equations (15.2.22) and (15.2.23) must allow for static (elastic) stable states. This last requirement implies that P must be proportional to dislocation flux ρV. A physically plausible form of P, satisfying this requirement, will be discussed elsewhere.[24] For the moment we justify our statement by reminding the reader that all accepted models for dislocation sources (such as that of Frank and Read[30]) imply the interaction of fixed obstacles with *mobile* dislocations. In preparing the subsequent analysis, equations (15.2.22) and (15.2.23) are written in nondimensional form as

$$v(\xi,\tau)=S^e(\xi,\tau)-\int_{A_1(\tau)}^{A_2(\tau)} d\xi'\,\frac{n(\xi',\tau)}{\xi'-\xi} \tag{15.2.24}$$

$$\frac{\partial n}{\partial \tau}=-\frac{\partial}{\partial \xi}(nv) \tag{15.2.25}$$

where intrinsic units of time, length, and stress have been introduced:

$$x^*=b,\qquad t^*=2\pi fB/\mu,\qquad V^*=x^*/t^*,\qquad \sigma^*=\mu/2\pi f$$

$$n(\xi,\tau)=b\rho(x^*\xi,t^*\tau),\qquad v(\xi,\tau)=V(x^*\xi,t^*\tau)/V^*$$

$$S^e(\xi,\tau)=\sigma^e(x^*\xi,t^*\tau)/\sigma^*,\qquad A_{1,2}=a_{1,2}/x^*,\qquad L\equiv[a_1,a_2]$$

In the remainder of this chapter we shall again allow $\xi\to x$ and $\tau\to t$, but the above transformations are to be understood.

15.2.3.2. The Statics in the Continuum Approximation

We first look for the existence conditions and spatial structures of "elastic" reference states with $v = 0$ and, consequently, $S^e = S^e(x)$. In order to retain the same notation as in Balakrishnan and Bottani,[35] our basic nondimensional equations for the time-independent case are written in a slightly different form:

$$\frac{1}{\pi}\int_{a_1}^{a_2} dx' \frac{n(x')}{x' - x} = S^e(x) \tag{15.2.26}$$

and

$$\int_{a_1}^{a_2} n(x)\, dx = 2\pi N = 2\pi(N^+ - N^-) \tag{15.2.27}$$

The latter equation is a conservation law ($P = 0$) that specifies the total Burgers vector of the distribution.

In equation (15.2.26) $S^e = \sigma + \tau$, where σ is the uniform applied stress (control parameter) and τ is the stress field of the fixed dislocations or obstacles in the medium.

We are interested in *confined* distributions of compact support $a_1 \leq x \leq a_2$. This requires the solution of the singular integral equation (15.2.26) with unknown integration limits a_1 and a_2. A unique solution can be obtained by requiring that the function

$$\Psi(z) = \frac{1}{\pi}\int_{a_1}^{a_2} dx' \frac{n(x')}{x' - z} \tag{15.2.28}$$

which is analytic in the cut z-plane, vanishes asymptotically as $|z| \to \infty$. This yields the condition[39]

$$\int_{a_1}^{a_2} \frac{S^e(x)}{\sqrt{(a_2 - x)(x - a_1)}}\, dx = 0 \tag{15.2.29}$$

The physical meaning of this condition lies in the fact that the stress field produced by the dislocation distribution has just the form (15.2.28) at the location $z(x, y)$[23] and must itself vanish at infinity. Equation (15.2.29) and the normalization condition (15.2.27) determine the boundary points a_1 and a_2. Equation (15.2.26) then has the unique solution[38,39]

$$n(x) = \frac{1}{\pi}\sqrt{(a_2 - x)(x - a_1)} \fint_{a_1}^{a_2} \frac{dx'}{x - x'} \frac{S^e(x')}{\sqrt{(a_2 - x')(x' - a_1)}} \tag{15.2.30}$$

which is valid when $S(x)$ is *not* singular in $[a_1, a_2]$, the case of current interest.

What remains is to specify the internal stress barrier $\tau(x)$, which must be nonsingular on the x-axis, capable of confining dislocations of either sign, physically realizable, and a reasonable simulation of the stress field of the defects usually present. All these criteria are satisfied by the form $-x(x^2+\varepsilon^2)^{-1}$. (The limit $\varepsilon \to 0$ yields the known[30] singular case of a locked dislocation at the origin.) This barrier can be regarded as arising from two screw dislocations located at the points $+\varepsilon$ and $-\varepsilon$ on the y-axis. The form also roughly simulates the stress due to an infinite array (along the y-axis) of edge dislocations parallel to the z-axis—a model "grain boundary." On setting $S^e(x) = \sigma - x(x^2+\varepsilon^2)^{-1}$, equations (15.2.26)-(15.2.29) finally yield the following result: if we define $\theta_1 = \pi - \tan^{-1}(a_1/\varepsilon)$, $\theta_2 = \tan^{-1}(a_2/\varepsilon)$, and $\phi = \frac{1}{2}(\theta_1 - \theta_2)$, the integral expression (15.2.30) yields

$$n(x) = \left(\frac{(a_2 - x)(x - a_1)}{\sqrt{(a_1^2+\varepsilon^2)(a_2^2+\varepsilon^2)}}\right)^{1/2} \left(\frac{\varepsilon \cos\phi - x \sin\phi}{x^2+\varepsilon^2}\right) \tag{15.2.31}$$

where a_1 and a_2 are the roots of the system

$$(\sin\theta_1 \sin\theta_2)^{1/2} \sin\phi = \varepsilon\sigma \tag{15.2.32}$$

and

$$1 - (\sin\theta_1 \sin\phi_2)^{1/2} \cos\phi - \tfrac{1}{2}\sigma(a_1 + a_2) = N \tag{15.2.33}$$

In the "phase space" (N, σ) the above system may or may not have roots (namely, possible elastic equilibrium configurations). The phase space in which these solutions are to be analyzed is the region $-\infty < \sigma < +\infty$, $N^+ \geq 0$, $N^- \geq 0$. For definiteness, we restrict ourselves to $\sigma \geq 0$ and $N = N^+ - N^- \geq 0$. (The other regions are analyzed similarly.) We begin with the special case $\sigma = 0$, in which the solution is the modulated Lorentzian

$$n(x) = \frac{\varepsilon}{x^2+\varepsilon^2}\left(\frac{a^2-\varepsilon^2}{a^2+\varepsilon^2}\right)^{1/2} \qquad (a_1 = a_2 = a)$$

for $|x| \leq a$, and $n(x) = 0$ for $|x| > a$. The spread $2a$ is determined for a given N from $N = 1 - \varepsilon(a^2+\varepsilon^2)^{-1/2}$. Thus, even if $\sigma = 0$, the barrier can support a compact distribution of positive dislocations. (No negative dislocations can remain in equilibrium when $\sigma = 0$ in this particular case.) Moreover, $N_{max} = 1$, corresponding to $a = \infty$, at which stage the distribution spreads out to the usual Lorentzian form (see Figure 15.5). If we further let $\varepsilon \to 0$, so that $S(x) = -1/x$, then $N = 1$ and $n(x) = \pi\delta(x)$ are obtained, thus recovering the case of a single locked dislocation of unit strength at $x = 0$. We may also let $\varepsilon \to 0$ with $\sigma > 0$ (a lock at $x = 0$ plus an applied stress). The general

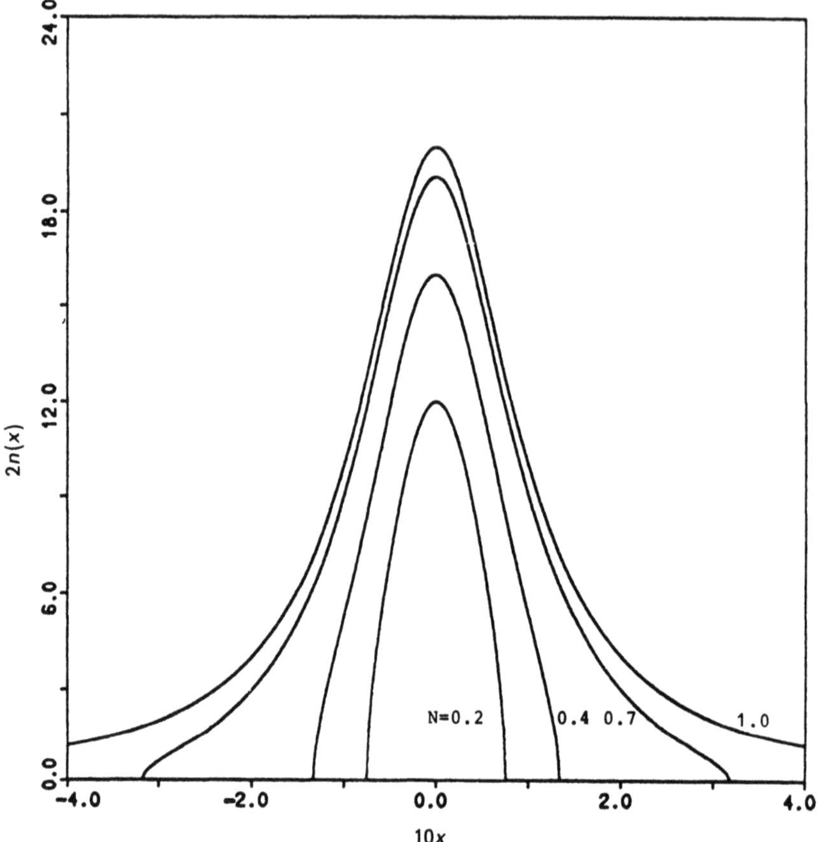

Figure 15.5. Function $n(x)$ for $\sigma = 0$ as N increases to $N_{\max} = 1$ ($\varepsilon = 0.1$ in all the plots). For $N < 1$, $|dn/dx| = \infty$ at $|x| = a$.

solution is now of the form $n(x) = c_1 x^{-1}[c_2 - (x - c_3)^2]^{1/2}$, where the constants c_i depend on σ and N, with an additional δ-function at $x = 0$ in some cases.

Next, the subspace $N = 0$ ($N^+ = N^-$), $\sigma > 0$, is examined. We find a compact distribution of dislocations of both signs, with a crossover at $x = \varepsilon \cot \phi$, as shown in Figure 15.6. As σ is increased this two-component pile-up shrinks, owing to the forced recombination of dislocations of opposite signs. Total annihilation occurs ultimately, because $N = 0$. This happens at the critical value $\sigma = (2\varepsilon)^{-1}$, when a_1 and a_2 reach the common value 0 from opposite sides. The two-component configuration is unstable against recombination by spontaneous fluctuations. We shall say more on this shortly.

Turning to values of N in the range $0 < N < 1$, $\sigma > 0$, we find *two* possible solutions for each σ and N. The first corresponds to a two-component localized distribution with $N^+ > N^- > 0$, while the second represents a one-component

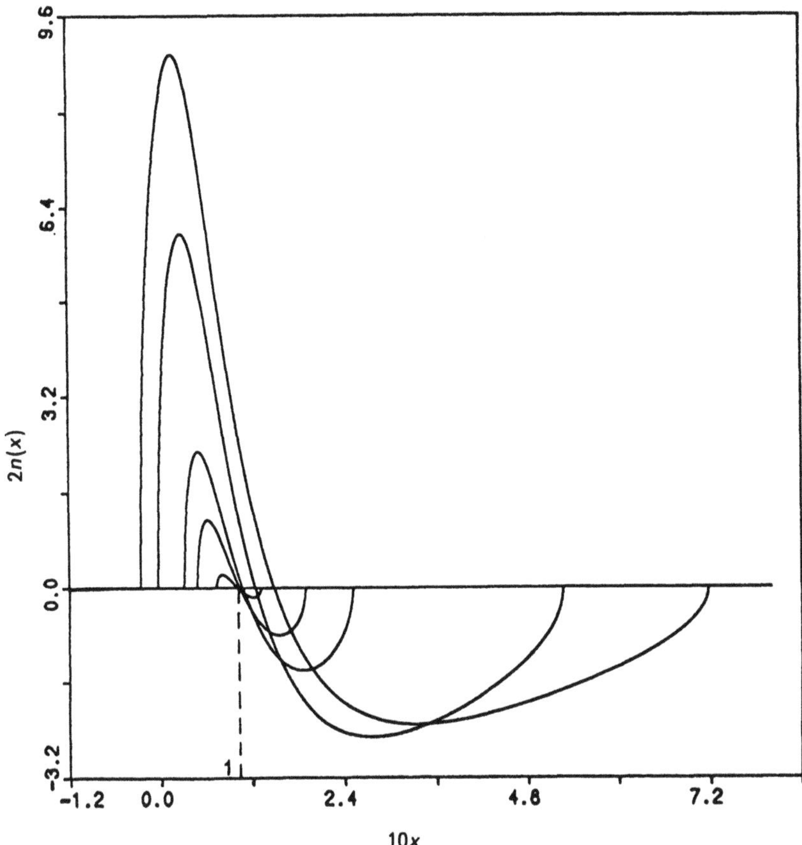

Figure 15.6. Recombination of dislocations as σ increases. The net area under each curve is $2\pi N = 0$ in this case. The graphs shown refer to $\sigma = 2.0, 3.0, 4.0, 4.5$, and 4.9.

distribution with $N = N^+$, $N^- = 0$. As σ is increased (keeping N fixed), the support of the two-component distribution shrinks (a_1 and a_2 move toward each other) owing to dislocation annihilation. Finally, at a threshold value $\sigma = \bar{\sigma}(N, \varepsilon)$ all the negative dislocations disappear, leaving behind a compact distribution of positive dislocations ($N^+ = N$, $N^- = 0$). This is a *limiting distribution*: the slope at its right extreme $x = a_2$ is no longer infinite but *zero* [$n \to 0$ like $(a_2 - x)^{3/2}$ as $x \to a_2$]. Further increase in σ causes a part of the dislocation population to spill over and breakaway (move to $x = \infty$). The subsequent equilibrium distribution is the saturated one appropriate to the stress value concerned (with zero slope at $x = a_2$).

The second solution represents a single-component compact distribution, shown in Figure 15.7. As σ is increased, both a_1 and a_2 are pushed to the right. At $\sigma = \bar{\sigma}(N, \varepsilon)$, the distribution attains exactly the same limiting form as described above: $dn/dx = 0$ at $x = a_2$, and a_1 reaches an accumulation

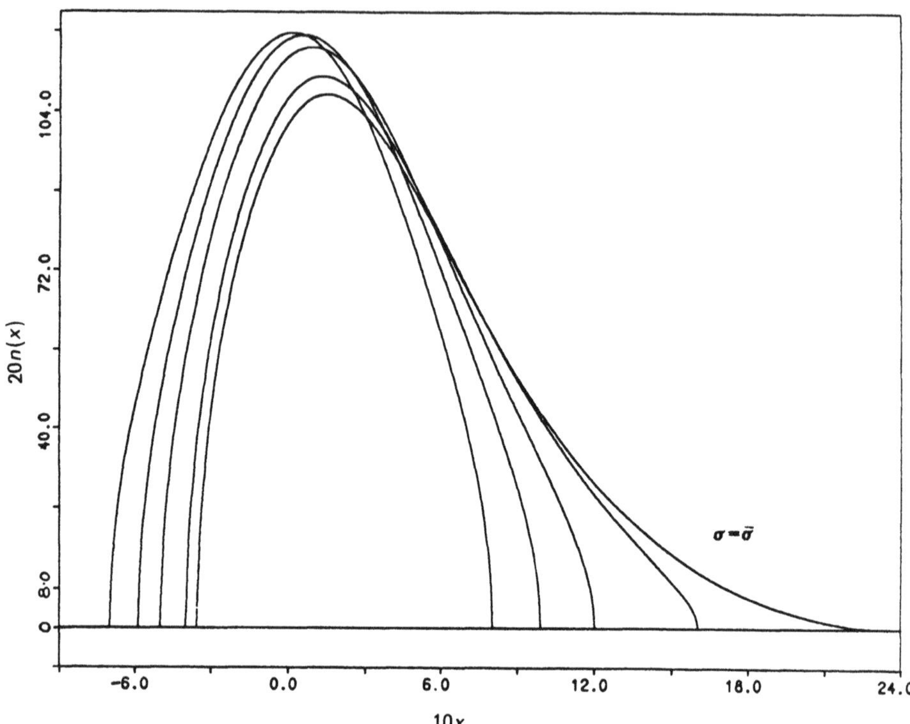

Figure 15.7. Distortion of the stable single-component distribution for $N = 0.2$ as σ increases through the values 0.4, 1.0, 1.6, 2.2, to $\bar{\sigma}(0, 2) \cong 2.45$.

point. A further increase in σ causes a spillover and breakaway of a part of the population, leaving a_1 unchanged and a limiting distribution with a reduced value of N given by the solution of $\bar{\sigma}(N, \varepsilon) = \sigma$.

The threshold stress $\bar{\sigma}$ has been found numerically for various values of N, and this enables us to draw the phase diagram of Figure 15.8 that summarizes our results. The full physical region is given by $-\infty < \sigma < \infty$, $-1 \leq N \leq 1$ (and of course $N_{\pm} \geq 0$), of which we have considered the portion $\sigma \geq 0$, $0 \leq N \leq 1$. The dark curve in the N^{+}-σ plane represents the threshold values $\sigma = \bar{\sigma}(N, \varepsilon)$. It ends at the "critical points" $(0, \frac{1}{2}\varepsilon)$ and $(0, 1)$, which have already been interpreted. Each point on the N^{+}-σ plane, within the triangular region enclosed by the threshold curve and the N^{+} axes, corresponds to a stable, confined, one-component distribution. The proof of the linear stability of this phase may be conducted with the aid of the dynamical equations (15.2.22) and (15.2.23) and is reported elsewhere. Translation of the threshold curve parallel to itself, as shown, yields the bifurcation surface. Points below it represent localized two-component distributions; above the surface, partial breakaway occurs (see the next section). However, the two-component configuration is unstable. Left to itself, the representative point of the system

www.ingramcontent.com/pod-product-compliance
Ingram Content Group UK Ltd.
Pitfield, Milton Keynes, MK11 3LW, UK
UKHW020151250726
13967UKWH00002B/997

* 9 7 8 1 4 8 9 9 2 0 5 9 1 *